Principles of Metal Refining

Principles of Metal Refining

T. Abel Engh
The Norwegian Institute of Technology
Trondheim, Norway

with contributions by

Christian J. Simensen
Centre for Industrial Research
Oslo, Norway

Olle Wijk
The Royal Institute of Technology
Stockholm, Sweden

Oxford New York Tokyo
OXFORD UNIVERSTY PRESS
1992

This book has been printed digitally and produced in a standard specification in order to ensure its continuing availability

OXFORD
UNIVERSITY PRESS

Great Clarendon Street, Oxford OX2 6DP

Oxford University Press is a department of the University of Oxford. It furthers the University's objective of excellence in research, scholarship, and education by publishing worldwide in

Oxford New York

Auckland Bangkok Buenos Aires Cape Town Chennai
Dar es Salaam Delhi Hong Kong Istanbul Karachi Kolkata
Kuala Lumpur Madrid Melbourne Mexico City Mumbai Nairobi
São Paulo Shanghai Singapore Taipei Tokyo Toronto

with an associated company in Berlin

Oxford is a registered trade mark of Oxford University Press in the UK and in certain other countries

Published in the United States
by Oxford University Press Inc., New York

Reprinted 2002

ISBN 0-19-856337-X

PREFACE

Dramatic changes and improvements have been made in the treatment of molten metals during the last thirty years. Thanks to continued advances, molten metals with a low content of impurities and the correct alloy composition are being produced more efficiently than ever before. Some effects of dissolved impurities and inclusions on the properties of the metal are discussed in Chapter 1, written by Christian J. Simensen.

Previously it was sufficient for the metallurgical graduate to know thermodynamics and to gain practical experience in handling molten metals. Today thermodynamics (Chapter 2) still plays a dominant part, but must be supplemented by a good understanding of kinetics (Chapter 4). Mass, heat, and momentum transfer (Chapter 7) are starting to play the same central role as in chemical engineering.

Fluid mechanics and turbulence theory are important not only for the kinetics concerning dissolved elements, but also to describe the removal of small particles (inclusions) from molten metals. This is treated in Chapter 5. Turbulent removal to the walls and top slag of non-metallic inclusions in steel and filtration of inclusions from aluminium is given particular attention. Hopefully, these results can stimulate workers concerned with other metals to develop methods for particle removal.

It should be noted that fluid dynamics does not push thermodynamics into the background. Thermodynamics is useful also to obtain badly needed data for physical and transport properties of molten metals, gases, salts, and slags.

A specialist in metal purification must know something about the origin and impurity content of the metal received. Also, it is necessary to have an understanding of what happens to the metal during solidification and ideally even later during the rolling, extrusion, and fabrication processes. Of course this is a near impossible task. The worker in refining metallurgy can hardly be both an expert in extractive and in physical metallurgy. Nevertheless, Chapters 3 and 9 attempt to give some background concerning the origin of the raw metal, what happens during casting, and the various uses of the final metal.

Removal of carbon from a molten metal is a typical example of a refining operation. In fact, decarburization of pig iron to produce steel is of enormous industrial importance. However, the literature concerning this field is vast and cannot be covered in this book. Instead, carbon removal from stainless steel is discussed in some detail both from a theoretical and practical point of view in Chapter 6 by Olle Wijk. The process illustrates some very important principles concerning removal of dissolved elements.

In Chapter 8, some of the problems in adding alloys to molten metals are reviewed and discussed. Here again heat and mass transfer effects are important. Heat diffusion in solids is treated in some (mathematical) detail. Due to the high thermal conductivity of metals, heat transfer relations for molten metals differ from what is found in chemical engineering systems.

Finally, in Chapter 10 refining and recycling are briefly considered. The main point is to stress the significance of the emerging field of recycling in order to conserve resources and the environment. Separation of the various parts of the recycled material plays a central role. This part of the process has much in common with the classical discipline of mineral beneficiation. The combination of beneficiation and refining arising in connection with recycling is new and interesting.

It is recognized that a number of disciplines and skills enter into the field of refining metallurgy. Perhaps a process metallurgist would be best equipped overall in this area. However, for instance, a chemical engineer also has a useful background in heat, mass, and momentum transfer. A physicist may have a good understanding of transport properties, thermodynamics, and fluid flow, while a mechanical engineer may be versatile in fluid flow and numerical calculations.

This book is written partly for undergraduates and partly for graduates. About half of the contents has been used as the curriculum for final year undergraduate students in process and physical metallurgy at the Norwegian Institute of Technology in Trondheim. The one-semester curriculum for these students was the first half of Chapter 2, all of Chapters 3 and 4, and part of Chapters 5 and 9. Students working for a Dr. Ing. (Ph.D.) degree in the field of refining metallurgy are also required to be familiar with the remaining chapters. At the Royal Institute of Technology in Stockholm, parts of the draft for the book have been given as course material for the Ph.D. students in process metallurgy.

Students who do not have a metallurgical background may be especially interested in the introductory Chapter 1, and Chapter 9 concerning solidification. For these students the first half of Chapter 2 that presents thermodynamics in a form tailor-made for describing low-alloy metals, should also be useful. Students should have at least some previous elementary training in thermodynamics. Chemical engineering students may find not only Chapter 4, but also Chapter 7 easy to master even at an undergraduate level.

In the study of Chapters 5, 7, and 8 some previous instruction in the areas of thermodynamics, transport phenomena, fluid flow, and turbulence theory is an advantage. Students who are not comfortable with their background in these areas may start by reading Sections 5.8–5.13. However, my advice to a student or worker in refining metallurgy who lacks background in some areas is not to worry. Study the difficult sections and chapters as well! Then do some extra homework by reading up on the basics.

Finally, I hope that even some of my university colleagues and co-workers in industry occupied in the field of refining metals will discover topics of interest. They will probably find that some points have been over-simplified, either due to sheer ignorance or in order to make life easier for the reader. I beg these co-workers to be tolerant and to take the time to send me their various comments.

T. A. E.

Norges Tekniske Høgskole
June 1992

ACKNOWLEDGEMENTS

Gradually, over the years, it has become evident that there is a very important field of endeavour that lies between extractive metallurgy and physical metallurgy. This book is an attempt to fill the gap. The importance of molten melt treatment was recognized early in Sweden, perhaps due to the long traditions in steelmaking. The author would like to acknowledge the inspiring influence of Professors Sven Eketorp and Krister Torssell at the Royal Institute of Technology, Stockholm. Also, I wish to mention the stimulating cooperation with Dr Hans Sandberg of SSAB and Dr Henrik Widmark of AB Sandvik Steel.

I would also like to acknowledge the excellent teaching I received in chemical engineering at the Norwegian Institute of Technology by Professors Aksel Lydersen, Sven Gregor Terjesen, and Olav Erga. It should be mentioned that my predecessor, the late Professor Marius Brostrup Müller, initiated formal instruction in refining steel ('secondary metallurgy') at our Division of Metallurgy.

In recent years it has been exciting and instructive to work in various areas of melt refining with Mr Karl Venås, Dr Surendra K. Saxena, Dr Geoffrey Sigworth, and Professors Madhu Nilmani and John Floyd, from the University of Melbourne.

A stay at MIT working with Professor John Frank Elliott has ensured that for me thermodynamics plays a decisive role in refining metallurgy. Later, Professors Harald Øye, Terkel Rosenqvist, Kai Grjotheim, and Ketil Motzfeldt have been my mentors in the field of thermodynamics. I am especially indebted to Ketil Motzfeldt for the extensive improvements he has made to the first part of Chapter 2.

I wish to acknowledge the encouragement and cooperation given to us by a number of graduates and graduate students. Dr. Ing. Birger Hoggen of the Department of Applied Earth Sciences and Metallurgy should be mentioned. From industry in recent years Dr. Ing. Terje Pedersen and Mr Erling Myrbostad of Hydro Aluminium have made important contributions in the field. Also, I wish to mention the assistance given to me concerning turbulence theory by Dr. Techn. Stein Tore Johansen. The present Dr. Ing. students Frede Frisvold, Dong Liang, Per Bakke, and Eirik Röhmen have all helped with the book.

Research has partly been conducted at the Norwegian Institute of Technology and at SINTEF (The Foundation of Scientific and Industrial Research at the Norwegian Institute of Technology), and some research and development has taken place in industry. Many of the research projects have been carried out in cooperation with Norsk Hydro a.s. and Elkem a.s. and with the Swedish steel industry. The author wishes to express gratitude for financial support given by the Royal Norwegian Council for Scientific and Industrial Research.

I am particularly happy that two of my colleagues, Professor Olle Wijk from the Royal Institute in Stockholm and Dr. Techn. Christian Simensen from the Centre for Industrial Research in Oslo have undertaken to write two of the chapters. The field of refining metallurgy is so wide that their special knowledge is vital. The Dr. Ing. students of Professor Olle Wijk are to be thanked for checking and proof-reading the manuscripts for the various chapters. Also, I wish to acknowledge the help and encouragement given to me on Chapter 9 by Dr. Ing. Halvard Tveit of Elkem.

Mr Roald Skjervø has assisted with the figures and diagrams. Mrs Asla Olufsen and Mrs Ingrid Page deserve the warmest thanks for typing the various parts of the typescript. My family and friends are to be thanked for their forbearance.

The following books have been very useful:

Iida, T. and Guthrie, R. I. L. (1988). *The physical properties of liquid metals.* Oxford University Press.

Turkdogan, E. T. (1980). *Physical chemistry of high temperature technology.* Academic Press, New York.

Lupis, C. H. P. (1983). *Chemical thermodynamics of materials.* North-Holland, New York.

The authors gratefully acknowledge permission to draw, or reset, the following list of figures and tables from previous publications.

Fig. 1.2	Talbot (1975)	*Int. Met. Rev.*
Fig. 1.3	Turner and Bryant (1967)	*J. Inst. of Metals*
Fig. 1.5	Allen and Hemitt (1933)	*J. Inst. of Metals*
Fig. 1.6	Mattson and Schückher (1958–59)	*J. Inst. of Metals*
Fig. 1.7	Rylski	Dept. of Mines and Technical Surveys, Ottawa
Fig. 1.8	Craighead, Lenning and Jafee	*J. of Metals*
Fig. 1.9	Higgins (1985)	Hodder and Stoughton, London
Fig. 1.11	(1978)	STU Information, Stockholm
Fig. 1.12	Kiessling (1981)	Symposium Uddeholm, Sweden
Figs 1.15, 1.16, 1.17	Leschiner and Kovalyov (1989)	Proc. Int. Workshop, Balatonfüred, Hungary
Fig. 1.20	Hilty (1985)	Bookcrafters Inc., Chelsea, MI
Fig. 1.21	Pickering (1975)	Union Carbide Corp.
Fig. 1.22	Pickering (1971)	Climax Molybdenum Corp.
Fig. 1.23	Dodd, Bai (1987)	Academic Press
Figs 1.24, 1.25	Lagneborg (1981)	Symposium Uddeholm, Sweden

Fig. 1.26	Nordberg (1980)	Swedish Inst. for Metal Research
Fig. 1.28	Emley (1966)	Pergamon Press
Fig. 1.29	Hanawalt, Nelson, Peloubet, (1942)	*Trans AIME*
Fig. 2.4	Gaskell (1973)	McGraw-Hill, New York
Fig. 2.9	Elliott, Gleisen, Ramakrishna (1963)	Addison Wesley, Reading
Fig. 2.10	Dewing (1972)	*Metall. Trans*
Fig. 2.12	Turkdogan (1980)	Academic Press, New York
Fig. 2.13	Selin (1987)	Doctoral thesis, Stockholm
Figs 2.14, 2.15, 2.20	Elliott, Gleiser, Ramakrishna (1963)	Addison Wesley, Reading
Fig. 2.17	Iida, Guthrie (1989)	Oxford University Press
Fig. 2.18	Chapman, Couling (1970)	Cambridge Univ. Press
Tables 2.2, 2.7	Lupis (1983)	North-Holland, New York
Tables 2.3, 2.4, 2.8, 2.9, 2.10	Sigworth, Elliott (1974, 1976)	*Can. Metall. Quarterly*
Tables 2.11, 2.12	Geiger, Poirier (1973)	Addison Wesley, Mass.
Table 2.13	Alward, Findlay (1974)	John Wiley, Hong Kong
Fig. 3.4	Hansen (1958)	McGraw-Hill, New York
Fig. 3.5	Davey (1986)	*Encyclopaedia of Materials Science and Engineering*, Oxford
Fig. 3.6	Goyal (1982)	*Journal of Metals*
Fig. 3.8	Kubaschewski and Alcock (1983)	Pergamon Press, Oxford
Fig. 3.9	Sears (1990)	*J. of Metals*
Fig. 3.10	Kuwabara, Umezawa, Mori, Watanabe (1988)	*Trans ISIJ*
Fig. 3.11	Feinman (1987)	*Plasma technology in metallurgical processing*, Warrendale, USA
Tables 3.1, 3.2, 3.5, 3.6, 3.7, 3.8, 3.9, 3.10, 3.11	Goetz (1986)	*The New Encyclopedia Britannica*, Chicago and
Table 3.12	Brandes (1983)	*Smithells Metals Reference Book*, Butterworths, London
Fig. 5.8	Asano, Nakano (1972)	*Trans ISIJ*
Fig. 5.19	Burghard (1970)	*Neue Hütte*
Fig. 5.10	Kawawa, Ohkabo (1968)	*Trans ISIJ*
Fig. 5.20	Nilmani (1984)	Symposium on extractive metallurgy, The Australian I.M.M. Branch, Melbourne
Figs 9.7, 9.9	Pfann (1966)	John Wiley, New York
Fig. 9.10	Hansen, Anderko (1958)	McGraw-Hill, New York
Fig. 10.1	Van Linden (1990)	Light Metals, Warrendale
Fig. 10.2	Schenker (1990)	*Lead-Zinc '90*, Warrendale

CONTENTS

NOTATION AND UNITS

Symbol	Description	Units
ΔA	interfacial contact area in control volume	m^2
A	interfacial contact area, cross-sectional area or integration constant, or constant	m^2
A_j	area in contact with phase j	m^2
A_M	Madelung constant	
A_p	surface area of particle or of injected powder	m^2
A_s	top surface area of bath	m^2
A_R	surface area of bubbles with radius R	m^2
A_v	vertical projection	m^2
A_{12}, A_{13}, A_{23}	contact areas between phases 1, 2, and 3	m^2
A_j^σ	interfacial area occupied by one mole of species j	m^2/mol
a	constant	
	or radius of inclusion	m
a_1, a_2, a_3	constants	
a_c	reference particle radius	m
a	activity coefficient	
a_i	activity of component i	
a_j	activity of component j	
a_1, a_2, a_3	constants	
a_s	specific surface area	m^{-1}
B	constant	
Bi, Bi*	Biot number	
b, b^*	constant, ratio of projected area to total surface area or thickness	m
b_j	constants	
C	constant	
	or number of components	
c	concentration, total concentration	$kmol/m^3$, kg/m^3,
	or volume fraction	or m^{-3}
c^*	concentration of nuclei	m^{-3}
$c(a)$	number of inclusions of diameter $2a$ per unit volume	m^{-4}
c_b	bulk concentration (of inclusions)	$number/m^3$
c_c	volume fraction cake	
c_E	eutectic composition	
c_{in}	inlet concentration	$kmol/m^3$
	or initial concentration	m^{-3}

Symbol	Description	Units
c_j, c_i	concentration of component j or i	kmol/m^3
c_e	concentration in liquid	kmol/m^3
c_n	concentration of nuclei	number/m^3
$c_{o'}$	outlet concentration	number/m^3
c_p	specific heat	J/(mol K)
c_r	concentration in reactor	kmol/m^3
c_s	concentration in solid or surface concentration	kmol/m^3
c_δ	concentration at distance δ_1 from wall	m^{-3}
[%C]	concentration of carbon	100 kg carbon/kg melt
[%C]e	hypothetical concentration of carbon in melt in equilibrium with gas phase	100 kg carbon/kg melt
c^*	number of nuclei per unit volume	m^{-3}
D	diffusion coefficient	m^2/s
D_j	diffusion coefficient of component j	m^2/s
D_E	eddy diffusivity for mass transfer	m^2/s
d	diameter	m
d_c	collision diameter	m
d_e	diameter of surface deformation	m
d_p	particle diameter	m
d_{pc}	critical particle diameter	m
E	filtration efficiency	
E	Young modulus	N/m^2
$\dot{E}$	stirring power	W
E_0	viscosity group or Eötvös number	
e	electronic charge	1.6×10^{-19}C
	or protrusion height (roughness)	m
	or factor	
e_i^j	interaction coefficient for effect of component j on i	
F	Helmholtz free energy	J/mol
	or degrees of freedom	
	or force	N
F_d	drag force	N
F_g	gravity force	N
f	function	
	or fraction liquid	
	or frequency	1/s
f_i	activity coefficient of component i	
f_i^σ	interfacial activity coefficient of component i	
$f_N(a)$	number-size distribution of inclusions	m^{-4}

f_R	number-size distribution of bubbles	m^{-4}
f_s	fraction solid	
f_x	activity coefficient for component x in melt	
G	Gibbs energy	J/mol, J
G_v	Gibbs energy per unit volume	J/m^3
G^σ	Gibbs energy of surface	J/mol, J
ΔG	change of Gibbs energy (free energy)	J/mol
ΔG_{mix}	change of Gibbs energy for mixture	J/mol, J
ΔG_{mix}^{id}	change of Gibbs energy for ideal mixture	J/mol, J
ΔG_{mix}^{xs}	excess Gibbs energy of mixing	J/mol, J
ΔG^0	standard Gibbs energy, Gibbs energy of adsorption	J/mol
ΔG_n	standard Gibbs energy for nucleus	J/mol
ΔG_n^*	critical Gibbs energy of nucleation	J
ΔG_v	Gibbs energy of nuclei	J/m^3
ΔG_{lens}^*	critical Gibbs energy of lens	J
$\dot{G}$	flow of gas	kmol/s, Nm^3/s
$\dot{G}_x$	removal rate of component x	kmol/s, Nm^3/s
g	Gibbs energy per unit area	$J/(mol\ m^2)$
	or acceleration of gravity	m/s^2
	or volume fraction	
	or function	
g^σ	Gibbs energy per unit area	J/m^2
$g(r)$	pair distribution function	
$[\%H]_v$	hydrogen percentage in melt in equilibrium with water vapour	per cent
H	distance from gas inlet to bath surface	m
	or thickness of filter	m
	or enthalpy	J
	or dimensionless group	
H_s	enthalpy that enters addition during shell period	J
H^σ	difference between molar enthalpy on the surface and in the bulk	J/mol
ΔH_m, ΔH_m^*	heat of melting	J/kg
h, h^*	heat transfer coefficient	$J/(m^2\ K\ s)$
ΔH	change of enthalpy	J/mol
$\overline{\Delta H^0}$	average values of enthalpy	J/mol
ΔH_v	enthalpy of evaporation	J/mol
i	component i or symbol for interface	
J	integral	
j	component j	

$[\%j]$	mass per cent of component j	100 kg j/kg melt
$[\%\mathrm{j}]^{\sigma}$	mass per cent of component j at surface	100 kg/kg melt
K, K'	equilibrium constants, inverse equilibrium constants or stress intensity factor	$N/m^{3/2}$
K_{Ic}	critical stress intensity factor	$N/m^{3/2}$
K_p	equilibrium constant for phosphorus equilibrium	
K_{p}	equilibrium constant for powder	
K_j	equilibrium constant for component j	
	or coefficient for electrical resistivity	Ωm
k	mass transfer coefficient for melt	m/s
	or thermal conductivity	J/(m K s)
	or number of components	
	or Boltzmann's constant 1.38×10^{-23}	J/K
	or partition ratio or coefficient	
	or turbulent kinetic energy per unit mass	m^2/s^2
k_{eff}	effective partition ratio or coefficient	
k_j	mass transfer coefficient for component j	m/s
k_{g}	mass transfer coefficient for gas	m/s
k_{H}	Henry's constant	
	or mass transfer coefficient for hydrogen	m/s
$k_{\mathrm{H_2}}$	mass transfer coefficient for H_2 in gas	m/s
$k_{\mathrm{H_2O}}$	mass transfer coefficient for H_2O in gas	m/s
k_{s}	thermal conductivity of shell	J/(m Ks)
k_{slag}	mass transfer coefficient for slag	m/s
k_{t}	total mass transfer coefficient for dissolved component or for inclusions	m/s
k_1	mass transfer coefficient for short contact time	m/s
k_2	mass transfer coefficient for long contact times	m/s
k_{tp}	total mass transfer coefficient for injected powder	m/s
k_{ts}	total mass transfer coefficient for top surface	m/s
L	characteristic vessel dimension, length of stirrer blades, distance	m
	or Lorenz number	$W\Omega/K^2$
l	Prandtl mixing length or length of molten zone	m
l_{e}	characteristic length of energy containing eddies	m
M	mass of bath	kg
	or symbol for alloying element	
	or Morton number	
M_j	molar weight of component j	g/mol or kg/kmol
M_{Na}	molar weight of sodium	g/mol or kg/kmol

M_s	mass of slag	kg
M_p	mass of powder	kg
M_1, M_2	molar weight of component 1, 2	g/mol or kg/kmol
$\dot{M}$	flow of metal into reactor	kg/s or kmol/s
$\dot{M}_p$	rate of injection of powder	kg/s
m	mass of atom or molecule or number of broken bonds	kg
$\dot{m}$	mass flux	kg/(m^2 s)
N	number of back-mix tanks or number of bubbles present in melt or number of atoms or particles	
N_A	Avogadro's number or ionic fraction of component A	mol^{-1}
N_B	ionic fraction of component B	
N_I	injection number	
N_R	number of bubbles of radius R	
N_V	dimensionless volume	
Nu	Nusselt number	
N_1, N_2	number of atoms or number of kmoles per second	kmol/s
$\dot{N}_R$	number of bubbles of radius/R created per second	1/s
n	integer or number or rotations per sec or number of kmoles or moles	1/s kmol, mol
n_i	number of molecules of component i or number of moles of component i	mol
n_i^σ	number of moles of component i at surface	mol
n_j	stoichiometric factor	
$\dot{n}$	molar flux	kmol/(m^2 s)
$\dot{n}_1$, $\dot{n}_2$, $\dot{n}_3$	molar flux from phases 1, 2, 3	kmol/(m^2 s)
$\dot{n}_a$	flux of inclusion of size a	1/(m^3 s)
[% O]$_i$	oxygen in inclusions	per cent
[% O]	dissolved oxygen in steel	per cent
P	number of phases	
p	pressure or probability	N/m^2
p_{bj}	partial pressure in bulk gas of component j	N/m^2
p_{H_2O}	partial pressure of H_2O in gas	N/m^2
p_i	pressure at interface or pressure of component i	bar

p_{in}	inlet pressure	bar
p_j	partial pressure of component j	N/m^2
p_{ji}	partial pressure of component j at interface	N/m^2
p_j^0	partial pressure of j that leaves metal j	bar
	or pressure over a pure melt of component j	N/m^2
p_{atm}	pressure above bath	bar
p	pressure	bar, N/m^2
p_{CO}, $p_{CO}^{(min)}$, $p_{CO}^{(max)}$	partial pressures of CO	bar
p_{inert}	partial pressure of inert gas	bar
p_o	outlet pressure	bar
p_0, p_i^0	vapour pressure of pure substance i	bar
p_{tot}	total pressure	bar
p_x	partial pressure of component x	bar
p_x^0	partial pressure of component x in bubbles when leaving melt	bar
Δp, Δp_c, Δp_f	total pressure drop, pressure drop in cake, and in filter, respectively	N/m^2
p_{inert}	partial pressure of inert gas	bar
p_1, p_2	pressure	N/m^2
Δp	pressure drop	N/m^2
Pe	Peclet number	
Pr	Prandtl number	
Q	heat	J
$\dot{Q}$	circulation flow of melt	m^3/s
$\dot{Q}_1$, $\dot{Q}_2$	volume flow of gas	m^3/s, Nm^3/s
$\dot{Q}_g$	flow of gas	m^3/s, m^3/hr, Nm^3/s
$\dot{Q}_{O_2}(j)$	volume of oxygen that reacts with element j per unit time	Nm^3/s
$\dot{Q}_{CO}$, $\dot{Q}_{Ar}$	volume flow of CO, argon respectively	Nm^3/s
$\dot{Q}_{O_2}$	total flow of oxygen	Nm^3/s
$\dot{q}$	flow of liquid	m^3/s
	or heat flux	J/(m^2 s)
R	radius of collector or addition, or bubble	m
	or universal gas constant	8.314 J/mol K
	or number of restrictions	
Re	Reynolds number	
r	radial distance	m
	or ratio	
r_c, r_f	resistance in cake and filter, respectively	m^{-2}

r_0	distance between ions	m
r^*	critical radius of nuclei	m
S	entropy or sum	J/(mol K), J/K
$S_1(x)$, $S_2(x)$	functions	
S_n	fatigue limit	N/m^2
S^σ	Difference between the molar entropy of the interface and the two bulk phrases	J/(mol K), J
ΔS	change of entropy	J/(mol K), J/K
ΔS_m	entropy of fusion	J/(mol K)
ΔS_n	entropy of nuclei	$J/(m^3 K)$
$\overline{\Delta S^0}$	average value of entropy	J/(mol K)
ΔS_{mix}	entropy of a mixture	J/(mol K), J/K
ΔS_{mix}^{id}	entropy of an ideal mixture	J/(mol K), J/K
Sh	Sherwood number	
s	length or position of solidification front	
Sc	Schmidt number	
T	temperature	°C, K
T_b	temperature of bulk	°C, K
T_{in}	initial temperature	°C, K
T_l	temperature of liquid	°C, K
T_W	temperature of wall	°C, K
T_m, T_m^*	melting temperature	°C, K
T_0	room temperature	°C, K
t	time, residence time	s
t_d	dissolution time	s
t_c	circulation time or critical time	s
t_a	melting down time	s
t_0	shell time	s
t_{tot}	total time	s
U	internal energy or total energy	J/mol, J J/m
U_e	elastic strain energy	J/m
U_s	surface energy	J/m
u	velocity of particle	m/s
u_r	velocity of particle relative to melt	m/s
u_{br}	velocity of particles relative to bubbles	m/s
V	volume or volume per mole	m^3, m^3/mol
V^*	critical volume of nucleus	m^3
V_m	volume at melting point or volume per mole at melting point	m^3, m^3/mol

V_3	volume of lens	m^3
$\dot{V}$	volume that bubbles sweep through per second	m^3/s
$\dot{V}_e$	effective volume purified per second	m^3/s
v	average velocity of melt or	m/s
	velocity of solidification front	m/s
v_b	velocity of bubble	m/s
	or velocity outside boundary layer	m/s
v_g	velocity of gas phase	m/s
v_m	mean velocity (in melt)	m/s
v_0	shear velocity along wall	m/s
	or eddy velocity in bulk melt normal to free surface	m/s
$\bar{v}_x$	average velocity along wall in x-direction	m/s
v_x	eddy velocity in x-direction	m/s
v_z	eddy velocity in z-direction	m/s
v_θ	velocity along meridian	m/s
v_{δ_1}	velocity at distance δ_1 from wall	m/s
W	thermodynamic probability	
	or work	J
We	Weber number	
Z	number of nearest neighbours	
	or number of electronic charges	
X	variable	
	or inclusion spacing	m
X_a	dimensionless group for addition	
x	component x in melt	
	or coordinate in axial direction	m
	or molar fraction	
	or variable	
x_i	mole fraction of component i	
x_j	mole fraction of component j	
x_n	mole fraction in plate n	
x_{in}	inlet mole fraction	
$[\% x]$	mass percentage of x in melt	100 kg x/kg melt
$[\% x]_e$	hypothetical percentage of x in melt	100 kg x/kg melt
$[\% x]_i$	concentration at interface	
$[\% x]_{in}$	initial mass percentage of x in melt	100 kg x/kg melt
$[\% x]_\infty$	final mass percentage of x in melt	100 kg x/kg melt
$(\% x)_i$	mass percentage of x in slag at interface	100 kg x/kg slag
$(\% x)_{in}$	initial mass percentage of x in slag	100 kg x/kg slag
$(\% x)_p^0$	exit mass percentage of x in slag	100 kg x/kg slag

y_{exit}	exit mole fraction	
y_j	mole fraction of component j in gas phase	
y_n	mole fraction in gas phase above plate n	
Z, Z_1, Z_2	efficiency of gas purging for monoatomic gas or number of neighbours, or number of charges	
Z^2	efficiency of gas purging for two-atomic gas	
z	spatial coordinate normal to interface	m
α	thermal diffusivity	m^2/s
	or constant	
α_{ij}	dimensionless parameter	
α_s	thermal diffusivity of shell	m^2/s
$\dot{\alpha}_e$	constant	s^{-1}
β	constant	
β_1, β_2, β_3	constants	
γ_i	activity coefficient of component i in slag on mass per cent basis or Raoultian activity coefficient	
γ_i^0	Raoultian activity coefficient at infinite dilution	
γ_{ip}	activity coefficient of component i in powder on mass per cent basis	
δ	collision diameter	m
	or thickness of diffusion boundary layer	m
δ_b	thickness of turbulent wall boundary layer	m
δ_1	thickness of restricted turbulence boundary layer	m
δ_2	thickness of viscous (laminar flow) boundary layer	m
ϵ	void fraction, energy parameter (see page 88)	
ϵ_0	dielectric constant	$C^2/(Nm^2)$
$\dot{\epsilon}$	specific power input	m^2/s^3
η	collision efficiency	
η_i, η_r	collisions efficiency due to interception and relative velocity, respectively	
Γ_i	concentration at surface of component i	mol/m^2
μ	viscosity	Ns/m^2
μ_i	chemical potential of component i	J/mol, J
	or viscosity of component i	Ns/m^2
μ_i^0	chemical potential of reference state	J/mol, J
μ_t	turbulent viscosity	Ns/m^2
ν	kinematic viscosity	m^2/s
κ	constant	°C/s
	or viscosity ratio	
	or compressibility	m^2/N

λ	distance between collisions	m
	or smallest eddy size	m
	or relative resistance	
	or ratio	
ρ, ρ^*	density of melt or shell	kg/m^3
	or distance	m
	or resistivity	Ωm
ρ_b	density of particles	kg/m^3
ρ_i	density of inclusion	kmol/m^3
ρ_μ	density of solid alloy addition	kmol/m^3
ρ_0	density of oxygen in inclusion	kmol O/m^3
ρ_0	resistivity of pure metal	Ωm
ρ_{slag}	density of slag	kg/m^3
σ	surface tension	N/m
	or electrical conductivity	1/(Ωm)
	or stress	N/m^2
σ_F	fracture stress	N/m^2
σ_l	surface tension of pure solvent	N/m
σ_m	tensile strength	N/m^2
σ_p	plastic energy required to extend crack	J/m^2
σ_s	surface energy of solid	J/m^2
σ_{sl}	interfacial tension between solid and pure solvent	N/m
σ_{13}, σ_{12}, σ_{23}	interfacial, tensions between phases 1 and 3, 1 and 2, and 2 and 3 respectively	N/m
τ	shear stress	N/m^2
	or residence time	
	or time constant	s
	or dimensionless time	
τ_a	dimensionless shell time	
τ_b	residence time of bubbles	s
τ_e	lifetime of the smallest eddy	s
τ_m	time constant for mixing	s
τ_0	time constant	s
	or shear stress at wall	N/m^2
	or dimensionless shell time	
τ_p	time constant for injection of powder	s
	or 'relaxation time' after a collision	s
τ_R	residence time for bubbles of size R	s
τ_{xy}	shear force	N/m^2
θ	polar angle or angle; contact angle	

ϕ	mean potential energy or friction factor equal to v_0^2/v_b^2	J/mol
$\phi(x)$, $\phi^*(x)$	functions	
ϕ_A	dimensionless group for contact area of bubbles	
ϕ_p	dimensionless group for contact area of powder	
ϕ_s	dimensionless group for contact area of top slag	
$\psi/[\% \text{ H}]$	hydrogen dimensionless group for bubble contact area	
ψ	stream function	m^3/s
$\psi(x)$	function	
ζ	$= 3\pi\mu d$	N s/m
Ω_μ	collision integral for viscosity	

Useful SI (Système International) units, conversion factors, and physical constants

Fundamental units

Physical quantity	Unit	Abbreviation
Length	metre	m
Mass	kilogram	kg
Time	second	s
Electric current	ampere	A
Temperature	kelvin	K
Amount of substance	mole	mol

Other physical quantities

Physical quantity	Unit	Abbreviation	In fundamental units
Mass	gram	g	10^{-3} kg
Mass	tonne	–	10^3 kg
Volume	litre	l	10^{-3} m^3
Force	newton	N	m kg/s^2
Energy	joule	J	m^2 kg/s
Power	watt	W	m^2 kg/s^3
Pressure	pascal	Pa	$kg/(ms^2)$
Electric charge	coulomb	C	As
Electric potential	volt	V	W/A
Electric resistance	ohm	Ω	V/A
Viscosity	–	Pas	kg/(ms)
Surface tension	–	–	kg/s^2

Multiples or fractions of SI units

Fraction	Prefix	Symbol
10^{-2}	centi	c
10^{-3}	milli	m
10^{-6}	micro	μ
10^{-10}	-	Å
10^{3}	kilo	k
10^{6}	mega	M
10^{9}	giga	G

The volume of ideal gas contained in 1 m^3 at 0°C and 1 bar is 1 Nm^3.
The volume of ideal gas contained in 1 l at 0°C and 1 bar is 1 Nl.

Conversion factors

1 mol of gas	= 22.41 Nl
1 kmol of gas	= 22.4 Nm^3
1 atm	= $1.013\ 25 \times 10^5$ Pa = 1.013 25 bar = 760 torr (mm Hg)
1 calorie	= 4.184 J
1 kcalorie	= 4.184 kJ
0 K	= − 273.15°C
1 electron volt (eV)	= $1.602\ 18 \times 10^{-19}$ J

Physical constants

gravitational acceleration at sea level and 45° latitude	$g \approx 9.806$ m/s^2
gas constant	$R = 8.314\ 51$ J/(K mol)
Boltzmann's constant	$k = 1.380\ 66 \times 10^{-23}$ J/K
Avogadro's number	$N_A = 6.022\ 14 \times 10^{23}$ atoms/mol
electron charge	$e = 1.602\ 18 \times 10^{-19}$ C
Faraday's constant	$F = 96\ 485.3$ C/mol
Planck's constant	$h = 6.626\ 08 \times 10^{-34}$ J s
mass of electron	$m_e = 9.109\ 39 \times 10^{-31}$ kg
speed of light	$c = 2.997\ 924\ 6 \times 10^8$ m/s
dielectic constant (permittivity of free space)	$\epsilon_0 = 8.854 \times 10^{-12}$ C^2/(Nm^2)

1

THE EFFECT OF DISSOLVED ELEMENTS AND INCLUSIONS ON THE PROPERTIES OF METAL PRODUCTS

Christian J. Simensen

1.1 INTRODUCTION

Metals have been in service for mankind since the early days of civilization. Tools were made from meteorites as far back as 10 000 BC, and copper was in use by the Sumerians in around 4500 BC. The use of light metals is of a much more recent date. The mass production of aluminium started in 1886 with the invention of the Hall–Héroult process, while magnesium was mass-produced only after World War I. The range of products is vast. Metals are in use in high-technology environments in satellites and aircraft and in common products such as kitchenware, refrigerators, and vacuum cleaners. All these products have one thing in common: certain impurities are prohibited as they ruin the metal quality, while other elements are added in order to meet specific requirements. Some of the major effects of impurities on product quality are outlined in this chapter in order to emphasize the necessity of chemical control and refining of the solvent metal and additions.

The earliest reported alloy development which was indirectly caused by impurities is the development of bronzes in the Middle East around 3500 BC. The first copper produced contained large quantities of arsenic which increased the strength of the metal. However, strengthening was more readily obtained by adding 5–10 per cent tin. Therefore products where arsenic was removed and tin was added were developed over the centuries. The result may be seen even today in precious artefacts and weapons.

The oldest metal products contained large quantities of material foreign to and insoluble in the matrix—inclusions—typically in the range 1–20 per cent. Modern production of metals gives a much lower content of inclusions and unwanted dissolved elements. The new refining and production techniques have resulted in impurity levels ranging from 2–3 per cent to a few parts per million (p.p.m.). Extreme requirements on purity have led to the development of advanced refining processes such as vacuum techniques and zone refining.

Examples of impurities in various commercial metal products are summarized in Table 1.1.[1] It is evident that metals like magnesium, zinc, and copper are available with a low content of impurities, ≈ 0.1 per cent, while iron is difficult to make without some carbon, and aluminium contains some iron and silicon. Remelted metal has an even higher content of impurities.

The literature, especially concerning steel, is so vast that no attempt has been made to summarize the various effects of dissolved elements and inclusions on

TABLE 1.1 *A survey of common impurities in some metals*

Raw Material:	Mg	Al	Ti Electr.	Mn Electr.	Fe	Cu	Zn
Content (%) of solvent element:	99.8	99.7	99.8	99.93	96	99.9	99.9
Impurity	Typical concentration (mass p.p.m.)						
H	0.2–3	0.2–0.5		150			
C					0.3–1.5%		
Mg		10–30					
Al	50–150						
Si	20–60	300–700	20		0.6–1.2%		
S					100–350		
Ca	14	3–10					
Ti	<1	30–50					
V		100–200					
Mn	50–120	10–30			0.4–2%		
Fe	100–400	500–2000				6	500–2000
Ni	1–5	10–80	40	25		3	
Cu	14	5–30					
Cd	7						100–300
Other elements:		Ga		Co, As, Mo		Ag	
Carbides	1.5–7	1–5					
Oxides	1–30	0.3–3	370		*	50–300	*
Nitrides	10–60	0.1–2					
Salts	*						
Sulphates				140			
Sulphides				180			

*Magnesium contains some salt; iron and zinc contain oxides.

the properties. Only a rudimentary outline and some examples are given. However, the chapter should serve to give the student a feeling of how properties can be modified by refining and micro-alloying. Because inclusions and properties in steel are described at such length in the literature, many examples presented here concern other metals like aluminium and copper. Nevertheless, at least some of the examples should also be interesting for people working with steel.

Impurities can be classified into four main groups. These include volatile elements, reactive elements, non-reactive elements, and inclusions. Volatile elements are elements such as hydrogen and sodium with a high vapour pressure. Reactive elements are those which can be removed from melts by

TABLE 1.2 *A comparison between aluminium metals from electrolysis cells and three-layer electrolysis with respect to impurities*

Impurity	Typical concentration (mass p.p.m.)	
	Electrolysis cells	three-layer electrolysis
Mg	10–40	<1
Si	300–700	1–5
Ti	30–50	<1
V	100–200	<1
Fe	500–2000	1–10
Ni	10–80	<1
Cu	5–30	<1
Zn	20–200	<1
Ga	80–180	1

adding oxygen, oxides, chlorides, water, salts, boron, or other chemicals. These impurities include calcium, lithium, and titanium. Examples of non-reactive elements are iron in aluminium which is very difficult to remove by ordinary processes. One way to refine aluminium is to use three-layer electrolysis where the impurities are absorbed in another liquid (see also Section 3.4). This aluminium product has a very low content of non-reactive impurities, namely less than 10 p.p.m. as shown in Table 1.2.[1]

The inclusions (see Figs 5.1 and 5.2) can be in both the liquid state and the solid state. Impurities in metal products lead to all kinds of changes, from an increase in the modulus of elasticity to a decrease in fatigue strength and ductility, an increase in corrosion, and a reduction in electrical conductivity. Some of the major problems are dealt with in this chapter in order to underline the importance of molten melt treatment and chemical and physical control of metal quality.

1.2 POROSITY

When a metal is held at high temperatures, some elements will evaporate and oxidize in air. Typical examples are hydrogen and sodium in aluminium. However, if the surroundings (air, furnace walls, slag, dross, or electrolyte) have a high content of such elements, the metal may absorb these impurities instead. The result can be porosity in the metal, and a reduced ductility. This section concerns the formation of pores and the deterioration of metal quality.

Generally, metals have a higher solubility of hydrogen in the liquid state than in the solid state. For example, the equilibrium content of hydrogen in magnesium at the melting point and one bar H_2 is 50 p.p.m. in the liquid and 37 p.p.m. in the solid state (Fig. 1.1). 2–6. The corresponding values for

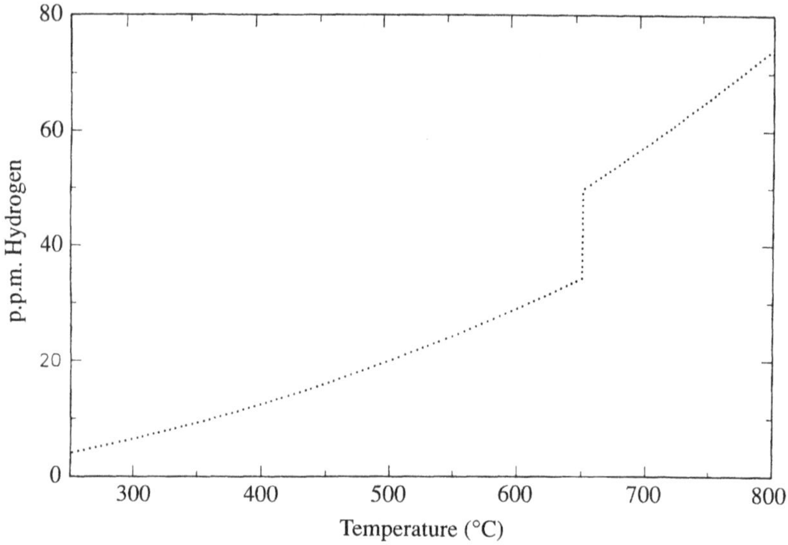

FIG. 1.1 Solubility of hydrogen in magnesium at 1 bar H_2.[2-6]

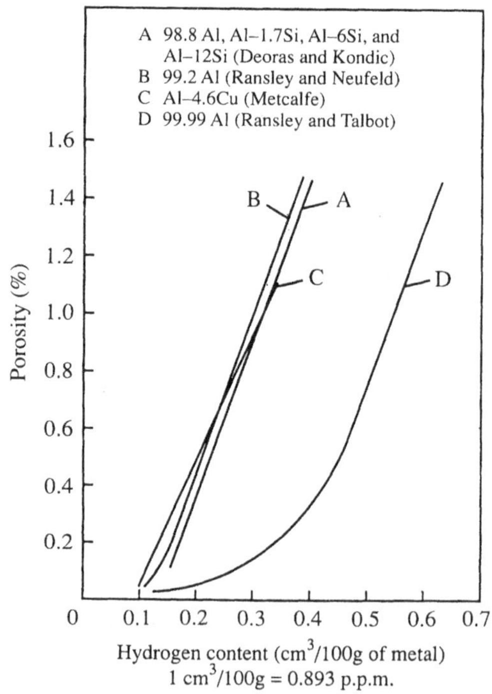

FIG. 1.2 Relation between hydrogen content and porosity for sand-cast bars, 25 mm diameter, of aluminium and aluminium alloys.[7]

aluminium are 0.64 p.p.m. in the liquid and as low as 0.03 p.p.m. in the solid state (see Fig. 4.14). When a metal with a high content of hydrogen is cast, hydrogen pores will be formed during solidification. The number of pores depends on the casting process and the content of hydrogen. An example is given for a series of aluminium alloys in Fig. 1.2.[7] It is evident that the porosity increases drastically with only small increases in the hydrogen content. Turner and Bryant[8] have shown that this porosity causes a reduction in short-transverse tensile properties of semicontinuously cast ingots of an Al–Cu–Mg–Si–Mn alloy (Fig. 1.3). Another effect is blistering of sheet metal during hot rolling or high-temperature annealing as revealed in Fig. 1.4. In this case hydrogen pores are nucleated on inhomogeneities such as grain boundaries and inclusions.[9] The pores grow by diffusion of hydrogen during high-temperature treatment. The result is blisters caused by hydrogen pores formed close to the metal surface.

In base metals, such as aluminium and magnesium, hydrogen alone creates the pores. In other metals pores may be caused by several dissolved elements. This is the case for copper alloys where both hydrogen and oxygen can be in solution. The content of the elements in molten copper at 1150°C is shown in Fig. 1.5.[10] It is evident that there is a relation between the solubility of the two elements. As shown by Allen and Hewitt,[10] water vapour bubbles may form in molten copper alloys by the reaction

$$H_2O(\text{gas}) = 2\underline{H} + \underline{O}. \tag{1.1}$$

The underlining indicates that the element is dissolved in the metal. Equation (1.1) is discussed in more detail in Chapter 3. Water vapour steam

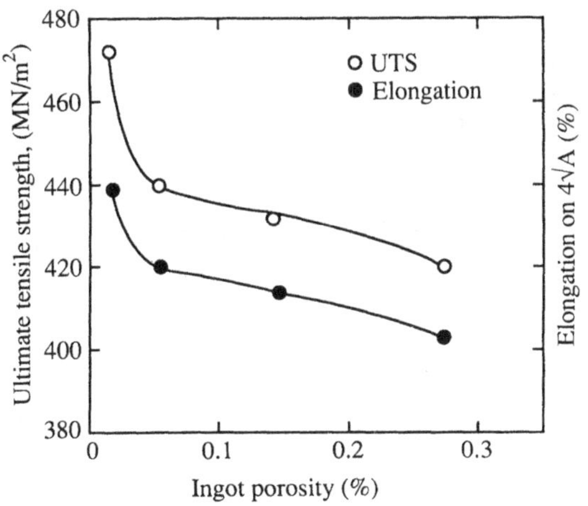

FIG. 1.3 Effect of ingot porosity on short-transverse tensile properties of 7.5 cm thick plate hot rolled from semicontinuously cast ingots of Al–4% Cu–0.6% Mg–0.6% Si–0.7% Mn with 28 × 112 cm cross-section.[8]

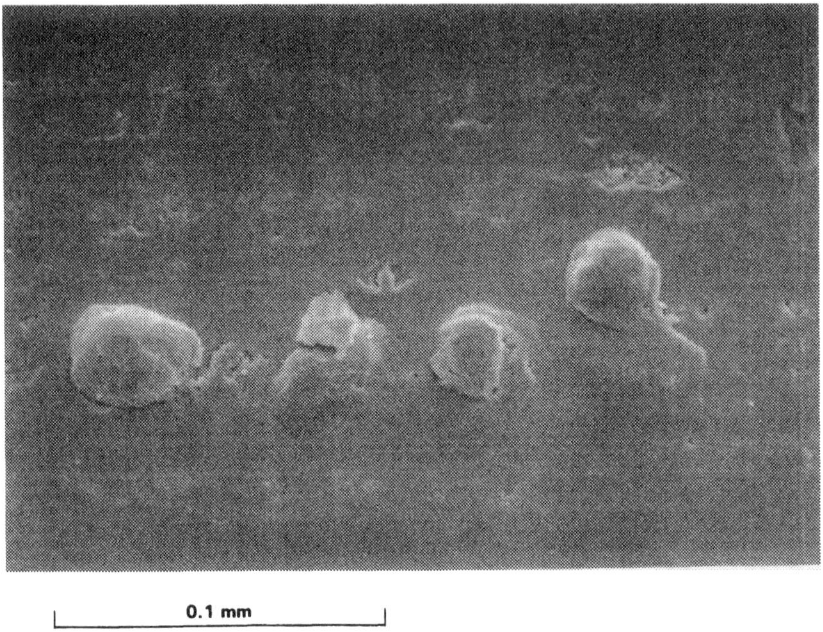

FIG. 1.4 Blisters in an aluminium sheet (photo: T. Ustad, Hydro Aluminium).

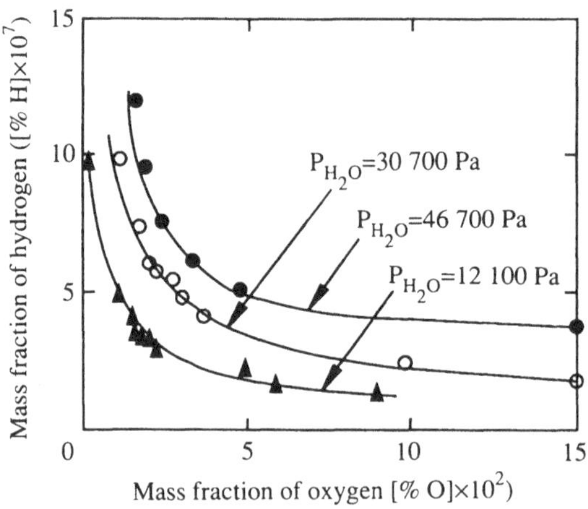

FIG. 1.5 Dissolved oxygen and hydrogen in molten copper at 1150°C.[10]

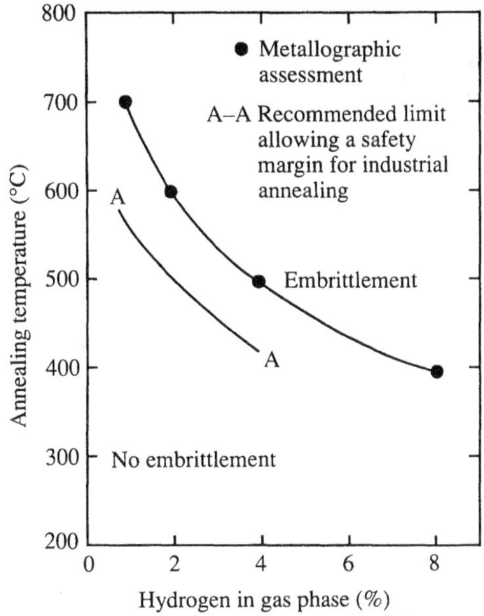

FIG. 1.6 Limiting conditions of temperature and gas-phase composition for steam embrittlement of tough-pitch copper containing 0.03 mass% O annealed for 1.5 hours.[12]

embrittlement can also occur in solid copper which contains oxygen. When the metal is heated above 400°C in an atmosphere containing hydrogen, the metal can be damaged beyond recovery by grain boundary cracking.[11] The mechanism is that hydrogen diffuses into the metal and reacts with cuprous oxide particles, generating steam according to

$$Cu_2O + 2\underline{H} = 2Cu + H_2O(\text{gas}). \tag{1.2}$$

Fortunately, however, even though hydrogen is a small atom and has a high diffusion coefficient in solid metals, it is possible to anneal copper before too much harm has been done. Mattsson and Schückher[12] have investigated the reaction and have given the limiting conditions for safe 1.5 hours annealing of tough-pitch copper containing 0.03 per cent oxygen. Their results are outlined in Fig. 1.6.

1.3 HYDROGEN EMBRITTLEMENT OF METALS

As already mentioned the solubility of hydrogen in solid metals increases with temperature. The solubility is high for metals like alpha-titanium which is 0.17

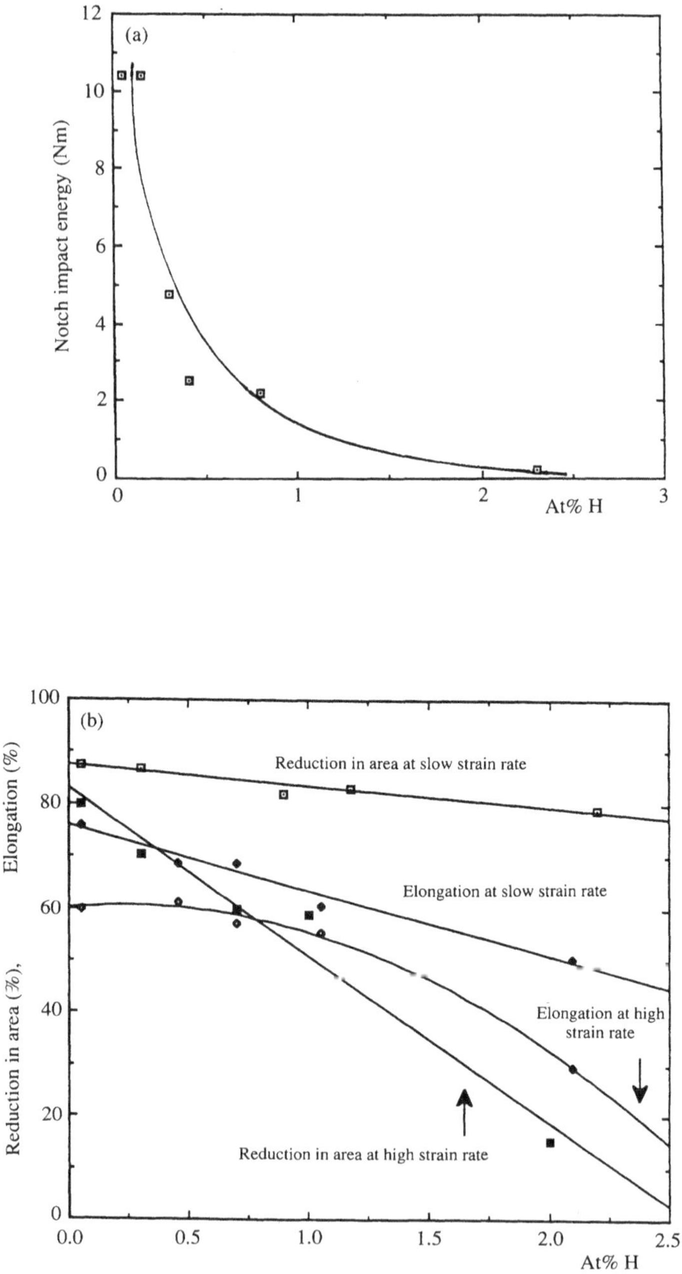

FIG. 1.7 The variation of the mechanical properties of pure titanium as a function of the hydrogen content and the testing speed at room temperature.[13] (a) Notch-impact energy; (b) elongation and reduction in area.

mass% (8 at%) at 325°C and low in iron (0.32 mass p.p.m. at 400°C). Metals such as iron, alpha-titanium, uranium, and alpha-zirconium do have one common property: they all become extremely brittle when the hydrogen content surpasses a critical value. Rylski[13] has measured the mechanical properties of pure titanium as a function of hydrogen at room temperature. As shown in Fig. 1.7 he found that both the notch impact energy and the elongation decreased rapidly with increased hydrogen content. The drop in notched impact strength was substantial even for contents of 0.005 mass%. Microstructure investigations have revealed that insoluble hydride platelets[14,15] are present when the hydrogen content is larger than 0.002 mass%. As long as this phase is present in a finely dispersed form, the mechanical properties can be improved. However, the phase grows quickly even at room temperature and forms massive particles. The result is a deterioration in impact and tensile properties as shown in Fig. 1.7. Therefore the brittleness in pure titanium is strongly related to the number and size of the hard hydride platelets.

Commercial titanium has a much lower solubility of hydrogen than high-purity material. The detrimental properties of hydrogen are then evident for far lower contents of hydrogen. It has been found[15] that beta-titanium is present at grain boundaries in such materials. Since the solubility of hydrogen is substantially higher in beta-titanium than alpha-titanium, most of the hydrogen is present in the beta phase, and hydrides are not present at low hydrogen levels (0.005 mass% H). For instance, Craighead, Lenning, and Jafee[16,17] tested the mechanical properties of Ti–8% Mn–H alloys. They measured a substantial drop in ductility at about 1 atomic% H (= 0.02 mass% H) as shown in Fig. 1.8. It has been found that the ductility is highly dependent upon the strain rate for this class of materials, and that the ductility is reduced for low strain rates. Furthermore, Koffila *et al.*[18] have shown that the effect is dependent upon temperature and disappears at about 90°C for a Ti–140 A alloy (which contains traces of iron, chromium, molybdenum, and carbon). According to Burte, the onset of deformation within the metal causes hydrogen to diffuse from an interstitial solution within the beta phase to certain preferred sites within the specimen. This microsegregation under influence of applied stresses, can then initiate microcracks. The propagation of these cracks may then cause brittle failure. This segregation of hydrogen, being a diffusion-controlled process, requires time in order to be damaging. Thus when the strain rate is low, a considerable degree of microsegregation of hydrogen may occur with the result that the material fractures in a brittle manner. The effect has been found to increase with increasing hydrogen content.

The embrittlement of hydrogen in steels shows similar features. However, in this case the ductility decreases progressively as the amount of hydrogen increases from infinitesimal values to a certain upper limit. Above this limit the ductility is unaffected by the hydrogen content and remains low. In steels, embrittlement is most likely to be caused by the segregation of hydrogen towards points in a triaxial stress state in the metal.[19] Furthermore, these

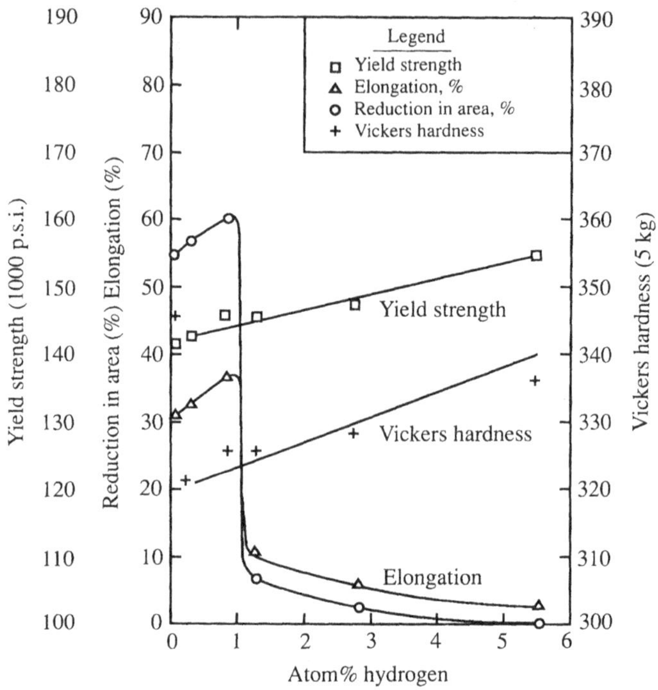

FIG. 1.8 Tensile properties of commercial Ti–8% Mn alloy as a function of hydrogen content.[16]

regions are created by a network of internal voids acting as internal stress raisers. Hydrogen is normally uniformly distributed throughout the metal lattice and is not damaging because its concentration is so low. But when hydrogen is segregated in highly stressed regions by stress-induced diffusion, the cohesive strength of the metal is reduced locally by the hydrogen atoms. The result is crack propagation at voids and an aggravated embrittlement.

1.4 ELECTRICAL CONDUCTIVITY

Both aluminium and copper are used in electrical devices and wires due to their high electrical conductivities. However, the impurities have a strong effect. For small concentrations of elements in solid solution in aluminium,[20] it has been found that the resistivity changes according to

$$\rho = \rho_0 + \sum_j K_j[\%\, j] \qquad (1.3)$$

TABLE 1.3 *The effect of impurity elements on the electrical resistivity of aluminium*

Element	Resistivity increase ($\Omega m \times 10^{-8}$) by adding 0.1% of an impurity element	
	In solid solution	As secondary phase
B	0.08	
Cr	0.40	0.02
Cu	0.032	0.003
Fe	0.26	0.006
Mg	0.05	0.02
Mn	0.29	0.03
Ni	0.08	0.006
Si	0.10	0.009
Ti	0.29	0.01
V	0.40	0.03
Zn	0.01	0.002
Zr	0.17	0.004

where ρ_0 is the resistivity of the pure metal, K_j is the change of electrical resistivity in the presence of 1 per cent of element j and [% j] is the concentration of element j. Values for the parameter K_j are given for a series of elements in Table 1.3. This demonstrates that elements like iron, titanium, vanadium, and chromium, all with several electrons in their valence band, have the largest effect. Willey[21] found that the effect of an element was a decade smaller if the element was present as intermetallic particles instead of in solid solution. This is the reason why boron is added to molten aluminium so that titanium and other elements can be removed as borides or are present only in very small quantities as particles in the solid.

The change in resistivity in copper by adding impurity elements is shown in Fig. 1.9.[22] The resistivity increases linearly for most elements in copper in agreement with eqn (1.3). Exceptions are iron, aluminium, and tin which may exhibit some deviation from linearity. The effect of iron, phosphorous, and silicon on electrical resistivity is surprisingly large. Concentrations of only 0.1 per cent cause an increase in resistivity or a reduction in conductivity of more than 20 per cent. Therefore copper in telephone wires is commonly alloyed with cadmium which improves the mechanical properties without reducing the conductivity more than 5–10 per cent.

Copper melts are often deoxidized with phosphorus in order to avoid water vapour in the products which are destined for welding or thermal treatment in hydrogen atmospheres. However, such materials will have an excess content of the order 0.04% P. The result is also a reduction of 25 per cent in conductivity which makes the material completely unsuitable for electrical purposes. It is necessary to employ high-purity electrolytic copper in such applications.

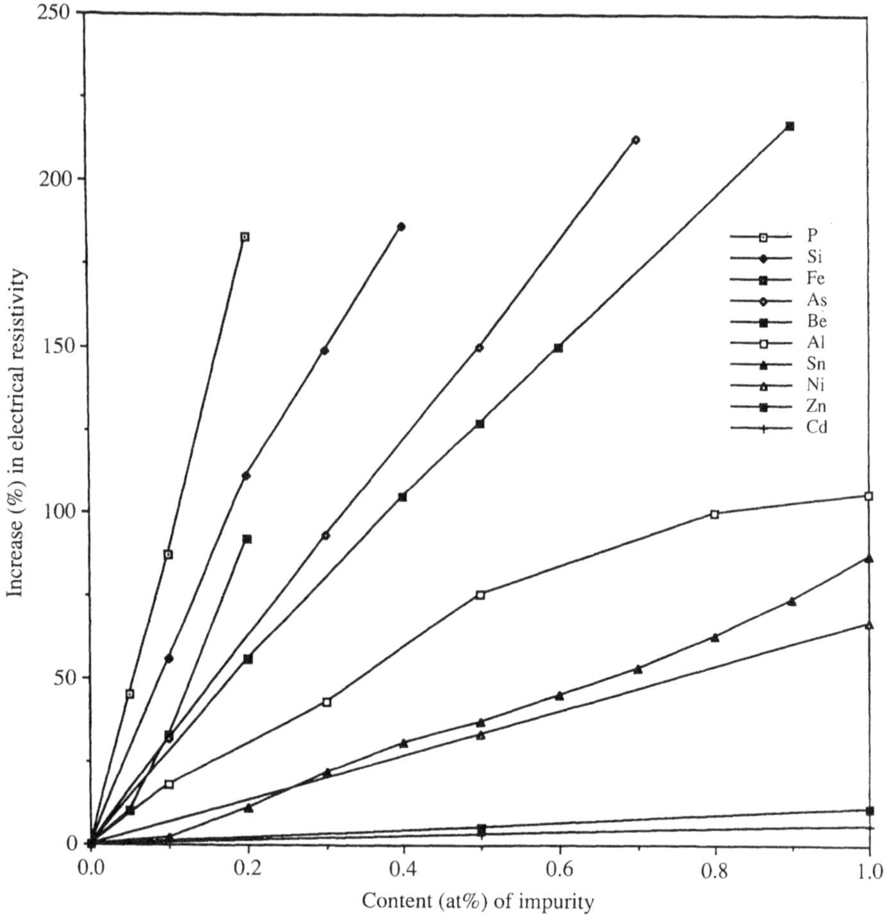

FIG. 1.9 The effect of impurities on the electrical resistivity of copper.[22]

1.5 MAGNETIC HYSTERESIS AND PARTICLES IN STEEL

Sheets of low-alloy steels are commonly used as material in electro-engines. One of the problems with these materials is that there is an energy loss, that is magnetic hysteresis, when the material is in a changing magnetic field. It is found that the material consists of magnetic domains. When a magnetic field is applied, two independent processes are operating as suggested by Becker and Döring.[23] There is an increase in the volume of domains which are favourably oriented with respect to the field at the expense of unfavourable ones. The other mechanism is that the direction of magnetization is rotated towards the direction of the field. Investigations of various steels have revealed that the hysteresis increases with the content of nitrides and carbides. An example is shown in

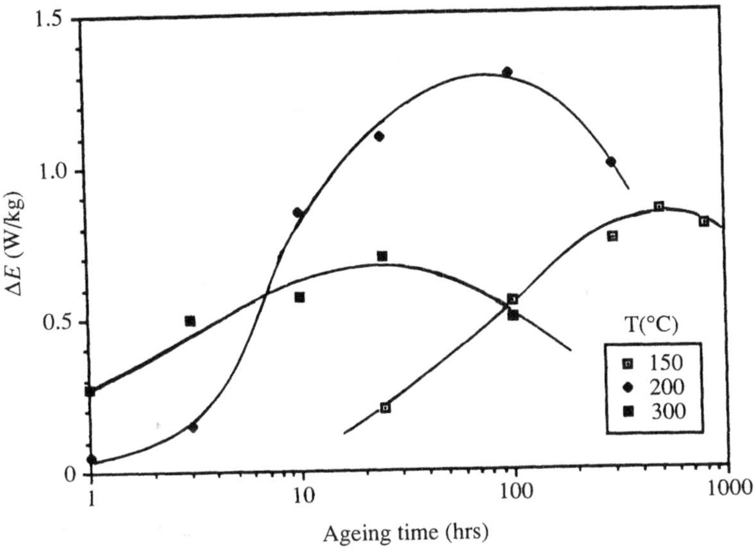

FIG. 1.10 Power loss, ΔE, in low-carbon steel as a function of preheat treatment between 150°C and 300°C. Magnetic flux density: 13 000 gauss.[24]

Fig. 1.10. Herø and Simensen[24] aged steels with coarse grains containing 0.01%C. The aging temperature was varied between 100°C and 300°C. The power loss in the aged materials was measured using a magnetic field of 13 000 gauss. It was found that the energy loss had a maximum after 1000 hours at 150°C and 30 hours at 300°C. Transmission electron microscopy investigations revealed that thin discs of cementite were formed on dislocations at high temperature, while ϵ-carbides were nucleated at low temperatures ($\leq$200°C). The latter were transformed to cementite during ageing. The maximum energy loss was related to thin discs, 1–1.5 μm in diameter. Leslie and Stevens[25] have obtained similar results for Fe–0.45% Mn–0.017% C and Nacken and Rahmann[26] for steels containing nitrides. The former showed that large carbides or nitrides obstruct the movement of the domain walls and secondary domains are formed if the discs are larger than 0.6 μm. The detrimental effects could be reduced by decreasing the carbon content to very low values of less than 0.003% C.

1.6 THE EFFECT OF IMPURITIES ON HOT DUCTILITY OF STEELS

Ljungström[27] has investigated the effect of impurities on the hot ductility of steels in the temperature range 1050–1250°C. He found that small amounts of lead, bismuth, tellerium, and tin in austenitic 17Cr–13Ni–3Mo steels reduced

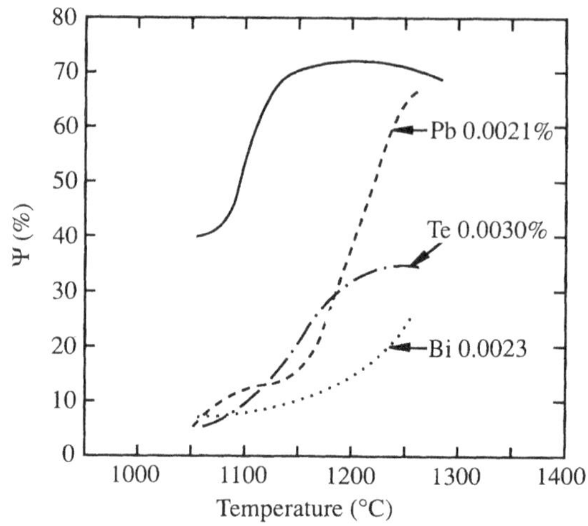

FIG. 1.11 Decrease in hot ductility at different temperatures of austenitic 17Cr–13Ni–3Mo steel caused by trace elements Pb, Te, and Bi.[28] (The ductility Ψ is defined as the percentage reduction of diameter at fracture. $\Psi = 100\,(d_0 - d)/d_0$.)

the hot ductility substantially as shown in Fig. 1.11. Comparison between the different elements showed that a lead equivalent could be defined empirically as[28]

$$\mathrm{Pb_{eq}} = \%\,\mathrm{Pb} + 4\%\,\mathrm{Bi} + 0.01\%\,\mathrm{Sn} + 0.02\%\,\mathrm{Sb} + 0.007\%\,\mathrm{As}. \quad (1.4)$$

Thus bismuth and lead have far the greatest effect. Even a content of 1 p.p.m. Bi or 21 p.p.m. Pb gives a severe reduction of the hot ductility at 1150°C. Microstructure investigations revealed that fractures were due to grain boundary embrittlement. Lead atoms segregated to the grain boundaries, and a detrimental effect was especially noticeable on those physical properties which depend on grain boundary conditions. The effects were accelerated by co-segregation of nickel, molybdenum, and chromium. A clean steel should therefore have a low content of the elements given by formula (1.4). The necessary cleanliness for stainless steels with a good hot ductility seems to be around 10 p.p.m. lead, but this depends upon the type of steel.

The detrimental effect of lead can, however, be counteracted by certain other elements. These are boron, zirconium, hafnium, cerium, and magnesium which are therefore deliberately added as trace elements. This is demonstrated in Fig. 1.12[29] showing how small additions of boron improve the hot ductility in a stainless ferritic-austenitic steel with 12–14 p.p.m. Pb. The effect is possibly caused by a segregation of boron to the grain boundaries which thereby prevents segregation of other elements. Additions of trace elements up to about 0.1 per cent are generally called micro-alloying. Micro-alloying has become an important part of material development during the last few decades.

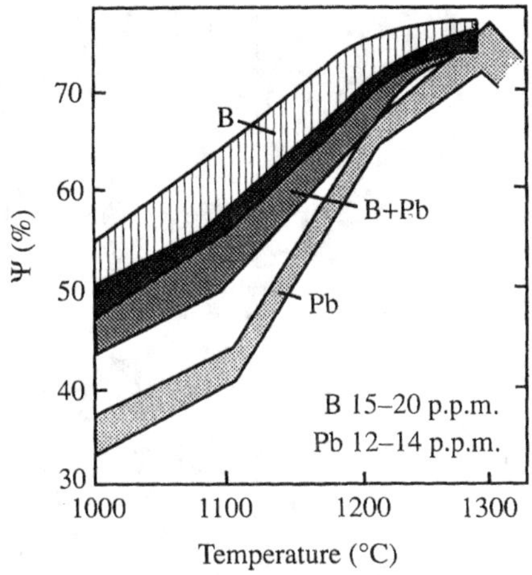

FIG. 1.12 Trace amounts of boron can counteract the adverse effects of lead on the ductility of 18Cr–15Ni–3Mo steel.[29] The cleaner steel has inferior ductility. (The ductility Ψ is defined as the percentage reduction of diameter at fracture. $\Psi = 100 (d_0 - d)/d_0$.)

1.7 THE EFFECT OF INTERMETALLIC PHASES ON MACROPROPERTIES

The level of impurities in most metals is low as shown in Table 1.1. The exceptions are iron and silicon in aluminium, iron in zinc and carbon, and manganese and silicon in iron. Another raw material which is rather impure, is commercial silicon. It contains commonly 0.1–1% Al, 0.1–1% Fe, 0.1–0.5% Ca, ≈0.1% Ti, and ≈0.1% C. Alloys which comprise one or several of these raw materials form intermetallic phases or carbides during solidification. An example is particles of Al_3Fe and α-AlFeSi and other phases in aluminium. The effect of intermetallic particles is rather similar to inclusions treated in Section 1.8. However, there is one main important difference. The intermetallic phases are chemically more active than most of the inclusions as they can go partly or totally into solid solution during annealing of the metal at high temperature.

Two important examples of the effect of intermetallic phases are presented in this section. One is the formation of *fir-tree zones* in aluminium. The other is the effect of large particles on fatigue of high-alloy aluminium.

1.7.1 The formation of fir-tree zones in aluminium

Certain products like offset plates and anodized panels require a homogeneous surface. This quality is rather difficult to attain for direct chill (DC) castings of commercial aluminium. When aluminium is DC-cast, sectioned and etched

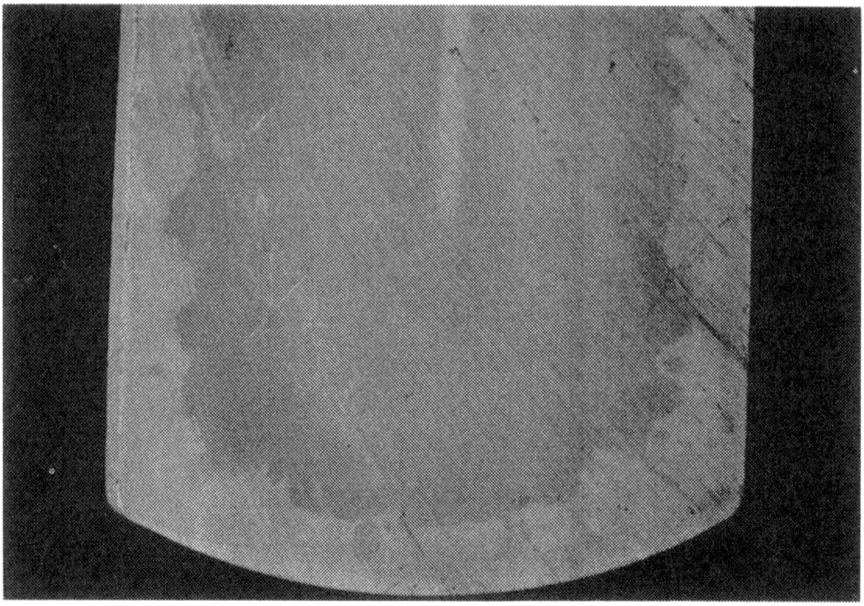

FIG. 1.13 A cross-section of an ingot showing a typical fir-tree structure. Etched in 10% NaOH at 70°C for 20 minutes.[32]

in NaOH solutions, or anodized its appearance can be as demonstrated in Fig. 1.13.[30] The surface shows a large variation in reflectivity with a bright outer area of the casting and a darker inner zone. This microstructure phenomenon has been called fir-tree zones. The result in this case is that a large section of the casting has to be scalped before it is sold to the customer.

Microstructure investigation by Westengen[30] has revealed that the dominant intermetallic phase present in the outer region is Al_mFe ($m \approx 4.2$)[31] and in the inner zone is Al_6Fe and Al_3Fe. When the material is annealed at a high temperature, the metastable Al_mFe dissolves and the equilibrium phase Al_3Fe is formed. Then the fir-tree structure also disappears. The effect is therefore due to the difference in chemical properties between Al_mFe and other Al–Fe phases.

When aluminium is DC-cast, the outer regions are solidified at a high cooling rate and the inner parts with a low cooling rate as shown in Fig. 1.14.[32] Both Westengen[30] and Miki *et al.*[33] have found that the phase Al_mFe is nucleated when the cooling rate is sufficiently large, that is, above 8–10°C/sec. The thickness of the fir-tree zone has a maximum value for materials with a Fe/Si ratio of about two and becomes rather small both for larger and smaller values. For instance, Brusethaug *et al.*[32] found for sheet ingots 0.6 m × 1.6 m in cross-section that the fir-tree zone was on average 0.25 m for material with

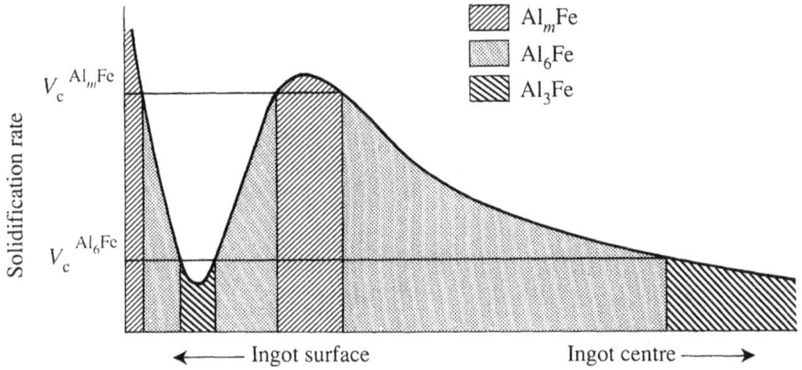

FIG. 1.14 Sketch of solidification rate over ingot cross- section and the occurrence of various zones.[30]

Fe/Si = 2, and 0 for materials with Fe/Si = 5. Therefore it is common to produce aluminium at a low solidification rate and to add iron to the melt in order to obtain a high Fe/Si ratio.

1.7.2 Fatigue in Al–Cu–Mg–Mn alloys

High-strength Al–Cu–Mg–Mn alloys (AA2024) have the best combination of toughness, fatigue, and strength properties in natural aged conditions. These materials are commonly used in dynamically loaded parts in aircraft. However, development of new commercial long-service aircraft, designed on the basis of the damage tolerance principle, has required further improvement in properties such as fatigue.

A series of Al–Cu–Mg–Mn alloys was studied by Lechiner and Kovalyov[34] in order to describe the effect of iron impurities. The content of iron was varied from 0.05 to 0.24 per cent, while the contents of silicon, manganese, and magnesium were kept at 0.2, 0.4, and 1.3 per cent, respectively. Since the insoluble Al–Fe–Si phases in the material also contain copper, the copper concentration was varied between 3.8 and 4.6 per cent. The analyses of the materials were carried out on 30 mm thick plates which had been commercially rolled and natural aged. Microstructure investigation revealed that the insoluble particles formed during casting were in the size range 5–20 μm. All the materials had around 75 per cent 5 μm particles, 20 per cent of 5–10 μm particles, and 5 per cent of particles larger than 10 μm. These large *excessive particles* were identified as Al–Fe–Mn–Cu–Si particles, while the small particles were Al–Cu–Mg phases without Fe. The content of the former phase increased rapidly with the copper content as shown in Fig. 1.15.

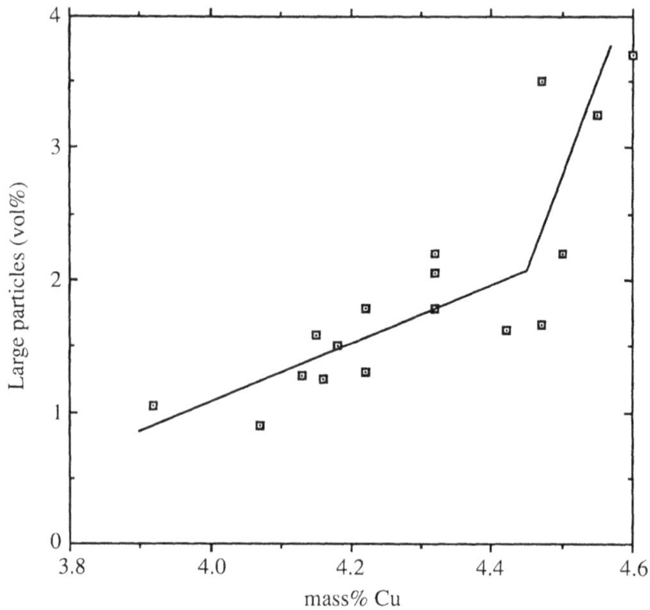

FIG. 1.15 Volume fraction of 'excessive particles' as a function of the copper content in Al–Cu–Mg–Mn alloys.[34]

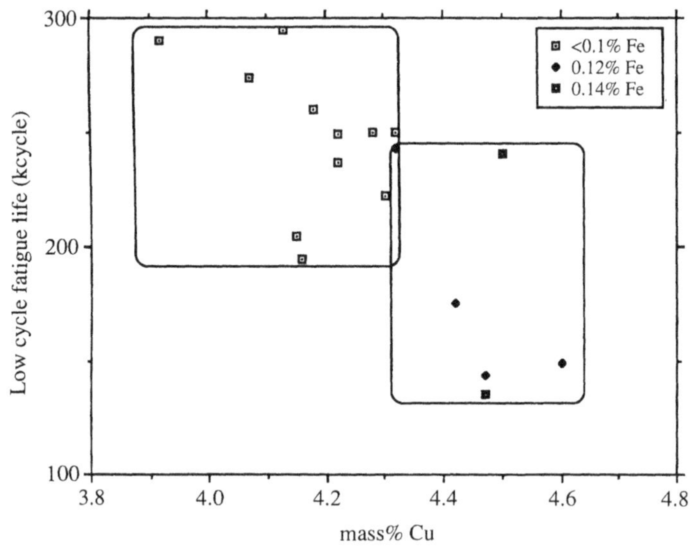

FIG. 1.16 Low cycle fatigue life versus the composition of Al–Cu–Mg–Mn plates. The materials with the lowest iron and copper content have the greatest fatigue life in these materials. (f = 3 Hz and axial tension stress of 155 MPa).[34]

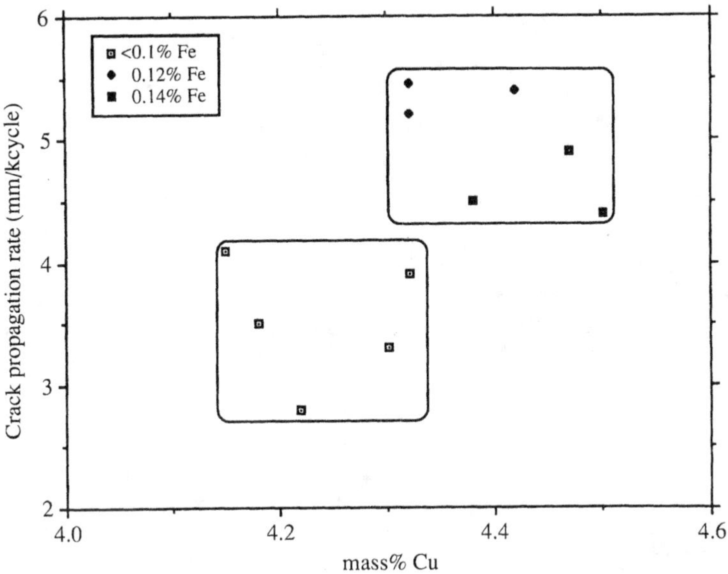

FIG. 1.17 Fatigue crack propagation rate versus the composition of Al–Cu–Mg–Mn plates. (Stress: 80 MPa, f = 5 Hz.)[34]

Low-cycle fatigue resistance was tested on notched specimens in axial tension with a frequency of 3 Hz and a maximum stress of 155 MPa. The results are summarized in Fig. 1.16. The low cycle life decreased for increasing volumes of large intermetallic particles and therefore also with increasing iron and copper content. The fatigue life was reduced from about 250 kc to 150 kc when the iron content increased from 0.07 to 0.15 per cent. Cracking resistance values were estimated for specimens with dimensions 10 mm × 500 mm × 1500 mm. Specimens were cut in the longitudinal direction from the central layers of the plates. Cracks with diameter 50 mm were made. Fatigue crack propagation was measured using a stress amplitude of 80 MPa and a frequency of 5 Hz. The crack propagation rate increased with increasing content of large Al–Cu–Fe–Si particles, that is, with the contents of copper and iron as shown in Fig. 1.17. These results clearly demonstrate the necessity of keeping the number of large intermetallic particles under control in advanced aluminium materials. Such investigations have led to the development of a new generation of alloys low in iron and also silicon with enhanced fracture toughness.

1.8 INCLUSIONS AND MECHANICAL PROPERTIES

There is a vast difference in the content of inclusions in commercial materials of various metals. Zinc and aluminium are typically in the range 0.1–10 p.p.m.,

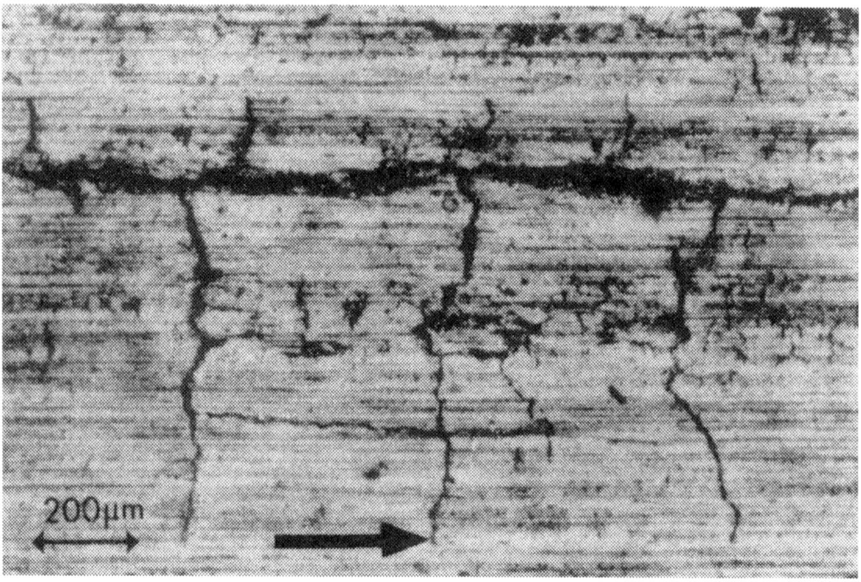

FIG. 1.18 Cracks along the rolling direction (arrow) and normal to the rolling direction in a Al–2% Mg sheet.[35]

magnesium 10–200 p.p.m., and steel 100–500 p.p.m. inclusions (Table 1.1). The effect on mechanical properties are correspondingly much more pronounced in steels than in the other materials.

However, even the very low number of inclusions in aluminium has an effect. For instance, they cause pinholes in thin foils with thickness less than 50 μm, and fracture of wires during the late stages of drawing. An example of harmful inclusions is shown in Figs. 1.18 and 1.19.[35] An Al–2% Mg alloy was cold-rolled, and large cracks were detected on the surface (Fig. 1.18). Microstructure investigation of the surface and of cross-sections through the defects revealed the presence of MgO and Al_2MgO_4 particles containing traces of chlorides (Fig. 1.19). These particles together with the common α-AlFeSi particles were situated in the cracks 0.1–0.5 mm below the rolled sheet. A simple explanation for the defect is that microvoids are formed at the oxide inclusions situated close to the surface. When the strain becomes too large for the material between the defect and the metal surface, the cracks will grow along the intermetallic particles causing small thin sheets of metal to peel off.

The cleanliness of steel with respect to non-metallic inclusions has always been of great concern to steelmakers. Very early the observation was made that inclusions had detrimental effects on properties and in particular ductility, fatigue, toughness, and machinability. The inclusions are mainly oxides and sulphides. Most of the oxides are small particles less than 0.2 μm in diameter,

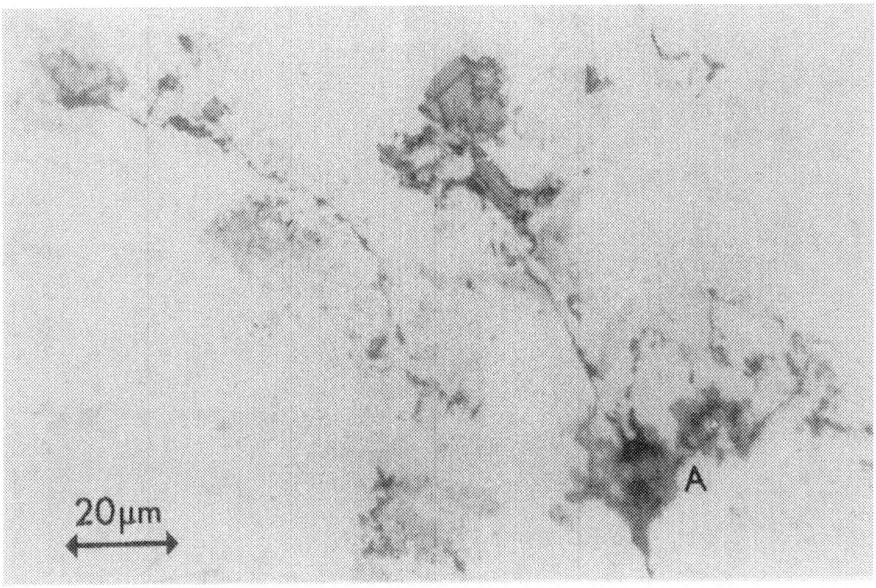

FIG. 1.19 Large oxide films and pores (A) at the cracks in the cold rolled sheet of Al–2% Mg alloy.[35]

but it is the large particles with diameters above 20 μm which give the harmful effects. Since the inclusions originate from the raw materials or are formed during melt operations, the primary objectives for steelmakers have been to develop processes for removal of inclusions from melts, and to introduce methods for controlling their cast product. This development has advanced significantly, and very impressive low levels of impurity content have been attained. One might ask why inclusions cannot be removed completely. In chapters 2 and 5 it will become apparent that this ideal is very difficult to attain in practice.

The shape of inclusions in wrought products is largely controlled by their behaviour during steel processing, that is, whether they are harder or softer than the steel matrix. The behaviour of different types of inclusions are schematically illustrated in Fig. 1.20.[36] Examples of hard inclusions are globular vitreous silicates (type (a)), crystalline agglomerates of high alumina calcium aluminates (type (b)), and 'loose' clusters of alumina (type (c)). In each case the matrix has to deform around the inclusions during rolling, creating cavities and deformation damage. The inclusions which have the most deleterious effect on toughness and ductility, are those which deform with the matrix (types (d) and (e) in Fig. 1.20). It is especially the properties in the transverse direction which become rather poor. Mixtures of MnS and oxides belong to type (d) while pure MnS inclusions give type (e). When an attempt is made to

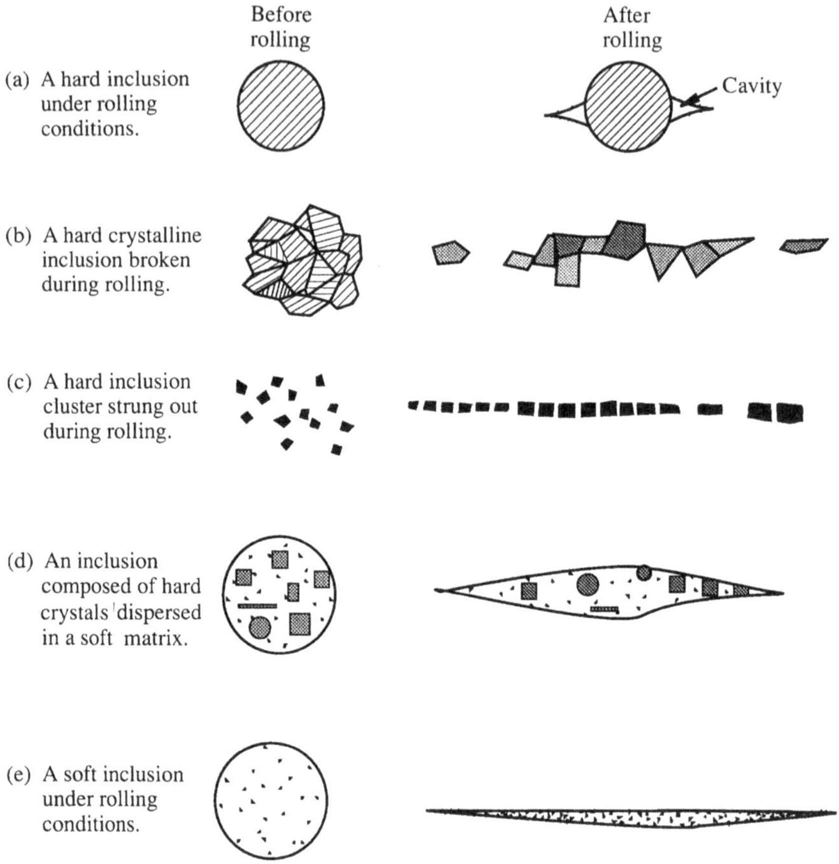

FIG. 1.20 Schematic representation of inclusion shapes before and after deformation.[36]

modify the inclusions in the refining of steel, the aim is to replace the soft inclusions with hard inclusions. In applications such as wire drawing, however, the soft inclusions (type (e)) are preferred.

It has been noticed that inclusions can also have a positive effect. The first development in this area was free-cutting steels where, for instance, sulphur was added to improve their machinability. This section, however, is focused on the more severe and damaging properties of common inclusions. The survey is partly based upon the work by Lagneborg.[37]

1.8.1 Ductile fracture

Fracture in bulk deformation processes such as forging, extrusion, and rolling usually occurs as ductile fracture, rarely as brittle fracture. However, depending on temperature and strain rate, the details of the ductile fracture mechanism vary. At temperatures below about half of the melting point of a given material (below the hot working region), a typical dimpled rupture type of mechanism occurs.[38]

It is well established that ductile fracture in steels is due to inclusions. Fracture is initiated by the formation of voids at the inclusions. Voids appear because particles do not deform, which forces the ductile matrix around the

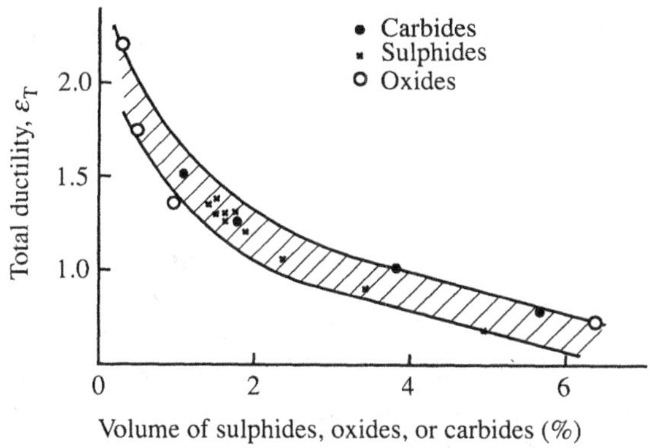

FIG. 1.21 Effect of second-phase particles on total ductility of steels.[39]

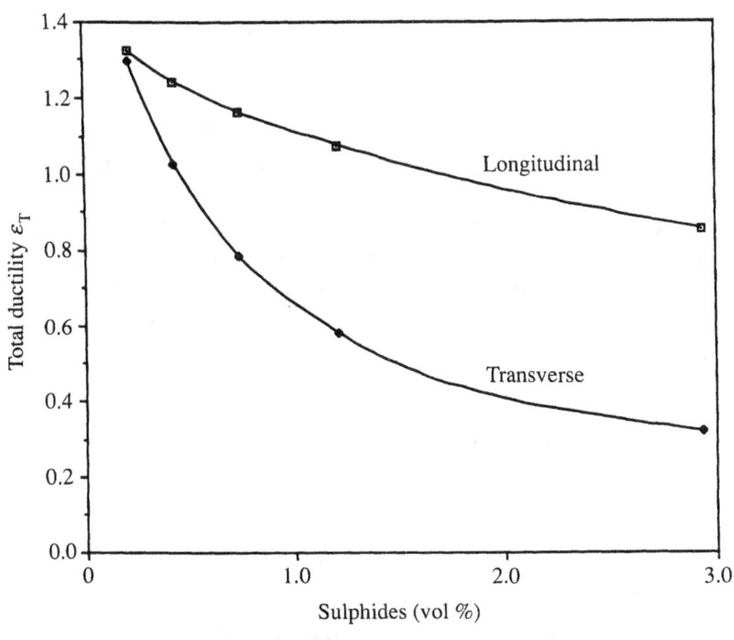

FIG. 1.22 Effect of the volume fraction of sulphides on the total ductility of steels for longitudinal and transverse testing.[40]

particle to deform more than normal. This in turn produces more strain hardening, thus creating a higher stress in the matrix near the particles. When the stress becomes sufficiently great, the interface may separate (decohesion) or the particle may crack. During continued straining these voids grow, develop cavities elongated in the straining directions, and the ligaments of matrix material between the voids become thinner. Finally the voids coalesce by localized deformation of the bridges between the voids, and the material fractures. Hence, we can distinguish three stages of the fracture process: void formation, void growth, and void coalescence. Experimentally it has been found that the fracture strain decreases strongly with increasing volume fraction of inclusions or content of carbon, oxygen, and sulphur as shown in Figs. 1.21[39] and 1.22.[40] The shape of the inclusions also has a strong effect. The elongation of inclusions during hot working, which takes place for many phases, has proved to be of particular significance. Thus, inclusions elongated in the tensile directions are far less harmful for ductility than discs and rods which are oriented perpendicular to the tensile direction. It should be noted that the content of inclusions in Figs 1.21 and 1.22 is very high compared to the mean level of inclusions encountered in commercial steels. However, in the steel products there may be areas having an excessive amount of inclusions. If these inclusions are located in a region of high stress, failure can take place.

1.8.2 Toughness

Toughness is the ability to absorb energy before fracture occurs. With respect to toughness, inclusions in metals behave like small cracks. The reason is that the bonding between the surrounding metal matrix and the inclusions is commonly weak, and that there exist small cracks at the interface even after casting. Thus theories concerning the dependence of toughness on cracks can readily be applied to inclusions.

When a material is brittle, Griffith's energy criterion can be used to describe catastrophic propagation of pre-existing cracks in the material.[40] According to this criterion, a crack will propagate in a brittle material when the decrease in the elastic strain energy is at least equal to the energy required to create the new crack surface. Using this criterion, it is possible to determine the tensile stress necessary for crack propagation. For a thin sheet containing an edge crack of length a (or equivalently an internal crack or an inclusion of length $2a$) which is subjected to an axial stress σ as illustrated in Fig. 1.23, the decrease in the elastic strain energy per unit thickness due to the crack is[41]

$$U_e = -\frac{\pi a^2 \sigma^2}{E} \tag{1.5}$$

where E is Young's modulus (modulus of elasticity). The increase in the surface energy due to the crack is $U_s = 4a\sigma_s$, where σ_s is the surface energy per unit area. The surface energy per unit area is discussed in the next chapter. It should

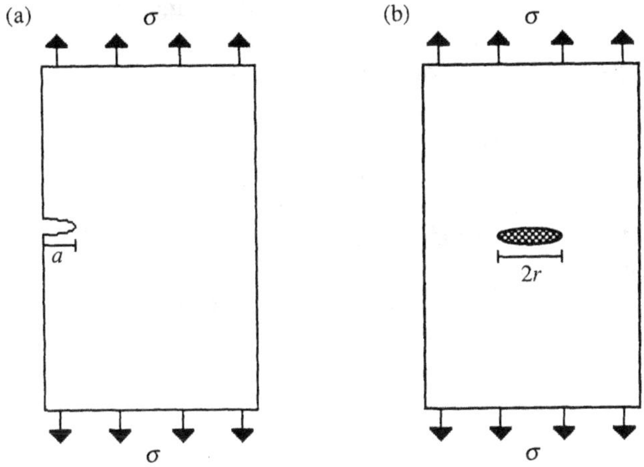

FIG. 1.23 A thin sheet with (a) an edge crack and (b) an inclusion subjected to an axial stress ($a = r$).[41]

not be confused with stress, σ, given as force per unit area. Young's modulus, E, gives the ratio between the stress and the elongation per unit length in a tensile test up to the elastic limit of the material. E has the same units as the stress σ. Values of E are included in tables in Chapter 3 for a number of metals.

The total change in energy is

$$U = U_e + U_s \tag{1.6}$$

According to the criterion an incremental increase in the crack length produces no change in the energy of the system. Then $dU/da = 0$. Here the case when the applied stress σ is constant is considered. From the above expressions one obtains

$$\sigma = \left(\frac{2E\sigma_s}{\pi a}\right)^{\frac{1}{2}}. \tag{1.7}$$

Orowan[42] has suggested that localized plasticity at the crack tip could be taken into account with an additional term σ_p, which is a measure of the plastic work required to extend the crack. Thus

$$\sigma = \left(\frac{2E(\sigma_s + \sigma_p)}{\pi a}\right)^{\frac{1}{2}}. \tag{1.8}$$

This equation can be generalized using linear elastic fracture mechanics as shown by Knott.[43] From linear elastic theory the stresses in the region of the crack tip are given by

$$\sigma_{ij} = \frac{K}{(2\pi\alpha_{ij}(\theta)r)^{\frac{1}{2}}} \tag{1.9}$$

where r and θ are the cylindrical coordinates of a point with respect to the crack tip, i and j indicate the directions, $a_{ij}(\theta)$ is in general a dimensionless parameter dependent upon the geometry of the specimen, and K is called the stress intensity factor. By comparison with eqn (1.9) it is seen that K is proportional to the square root of $E(\sigma_s + \sigma_p)$. It therefore depends on the physical properties of the metal, on Young's modulus, the surface energy, and the plastic work. The fracture toughness, K_{IC}, is the critical stress intensity factor in the opening mode, and this may be measured. The dimensions of the specimen are in this case selected so that there is plane strain at the crack tip. Then for the precracked specimen the fracture stress, σ_F, is given by

$$\sigma_F = \frac{K_{IC}}{(\pi \alpha r)^{\frac{1}{2}}} \tag{1.10}$$

where the factor α depends on the geometry of the specimen.

According to Lagneborg,[37] inclusions may promote fracture by two fundamentally different mechanisms. The first mechanism is that an inclusion of diameter $2r$ acts as a crack with diameter $2r$ from which a fracture develops. It has been verified experimentally that this equivalence between an inclusion and a crack is usually a good approximation. The second mechanism is that K_{IC} depends not only on the metal matrix but also on the inclusions.

Kiessling and Nordberg[44] have shown that cracks are initiated at inclusions and that spherical or cylindrical inclusions may be regarded as sharp cracks. A prerequisite for this is that small border cracks or cracks in the inclusion are formed at the early stages of loading the material. When such a sharp crack has been formed, the type of inclusion and its properties should have little influence. In fact it has been shown experimentally for rolled steels that small cracks are frequently observed at hard inclusions, such as silicates, as well as at plastic inclusions like manganese sulphides. According to the second mechanism K_{IC} also depends on the inclusions if cracks are not present at the interface between inclusion and matrix. It is pointed out that the tendency for borderline cracks to form depends both on cleavage strength and the strength of the inclusion–matrix interface.

Knowing the fracture toughness, K_{IC}, the critical inclusion size, r, can be determined by using eqn (1.10). The critical defect size for a design stress which has for simplicity been put equal to the yield stress, has been calculated by Lagneborg.[37] He found that the critical inclusion size decreased with the yield stress both through the direct influence of this stress and the decreasing fracture toughness, K_{IC}. The critical size varied from 0.18 mm to 1 mm for different steels. When the temperature was decreased, the yield strength increased and the fracture toughness decreased and resulted in a decrease of the critical inclusion size. Lagneborg found that of all the measured inclusion parameters, it was only the size of the largest ones which counted and which controlled the brittle fracture.

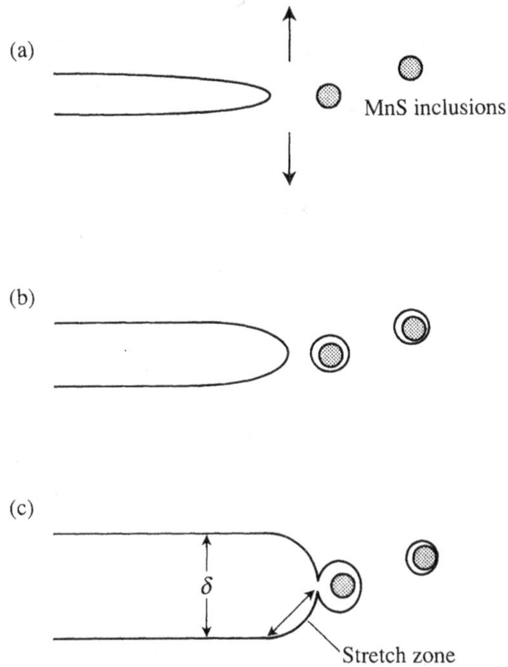

FIG. 1.24 Schematic illustration of the internal necking process at MnS inclusions in steel: (a)–(c) represent increasing strains which cause crack blunting, void growth, and eventual coalescence of crack tip and void. δ = COD at initiation of the coalescence.[37]

Inclusions also have an influence upon growth propagation which leads to ductile fracture. Assume that a load is applied to a steel containing a sharp precrack. As the load increases and the crack opening displacement, COD, increases as illustrated in Fig. 1.24, the inclusions near the crack tip experience increasing strains. Then voids are formed at low strains around the inclusions as the bonds between the matrix and the inclusions are weak. Void growth is limited until the voids are subjected to high plastic strains and stresses. Initiation of fracture may then take place at the points were the blunting tip first coalesces with the growing void nearest to the crack tip.

The initiation process of ductile fracture at the tip of a precrack has been modelled by several investigators. One of the main problems has been to describe conditions near the crack tip. However, this difficulty can be avoided by considering the local stress field around the crack tip. Thus the J integral[45] for stress energy is a surface integral that encloses the crack tip from one side of the crack to the other.

The critical value of J required to initiate crack extrusion[46] is

$$J = 0.46\tau X \tag{1.11}$$

where τ is the shear yield stress and X is the inclusion spacing. The crack opening displacement, COD, may be calculated from the J integral. Figure 1.25 shows results of calculations of the dependence of the COD as a function of the spacing and radius of the particles (Rice and Johnson[47]). Experimental results for steels tabulated in Table 1.4. are inserted in the diagram. There is fairly satisfactory agreement for steels with a low strength and a high work hardening capacity.[37] It is also evident that the theory is less applicable for high-strength steels. It is, for instance, typical that the critical COD decreases with decreasing tempering temperature in steel En25. Knott[48] has shown that these low COD values appear to be related to shear cracks emerging from the tip of the precrack before a void has fully coalesced with the crack tip. He has furthermore demonstrated that these shear cracks in quenched and tempered steels seem to run from carbide to carbide by decohering the carbide–matrix

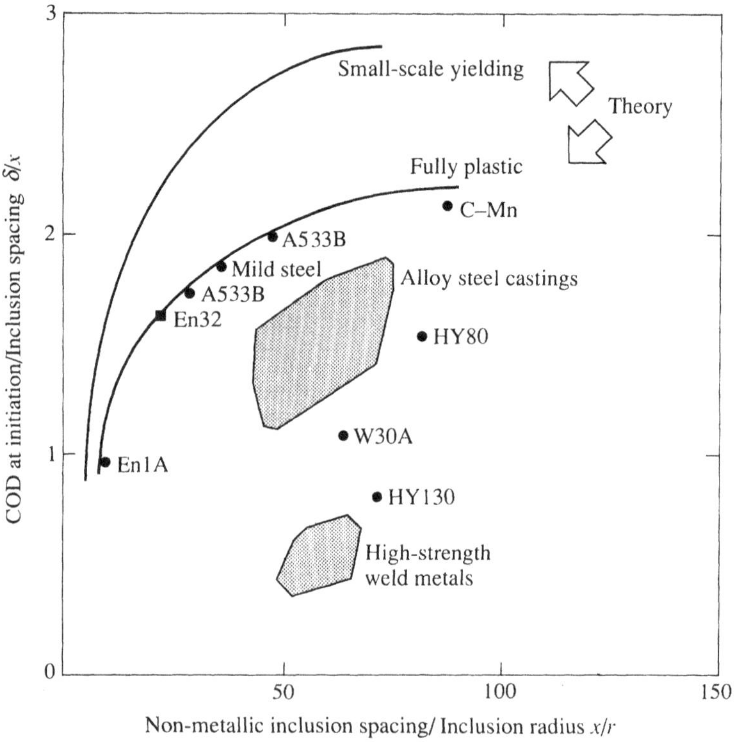

FIG. 1.25 Crack tip ductility for a series of steels[37] listed in Table 1.4. The results are compared with the theoretical curves determined by Rice and Johnson.[47] There is a reasonable agreement for the transverse directions of steels with high working capacity: En1A, En32, A533B, mild steel, and C–Mn steel, but other steels fail to demonstrate the predicted ductility.

TABLE 1.4 *Composition of steels used in toughness measurements illustrated in Fig. 1.25*

Name	Concentration (mass%)				
	C	Mn	Ni	Cr	Others
En1A	0.12	0.90			S: 0.24
En32	0.17	0.74	0.02	0.01	S: 0.03
A33B	0.24	1.42	0.70		Mo: 0.48
C–Mn	0.10	1.38			S: 0.009
Alloy castings	0.23	0.9	1.8–2	0.9	Mo: 0.4
HY80	0.17	0.34	2.36	1.23	Mo: 0.34
Ducol W30A	0.16	1.3	0.2	0.6	Mo: 0.23
Hy130	0.10	0.83	4.93	0.50	Mo: 0.53
Weld metals	0.1	1–1.6	2.3–3	0.5–1	Mo: 0.5
En25	0.34	0.72	2.72	0.62	Mo: 0.55

interface. From these observations it is evident that the spatial distribution as well as the inclusion spacing and the size distribution of inclusions, has a significant effect upon the the ductile fracture. However, these effect are not satisfactorily accounted for in the present theories.

1.8.3 Fatigue

Failure in fatigue of materials is very often due to inclusions. Inclusions reduce the fatigue strength substantially. This is the reason why inclusion studies have always been necessary to rate the quality of steels subjected to fatigue in service, such as ball-bearing, flapper valve, wheel, and shaft steels. It has been established that the fatigue strength can be predicted from the tensile strength and the type, content, and size of inclusions.

For materials without inclusions it has been found empirically that the fatigue limit, S_n, is related to the tensile strength, σ_m:[47]

$$S_n \approx 0.5\sigma_m. \tag{1.12}$$

This equation is obeyed for low-strength steels, while high-strength steels have a tendency to give lower values.

The first, more quantitative, treatment of the relationship between inclusions and fatigue properties has been given by Peterson.[49,50] He considered the inclusions as internal holes and calculated the stress concentrations by elastic theory. He stated that if a volume around the inclusion larger than a critical value was stressed beyond the ideal fatigue strength as a result of the stress concentration, the fatigue limit of the whole material is reached. The inclusion will in this way influence the fatigue strength through its size, shape, and orientation relative to the applied stress. On this basis it may be shown that a steel containing a spherical inclusion of radius r would have a fatigue limit[51]

$$S_n = \frac{0.5\sigma_m}{1 + (1 + A/r)^{-1}} \tag{1.13}$$

where 0.5 σ_m is the ideal fatigue limit and A represents a measure of a zone around the inclusion where the material is stressed beyond the ideal fatigue strength. Peterson found, as expected, that A decreases with increasing strength of the steel. He proposed the relation

$$A = 0.025\ (2070/\sigma_m)^{1.8} \tag{1.14}$$

where A is in mm and σ_m is in N/mm^2.

Thus, provided that it is the size of the inclusions that will cause failure, eqns (1.13) and (1.4) are able to predict the fatigue strength. Nordberg[51] has compared the prediction by these equations with actual measurements of fatigue strength and the size of the defect that initiated failure. The defect represented a series of various types of inclusions as well as pores. As shown in Fig. 1.26 the agreement between Nordberg's measurements and the theory is very good, especially since eqns (1.13) and (1.14) are only a first approximation.

Despite the success of the theory outlined, there is clear evidence that the internal properties of inclusions have an effect as well. Hence Brooksbank and Andrews[52] have shown that inclusions with smaller thermal expansion that the steel matrix, typically most oxide inclusions, affect the fatigue strength detrimentally and that inclusions with larger thermal expansion, like many sulphides, are not detrimental. These authors argue that the effect is due to the fact that the former inclusions set up tensile stresses. This hypothesis has been supported by the measurements of Hauser and Wells[53] who have found that initiation of cracks in transverse fatigue testing of high-strength steels always

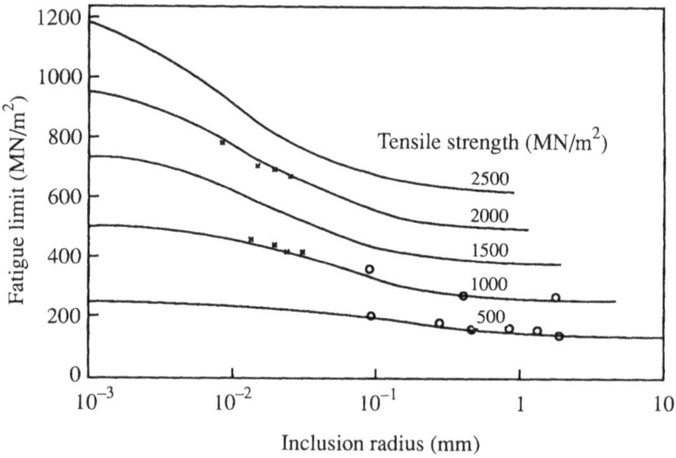

FIG. 1.26 Comparison between the experimental and predicted fatigue limit in steel.[51]

starts at inclusions other than manganese sulphides, in spite of the fact that these are longer than the initiating inclusions.

1.8.4 Machinability

Machining of metal pieces is a common standard industrial process. An example of machining of a metal workpiece is shown in Fig. 1.27. The purpose of such an operation is to remove the outer part and produce a metal piece that has a smooth surface of specified geometry and size. It is important to avoid any abrasion of the tool. When the process is running with low maintenance costs and with a low percentage of workpiece failures, the material is said to have a good machinability.

It is well established that non-metallic inclusions have a strong influence upon the machinability of steels. The effect is closely related to the inherent properties of the inclusions. As a result of this, free-cutting or free-machining grades of steel have been developed where advantage has been taken of the effect of special additions. One classical example is resulphurized steels with sulphur additions of up to about 0.35 per cent sulphur, containing mainly manganese sulphides. The non-metallic inclusions affect machinability in a complex manner. Kiessling[54] has summarized the conditions that should be met in order to improve the machinability as follows:

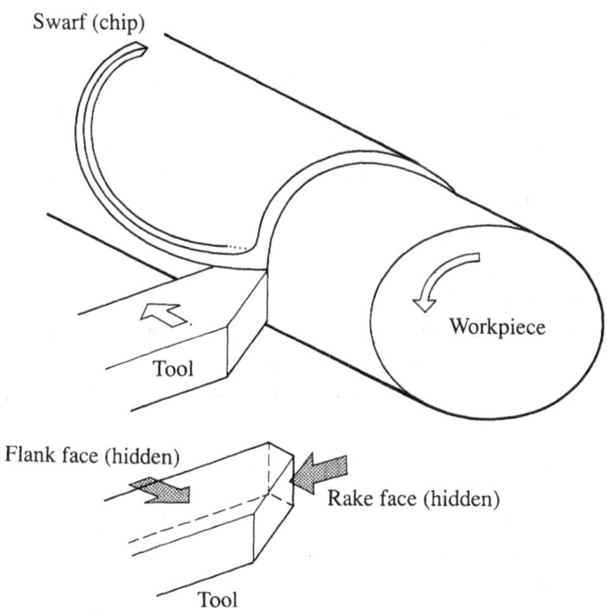

FIG. 1.27 Machining of a cylindrical metal piece. The tool is drawn twice to illustrate where the flank face and the rake face are on the tool. Chip is a short swarf.

1. The inclusions should act as stress raisers in the shear plane of the swarf so initiating crack formation and embrittling chips. The chip–tool contact length then decreases, which is beneficial. Inclusions should not, however, be such strong stress raisers that the work piece cracks.
2. The inclusions should participate in the flow of metal in the flow zone, increasing the shear of the metal. They should not cut through the plastic flow of metal and thus damage the tool surface.
3. The inclusions should form a diffusion barrier on the rake face (see Fig. 1.27) of the tool at the temperature of the tool–chip interface. This temperature depends on several variables, especially the cutting speed.
4. The inclusions should give a smooth piece surface and not act as abrasives on the flank face of the tool.

An example is MnS particles in steels. These inclusions act as stress raisers in the shear plane of the swarf and thereby improve the machinability. Large inclusions have a better effect than the smaller ones. Ollilainen[55] has improved the machinability of steels by adding calcium and forming oxides with a low melting point and a low hardness.

1.9 PITTING CORROSION

Parts made of steel and aluminium, copper, titanium, and magnesium alloys are common in products such as cars, buildings, and oil rigs. These devices are designed to have a long service life and therefore are highly resistant to corrosion. There are a variety of ways in which products of metal alloys can corrode. This section concerns only corrosion related to impurities. In most cases it turns out that corrosion is associated with impurity phases formed either in the melt or during casting. However, since the mechanisms are not known in detail in most cases, only a few examples are presented here.

Magnesium is a metal which corrodes rather easily. The presence of more than 0.016% Fe or 0.004% Ni aggravates the corrosion rate of magnesium considerably as shown in Fig. 1.28.[56] These tests were carried out in sodium chloride solutions. Microstructure investigations revealed the presence of intermetallic phases. Simensen and Oberländer[57] have shown that these are α-Fe and Fe_3Si in commercial Mg–Fe alloys, and that these phases are formed in the melt for iron contents as low as 0.01% Fe. The solubility of nickel in solid magnesium is also extremely low, and Mg_2Ni is formed. Hanawalt *et al.*[58] have shown that this corrosion can be reduced substantially by adding manganese (Fig. 1.29) which causes the formation of other intermetallic phases high in manganese. In both examples the intermetallic phases are the origin of pitting corrosion. Pitting corrosion caused by inclusions in steel is discussed in some detail in the following example.

It has been known for a long time that non-metallic inclusions may have a

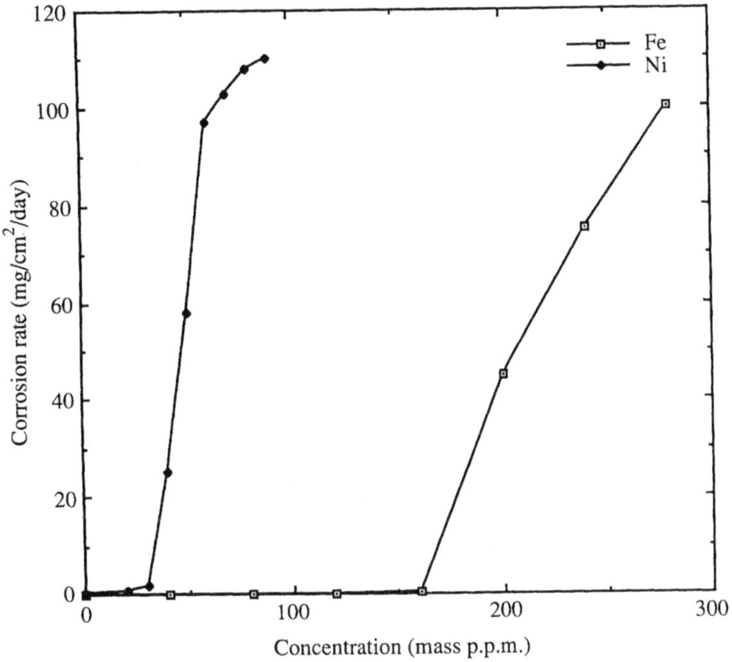

FIG. 1.28 The corrosion rate as a function of the nickel content in a Mg–0.6% Zr alloy and iron in magnesium when tested in a 3% NaCl solution.[56]

harmful effect on the corrosion resistance of steel. The correlation between inclusions and corrosion is particularly obvious in localized corrosion, such as pitting. As one might expect, this effect strongly depends on the chemistry of the inclusions. It was found that various oxide inclusions do not appear to have any effect. Microstructure investigations by means of scanning electron microscopy have revealed that sulphide inclusions are the most likely initiation sites for pitting in stainless steels. These sulphides are MnS with varying amounts of dissolved iron and chromium. On the basis of such observations, Eklund[59,60] has proposed the following mechanism: sulphides are in general good electronic conductors, and accordingly the manganese sulphides in stainless steel can be polarized to the potential of the passive steel surface. At this potential MnS, as shown both experimentally and theoretically by Eklund, is not thermodynamically stable and will tend to dissolve in a surrounding salt solution. By the dissolution of sulphides, ions of manganese as well as iron and chromium form locally. They form, in turn, hydrogen ions, hydroxy and chloride complexes. One of the important reactions which takes place is

$$MnS + 4H_2O \Rightarrow Mn^{2+} + SO_4^{2-} + 8H^+ + 8e^-. \quad (1.15)$$

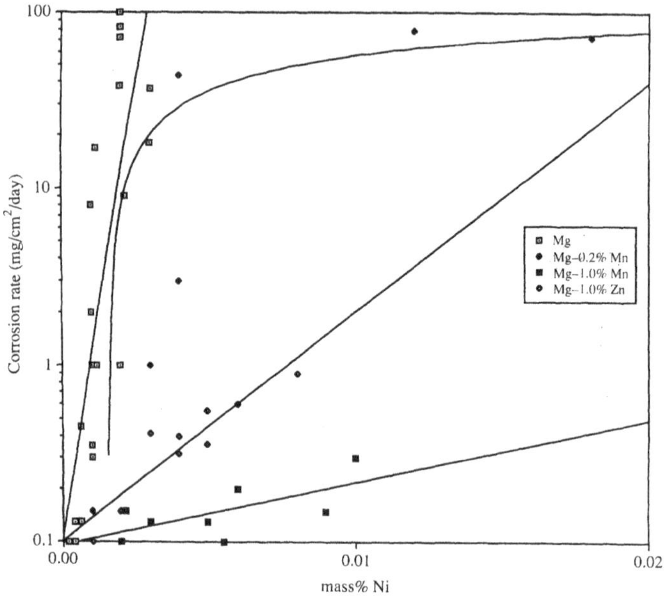

FIG. 1.29 The corrosion rate of pure Mg, Mg–Mn, and Mg–Zn alloys as a function of the nickel content.[58] Corrosion is drastically reduced by alloying with Mn or Zn.

From the reaction in eqn (1.15) it is evident that the solution in the micropit becomes more and more acid. When a certain degree of acidity is obtained, the uncovered metal surface around the sulphide will no longer be able to passivate, and metal ions from the steel will enter the solution. At this moment the electrochemical potential drops dramatically. Then the dissolution of sulphide will be terminated as the potential will now be within the region of thermodynamic stability of the sulphide. If the formation of hydrogen ions from the hydrolysis of the increased metal ion concentration is high, the dissolution of the metal matrix will continue as the solution remains quite acid and a self-sustaining pit has been developed. Otherwise, the solution will be neutralized by the formation of H_2S gas and the uncovered metal will passivate by the formation of a hydroxide layer.

An interesting alloy development has been made based on this model for pitting corrosion. Steels low in manganese have been produced, containing chromium sulphides instead of manganese sulphides.[61,62] Compared with manganese sulphides, this phase is much less soluble in acids. The result is a steel which is far more resistant to pitting corrosion.

1.10 CONCLUDING REMARKS

There is no doubt that manufacturers of metal products are keenly aware of impurities and their effect on properties. It is common for them to use advanced methods in product control or to require a certificate with respect to impurities when they buy raw materials. This chapter has revealed some of the reasons. The examples selected have been from the production of steel, aluminium, titanium, copper and magnesium. However, similar effects are also found for other materials. For instance, hot shortness in zinc alloys is caused by local melting of the eutectic ternary Zn–Al–Mg phase. Accelerated corrosion along grain boundaries in zinc is caused by traces of lead, cadmium, tin, and thallium. Even small amount of zinc in lead is a problem as it causes brittleness, intergranular weakness, and dimensional instability. Phosphorous causes grain boundary embrittlement in tungsten materials formed by powder metallurgy. Thus it is apparent that the list of the effects of impurities is long. It should be noted that most properties depend to some degree on dissolved impurities and inclusions and thus refining metallurgy is an important field of endeavour. There are numerous problems that can be solved in a variety of ways. Refining metallurgy is exciting; knowledge and imagination are needed! It is hoped that this and the following chapters will contribute to our understanding of the problems and their solution.

REFERENCES

1. Simensen C. J. (1989). Sources of impurities in aluminium melts and their control. In *Aluminium melt refining and alloying C* 1-*C* 20, (ed. M. Nilmani), University of Melbourne, Australia.
2. Watanabe T., Huang Y. C., and Komatsu R. (1976). Solubility of hydrogen in magnesium. *J. Jpn. Inst. Light Metals*, **26**, 76–81.
3. Shapavalov V. I., Serdyuk N. P., and Semik O. P. (1981). Magnesium–hydrogen and aluminium–hydrogen phase diagrams. *Dop. Akad. Nauk Ukr. RSR. Ser. A Fiz.-Mat. Tekh. Nauki*, **6**, 99–101 (in Russian).
4. Popovic Z. D. and Piercy G. R. (1976). Measurement of the solubility of hydrogen in solid magnesium. *Metall. Trans. A*, **6**, 1915–7.
5. Koeneman J. and Metcalfe A. G. (1959). The solubility of hydrogen in magnesium. *Trans. ASM*, **51**, 1072–82.
6. Chernega D. F., Gotvjanskij, U. A., and Prisjasznjok, T. N. (1979). Diffusion and solubility of hydrogen in liquid magnesium. *Vestn. Kiev. Politekh. In-ta. Kiev*, **16**, 56–59.
7. Talbot D. E. J. (1975). Effects of hydrogen in aluminium, magnesium, copper and their alloys. *Int. Met. Rev.*, **20**, 166–84.
8. Turner A. N. and Bryant A. J. (1967). The effect of ingot quality on the short-transverse mechanical properties of high-strength aluminium alloy thick plate. *J. Inst. Metals*, **95**, 353–9.
9. Kostron H. (1952). Aluminium and gas. *Z. Mkde*, **43**, 269–84 and 373–87.

10. Allen N. P. and Hewitt T. (1933). The equilibrium of reaction between steam and molten copper. *J. Inst. Metals*, **51**, 257–75.
11. Harper S., Callcut V. A., Townsend D. W., and Eborall R. (1961–62). The embrittlement of tough-pitch copper during annealing or preheating. *J. Inst. Metals*, **90**, 422–9.
12. Mattsson E. and Shückher F. (1958–59). An investigation of hydrogen embrittlement in copper. *J. Inst. Metals*, **87**, 241–7.
13. Rylski O. Z. P.M. 203. Dept. of Mines and Technical Surveys, Mines Branch, Ottawa, Canada.
14. Liue T. S. and Steinberg M. A. (1958). The mode of hydride precipitation in alpha titanium and alpha titanium alloys. *Trans. Amer. Soc. Metals*, **50**, 455–77.
15. Cotterill P. (1961). The hydrogen embrittlement of metals in Progress in materials science. **9**, 201–304, Pergamon Press, Oxford.
16. Craighead C. M., Lenning G. A., and Jafee R. I. (1956). Hydrogen embrittlement of beta-stabilized titanium alloys. *J. Metals N. Y.*, **8**, 923–8.
17. Jafee R. I., Lenning G. A., and Craighead C. M. (1956). Effect of testing variables on the hydrogen embrittlement of titanium and a Ti-8pct Mn alloy. *J. Metals N. Y.*, **8**, 907–13.
18. Koffila, R. J. and Burte H. M. (1955). Hydrogen contamination in titanium and titanium alloys. Part I: Hydrogen embrittlement on alpha-beta titanium alloys. USAEC Report No. WADC-TR-54-616 (unclassified).
19. Troiano A. R. (1960). The role of hydrogen and other interstitiels in the mechanical behaviour of metals, Cambell memorial lecture. *Trans. ASM*, **52**, 54–80.
20. Mondolfo L. F. (1976). *Aluminium alloys, structure and properties*. Butterworths, London.
21. Willey L. A. as quoted by W. A. Dean (1967). In *Aluminium* 1, (ed. K. R. Horn), p. 174. ASM, Metals Park, Ohio.
22. Higgins R. A. (1981). *Engineering metallurgy* 1. *Applied physical metallurgy*. Hodder and Stoughton, London.
23. Becker R. and Döring W. (1939). *Ferromagnetismus*. J. Springer, Berlin.
24. Simensen C. J. and Herø H. (1974). Magnetic ageing of sheet metals of Fe-Si-C alloys. SI report 730112 (in Norwegian).
25. Leslie W. C. and Stevens D. W. (1964). The magnetic ageing of low-carbon steels and silicon irons. *Trans. of the ASM*, **57**, 264–83. Also Leslie W. C. (1961). The quench-ageing of low carbon iron and iron-manganese alloys. *Acta Met.*, **9**, 1004–22.
26. Nacken M. and Rahmann J. (1962). Untersuchung zur Abschreckalterung stickstoffhaltiger Stähle durch Messunge der Dämpung, Koerzitivfeldstärke und Rückbildung. *Archiv für das Eisenhüttenwesen*, **33**, 131–40.
27. Ljungström L. G. (1977). The influence of trace elements on the hot ductility of austenitic 17 Chromium –13% Nickel – molybdenum – steel. *Scand. J. Metall.*, **6**, 176–84.
28. Effects of impurities on hot ductility of austenitic stainless steels. STU information, 1978, No. 105, Stockholm.
29. Kiessling R. (1981). Clean steel – a debatable concept. In *Swedish Symposium on Non-Metallic Inclusions in Steel*, (ed. H. Nordberg and R. Sandberg) pp. 7–18. Uddeholms AB, Hagfors, Sverige.
30. Westengen H. (1981). Structure inhomogeneities in direct chill cast sheet ingots of commercial purity aluminium, 7. Internationale Leichtmetalltagung Leoben-Wien, Sekretariat der 7. ILMT, Leoben, Austria, pp. 103–5.

31. Skjerpe P. (1988). Structure of Al_mFe. *Acta Cryst.*, **B44**, 480–6.
32. Brusethaug S., Porter D., and Vorren O. (1987). The effect of process parameters on the fir-tree structure in DC-cast rolling ingots, 8. Internationale Leichtmetalltagung Leoben-Wien, Sekretariat der 8. ILMT, Leoben, Austria, pp. 472–6.
33. Miki I., Kosuge H., and Nagahama K. (1975). Supersaturation and decomposition of aluminium–iron alloys during solidification. *J. Jap. Inst. L. Met.*, **25**, 1–9 (in Japanese).
34. Leschiner L.N. and Kovalyov V.G. (1989). Effect of Fe and Si on properties and structure of Al–Cu–Mg alloys. In *Proc. Int. Workshop in Balatonfüred, Hungary,* (ed. I. Kovács), Trans. Tech. Publ., Switzerland.
35. Simensen C.J. (1974). Investigations of cracks in rolled materials, SI report 720717-7 (in Norwegian).
36. Hilty D.C. (1985). Inclusions in steel. In *Electric furnace steelmaking*, (ed. C.R. Taylor), pp. 237–51. Iron and Steel Soc. Inc., Bookcrafters Inc., Chelsea, MI.
37. Lagneborg R. (1981). The influence of non-metallic inclusions on properties in steel—a review. In *Swedish symposium on non-metallic inclusions in steel*, (ed. H. Nordberg and R. Sandberg), Sweden, pp. 285–352. Uddeholms AB, Hagfors.
38. Dieter, G.E. (1985). Bulk workability testing. In *Metals Handbook*, (9th edn) Vol. 8, pp. 571–3. Metals Park, Ohio.
39. Pickering F.B. (1975). High-strength low-alloy steels—a decade of progress. Microalloying 75, Int. Symp. by Union Carbide Corp., p. 1.
40. Pickering F.B. (1971). Towards improved ductility and toughness. Climax Molybdenum Comp. Conf., Kyoto, p. 9.
41. Dodd B. and Bai Y. (1987). *Ductile fracture and ductility*. Academic Press, London.
42. Orowan E. (1950). Fatigue and fracture of metals. MIT Symp., Wiley, New York.
43. Knott J.F. (1973). *Fundamentals of fracture mechanics*. Butterworths, London.
44. Kiessling R. and Nordberg H. (1971). Influence of inclusions on mechanical properties in steel. Critical inclusion size. *Rep. Royal Swedish Acad. Eng. Sci.*, **169**, 1, 159–69.
45. Andersson H. (1977). Analysis of a model for void growth and coalescence ahead of a moving crack tip. *J. Mech. Phys. Solids*, **25**, 217–33.
46. Rice J.R. (1968). A path independent integral and the approximate analysis of strain concentration by notches and cracks. *J. Appl. Mech.*, **35**, 379–86.
47. Rice J.R. and Johnson M.F. (1970). In *Inelastic behaviour of solids*, (ed. M.F. Kanninen *et al.*), p. 641, McGraw-Hill, New York.
48. Knott J.F. (1980). Micromechanisms of fibrous crack extension in engineering alloys. *Metal Science*, **14**, 327–36.
49. Peterson R.E. (1959). Proc. symp. fatigue of aircraft structure, WADC TR 59-507, p. 273.
50. Peterson R.E. (1959). In *Metal fatigue*, (ed. G. Sines). McGraw-Hill, New York.
51. Nordberg H. (1980). The effect of notches and non-metallic inclusions on fatigue properties of high strength steel. Internal Report, IM-1450, Swedish Institute for Metal Research.
52. Brooksbank D. and Andrews K.W. (1970). Stress fields around inclusions and their relation to mechanical properties; production and application of clean steels. *Int. Conf. at Balatonfüred*, pp. 186–198. Iron and Steel Institute, Publ. 134.
53. Hauser J.J. and Wells M.G.H. (1976). The effect of inclusions on fatigue properties in high-strength steels. In *Mechanical Working and Steel Processing*, p. 186. AIME, New York.
54. Kiessling R. (1978). *Non-metallic inclusions in steel*. Metals Society, London.

55. Ollilainen V. (1981). The effect of Ca treatment on the machinability of steel. In *Swedish symposium on non-metallic inclusions in steel*, (ed. H. Nordberg and R. Sandberg), pp. 429–49. Uddeholms AB, Hagfors, Sverige.
56. Emley E. F. (1966). *Principles of magnesium technology*. Pergamon Press, Oxford.
57. Simensen C. J. and Oberländer B. C. (1980). A survey of inclusions in magnesium. *Z. Prakt. Metallographie*, **17**, 125–36.
58. Hanawalt J. D., Nelson C. E., and Peloubet J. A. (1942). Corrosion studies of magnesium and its alloys. *Trans. AIME*, **147**, 273–99.
59. Eklund G. (1971). Relation between pitting and non-metallic inclusions. *Jernkont. Ann.*, **155**, 637–42.
60. Eklund G. (1974). Initiation of pitting at sulphide inclusions in stainless steel. *J. Electro-Chem. Soc.*, **121**, 467–73.
61. Szummer, A., Lublinska, K., and Janik-Czachov, M. (1991). Low manganese corrosion resistant steels. *ISIJ*, **31**, No. 2, 239–40.
62. Konig, U., Bier, M., and Schultze, J. W. (1991). The passivation of iron and iron/chromium alloys in acid and neutral solution. *ISIJ*, **31**, No. 2, 142–8.

2

THERMODYNAMICS AND TRANSPORT PROPERTIES

2.1 THERMODYNAMICS

2.1.1 Introduction

Thermodynamics and kinetics are the tools we use to describe how impurities enter the melt and how they can be removed. Refining metallurgy is a never-ending battle of trying to remove impurities and inclusions from the molten metal. These impurities come from various sources. Partly, they stem from the raw materials (ores) used in extractive metallurgy, and partly they enter the melt during the alloying and refining treatment. The melt picks up impurities from the containing walls (refractories), from the atmosphere, and even from the phase used to extract the impurities (slag). The slag and the walls must stand up to the high temperatures and chemical attack from the melt. The walls are therefore composed of refractory materials such as Al_2O_3 or MgO in the case of steelmaking. These refractory materials must be relatively inexpensive; they are therefore never 'pure'. They also contain binding materials such as SiO_2, B_2O_3, carbon-based binders, and so on.

Finally, oxygen, hydrogen, and nitrogen enter the melt due to contact with the atmosphere. This problem of 'pick-up' is illustrated in Fig. 2.1

When an alloy is produced one starts with a more or less pure molten metal and then adds the various alloying elements in controlled amounts. Thus for instance, Si, Mn, and Zn are added to molten aluminium to create the specified aluminium alloy composition. Similarly, iron is the starting point for iron-based alloys and copper for copper-based alloys. The alloy additions may contain a number of undesired impurities and inclusions. Production of iron, aluminium, copper, and so on is the subject of extractive metallurgy. In Chapter 3 production of the metals and alloy addition is discussed only in so far as light is thrown on the origin and amounts of dissolved impurities and inclusions.

In high-temperature metallurgy (pyrometallurgy) thermodynamics tells us what processes or reactions are possible. Furthermore, thermodynamics give us the 'driving force' for the description of the rates (kinetics). From thermodynamics we may map out both the positive and negative aspects of a refining process. Usually, we assume that the pressure and temperature are constant and apply thermodynamics to study equilibria at the interface between two or more phases. The interfaces may be slag–metal, refractory–metal, inclusion–metal, or atmosphere–metal. In pyrometallurgy, due to the high temperatures and correspondingly high reaction rates, we often assume that the

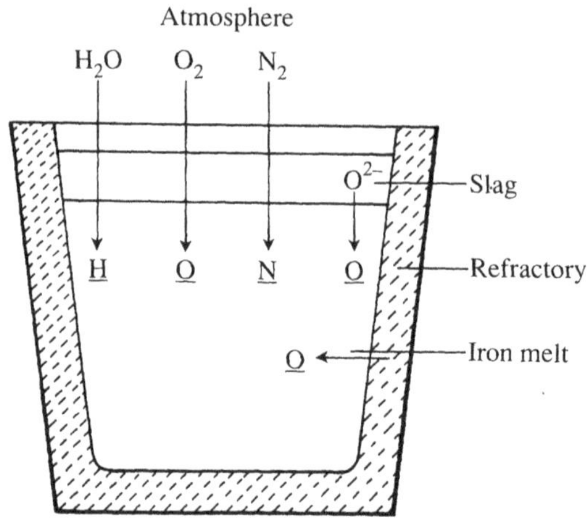

FIG. 2.1 Pick-up of impurities in steel from the atmosphere and refractory.

chemical reaction rates are not rate-limiting (see Chapter 4). Mass and heat transfer to and from the interfaces determine the kinetics. It is then very important that the thermodynamics (the equilibria) are given correctly for the interfaces.

Even at high temperatures it is possible that the equilibria are not attained at the interfaces. This may be the case in problems concerning contact between more than two interfaces or when contact times are very short. Examples of this are removal of inclusions to walls, bubbles, or filters, wetting problems, and rapid solidification. Towards the end of this chapter, surface and interfacial tension problems are discussed. Wetting is partly treated in Chapter 5.

In the following, thermodynamics is presented in a form tailor-made for the description of molten low-alloy metals. It is assumed that the student has some previous knowledge of thermodynamics.

2.1.2 Enthalpy, entropy, and Gibbs energy

As is well known, Thomson and Berthelot believed that the tendency of a reaction to take place is determined by the heat of reaction. For a reaction to occur, the heat given off from the system had to be positive. In our present terminology we consider heat absorbed by the system as positive, and we would say that the *enthalpy change* of the reaction, ΔH, would have to be negative. While this is true for a number of spontaneous and well-known reactions, it is also easy to find examples of spontaneously occurring processes where the system absorbs heat, that is, ΔH of the reaction is positive. Thus, there must be another and more subtle quantity which plays a part in determining the tendency for a reaction to take place, or, more generally, the direction of a spontaneous process. This quantity is the *entropy*, or rather the *entropy change* ΔS of the process.

The simplest definition of the entropy content of a body is the reversibly absorbed heat (or enthalpy) divided by the absolute temperature. When a body or a system absorbs heat at a constant temperature, such as, for example, when a chunk of ice is melting, the entropy change is simply

$$\Delta S = \frac{\Delta H}{T}. \tag{2.1}$$

However, there is also another aspect of entropy: it is related to probability, or rather to disorder in the system. The relation to the *thermodynamic probability W* was first proposed by Boltzmann:

$$S = k \ln W. \tag{2.2}$$

Here $k = 1.3806 \times 10^{-23}\ \mathrm{J\,K^{-1}}$, which is identical to the gas constant divided by the number of particles in a mole. The thermodynamic probability, W, may be defined as the total number of possible and distinguishable microstates of a system, subject to the condition of constant total energy, volume, and amount of substance. In a perfect crystal at absolute zero, the atoms can be arranged in one way only, that is, $W = 1$, and $S = k\ln 1 = 0$. (This, in fact, is the essential content of the third law of thermodynamics.) As the crystal is heated, the atoms start to vibrate, and according to the quantum theory they are distributed over a certain set of energy levels. With increasing temperature, which means increased energy, the atoms may be distributed over a wider range of energy levels, which means increasing disorder, which is turn means increasing entropy. With precise knowledge of these energy levels, both the entropy and the enthalpy of pure substances may, in principle, be calculated from statistical thermodynamics. In practice, this is generally not possible for solids and liquids, whereas the data for gases is usually acquired in this way. Thus, statistical thermodynamics has contributed significantly to our store of thermodynamic data. It does not, however, have any significant impact on our use of this data and hence, for our purposes, we will not pursue the subject of statistical thermodynamics. Nevertheless, it is important to have at least a qualitative understanding of the connection between the disorder in a system and its entropy.

The entropy resulting from heating is called *thermal entropy*. But increased disorder may also result when particles (atoms or molecules of different kinds) exchange positions in space. This gives rise to *configurational entropy*. We will give a simple example of the statistical calculation of the latter. Suppose we have N_1 atoms of substance A and N_2 atoms of substance B, and that these substances are in liquid or gaseous form so that they may mix readily. We assume furthermore that there is no specific interaction between atoms A and B (or more properly, that the bond energy of the bond A–B is equal to the average of the bonds A–A and B–B). In this case the most probable state is that of random mixing of A and B atoms. The total number of atoms is $N_1 + N_2$ and the total number of different ways these atoms may be arranged in a quasi-lattice is $(N_1 + N_2)!$ Permutation of identical atoms does not, however, lead to a new and distinguishable microstate; hence, we have to reduce the total number by $N_1!N_2!$ From this the entropy change is given by

$$\Delta S = k \ln W = k \ln \frac{(N_1 + N_2)!}{N_1!N_2!}. \tag{2.3}$$

When N_1 and N_2 are large numbers we may use Stirling's approximation: $\ln(N!) = N\ln N - N$. After some rearrangement, this leads to

$$\Delta S = -(N_1 + N_2)k\left[\frac{N_1}{N_1 + N_2}\ln\frac{N_1}{N_1 + N_2} + \frac{N_2}{N_1 + N_2}\ln\frac{N_2}{N_1 + N_2}\right].$$

The fractions are nothing but the mole fractions. Dividing through by the number of atoms in a mole, we get the familiar equation for the entropy change when n_1 moles of substance A and n_2 moles of substance B form an ideal mixture:

$$\Delta S^{\mathrm{id}}_{\mathrm{mix}} = -(n_1 + n_2)R(x_1 \ln x_1 + x_2 \ln x_2). \tag{2.4}$$

Now let us suppose that the mixture consists essentially of substance A, with only a little of substance B dissolved. It is then realized that when the added amount of substance B tends towards zero, then the entropy change per mole of B added tends towards infinity because of the logarithmic relationship. Due to this peculiar property of the entropy, a pure metal is highly improbable, and in fact does not exist. Nature tends to create disorder, to mix up the elements, and this behaviour generates an abundance of difficulties for the worker in refining metallurgy.

Let us now return to the general question of 'the driving force' of a process or a reaction. From what has already been said about entropy and disorder we may assume that a spontaneous process will always occur in the direction of increasing entropy (which is one way of stating the second law of thermodynamics). That is, the *total* change in entropy must be positive. This total is the sum of the change in the system under consideration and the change in the environment. Denoting the environment by e, we have

$$\Delta S_{\mathrm{tot}} = \Delta S + \Delta S_{\mathrm{e}} \geqslant 0 \tag{2.5}$$

where the inequality sign denotes a spontaneous process while the equality stands for equilibrium.

We usually prefer to take into account only the changes taking place within our system, leaving the rest of the universe out of consideration; hence we would like to eliminate ΔS_{e}. Let us consider a system in which a process, for example a chemical reaction, takes place at constant temperature and pressure. In the system, then, we have the enthalpy change ΔH and the entropy change

ΔS. The environment essentially plays the part of a heat reservoir; thus by the first law of thermodynamics, $\Delta H_e = -\Delta H$. Furthermore, from eqn (2.1), the entropy change in the environment is $\Delta S_e = \Delta H_e/T = -\Delta H/T$. Inserted into eqn (2.5) this gives

$$\Delta S - \frac{\Delta H}{T} \geqslant 0.$$

By a slight rearrangement

$$\Delta H - T\Delta S \leqslant 0.$$

We introduce another function, the *Gibbs energy* (formerly known as the Gibbs free energy) $G = H - TS$, named in honour of J. Willard Gibbs (1839–1903). The criterion for spontaneous change, respectively equilibrium, in a system is then[1]

$$\Delta G = \Delta H - T\Delta S \leqslant 0. \qquad (2.6)$$

Again the inequality refers to a spontaneous change and the equality refers to equilibrium (a reversible change). The change in Gibbs energy thus has replaced the *heat of reaction* as the correct measure of the driving force (strictly speaking, for reactions taking place at constant temperature and external pressure, which is usually the case).

The change in Gibbs energy may most often be considered the net outcome of two opposing tendencies, as appears from the equation for ΔG. We may illustrate this with the freezing of water as a very simple example. In order for freezing to occur heat must be given off and ΔH is negative. But freezing leads to increased order on a molecular scale, which means that ΔS is also negative. Thus the two terms ΔH and $T\Delta S$ work in opposite directions. Which of the two terms will 'win' is then determined by the temperature. At $T < 273$ K, ΔH wins, and water freezes. At $T > 273$ K, $T\Delta S$ wins, and ice melts.

2.1.3 The effect of high temperature on molten metals

Usually in a molten metal alloy we assume that the various elements are present only as atoms. Since the dissolution of one substance in another can occur only if there is a decrease in the Gibbs energy, it follows that, generally speaking, gases and refractory compounds do not dissolve in molten metals as readily as do other metals. To understand this, the dissolution of a solid compound can be visualized as occurring in three steps as follows.

1. The compound is broken down into its various elements.
2. The elements are brought into liquid form, which means that they must be melted if they are solid, or condensed if they are gaseous.
3. The liquid elements are dissolved in the liquid metal.

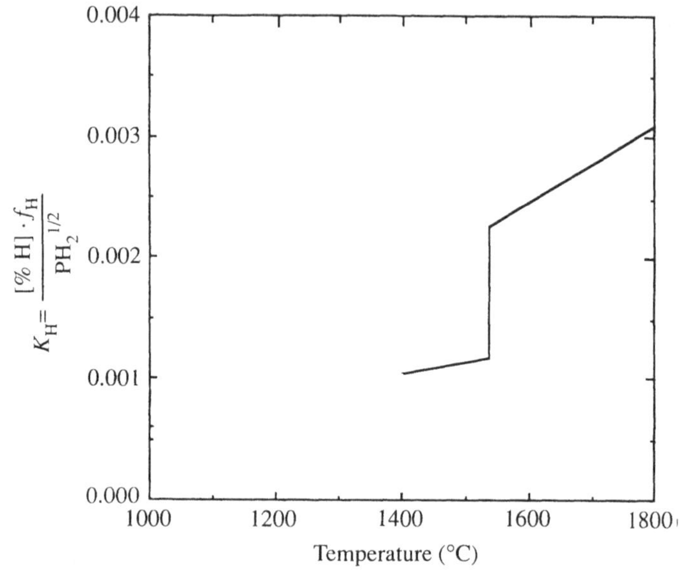

FIG. 2.2 The solubility of hydrogen in iron.

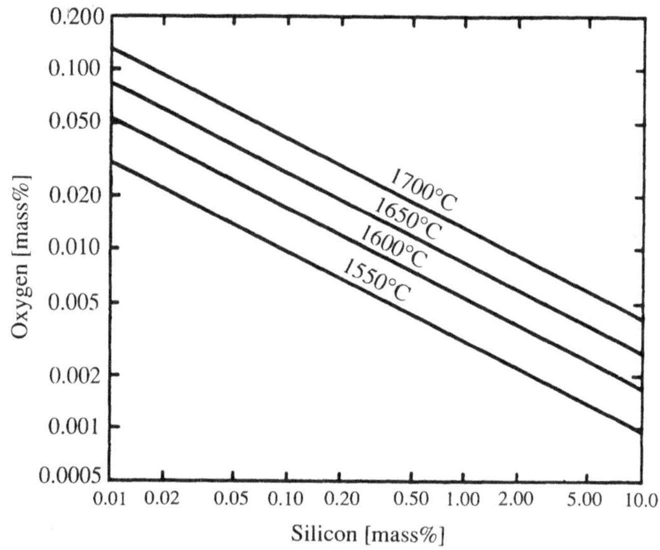

FIG. 2.3 Oxygen in equilibrium with silicon in steel at various temperatures.

Steps 1 and 2 will always have positive ΔG values, therefore gases and chemical compounds are generally less soluble than metals.

For solid compounds, the positive Gibbs energy of melting and of atomization decreases with rising temperature; at the same time the term $T\Delta S_{mix}$ tends to decrease the Gibbs energy of step 3. Thus the solubility of a solid in a liquid metal rises with temperature. The same is true for the solubility of the two-atomic gas H_2. Two examples are given of this effect in Figs 2.2 and 2.3. This behaviour is very important. When the temperature in a metal is increased, various elements and gases from the refractory, atmosphere, slag, and so on are dissolved in the melt. Conversely, during solidification solubility decreases, the processes are reversed and inclusions (particles) and gas (H_2 pores) are precipitated. This effect is one of the major problems in refining metallurgy.

It should be mentioned that if the process of mixing produces heat (ΔH_{mix} negative), this may compensate for steps 1 and 2 and give a high solubility. In this case the solubility may in fact decrease with increasing temperature.[11]

2.1.4 Chemical potentials and activities

Let U denote the internal energy of a system, Q the heat, and W the work supplied to the system. The first law of thermodynamics in differential form may then be written

$$dU = \partial Q + \partial W$$

where the symbols indicate that ∂Q and ∂W are not exact differentials, but merely small amounts of heat and work. If we consider an equilibrium process in a closed system, with volume work as the only form of work, these two quantities may be replaced by

$$dU = TdS - pdV. \tag{2.7}$$

Now let us extend this fundamental equation of thermodynamics to include changes in the chemical composition of the system. We note that various forms of work and energy may be written as the product of an intensive quantity (such as electrical potential, pressure, mechanical force, and soon) and an extensive quantity (such as electrical charge, volume, distance). By analogy we extend this to the chemical energy of a substance. The extensive quantity in this case is the amount of substance, usually given in terms of the number of moles, n_i. The intensive quantity is simply named the *chemical potential*, with the symbol μ_i for substance i. Thus eqn (2.7) is extended to

$$dU = TdS - pdV + \sum_i \mu_i dn_i \tag{2.8}$$

where the dn_i's may be either positive or negative. In order to gain insight into the nature of this chemical potential, we again write the definition of Gibbs energy:

$$G = U + pV - TS.$$

Taking the total differential, the following is obtained:

$$\mathrm{d}G = \mathrm{d}U + p\mathrm{d}V + V\mathrm{d}p - T\mathrm{d}S - S\mathrm{d}T.$$

Introducing $\mathrm{d}U$ from eqn (2.8) gives

$$\mathrm{d}G = V\mathrm{d}p - S\mathrm{d}T + \sum_i \mu_i \mathrm{d}n_i. \tag{2.9}$$

For a process occurring at constant pressure and temperature, $\mathrm{d}p$ and $\mathrm{d}T$ are zero, and the equation reduces to $\mathrm{d}G = \Sigma_i \mu_i \mathrm{d}n_i$. From this we have

$$\mu_i = \left(\frac{\partial G}{\partial n_i}\right)_{p,T}. \tag{2.10}$$

In other words, at constant temperature and pressure, the chemical potential of a substance is the *partial molar Gibbs energy* of that substance. It could be written $\bar{G}_i$ as for other partial molar molar quantities, but because of its significance in chemical thermodynamics, the special symbol μ_i is used. (For a pure substance, the term 'partial' is superfluous, and the chemical potential is simply the Gibbs energy per mole.)

As is well known, the internal energy, U, of a substance cannot be stated as an absolute quantity; it can only be stated relative to some chosen reference state. The same is then true for the Gibbs energy, and hence for the chemical potential. The chemical potential of a substance in its reference state is usually designated μ_i^0. Thus, when substance i is transformed from its reference state to a mixture, the change in its chemical potential is $(\mu_i - \mu_i^0)$.

Next we introduce yet another concept, the *activity* a_i, defined by the equation

$$a_i = \exp\left(\frac{\mu_i - \mu_i^0}{RT}\right)$$

or more conveniently

$$RT \ln a_i = \mu_i - \mu_i^0. \tag{2.11}$$

Notice that in the reference state where $\mu_i = \mu_i^0$ the actvity a_i becomes equal to 1. It is clear from the defining equation that the physical significance of the activity depends on the choice of reference state. This will be discussed in the following sections.

2.1.5 The pure substance as reference state, and Raoult's law

Before proceeding, it is necessary to make a semantic note: 'mixture' and 'solution' are synonyms in the field of chemical and metallurgical thermodynamics. We often talk of mixtures when the components are completely or nearly completely miscible, and of solutions when some components (the solutes) are present only in low concentrations in the main component (the solvent). In principle, however, there is no difference and we will use one or other of the two terms without discrimination.

We will now consider a liquid mixture of the two substances A and B. As reference states we choose the pure liquid substances A and B at the temperature of the mixture. Now it is common knowledge from phase diagrams that one or both of the components in pure form may be solid at a temperature when their mixture is liquid. Thus the pure, liquid component may not be the thermodynamically stable form at the relevant temperature; nevertheless, this is the reference state we have chosen.

For mixtures in which both components are present in comparable amounts, the mole fractions x_1 and x_2 are the most reasonable expressions for the composition. We will assume, as in Section 2.1.2, that the two substances mix in all proportions without any specific interaction. This means that we assume the enthalpy of mixing $\Delta H_{\text{mix}} = 0$. The entropy of mixing in this case is the same as that given in eqn (2.4) from statistical arguments. It follows that

$$\Delta G_{\text{mix}} = \Delta H_{\text{mix}} - T\Delta S_{\text{mix}} = RT(n_1 \ln x_1 + n_2 \ln x_2). \qquad (2.12)$$

This sort of mixture, in which the enthalpy of mixing is zero and the entropy corresponds to completely random mixing, is called an *ideal mixture*. Focusing on, for example, component B, we get from eqn (2.12) by a differentiation

$$\frac{\partial \Delta G_{\text{mix}}}{\partial n_2} = \mu_2 - \mu_2^0 = RT \ln x_2. \qquad (2.13)$$

By comparison with eqn (2.11) it is seen that the activity of a component in an ideal solution is equal to its mole fraction. When $x_2 = 1$, then $a_2 = 1$. The pure component is the standard state. This is the virtue of the concept of activity in general: it becomes equal to the chosen measure of concentration in the limiting case.

Most mixtures do not show ideal behaviour, and the activity of a component in a mixture generally must be found by experiment. When it is found that the activity departs from the mole fraction, this is expressed by an *activity coefficient* γ_i such that

$$a_i = \gamma_i x_i. \qquad (2.14)$$

In a system with several phases at equilibrium, the chemical potential of any component of the system must be the same in all the phases. It follows that the activity of the component must also be the same, provided that the activity in the different phases is referred to the same reference state.

In the above presentation, the ideal solution and the activity were introduced on the basis of a statistical picture of mixing. Textbooks on thermodynamics usually follow a slightly different course by first considering the isothermal expansion of an ideal gas, and next the vapour pressure of substances. Thus, if the vapour pressure of the pure substance (the reference state) is p_i^0, and the vapour pressure of that substance in a mixture is p_i, then its activity in the mixture is $a_i = p_i/p_i^0$, provided the vapour behaves as an ideal gas. The identity with the activity as defined above comes about because the chemical potential of the substance must be the same in the mixture and in the vapour at equilibrium, and because ideal gases form ideal mixtures as defined above.

The French chemist F.-M. Raoult (1830–1901) found for a number of liquid mixtures that the vapour pressure of a component is proportional to the mole fraction of that component in the mixture, that is, $p_i = x_i p_i^0$. This is known as *Raoult's law*. It is strictly valid for an ideal mixture as defined above. In modern usage it is not only restricted to vapour pressures; whenever the activity equals the mole fraction we say that Raoult's law is obeyed.

As indicated above, most mixtures do not show ideal behaviour. Thus for activity greater than the mole fraction we talk about positive deviation from Raoult's law; when it is smaller than the mole fraction we call it negative deviation. It is important to note, however, that regardless of deviations over most of the composition range, Raoult's law always holds as a limiting law for the main component of a mixture (the solvent) when its mole fraction tends towards unity. This is stated here without proof and shown graphically in Fig. 2.4. We will return to it in Section 2.1.7.

2.1.6 The dilute solution and Henry's law

The English physician and chemist W. Henry (1774–1836) studied the solubility of gases in liquids, and he found that the pressure of the gas over the solution was proportional to its concentration in the solution. This is called *Henry's law*. With concentration expressed in terms of mole fraction we may write it

$$P_i = k_H x_i \tag{2.15}$$

where k_H is an empirical constant. We will say that Henry's law is valid for a solute whenever the activity of that solute is proportional to its concentration. This is true for any solute at very low concentration (that is, at 'infinite dilution') in any solvent.

The mole fraction, however, may not be the best or most practical choice of expression for concentration in dilute solutions. For example, in salts and other substances in aqueous solution, concentration is usually expressed in terms of moles per litre. Correspondingly, the reference state for a solute is chosen as its solution in water at 1 mol/l. For solutions in liquid metals, concentration is more conveniently expressed as mass per cent (mass%) solute, with standard state 1 mass% in the relevant solvent metal. Then Henry's law becomes:

$$a_i = c_i. \tag{2.16}$$

The proportionality constant in Henry's law has disappeared, and it says simply that the activity of a component at low concentration is equal to its concentration. In some solutions, however, the activity may deviate from the concentration even at concentrations lower than that of unit activity. This is definitely the case, for example, in aqueous solutions of electrolytes, where marked deviations from Henry's law appear already at concentrations below 0.1 mol/l. In such cases, at the chosen reference state of unit concentration the unit activity only represents a hypothetical state.[2] This near-trivial problem is not discussed here because for solutions in liquid metals the reference state of 1 mass% usually represents such a low concentration that Henry's law is reasonably well obeyed. At larger concentrations, deviations may occur. This is discussed in the following.

When deviations occur, the activity may be expressed in terms of the concentration multiplied by an activity coefficient:

$$a_i = f_i[\%\, i]. \tag{2.17}$$

In this equation, the symbol [% *i*] stands for the concentration expressed as mass% of component *i*. Note that different symbols are used for the activity coefficient, depending upon whether the activity is expressed 'according to Raoult' or 'according to Henry'. The use of γ and f as introduced here is opposite to convention in chemistry where f is used for the activity coefficient with reference to Raoult's law, and γ with reference to Henry's law. The use of

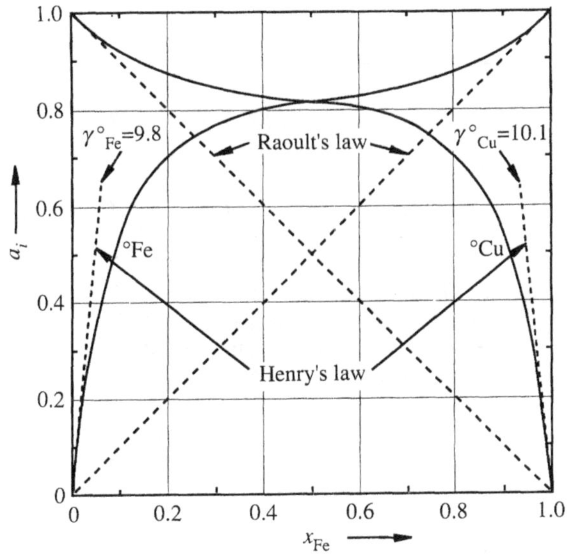

FIG. 2.4 Activities in liquid mixtures of copper and iron at 1550°C. The Henry's law tangent is indicated for copper dissolved in iron and iron dissolved in copper.[2]

γ and f in the present book, however, has a long standing in the field of metallic solution. For the activity itself, however, the same symbol a is commonly used regardless of the reference. For this reason, it is important to emphasize the reference state that is used, whenever doubt may occur.

The relation between Raoult's and Henry's laws is shown graphically in Fig. 2.4. It is seen that, on a mole fraction scale, Raoult's law says that the activity as a function of concentration has a slope of unity. Henry's law, on the other hand, says that, starting from zero or near zero, the activity curve has a constant slope, but it says nothing about the magnitude of this slope.

From Fig. 2.4. it is also seen that a mixture which shows positive deviation according to Raoult shows negative deviation according to Henry. Correspondingly, negative deviation from Raoult's law is accompanied by positive deviation from Henry's law.

In most tables of thermodynamic data, the pure component is the reference state. To allow us to use this data with our reference state of 1 mass%, the Gibbs energy change for transfer between the two reference states is given in Tables 2.1–2.4 for the solvent metals aluminium,[4] iron,[5] copper,[6] and cobalt.[7] That is, the tables give the change in Gibbs energy, ΔG_i^0, that occurs when 1 mole

TABLE 2.1 *Standard Gibbs energies* ΔG_i^0 *for solution of elements to 1 mass% solution in liquid aluminium.*[4] *s denotes solid, l denotes liquid, and g denotes gas*

Element, i		ΔG_i^0 (joule/mole)
Be	(s)	$39\,360 - 40.17\ T$
Bi	(l)	$22\,210 - 48.90\ T$
Ca	(l)	$(-43\,510 - 41.55\ T)$
Cd	(l)	$29\,710 - 52.29\ T$
Cu	(s)	$-4750 - 68.555\ T$
Fe	(s)	$-112\,970 - 14.272\ T$
Ga	(*l*)	$3480 - 48.33\ T$
Ge	(l)	$-11\,550 - 51.413\ T$
$\frac{1}{2}H_2$	(g)	$48\,800 + 27.29\ T$
In	(l)	$28\,450 - 55.32\ T$
Li	(l)	$-24\,270 - 12.61\ T$
Mg	(l)	$-14\,550 - 38.66\ T$
Na	(l)	$34\,430 - 21.05\ T$
Ni	(s)	$-118\,320 - 54.95\ T$
Pb	(l)	$41\,710 - 53.69\ T$
Sb	(l)	$54\,810 - 90.33\ T$
Si	(s)	$40\,160 - 85.84\ T$
Sn	(l)	$24\,460 - 59.97\ T$
Zn	(l)	$10\,620 - 49.84\ T$

TABLE 2.2 *Standard Gibbs energies for solution of elements to 1 mass% solution in liquid iron*[5, 24]

Element, i		ΔG_i^0 (joule/mole)
Ag	(l)	$82\,430 - 43.76\,T$
Al	(l)	$-63\,180 - 27.91\,T$
B	(s)	$-65\,270 - 21.55\,T$
C	(graphite)	$22\,590 - 42.26\,T$
Ca	(g)	$-39\,460 + 49.37\,T$
Co	(l)	$1000 - 38.74\,T$
Cr	(l)	$-37.70\,T$
Cr	(s)	$19\,250 - 46.86\,T$
Cu	(l)	$33\,470 - 39.37\,T$
$\frac{1}{2}H_2$	(g)	$36\,480 + 30.46\,T$
Mn	(l)	$4080 - 38.16\,T$
Mo	(l)	$-42.80\,T$
Mo	(s)	$27\,610 - 52.38\,T$
$\frac{1}{2}N_2$	(g)	$3600 + 23.89\,T$
Nb	(l)	$-42.68\,T$
Nb	(s)	$23\,010 - 52.30\,T$
Ni	(l)	$-20\,920 - 31.05\,T$
$\frac{1}{2}O_2$	(g)	$-117\,150 - 2.89\,T$
$\frac{1}{2}P_2$	(g)	$-122\,170 - 19.25\,T$
Pb	(l)	$212\,550 - 106.27\,T$
$\frac{1}{2}S_2$	(g)	$-135\,060 + 23.43\,T$
Si	(l)	$-131\,500 - 17.24\,T$
Sn	(l)	$15\,980 - 44.43\,T$
Ti	(l)	$-46\,020 - 37.03\,T$
Ti	(s)	$-31\,130 - 44.98\,T$
U	(l)	$-56\,070 - 50.21\,T$
V	(l)	$-42\,260 - 35.98\,T$
V	(s)	$-20\,710 - 45.61\,T$
W	(l)	$-48.12\,T$
W	(s)	$31\,380 - 63.60\,T$
Zr	(l)	$-51\,050 - 42.38\,T$
Zr	(s)	$-34\,730 - 50.0\,T$

of the pure substance i, initially in its reference state at 1 bar pressure and at temperature T, is dissolved in the solvent metal to give a solution with $[\%\,i] = 1$ at the same temperature T and 1 bar pressure. It may be added that this ΔG_i^0 is essentially the change in partial molar Gibbs energy of component i as a result of the dissolution process, since the change in partial molar Gibbs energy of the solvent metal as a result of the small change in concentration from 100 mass% to 99 mass% is essentially zero.

TABLE 2.3 *Standard Gibbs energies for solution of elements to 1 mass% solution in liquid copper*[6, 24]

Element, i		ΔG_i^0 (joule/mole)
Ag	(l)	$16\,320 - 44.02\,T$
Al	(l)	$-\ 36\,110 - 57.91\,T$
As	(g)	$-\ 93\,510 - 39.50\,T$
Au	(l)	$-\ 19\,370 - 50.58\,T$
Bi	(l)	$24\,940 - 63.18\,T$
C	(graphite)	$35\,770 + 50.21\,T$
Ca	(l)	$-\ 92\,890 - 34.31\,T$
Cd	(g)	$-107\,530 + 53.14\,T$
Cd	(l)	$-\ 7\,780 - 42.68\,T$
Co	(s)	$33\,470 - 37.66\,T$
Cr	(s)	$46\,020 - 36.48\,T$
Fe	(s)	$54\,270 - 47.45\,T$
Fe	(l)	$38\,910 - 38.79\,T$
Ga	(l)	$-\ 45\,190 - 36.32\,T$
Ge	(l)	$-\ 66\,940 - 32.89\,T$
$\frac{1}{2}$ H_2	(g)	$43\,510 + 31.38\,T$
In	(l)	$-\ 39\,960 - 23.35\,T$
Mg	(g)	$-168\,200 + 63.18\,T$
Mg	(l)	$-\ 36\,280 - 31.51\,T$
Mn	(l)	$(-\ 8\,160 - 36.94\,T)$
Mn	(s)	$(6\,490 - 46.61\,T)$
Ni	(l)	$9\,790 - 37.66\,T$
Ni	(s)	$27\,410 - 48.12\,T$
$\frac{1}{2}$ O_2	(g)	$-\ 85\,350 + 18.54\,T$
Pb	(l)	$36\,070 - 58.62\,T$
Pd	(s)	$3\,350 - 42.26\,T$
Pt	(s)	$(-\ 42\,680 - 43.81\,T)$
$\frac{1}{2}$ S_2	(g)	$-119\,660 + 25.23\,T$
Sb	(l)	$-\ 52\,300 - 43.51\,T$
Se	(g)	$-\ 76\,150 - 40.08\,T$
Si	(l)	$-\ 62\,760 - 31.38\,T$
Si	(s)	$-\ 12\,130 - 61.42\,T$
Sn	(l)	$-\ 37\,240 - 43.52\,T$
Te	(g)	$-\ 41\,840 - 44.06\,T$
Tl	(l)	$28\,160 - 49.12\,T$
V	(s)	$117\,570 - 75.73\,T$
Zn	(l)	$-\ 23\,600 - 38.37\,T$

TABLE 2.4 *Standard Gibbs energies for solution of elements to 1 mass% solution in liquid cobalt*[7, 24]

Element, i		ΔG_i^0 (joule/mole)
Al	(l)	$-91\,210 - 27.15\,T$
C	(graphite)	$20\,880 - 20.42\,T$
Fe	(l)	$-12\,050 - 27.49\,T$
$\frac{1}{2}$ H_2	(g)	$35\,560 + 31.51\,T$
Mn	(l)	$-37.70\,T$
$\frac{1}{2}$ N_2	(g)	$41\,840 + 21.92\,T$
Ni	(l)	$1\,260 - 44.14\,T$
$\frac{1}{2}$ O_2	(g)	$-61\,630 - 9.71\,T$
$\frac{1}{2}$ S_2	(g)	$-135\,140 + 30.25\,T$
Si	(l)	$-146\,440 - 7.03\,T$
Ti	(s)	$-94\,060 - 44.43\,T$
V	(s)	$-25\,690 - 44.61\,T$

To show the conversion from mole fraction to mass per cent, we consider a binary mixture of A and B, with mole fractions x_1 and x_2. We then have

$$[\%\,\mathrm{B}] = \frac{100x_2M_2}{x_1M_1 + x_2M_2} \tag{2.18}$$

where M_1 and M_2 are the atomic weights of the two components. At infinite dilution the equation reduces to

$$[\%\,\mathrm{B}] = 100x_2\frac{M_2}{M_1}. \tag{2.19}$$

The ratio between the Henrian activity a_2^{H} and the Raoultian activity a_2^{R} at infinite dilution is then given by

$$\frac{a_2^{\mathrm{H}}}{a_2^{\mathrm{R}}} = \frac{[\%\,\mathrm{B}]}{\gamma_2^0 x_2} = \frac{100M_2}{\gamma_2^0 M_1}. \tag{2.20}$$

In the latter equation, γ_2^0 is the Raoultian activity coefficient for component B at infinite dilution, which is the same as the slope of the Henry's law tangent at $x_2 = 0$.

As stated in the introduction to this section, Henry's law in its original form says that the concentration of a dissolved gas is proportional to its pressure over the solution, cf. eqn (2.15). This is not always true, however. When a diatomic gas such as hydrogen dissolves in a liquid metal, it simultaneously dissociates:

$$\tfrac{1}{2}\mathrm{H_2(g)} = \mathrm{H(in\ metal)}.$$

The equilibrium constant has the form

$$K = \frac{a_{\mathrm{H}}}{p_{\mathrm{H}_2}^{\frac{1}{2}}}.$$

At low concentrations

$$[\%\,\mathrm{H}] = p_{\mathrm{H}_2}^{\frac{1}{2}} K. \tag{2.21}$$

Equation (2.21) is commonly referred to as Sievert's law, generally valid for the dissolution of diatomic gases in liquid metals.

In the cases where Henry's law does not hold for the solute, Raoult's law does not hold for the solvent. This is the case whenever the solute dissociates in the solvent. Hydrogen in a liquid metal is one example; another is the dissolution of a salt in water. Both 'laws' depend on the condition that the dissolved substance is present in the same molecular form in the solution and in the standard state.

For mass transfer between metal and a second phase, resistance to transfer is often on the metal side. Then it will be seen in Chapter 4 that the notation is simpler if the inverse equilibrium constant is employed. In this case, the symbol K is also used for the inverse equilibrium constant.

2.1.7 Gibbs–Duhem's law

From eqn (2.9), for a system at constant pressure and temperature, we have

$$\mathrm{d}G = \sum_i \mu_i \mathrm{d}n_i. \tag{2.22}$$

This equation is valid for an infinitesimal amount of a system containing a number of components. Now we may conceptually increase the amount of the system while keeping its composition constant, which means that all of the μ_i's remain constant. This sort of 'chemical integration' gives

$$G = \sum_i \mu_i n_i. \tag{2.23}$$

Differentiation of eqn (2.23) gives

$$\mathrm{d}G = \sum_i \mu_i \mathrm{d}n_i + \sum_i n_i \mathrm{d}\mu_i. \tag{2.24}$$

By comparison with eqn (2.22) it is seen that we must have

$$\sum_i n_i \mathrm{d}\mu_i = 0. \tag{2.25}$$

This is the basic form of the Gibbs–Duhem equation which may take various forms depending upon the application. The equation gives a relation between the chemical potentials or activities of the various components. We will apply it to one mole of a binary mixture. The combination of eqns (2.25) and (2.11) gives

$$x_1 \mathrm{d}\ln a_1 + x_2 \mathrm{d}\ln a_2 = 0. \tag{2.26}$$

Equation (2.26) may be used to evaluate the activity of a component in a solution when that of the other is known over a range of compositions. The integration meets with some difficulty, however, because in the limit when $a_i \to 0$, $\ln a_i \to -\infty$. To avoid this difficulty, we note that $x_1 + x_2 = 1$, $\mathrm{d}x_1 + \mathrm{d}x_2 = 0$, and hence also

$$x_1 \mathrm{d}\ln x_1 + x_2 \mathrm{d}\ln x_2 = 0. \tag{2.27}$$

By subtraction of eqn (2.27) from eqn (2.26) we get

$$x_1 \mathrm{d}\ln \gamma_1 + x_2 \mathrm{d}\ln \gamma_2 = 0. \tag{2.28}$$

Thus the Gibbs–Duhem equation may also be applied to the activity coefficients, and these do not go to zero at any point. We will not, however, pursue the integration of Gibbs–Duhem's equation; this is discussed in textbooks on thermodynamics such as that of Gaskell.[2]

We will instead use eqn (2.26) to demonstrate the connection between Raoult's and Henry's laws. For this purpose, we will regard component B as the minor constituent, and we will assume that Henry's law is valid for it. Composition is expressed in terms of mole fractions; thus Henry's law is written

$$a_2 = k_{\mathrm{H}} x_2$$

which gives

$$\mathrm{d}\ln a_2 = \mathrm{d}\ln x_2 = \frac{\mathrm{d}x_2}{x_2}.$$

It is noted also that $\mathrm{d}x_2 = -\mathrm{d}x_1$. With these substitutions, eqn (2.26) yields

$$x_1 \mathrm{d}\ln a_1 - \mathrm{d}x_1 = 0$$

or

$$\mathrm{d}\ln a_1 = \mathrm{d}\ln x_1.$$

Integrated, it becomes

$$\ln a_1 = \ln x_1 + \ln \text{const.}$$

or

$$a_1 = \text{const.}\, x_1. \tag{2.29}$$

But by definition, $a_1 = x_1$ for $x_1 = 1$, hence we must have const. $= 1$, and thus $a_1 = x_1$. In other words, *in the concentration range where Henry's law is valid for the solute, Raoult's law is valid for the solvent*, which is what we set out to prove.

For metals there is always a range at high dilution (low concentration) where Henry's law is valid for the solute, and as a consequence Raoult's law must always be valid for the solvent when its mole fraction approaches unity. This was stated without proof at the end of section 2.1.5, and it is demonstrated graphically in Fig. 2.4. If it is found experimentally for some systems that Raoult's law is not obeyed as $x_1 \to 1$, this means that the calculation is based on an erroneous assumption regarding the formula of the dissolved species. This would be the case if one assumes, for example, that hydrogen dissolves in a molten metal as H_2, or that sodium chloride dissolves in water as NaCl molecules. However, in the example, if H is selected as the species present in the metal, eqn (2.29) applies and Raoult's law is valid. Conversely, agreement with Raoult's law for the solvent may be used to determine the molecular identity of the solute. This, is essence, is the basis for the classical method of determining molecular weights by freezing-point depression.

2.1.8 Gibbs energies of solution

To illustrate the use of the Gibbs energy of solution tables the equilibrium constants for the dissolution of H_2 and Na (gas) in molten aluminium are worked out. These equilibrium constants are employed in Chapter 4.

For the reaction from Table 2.1:

$$\tfrac{1}{2}\,H_2 = \underline{H}$$

$$K_H = \frac{P_{H_2}^{\frac{1}{2}}}{a_H} = \exp\left(\frac{\Delta G_H^0}{RT}\right) = \exp\left\{\left(\frac{\Delta H^0}{T} - \Delta S^0\right)\Big/R\right\}.$$

At 760°C from Table 2.1

$$\frac{\Delta G_H^0}{RT} = \left(\frac{48\,800}{760+273} + 27.29\right)\Big/8.314.$$

Then

$$a_H = p_{H_2}^{\frac{1}{2}}/K_H \qquad \text{and} \qquad [\%\,H] = \frac{P_{H_2}^{\frac{1}{2}}}{K_H f_H}$$

where

$$K_H = 0.78 \times 10^4.$$

The solubility in aluminium is given in Fig. 4.14. Hydrogen solubility curves usually have the same general shape for the various metals. Figure 2.2 shows the hydrogen solubility in iron. For low-alloy steels $f_H \approx 1$. This will be discussed in the following.

Next, for

$$\mathrm{Na(l)} = \underline{\mathrm{Na}}$$

$$K_{\mathrm{Na}} = \frac{1}{a_{\mathrm{Na}}} = \exp\left(\frac{\Delta G^0_{\mathrm{Na}}}{\mathrm{RT}}\right).$$

At 1100 K

$$\frac{\Delta G^0_{\mathrm{Na}}}{\mathrm{RT}} = \left(\frac{34\,430}{1100} - 21.05\right) \Big/ 8.3145 = 1.23$$

and $K_{\mathrm{Na}} = 3.43$. Thus, in equilibrium with liquid sodium, the sodium activity in aluminium, given in per cent, is $0.291 = 1/3.43$.

$$[\%\mathrm{Na}] = a_{\mathrm{Na}}/f_{\mathrm{Na}}.$$

If the content of alloy additions is low, $f_{\mathrm{Na}} \sim 1$.

The thermodynamic data given in the tables is valid only in a limited temperature range. This range is found in the references. Usually, the tables must be used in conjunction with tables of standard Gibbs energy data.[8,9,10]

Appendix 1 contains a selection of the Gibbs energy of reaction. As an example of the use of Appendix 1 and the tables, let us look at the dissolution of SiO_2 in molten iron.

From Appendix 1:

for:	ΔG^0(J/mol)
SiO_2 (s) = Si(s) + O_2	906 000 − 176 T
Si(s) = Si(l)	50 540 − 30.0 T
and from Table 2.2:	
Si(l) = $\underline{\mathrm{Si}}$	−131 500 − 17.24 T
O_2(g) = 2$\underline{\mathrm{O}}$	−234 300 − 5.78 T
Addition of the above gives:	
SiO_2(s) = $\underline{\mathrm{Si}}$ + 2$\underline{\mathrm{O}}$	590 740 − 229.02 T

Then

$$\frac{a_{\mathrm{Si}} a_{\mathrm{O}}^2}{a_{\mathrm{SiO_2}}} = \exp\left\{\left(-\frac{590\,740}{T} + 229.02\right) \Big/ 8.3145\right\}$$

$$= \exp\left(-\frac{71\,049}{T} + 27.54\right).$$

This equation is presented in Fig. 2.3, where corresponding values of [% Si] and [% O] are plotted for various steelmaking temperatures. The difference between activities and concentrations is disregarded in Fig. 2.3. The ratios between activity and concentration can be calculated from the interaction coefficients given later. It is found that the differences are small for Si and O in iron.

2.1.9 Interaction coefficients

Henry's rule applies only in a limited low concentration range. As the concentration of component i increases, the activity a_i deviates from the line given by Henry's rule. In the following it will be seen how corrections can be introduced to describe the deviations. When Henry's rule was discussed, it was assumed that there was no interaction between the solute atoms i. Also it was assumed that no other solute atoms j were present in the melt.

However, the i atoms will in fact interact with each other and with the j atoms. This will change the enthalpy of solution for the i atoms and also the structure around the i atoms and thus effect the entropy of the i atoms. Then we write for the chemical potential

$$\mu_i \approx \mu_i(\text{no interaction}) + \sum_{j=2}^{k} \frac{\partial \mu_i}{\partial x_j} x_j. \tag{2.30}$$

Here μ_i (no interaction) is the chemical potential if there had been no interaction between the solute components. $\partial\mu_i/\partial n_j$ is the effect of component j on i. $\partial\mu_i/RT\partial x_j = \varepsilon_i^j$ is called the first-order interaction coefficient. From eqns (2.9) and (2.10) we see that

$$\frac{\partial \mu_i}{\partial n_j} = \frac{\partial G}{\partial n_i \partial n_j} = \frac{\partial \mu_j}{\partial n_i}.$$

Thus, it is realized that

$$\varepsilon_i^j = \varepsilon_j^i. \tag{2.31}$$

From eqns (2.11) and (2.14) it is seen that eqn (2.30) becomes

$$\ln\gamma_i = \ln\gamma_i^0 + \sum_{j=2}^{k} \varepsilon_i^j x_j \tag{2.32}$$

where γ_i^0 is the Raoultian activity coefficient if there had been no interaction between the solute components. It therefore corresponds to Henry's law (cf. Fig. 2.4).

In practical applications, as mentioned previously, it is usually preferable to employ concentration units of mass per cent and the corresponding activity coefficient

$$f_i = \frac{a_i}{[\%\, i]}. \tag{2.33}$$

The interaction coefficients e_i^j in this case are defined by the equation

$$\log f_i = \sum_{j=2}^{k} e_i^j \, [\%\, j]\,. \tag{2.34}$$

Often we wish to convert from the ε_i^j to the e_i^j coefficients. To accomplish this it is pointed out that the concentration units are related by

$$[\%\, i] = \frac{100 x_i M_i}{x_1 M_1 + \sum_{j=2}^{k} x_j M_j}\,. \tag{2.35}$$

Furthermore, the definitions of the Henrian activity a_i^{H} and Raoultian activity, a_i^{R}, are not identical. To find the conversion it is recalled that at infinite dilution $a_i^{\mathrm{H}} = [\%\, i]$. Then from eqns (2.35) and (2.14) the conversion is obtained by comparing two calculations of the mole fraction x_i at infinite dilution.

$$x_i = \frac{M_1 a_i^{\mathrm{H}}}{100 M_i} = \frac{a_i^{\mathrm{R}}}{\gamma_i^0}\,.$$

Thus the relation between the Henrian and Raoultian activities is given by

$$\frac{M_1 a_i^{\mathrm{H}} \gamma_i^0}{100 M_i a_i^{\mathrm{R}}} = 1. \tag{2.36}$$

From the definitions of the activity coefficients we get

$$1 = \frac{M_1 a_i^{\mathrm{H}} \gamma_i^0}{100 M_i a_i^{\mathrm{R}}} = \frac{M_1 f_i \, [\%\, i]\, \gamma_i^0}{100 M_i \gamma_i x_i}\,. \tag{2.37}$$

Finally, insertion of eqn (2.35) into the right-hand side of eqn (2.37) gives:

$$1 = \frac{f_i \gamma_i^0}{\gamma_i \left(1 + \sum_{j=2}^{k} x_j (M_j/M_1 - 1)\right)}\,. \tag{2.38}$$

If we now take the logarithm of the two sides of the equation, apply eqns (2.34) and (2.32) for $\log f_i$ and $\ln \gamma_i/\gamma_i^0$, respectively, and insert x_j instead of $[\%\, j]$ using eqn (2.35), the first-order x_j terms in the Taylor expansion should vanish. This gives[11]

$$230 e_i^j M_j/M_1 + \left(1 - \frac{M_i}{M_1}\right) = \varepsilon_i^j\,. \tag{2.39}$$

Equation (2.39) is derived in Problem 2.6. The term $1 - M_j/M_1$ has mistakenly often been omitted in the literature. Orders of magnitude for the interaction coefficients should be kept in mind. The ε_i^j coefficients usually vary from -10 to $+10$. The corresponding e_i^j coefficients are 100 times smaller. Ohtani[12] and others suggest that the ε_i^j change in a systematic manner with the atomic number. For instance, for iron as the solvent it has been possible to construct graphs of ε_j^i as a function of atomic number and position in the periodic table of elements.[13] Values of ε_j^i not available experimentally can then be obtained by interpolation or extrapolation of the graphs. In this book, only a selection of experimentally determined e_j^i are presented for aluminium, iron, copper, and cobalt as solvent metals. Much of the data is tentative. It is therefore only offered as survey values which the reader may find convenient to employ.

In Tables 2.5 and 2.8 some interaction coefficients are given for binary aluminium[14] and copper-based alloys.[15] Tables 2.6, 2.7, 2.9, and 2.10 contain interaction coefficients for ternary alloys with aluminium, iron, copper, and cobalt as solvent metals.

To illustrate the use of the interaction coefficients for ternary systems we calculate the activity coefficient for hydrogen, f_H, for one of the aluminium alloys given in Table 4.1. With 6% Si and 3.5% Cu we obtain from Table 2.6, $e_H^{Si} = 0.03$ and $e_H^{Cu} = 0.03$, then $\log f_H = 0.030 \times 6 + 0.030 \times 3.5 = 0.285$, and $f_H = 1.93$.

We notice that only iron, aluminium, copper, and cobalt-based melts have been included. For other solvent metals the data is even more limited.

Second-order effects that give corrections proportional to the second power of the solute concentrations have been obtained in the literature.[11] Such data is included in this book only in Chapter 6 for stainless steel where the alloy content is high.

2.1.10 Equilibria between particles (inclusions) and melts; precipitation deoxidation

Precipitation deoxidation of molten steel is described here partly due to its importance and partly to illustrate the use of thermodynamic data to calculate

TABLE 2.5 *Interaction coefficients in liquid Al–i binary alloys for components i on itself*[4] *(for ε_i^i see eqn (2.32) with j = i, for e_i^i see eqn (2.34) with j = i)*

Element, *i*	ε_i^i	e_i^i	*T(K)*
Cd	− 5.0	−0.0019	1373
Cu	2.2	0.0066	1373
Ga	− 0.3	0.0022	1023
Ge	+ 3.0	0.0076	1200
In	− 4.5	−0.0013	1173
Mg	3.0	0.014	1073
Si	16.0	0.067	1100
Sn	6.0	0.0093	1000
Zn	− 0.9	0.0009	1000

TABLE 2.6 *Interaction coefficients in ternary aluminium alloys*[4] *(see eqns (2.32) and (2.34))*

Part A	Metallic alloys	Temperature (K)
$\varepsilon_{Zn}^{Si} = 2.2$	$e_{Zn}^{Si} = 0.0094$	963–1053
$\varepsilon_{Si}^{Zn} = 2.2$	$e_{Si}^{Zn} = 0.0065$	
$\varepsilon_{Li}^{Sn} = -16.0$	$e_{Li}^{Sn} = -0.012$	949
$\varepsilon_{Sn}^{Li} = -16.0$	$e_{Sn}^{Li} = -0.28$	
$\varepsilon_{Na}^{Si} = (-12)$	$e_{Na}^{Si} = (-0.05)$	973
$\varepsilon_{Si}^{Na} = (-12)$	$e_{Si}^{Na} = (-0.062)$	
$\varepsilon_{Mg}^{Si} = (-9.0)$	$e_{Mg}^{Si} = (-0.037)$	973–1073
$\varepsilon_{Si}^{Mg} = (-9.0)$	$e_{Si}^{Mg} = (-0.044)$	
$\varepsilon_{Sn}^{Pb} = (-1.5)$	$e_{Sn}^{Pb} = (0.003)$	973–1073
$\varepsilon_{Pb}^{Sn} = (-1.5)$	$e_{Pb}^{Sn} = (0.0019)$	

Part B Al–H–j alloys

Element, j	ε_H^j	e_H^j	Temperature (K)
Ce	(−100)	(−0.08)	800
Cu	15	0.030	700–800
Cr	(−1)	(0)	800–900
Fe	(−1)	(0)	800–900
Mg	(−2)	(−0.01)	700–800
Mn	27	(0.06)	800
Ni	(19)	(0.04)	700–800
Th	−20	−0.006	800–900
Ti	(−42)	(−0.1)	800–900
Si	7.1	0.03	700–800
Sn	0.6	0.004	800–900

equilibria between two immiscible phases. Oxygen dissolved in steel increases with temperature at equilibrium with atmosphere, slag, or refractories. When the temperature is decreased during teeming and solidification, this process is reversed; the solubility of oxygen decreases. Then oxygen may react with an alloying element, M, in the steel

$$x\underline{O} + y\underline{M} = M_yO_x \qquad (2.40)$$

where the reaction product M_yO_x is a solid, liquid, or gaseous phase. For instance, M may be Si, Mn, C, or the solvent Fe. Oxygen often reacts

TABLE 2.7 *Interaction coefficients* e_i^j *in ternary iron-based alloys*[11,5] *(see eqn (2.34))*

i \ j	Al	B	C	Ca	Co	Cr	Cu	i \ j	H	Mn	Mo	N	Nb	Ni
Al	0.045		0.091	−0.047				Al	0.24			−0.058		
B		0.04	0.22					B	0.49			0.074		
C	0.043	0.24	0.124	−0.097	0.008	−0.024	0.016	C	0.67	−0.012	−0.008	0.09	−0.06	0.012
Ca	−0.072		−0.34	−0.002				Ca						−0.044
Co			0.021		0.0022	−0.022		Co	−0.14					
Cr			−0.12		−0.02		0.02	Cr	−0.33		−0.002	−0.18		0.0002
Cu			0.066			0.018	0.023	Cu	−0.24			0.026		
H	0.013	0.05	0.06		0.002	−0.002		H	0.0	−0.0014			−0.0023	
Mn			−0.07					Mn	−0.31			−0.15		
Mo			−0.10			−0.0003		Mo	−0.20			−0.10		
N	−0.03	0.094	0.103		0.011	−0.046	0.009	N		−0.04	−0.011		−0.07	0.0063
Nb			−0.49					Nb	−0.61			−0.47		
Ni			0.042	−0.067		−0.0003		Ni	−0.25			0.013		0.0009
O	−3.9	−2.6	−0.45		0.008	−0.04	−0.013	O	−3.1	−0.021	0.0035	0.057	−0.14	0.006
P			0.13			−0.03	0.024	P	0.21			0.094		0.0002
Pb	0.021		0.066			0.02	−0.028	Pb		0.023				−0.019
S	0.035	0.13	0.11		0.0026	−0.011	−0.008	S	0.12	−0.026	0.003	0.01		
Si	0.058	0.02	0.18	−0.067		−0.0003	0.014	Si	0.64			0.09		0.005
Sn			0.37			0.015		Sn	0.12			0.27		
Ti						0.055		Ti	−1.1			−1.8		
V			−0.34					V	−0.59			−0.35		
W			−0.15					W	0.088			−0.07		
Zr								Zr				−4.10		

TABLE 2.7 (*continued*)

i \ *j*	O	P	Pb	S	Si	Sn	*i* \ *j*	Ti	V	W	Zr
Al	−6.6		0.0065	0.03	0.0056		Al				
B	−1.8			0.048	0.078		B				
C	−0.34	0.051	0.008	0.046	0.08	0.041	C		−0.077	−0.0056	
Ca					−0.1		Ca				
Co	0.018		0.003	0.0011			Co				
Cr	−0.14	−0.053	0.0083	−0.02	−0.0043	0.009	Cr	0.059			
Cu	−0.065	0.044	−0.0056	−0.021	0.027		Cu				
H	−0.19	0.011		0.008	0.027	0.0053	H	−0.019	−0.0074	0.0048	
Mn	−0.083	−0.0035	−0.0029	−0.048			Mn				
Mo			0.0023	−0.0005			Mo				
N	0.05	0.045		0.007	0.047	0.007	N	−0.53	−0.093	−0.0015	−0.63
Nb	−0.83			−0.047			Nb				
Ni	0.0	−0.0035	−0.0023	−0.0037	0.0057		Ni				
O	−0.2	0.07		−0.133	−0.131	−0.0111	O	−0.6	−0.3	0.0085	−3.0
P	0.13	0.062	0.011	0.028	0.12	0.013	P				
Pb		0.048		−0.32	0.048	0.057	Pb				
S	−0.27	0.29	−0.05	−0.028	0.063	−0.0044	S	−0.072	−0.016	0.01	−0.052
Si	−0.23	0.11	0.01	0.056	0.11	0.017	Si		0.025		
Sn	−0.11	0.036	0.035	−0.028	0.057	0.0016	Sn				
Ti	−1.80			−0.11			Ti	0.013			
V	−0.97			−0.028	0.42		V		0.015		
W	−0.052		0.0005	0.035			W				
Zr				−0.16			Zr				

TABLE 2.8 *Interaction coefficients in binary liquid copper alloys*[6]

Element, i	e_i^i
Ag	-0.005
Al	0.14
Au	0.008
Bi	$-9/T + 0.0052$
Ca	(0.14)
Fe	-0.029
Ga	0.028
Ge	0.051
H	(0)
Mg	(0.1)
Mn	0.03
O	$-414/T + 0.122$
Pb	-0.0006
S	$-179/T - 0.004$
Sb	(0.04)
Sn	0.025
Ti	-0.008
Zn	0.017

TABLE 2.9 *Interaction coefficients in ternary alloys of Cu*[6]

Element, j	e_H^j	e_O^j	e_S^j
Ag	0.0006	0	
Al	0.0058		
Au	0.0003	0.015	0.012
Co	0.015	-0.32	-0.023
Cr	0.0092		
Fe	-0.015	$-20\,000/T + 10.8$	$-125/T + 0.042$
Mn	-0.006		
Ni	-0.026	$-169/T + 0.079$	$-159/T + 0.069$
P	0.088	$6230/T + 3.43$	
Pb	0.031	-0.007	
Pt	-0.0084	0.057	0.019
S	0.073	-0.164	
Sb	0.031		
Si	0.042	(-62)	0.062
Sn	0.016	-0.009	
Te	-0.012		
Zn	0.029		

TABLE 2.10 *Interaction coefficients in ternary cobalt-based alloys for H, N, O, and S at 1600°C.*[7, 24]

Element j	e_H^j	e_N^j	e_O^j	e_S^j
Al	0.014	0.04		
Au			0.0026	
B	0.09			
Cr	0	−0.043	−0.07	
Cu	−0.0031	−0.009	−0.009	−0.016
Fe	0.0014	−0.01	−0.019	−0.0034
Mn	−0.0035		−0.20	
Mo	0.0041	−0.008	−0.0011	
Nb	0	−0.042		
Ni	−0.002	0.024		−0.007
S			−0.133	
Si	0.021	0.12		
Sn	0.006			
Ta	0.0054	−0.029		
Ti		−0.45		
V		−0.1	−0.28	
W	0.0064	−0.007		

simultaneously with the various components and complex inclusions may form. In the case of carbon, CO bubbles are created. If the inclusions or CO bubbles are trapped in the solidified metal, they may be harmful to the mechanical properties. This was discussed in Chapter 1.

How do we try to reduce the amount and size of such inclusions? Before casting, a deoxidation element is introduced with a much greater affinity for oxygen than the alloying elements. Then ideally the content of dissolved oxygen is lowered to such low levels that it does not react with the melt components during solidification. However, there is a catch to this procedure; the deoxidation product must not remain in the melt! For instance if Al is used to precipitate oxygen as follows,

$$2\underline{Al} + 3\underline{O} = Al_2O_3,$$

the hard and brittle Al_2O_3 particles must be removed. In Chapter 5 this removal of inclusions is treated in detail.

From standard Gibbs energy data[5, 11] and the Gibbs energies of solution given in Table 2.2 one can calculate the diagrams presented in Fig. 2.5 that relate oxygen activity and activity of the various deoxidation elements. An example of the calculations will be given. The activity of calcium is so low that it is not included in the diagram. Zirconium and calcium are the most efficient deoxidation elements. The reason why Al is preferred in practice is partly the lower price of Al and partly problems of separating the ZrO_2 inclusions from

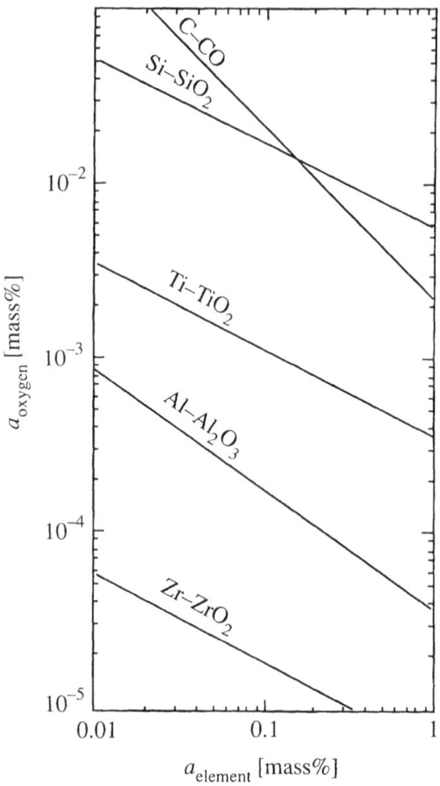

FIG. 2.5 The activity of oxygen in steel at 1600°C in equilibrium with various deoxidation elements.

the melt. The use of calcium will be discussed in the next section.

Carbon and silicon seem to be poor deoxidizers according to the diagram. However, it should be remembered that Fig. 2.5 is calculated with the activity equal to 1 for the deoxidation product and $p_{CO} = 1$ bar for carbon. If p_{CO} is much less than 1 bar, the line for carbon moves downwards (see Fig. 2.6) and carbon may become a useful deoxidant. How can we reduce the partial pressure of CO? One way is to dilute the CO gas with an inert gas such as argon or nitrogen. Another solution is to evacuate the gas above the melt so that the total pressure is less than 1 bar. In practice there is a limit to how low we can reduce oxygen levels by this approach. The explanation lies in the realm of kinetics treated in Chapter 7.

As for carbon, it is observed from the equilibrium

$$\underline{Si} + 2\underline{O} = SiO_2$$

that silicon can become an effective deoxidant if the activity of SiO_2 is

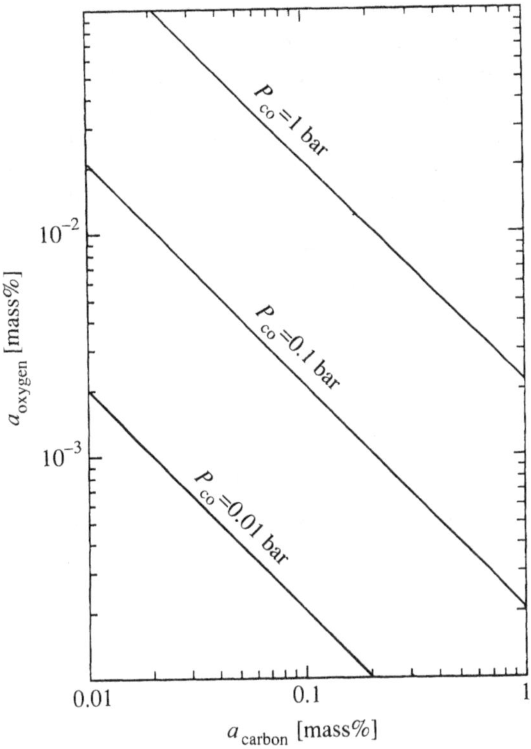

FIG. 2.6 Activity of oxygen in steel in equilibrium with carbon at 1600°C for various pressures of CO.

reduced. As will be seen from Fig. 2.14, a_{SiO_2} is far less than 1 in a CaO-Al_2O_3-SiO_2 slag high in lime (basic slag).

It will be shown how the Al line in Fig. 2.5 is obtained. From Appendix 1 and Table 2.2:

	ΔG^0(J/mol)
$\frac{1}{3} Al_2O_3 = \frac{2}{3} Al(l) + \frac{1}{2} O_2$	$562\,433 - 109\,T$
$\frac{1}{2} O_2 = \underline{O}$	$-117\,150 - 2.89\,T$
$\frac{2}{3} Al(l) = \frac{2}{3} \underline{Al}$	$-42\,120 - 18.6\,T$
$\frac{1}{3} Al_2O_3 = \frac{2}{3} \underline{Al} + \underline{O}$	$403\,163 - 130.5\,T$

or at $T = 1873$ K and $a_{Al_2O_3} = 1$

$$\frac{a_{Al}^{\frac{2}{3}} a_O}{a_{Al_2O_3}^{\frac{1}{3}}} = K = 3.74 \times 10^{-5}. \tag{2.41}$$

Notice that the scales in Fig. 2.5 are logarithmic and that the slope is $-\frac{2}{3}$.

From the figure it might be imagined that dissolved oxygen could be brought down to whatever levels we desire by the addition of a strong deoxidizer. However, this is not the case. A deoxidizer, M, has a strong affinity for oxygen and this results in an interaction coefficient e_O^M which is large and negative. This gives a minimum for [% O] at low values of M. We will show this for M = Al. The logarithm of the equilibrium equation gives

$$\tfrac{2}{3}\log a_{Al} + \log a_O = \log K(T)$$

or, introducing activity coefficients,

$$\tfrac{2}{3}(\log f_{Al} + \log[\%\,Al]) + \log f_O + \log[\%\,O] = \log K(T). \qquad (2.42)$$

The minimum is found from

$$\frac{\partial[\%\,O]}{\partial[\%\,Al]} = O.$$

From the definition of the activity coefficients (see eqn (2.34))

$$e_O^{Al} = \frac{\partial \log f_O}{\partial[\%\,Al]}$$

and

$$e_{Al}^{Al} = \frac{\partial \log f_{Al}}{\partial[\%\,Al]}.$$

The minimum is given by differentiating eqn (2.42) with respect to [% Al]. Notice that log[% Al] = ln [% Al]/ln 10 = 0.434 ln [% AL]. Then

$$\tfrac{2}{3}\left(e_{Al}^{Al} + \frac{0.434}{[\%\,Al]}\right) + e_O^{Al} = O$$

$$[\%\,Al] = -\frac{0.434}{\frac{3}{2}e_O^{Al} + e_{Al}^{Al}}.$$

This determines the minimum for [% O]. From Table 2.7 of interaction coefficients $e_O^{Al} = -3.90$ and $e_{Al}^{Al} = 0.045$ so that [% Al] = 0.075. Furthermore, since $e_O^O = -0.20$ and $e_{Al}^O = -6.6$

$$\log f_O = e_O^O[\%\,O] + e_O^{Al}[\%\,Al] \approx e_O^{Al}[\%\,Al] = -3.90 \times 0.075$$

which gives $f_O = 0.5$ and

$$\log f_{Al} = e_{Al}^{Al}[\%\,Al] + e_{Al}^O[\%\,O] \approx e_{Al}^{Al}[\%\,Al] = 0.045 \times 0.075$$

so that $f_{Al} \approx 1$.

Finally we find from the equilibrium that the lowest oxygen value attainable by Al deoxidation with Al_2O_3 as the reaction product is [% O] = 0.0004 or 4 p.p.m. However, if Al_2O_3 is mixed with other compounds so that the

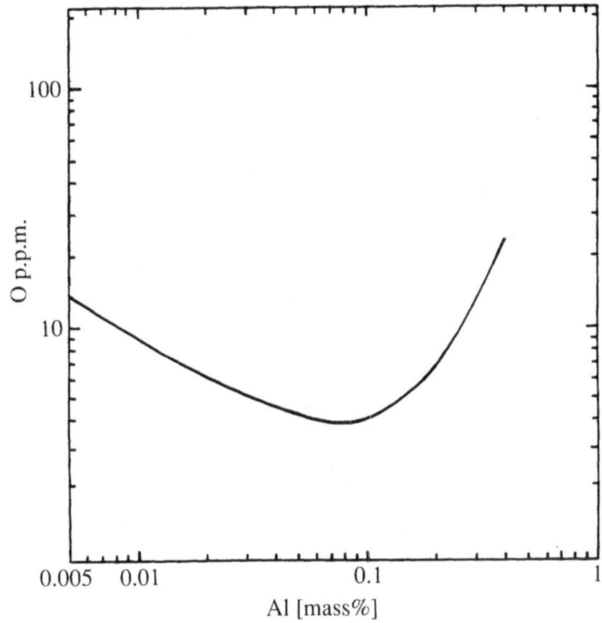

FIG. 2.7 Oxygen in steel in equlibrium with aluminium. Only first-order interaction coefficients are taken into account.

activity is less than 1, lower oxygen concentrations can be attained. Figure 2.7 gives [% O] as a function of [% Al] assuming $a_{Al_2O_3} = 1$. The oxygen values must not be confused with total oxygen which also includes oxygen in the form of Al_2O_3 inclusions. In steelmaking it is usually difficult to attain total oxygen below 10–20 p.p.m. Removal of inclusions of Al_2O_3 will be discussed in Chapter 5.

Due to pick-up of oxygen from the surroundings of the melt, there will be a steady loss of deoxidizer. Thus the Al is often added to the melt in excess of the amount corresponding to 0.075%.

In equilibrium with a slag inclusion where the Al_2O_3 activity is less than 1, eqn (2.41) may be used to obtain a_O. For instance, $a_{Al_2O_3} = 0.085$. At a temperature of 1823 K, the equilibrium constant in eqn (2.41) is reduced to $K = 1.85 \times 10^{-5}$. Then $a_O = 0.0002 \times 1.85/3.7 \times 0.085^{1/3} = 0.000\,04\%$. Thus by dilution of the Al_2O_3 reaction product we can attain very low oxygen levels. This last result is used in the following section on Ca addition.

2.1.11 Modification of inclusions; Ca additions to steel

In precipitation deoxidation, even if removal of the inclusions to the top slag and walls proceeds at a high rate, there will always be present a certain number of particles. The various removal, pick-up, and nucleation processes that may

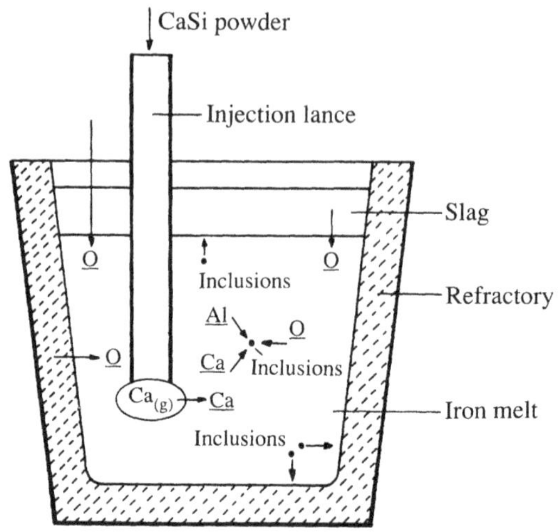

FIG. 2.8 The various removal, pick-up, and nucleation processes that occur when Ca is added to steel.

occur are illustrated in Fig. 2.8. Harmful effects of the inclusions were reviewed in Chapter 1. In rolling and other deformation processes it is found that the particles should not be too hard. The inclusions will then destroy the matrix surrounding them, and voids are created and grow during the deformation process. On the other hand, if the inclusions are too soft, as may be the case with MnS, for instance, they elongate and give poor mechanical properties in the transverse direction.

Furthermore, it has been found that solid oxide or sulphide inclusions tend to clog teeming or casting nozzles.[14,15] Figure 2.9 gives the temperature–composition diagram for the system CaO–Al_2O_3. Typically, there is a molten region at about equal amounts of CaO and Al_2O_3. Silica also increases the extent of the molten region. Clogging is reduced when the inclusions contain roughly equal amounts by mass of CaO, Al_2O_3, and SiO_2.

How do we change the composition of Al_2O_3 inclusions to inclusions that also contain CaO? One approach is to add calcium to the melt, in spite of the fact that calcium is an expensive metal and also has a high vapour pressure at steelmaking temperatures. Calcium is dissolved into the melt by injection of Ca and CaSi[16,17] or by wire feeding. This calcium together with aluminium may precipitate liquid calcium aluminate inclusions. At the same time precipitation of solid CaS should be prevented.

The objective outlined above presents a formidable problem both from a practical and a theoretical point of view. In the following we will look at the various equilibria to outline the thermodynamics involved in creating liquid calcium aluminate inclusions with no CaS precipitates.

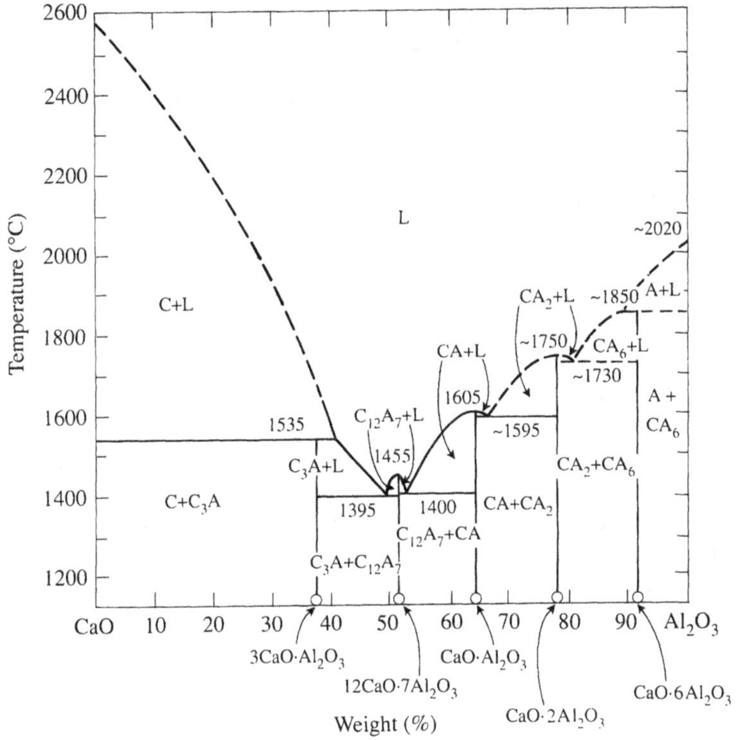

FIG. 2.9 Temperature–composition diagram for the system $CaO–Al_2O_3$.[27]

The phase rule is a useful tool in the treatment of the problem. Due to its general importance it is discussed in the following section.

2.1.12 The phase rule applied to the problem of calcium addition

When dealing with a system comprising several phases and a number of components, application of the phase rule may give useful information about the behaviour of the system. We will briefly review the background of this rule.

Assume that we have a system consisting of a number C of different components, distributed over a number P of phases. At equilibrium, the chemical potential of each component must be the same in all of the phases. For each component this gives $P - 1$ relations regarding the chemical potentials, and thus a total of $C(P - 1)$ equations for the whole system.

In each phase there are $C - 1$ composition variables, and thus a total of $P(C - 1)$ composition variables, plus the two variables defining temperature and pressure. Thus we have

Number of variables	$PC - P + 2$
− number of equations	$-PC + C$
Number of independent variables: $F = C - P + 2$	

This is Gibbs' phase rule. The equation looks simple, but its application quite often requires some thought. A main question is this. How do you count the components? To cut short the discussion, we may state that in metallurgy it is basically a good idea to take the number of elements in the system as the number of components. In so doing, however, one has to look out for possible fixed relations between the amounts of different elements, which eventually represent restrictions on the number of free variables. This is notably the case in systems in which each element is present in one oxidation state only, since there must be a balance between positive charges (cations) and negative charges (anions) in such a system. Thus the more complete phase rule may be written $F = C - P + 2 - R$, where R stands for the number of restrictions. It may have the values 0 or 1, and rarely 2.

Turning to the system presented in the previous section, we disregard silicon, and thus we have $C = 5$ (Fe, O, S, Al, Ca). There is a metal phase, an aluminate phase, and a calcium sulphide phase; thus $P = 3$. There is no restriction, $R = 0$, since several of the elements occur both in their elemental forms and in compounds. Thus we have the number of free variables.

$$F = 5 - 3 + 2 = 4.$$

If pressure and temperature are given, this leaves two degrees of freedom. To use these we can adjust $\underline{\text{Al}}$ and $\underline{\text{O}}$. Finally, hopefully, $\underline{\text{S}}$ can be kept at such low levels that CaS does not form. This last point means, in terms of the phase rule, that P is reduced by one. Then there is an additional degree of freedom and we can adjust for instance, $\underline{\text{S}}$.

Next we will turn our attention to the aluminate inclusions by considering separately the system $CaO–Al_2O_3$.[18] Using the 'rule of elements' this is a three-component system, but there is also a restriction (fixed oxidation numbers) so that $F = 3 - P + 2 - 1 = 4 - P$. (We would of course get the same result if we considered the system as composed of the two components CaO and Al_2O_3.)

With pressure and temperature given, one finds $F = 0$ for $P = 2$ in this system. That is, equilibrium at a fixed temperature between a solid and a liquid, or two solid phases, requires a definite composition in each of the phases, as is readily apparent from the phase diagram in Fig. 2.9. At temperatures around 1500°C and about equal mass fractions of alumina and calcium oxide, the system comprises only a liquid phase in which the composition may be varied. On the CaO-rich side the melt eventually precipitates $3CaO{\cdot}Al_2O_3$, while on the Al_2O_3-rich side $CaO{\cdot}Al_2O_3$ is precipitated. In the following we will discuss whether it is practical to deoxidize with a combination of aluminium and calcium, to remain within the region where the oxide inclusions are entirely liquid at 1500°C, and simultaneously prevent precipitation of CaS.

To give inclusions that contain CaO, calcium must be able to compete with aluminium as a deoxidizer. At the same concentration (activity) levels, Ca is in fact a better deoxidizer than Al. The problem in practice may become one of reaction rates and mechanisms involved, but still thermodynamics yields important information. What will happen when solid calcium is added to molten steel at 1500°C? First of all it melts (m.p. 838°C). Next we may find its boiling point from the equation in Appendix 1:

$$\mathrm{Ca(l)} = \mathrm{Ca(g)}; \qquad \Delta G^0 = 156\,800 - 89.5\,T(\mathrm{J/mol}).$$

That is, the boiling point is $156\,800/89.5 = 1752\,K = 1479°\mathrm{C}$. More accurate data gives 1755 K, or 1482°C. (The discrepancy is due to the use of a simple linear function for ΔG^0 as given in Appendix 1.) In fact, calcium is added to molten steel at even higher temperatures, and it is then realized that severe losses by volatilization of calcium is to be expected. How should this problem be reduced and the yields of calcium be increased? In practice two approaches are in use. If Ca is injected deep down in the melt, pressure is higher and boiling is prevented. Furthermore, one may employ an alloy of calcium where the activity relative to pure calcium is less than 1. Such an alloy is CaSi. Use of this alloy also gives the modified inclusions containing SiO_2.

Let us assume that a vapour pressure of calcium higher than 1 bar cannot be sustained. What is then the activity of calcium in the melt? For the reaction according to Table 2.2.

$$\mathrm{Ca(gas)} = \underline{\mathrm{Ca}}; \qquad \Delta \mathrm{G}^0 = -39\,460 + 49.37\,T\,\mathrm{J/mol}.$$

Then at 1482°C

$$\frac{a_{\mathrm{Ca}}}{p_{\mathrm{Ca}}} = 0.039$$

so that the activity of calcium is 0.039% at 1 bar Ca pressure. This value is roughly the same as the activity of the aluminium in the melt. Then mass transfer rates by diffusion to the new modified inclusions created during teeming and solidification should be roughly the same for calcium and aluminium. Thus conditions are present for producing new $CaO{\cdot}Al_2O_3$ inclusions.

Mass transfer problems are studied further in Chapters 5 and 7. However, there is one point remaining. What are the conditions for avoiding precipitation of CaS? The relevant equilibrium is

$$\mathrm{CaO} + \underline{\mathrm{S}} = \mathrm{CaS} + \underline{\mathrm{O}}.$$

From the thermodynamic data (Appendix 1 and Table 2.2):

	ΔG^0 (J/mol)
$CaO(s) = Ca(l) + \frac{1}{2} O_2$	$640\,000 - 108.6\,T$
$\frac{1}{2} O_2 = \underline{O}$	$-117\,150 - 2.89\,T$
$Ca(l) + \frac{1}{2} S_2(g) = CaS(s)$	$-548\,000 + 104\,T$
$\underline{S} = \frac{1}{2} S_2(g)$	$135\,060 - 23.43\,T$
$CaO(s) + \underline{S} = CaS(s) + \underline{O}$	$109\,910 - 30.92\,T$

Then at 1550°C

$$\frac{a_S}{a_O} = 34\frac{a_{CaS}}{a_{CaO}}.$$

For equal amounts by mass of CaO and Al_2O_3 at 1550°C it is found from Elliott *et al.*[27] that $a_{CaO} = 0.4$ and $a_{Al_2O_3} = 0.085$. Previously it was shown for Al deoxidized steel in equilibrium with this slag that $a_O = 0.000\,04\%$. Then with $a_{CaS} = 1$

$$a_s = 0.000\,04 \times 34/0.4 = 0.0034.$$

Thus a_S should be less than 0.0034% or 34 p.p.m. in order to prevent precipitation of CaS. Due to the very low oxygen activity, the steel must be desulphurized to low levels prior to the calcium treatment in order to prevent formation of CaS.

2.1.13 The regular solution model and molten salts

In the preceding sections emphasis has been placed on activities in dilute metallic solutions, where concentration is expressed in terms of mass per cent and Henry's law forms the basis for the definition of activity. On the other hand, when the solution can no longer be regarded as dilute, as, for example, in mixtures of molten salts, composition is better expressed in terms of mole fractions with activities related to Raoult's law.

In Section 2.1.5 an ideal solution was defined as one which obeys Raoult's law over the whole composition range. In practice, however, this is a rare exception; most mixtures show deviations from ideality. In order to discuss these deviations, we introduce the concept of *excess* quantities. An excess quantity is the difference between the real quantity and the corresponding ideal value. We will look at the excess Gibbs energy of mixing, ΔG^{xs}_{mix}.

For the Gibbs energy of mixing for a binary solution we have in general:

$$\Delta G_{mix} = RT(n_1 \ln a_1 + n_2 \ln a_2). \tag{2.43}$$

The corresponding expression for an ideal solution is

$$\Delta G^{id}_{mix} = RT(n_1 \ln x_1 + n_2 \ln x_2). \tag{2.12}$$

Taking the difference, and noting that $\ln a_1 - \ln x_1 = \ln(a_1/x_1) = \ln\gamma_1$, we have

$$\Delta G^{xs}_{mix} = RT(n_1 \ln\gamma_1 + n_2 \ln\gamma_2).$$

From this follows

$$\frac{\partial \Delta G^{xs}}{\partial n_1} = RT\ln\gamma_1; \qquad \frac{\partial \Delta G^{xs}}{\partial n_2} = RT\ln\gamma_2. \tag{2.44}$$

Equations (2.44) are quite general. Now we will introduce a class of solutions in which the enthalpy of mixing deviates from the ideal value (which is zero), while the entropy of mixing still has very nearly the ideal value (as given by eqn (2.4)). This class, introduced by J.H. Hildebrand in 1929, is called *regular solutions*. Because of the ideal entropy of mixing, we have for regular solutions

$$\Delta G^{xs}_{mix} = \Delta H_{mix}. \tag{2.45}$$

Next we will go one step further, and assume that the enthalpy of mixing is symmetrical with respect to the two components. We also know that we must have $\Delta H_{mix} = 0$ for $x_1 = 0$ and for $x_2 = 0$. Furthermore because of the symmetry it must have a maximum for $x_1 = x_2 = 0.5$. A simple function which meets these criteria is

$$\Delta G^{xs}_{mix} = \Delta H_{mix} = (n_1 + n_2)\alpha x_1 x_2 = \alpha \frac{n_1 n_2}{n_1 + n_2}$$

where α is an empirical constant. By differentiation as in eqn (2.44), this gives the expressions for the activity coefficients:

$$\ln\gamma_1 = \frac{\alpha}{RT}x_2^2; \qquad \ln\gamma_2 = \frac{\alpha}{RT}x_1^2. \tag{2.46}$$

These equations are representative of a model which we will call the *symmetrical regular solution*.

For molten salt mixtures it is often assumed that the different types of cations A^+ and B^+ are randomly distributed among the anions and the anions X^-, Y^- are randomly distributed among the cations (the Temkin model). For a binary salt mixture AX–BX with a common ion X, eqn (2.46) then becomes

$$\ln\gamma_{AX} = \frac{\alpha}{RT}N_B^2 \tag{2.47}$$

$$\ln\gamma_{BX} = \frac{\alpha}{RT}N_A^2 \tag{2.48}$$

where the N_A and N_B are the ionic fractions

$$N_A = \frac{x_A}{x_A + x_B}$$

$$N_B = \frac{x_B}{x_A + x_B}. \tag{2.49}$$

The cryolite melt employed in aluminium electrolysis cell may disregarding the presence of the AlF_6^{3-} ion, be regarded as composed of NaF and $NaAlF_4$ in the molar ratio 2:1, with Na^+ as the common ion. Then[19]

$$RT\ln\gamma_{NaF} = \alpha(1 - x_{NaF})^2 \tag{2.50}$$

where

$$\frac{\alpha}{RT} = -\frac{51\,000}{8.3145 \times 1285} = -4.77.$$

The validity of this equation is tested in Fig. 2.10. The value 51 000 J is found from the slope of the experimental points.

To pursue this example further, Na contamination in the Hall–Héroult cell is studied. The content of sodium is determined in molten aluminium in equilibrium with cryolite of the stoichiometric composition Na_3AlF_6 at the

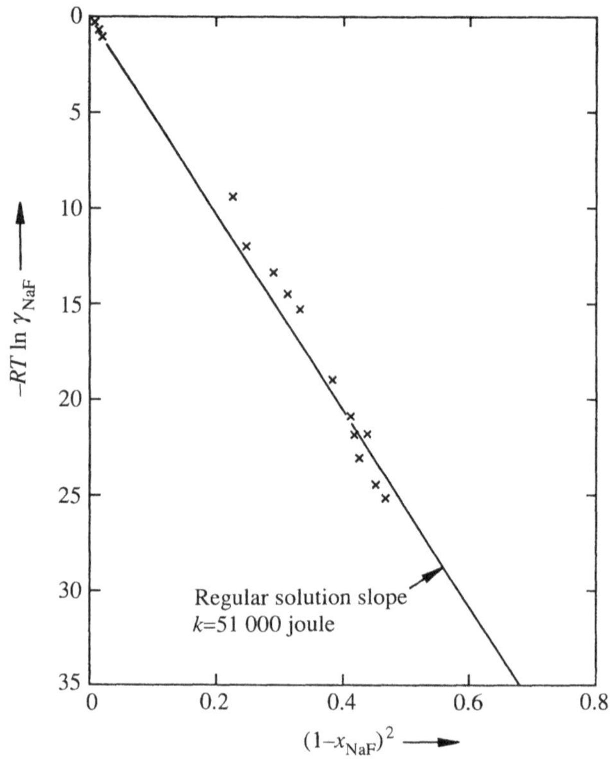

FIG. 2.10 Example of regular solution behaviour. The system is Na_3AlF_6 at the liquidus temperature.[19], 1285 K.

melting point 1285 K. (Sodium causes edge cracks in rolled products. For some qualities sodium levels below 1 p.p.m. are required.)

$$x_{\mathrm{NaF}} = 0.67 \quad \text{and} \quad x_{\mathrm{NaAlF_4}} = 0.33$$

$$\ln\gamma_{\mathrm{NaF}} = -4.77 \times 0.33^2 \quad \text{and} \quad \gamma_{\mathrm{NaF}} = 0.595.$$

From the free energy tables, ΔG^0 for the exchange reaction between Na and Al may be obtained:

	ΔG^0 (J/mol)
$2\mathrm{NaF(l)} = 2\mathrm{Na(g)} + \mathrm{F_2(g)}$	$1\,248\,000 - 296.0\,T$
$\mathrm{Na(g)} = \mathrm{Na(l)}$	$-101\,300 + 87.9\,T$
$\mathrm{Na(l)} = \underline{\mathrm{Na}}$	$34\,434 - 21.04\,T$
$\mathrm{F_2} + \frac{1}{3}\mathrm{Al(l)} + \mathrm{Na(g)} = \frac{1}{3}\mathrm{Na_3AlF_6(l)}$	$-1\,126\,667 + 207.8\,T$
$2\mathrm{NaF(l)} + \frac{1}{3}\underline{\mathrm{Al}} = \underline{\mathrm{Na}} + \frac{1}{3}\mathrm{Na_3AlF_6(l)}$	$54\,463 - 21.35\,T$

Then at 1285 K:

$$\ln\left(\frac{f_{\mathrm{Na}}[\%\,\mathrm{Na}]}{a_{\mathrm{Al}}^{\frac{1}{3}}a_{\mathrm{Naf}}^{2}}\right) = \frac{-(54\,463 - 21.35 \times 1285)}{8.3145 \times 1285} = -2.53$$

or

$$f_{\mathrm{Na}}[\%\,\mathrm{Na}] = 0.080 \times 1 \times (0.67 \times 0.595)^2 = 0.0125 \quad \text{or} \quad 125\ \text{p.p.m.}$$

In practice the electrolyte is higher in AlF_3 than corresponds to Na_3AlF_6. This lowers the sodium content to about 100 p.p.m.

Removal of sodium from aluminium is discussed in detail in Chapter 4.

2.1.14 Slags

In the ores that serve as a basis for metal production, the metal or metal compound that we wish to extract is intimately mixed with various oxides and sulphides. Silicates, aluminates, and phosphates are examples of such components. In the pyrometallurgical processes they remain as a molten oxide phase insoluble in the primary metal.

Sometimes the slag also contains significant amounts of the metal we are producing, for instance due to oxidation of the metal. This is the case when phosphorus or carbon are removed from steel under oxidizing conditions.

Slags can be employed to extract unwanted elements from the metal, for instance $\underline{P}$ from steel as just mentioned. Another impurity transferred to slag from steel is sulphur. However, sulphur removal takes place under reducing conditions, that is, when the activity of oxygen in the steel is low. Then an exchange reaction takes place between sulphur in the steel and oxygen in the slag. (This is discussed later.)

The present knowledge of the structures of slags is based on X-ray and neutron diffraction studies of slags and glasses.[20,21] The structures of silicate, phosphate, and borate glasses are conceived as random networks in which the small atoms with a high positive oxidation number, B(+III), Si(+IV), and P(+V), the 'network formers', are tetrahedrally surrounded by oxygen atoms, O(−II), which form bridges between the positive atoms. The most typical example of such structures is afforded by silica and silicates. In pure silica, the SiO_4 tetrahedra share all four corners with neighbouring tetrahedra, forming a three-dimensional network. With the addition of a basic oxide such as Na_2O or CaO, oxygen bridges are broken, and the structure is changed to one in which the SiO_4 tetrahedra form sheets, and then chains (or rings) as illustrated in Fig. 2.11. With the addition of sufficient basic oxide the structure finally contains isolated SiO_4^{4-} ions, such as in calcium orthosilicate, Ca_2SiO_4. Along with the breaking of oxygen bridges, the viscosity of the molten slag or oxide mixture goes down.

Various attempts have been made to interpret the thermodynamic properties of slag melts in terms of the ionic models.[22,23] These models were applied to the equilibrium relations for slag–metal exchange reactions in terms of the concentrations of ionic species. A problem is that the activity coefficients of the various ionic species can only be determined experimentally. Therefore there is a limit to how far the theory can be put to use. Ultimately, for a given slag–melt system experiments must be performed, for instance to determine the partition coefficient for an impurity. This will be discussed shortly. The theoretical studies are helpful in a general way; they throw light on the concept of basic and acid slags. Also they suggest the variables to use in describing the results of the experimental studies.

An acid slag, containing a large amount of network formers, cannot be used

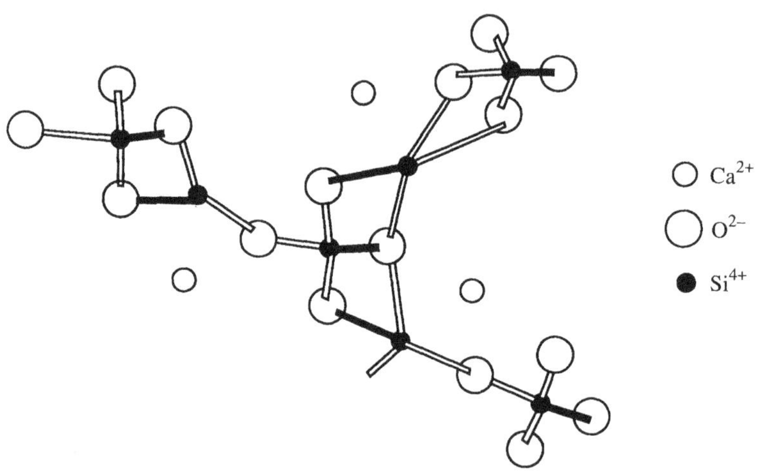

FIG. 2.11 Silicate chains in a slag partly broken by CaO addition.

to extract impurities such as oxygen or sulphur from a melt. One reason is thermodynamic: such a slag has a low capacity for these impurities. Furthermore, acid slags have high viscosity so that the mass transfer rates are low.

The typically basic oxides are essentially ionic compounds with relatively large cations of low charge. Examples, in order of increasing 'basicity', are MgO, FeO, MnO, CaO, PbO, BaO, Na_2O, and K_2O. Magnesium and iron oxides are not very effective in breaking oxygen bridges, while the oxides of sodium and potassium have a strong ability to break the network.

The action of the basic oxides depends upon their ability to contribute oxygen ions, which in turn react with the acid oxide. The reaction to complete the breaking of the network may be written as follows for the various acid oxides:

$$SiO_2 + 2O^{2-} = SiO_4^{4-}$$

$$P_2O_5 + 3O^{2-} = 2PO_4^{3-}$$

$$B_2O_3 + 3O^{2-} = 2BO_3^{3-}.$$

Further addition of basic oxide beyond this point results in free oxygen ions. It is realized that the content of oxygen ions plays an important role in determining the slag properties. The ionic model described above allows us to calculate the content of O^{2-} ions in the slag. Thus, if a slag is made up of w moles of SiO_2; x moles of P_2O_3, B_2O_5, or Al_2O_3; y moles of MgO, FeO, MnO, CaO, PbO, or BaO (we write M for these cations); and z moles of Na_2O or K_2O, we have the following equation for the number of moles of O^{2-}:

$$\begin{aligned} &wSiO_2 + x(P_2O_5, B_2O_3, Al_2O_3) + yMO + z(Na_2O, K_2O) \\ &= wSiO_4^{4-} + 2x(PO_4^{3-}, BO_3^{3-}, AlO_3^{3-}) \\ &\quad + (y + z - 2w - 3x)O^{2-} + yM^{2+} + 2z(Na^+, K^+). \end{aligned} \tag{2.51}$$

This equation fulfils the requirement that the number of each atom species is conserved and that electrical neutrality is retained. It is seen that the basic components increase O^{2-} while the acid components (including Al_2O_3) consume O^{2-} ions.

2.1.15 The equilibrium between sulphur in steel and in a basic slag

Removal of sulphur from steel to a basic slag is a very important part of steelmaking. Therefore the equilibrium between steel and slag is discussed here. Another reason is to illustrate the use and limitations of the ionic theories outlined above.

In the slag, sulphur is assumed to be present as S^{2-} similar to the O^{2-}. The sulphur, in moving from metal to slag, must acquire two electrons since the components in the metal are all in atomic form. Then the only manner in which

S transfer can take place is by an exchange reaction with a negatively charged ion from the slag. The only candidate seems to be O^{2-}:

$$\underline{S} + O^{2-} = S^{2-} + \underline{O}.$$

Sulphur is removed from the melt and in return we are getting oxygen in the metal! To drive the equilibrium to the right it is seen that O^{2-} should be high (basic slag) and $\underline{O}$ should be low (reducing conditions).

For the exchange equilibrium we may write

$$\frac{a_O a_{S^{2-}}}{a_S a_{O^{2-}}} = K_s(T)$$

where the equilibrium constant $K_s(T)$ depends not only on the temperature but also on the composition of the slag. The simplest manner of eliminating $a_{O^{2-}}$ is to assume that there is equilibrium between oxygen in the melt and in the slag:

$$\underline{O} + Fe = O^{2-} + Fe^{2+}.$$

If the two reactions are added, one obtains

$$\underline{S} + Fe = S^{2-} + Fe^{++}.$$

This equation probably does not reflect the mechanism of sulphur removal, but can be used to determine the equilibrium. Then

$$\frac{a_{s^{2-}} a_{Fe^{2+}}}{a_s a_{Fe}} = K_s'(T).$$

From a fundamental point of view one would wish to obtain the 'sulphur capacity' of the slag given as the ratio

$$\frac{x_{S^{2-}}}{[\%\,S]} = \frac{K_s'(T) f_s}{\gamma_{S^{2-}} \gamma_{Fe^{2+}} x_{Fe^{2+}}}$$

where $x_{S^{2-}}$ refers to the anion mole fraction and $x_{Fe^{2+}}$ refers to the cation mole fraction. However, the activity coefficients $\gamma_{S^{2-}}$ and $\gamma_{Fe^{2+}}$ are unknown and the factor $K_s'(T)/(\gamma_{S^{2-}}\gamma_{Fe^{2+}})$ must be determined experimentally. Furthermore, from a practical point of view, it is the ratio (% S)/[% S] that is of cardinal interest. Figure 2.12 shows some industrial results[24] for the sulphur capacity (partition coefficient) $K_s = (\%\,S)/[\%\,S]$, where (% S) is the sulphur content in the slag in mass per cent. $x_{Fe^{2+}}$ has been replaced by (% FeO*) where the asterisk indicates that in the chemical analysis of the slag small amounts of Fe^{3+} have also been included. The figure shows that (% S)/[% S](% FeO*) depends primarily on the content of network formers summed up as mass percentages. As this sum increases (as the slag becomes more acid) the sulphur capacity goes down. Also the sulphur capacity is reduced as the conditions become more oxidizing, that is, as (% FeO) and [% O] become higher.

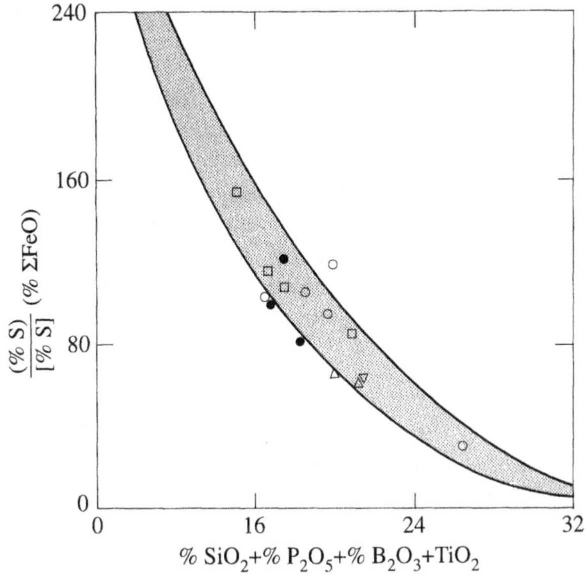

FIG. 2.12 The sulphur partition coefficient (%S)/[%S] between slag and steel as a function of acid components in slag at temperatures from 1550°C to 1650°C.[24]

One would expect that summation of mole percentages corresponding to the ions SiO_4^{4-}, TiO_4^{4-}, PO_4^{3-}, and BO_3^{3-}, respectively, weighted by the factors $2w$ and $3x$ from eqn (2.51) would give a better fit. However, a closer look shows that the improvement would be marginal especially since the content of SiO_2 dominates. Therefore preference has been given to the industrially more practical units of mass per cent.

2.1.16 The equilibrium between phosphorus in steel and in a basic slag

For many years it has been a problem to obtain a satisfactory description of the phosphorus equilibrium between steel and slag.[25] Here, the application of ionic theories to the problem will be outlined. Then a more direct approach will be given.

As previously mentioned, phosphorus is mainly present in a basic slag as PO_4^{3-}. To produce a phosphate ion, PO_4^{3-}, with its negative charge, $\underline{P}$ in the melt must unite with O^{2-} ions. If phosphorus, oxygen, and negative charge are balanced, the following equation for the phosphorus exchange is obtained

$$\underline{P} + \tfrac{5}{2}\underline{O} + \tfrac{3}{2}O^{2-} = PO_4^{3-}$$

and the equilibrium constant for the reactions becomes

$$K_p = \frac{a_{PO_4^{3-}}}{a_p a_O^{\frac{5}{2}} a_{O^{2-}}^{\frac{3}{2}}}$$

$$\frac{x_{PO_4^{3-}}}{[\%\,P]} = \frac{f_P K_P a_O^{\frac{5}{2}} a_{O^{2-}}^{\frac{3}{2}}}{\gamma_{PO_4^{3-}}}. \tag{2.52}$$

It is seen that the distribution of phosphorus between slag and melt (partition coefficient) is proportional to $a_{O2}^{\frac{5}{2}}$ (oxidizing conditions) and with $a_{O^{2-}}^{\frac{3}{2}}$ (basic slag, high lime content). Probably partly due to the complex form of the relation and the difficulties in determining $a_{O^{2-}}^{\frac{3}{2}}$ and $\gamma_{PO_4^{3-}}$, eqn (2.52) has not proved satisfactory in industrial applications. Instead a more direct empirical approach may be useful. Selin[26] has taken slags of the composition employed industrially, equilibrated them with steel melts, and measured the equilibrium phosphorus contents in the steel. The reference slag was

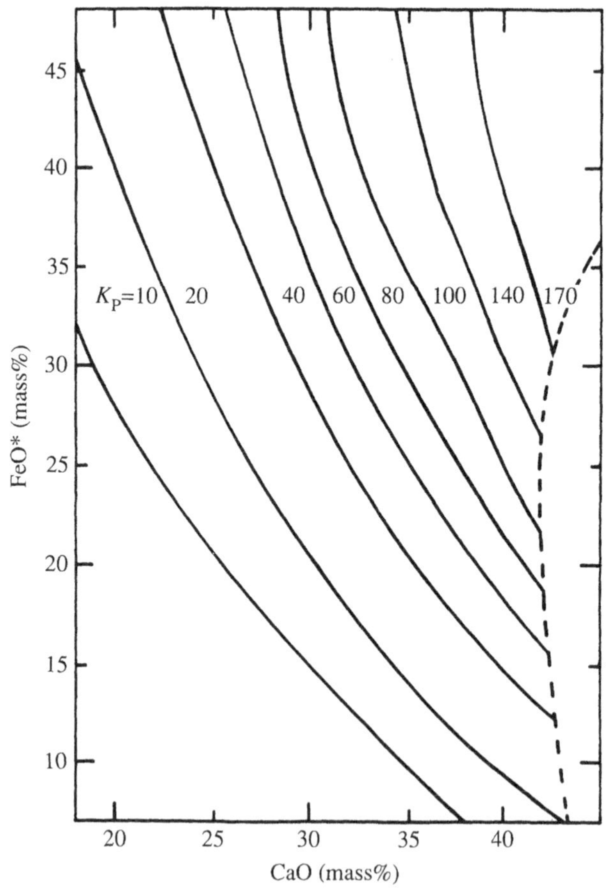

FIG. 2.13 Lines for constant partition coefficient (%P)/[%P] between slag and steel as a function of lime content and FeO* content in MgO-saturated slag (see text) at 1600°C.[26]

CaO–FeO–MgO + 0.4% P in slag in an MgO crucible. The properties of more complex slags containing additions of Al_2O_3, TiO_2, and VO_4 were determined relative to the corresponding properties of the reference system.

One may wonder about the number of variables involved in the description of the reference system. To get an idea of this, we invoke the Gibbs phase rule again:

$$F = C + 2 - P$$

where C is the minimum number of components required to create the system and P is the number of phases. The laboratory system consists of four phases: slag, metal, atmosphere, and MgO crucible so that $P = 4$. The minimum number of components is $C = 7$: Ar, Ca, Fe, Mg, Si, P, and O. This gives five degrees of freedom: $F = 7 + 2 - 4 = 5$. At fixed temperature and pressure (1 bar) and fixed phosphorus content in the slag there only remains two degrees of freedom. For these independent variables CaO and FeO* were chosen. Then on a diagram with CaO and FeO* on the ordinates one can describe properties of the system such as the phosphorus content of the melt. Figure 2.13 gives lines for constant phosphorus distribution:

$$K_p = \frac{(\%\,P)}{[\%\,P]}\,.$$

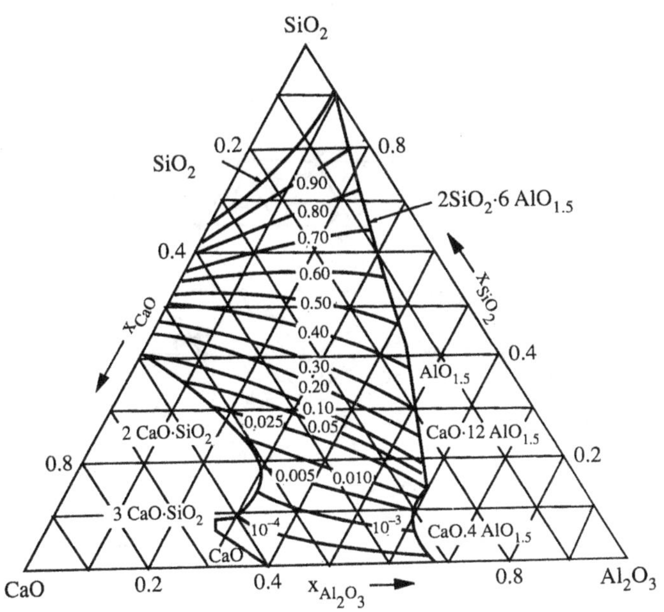

FIG. 2.14 SiO_2 activities in a CaO–Al_2O_3–SiO_2 slag at 1550°C.[27] Concentrations are given in mole fractions. Activity is referred to the pure solid oxide.

The results are in agreement with the predictions from eqn (2.52); the value of K_p increases with increasing FeO^* in the slag (oxidizing conditions) and with increasing CaO (basic slag).

2.1.17 Activities of slag components

It was seen in the previous section that it is difficult to determine the phosphorus capacity of steelmaking slags. To give a more balanced picture it is worth pointing out that there exists a wealth of experimental data concerning activities in slag system.[26-28] This is especially the case for 'blast furnace' type slags of SiO_2, Al_2O_3, and CaO. Figure 2.14 gives SiO_2 activities for this system while Fig. 2.15 provides the CaO and Al_2O_3 activities. In problem 2.8 an example is given of the use of these diagrams. In the example, Ca is removed from a FeSi melt by an exchange reaction with SiO_2 in CaO–Al_2O_3–SiO_2 slag.

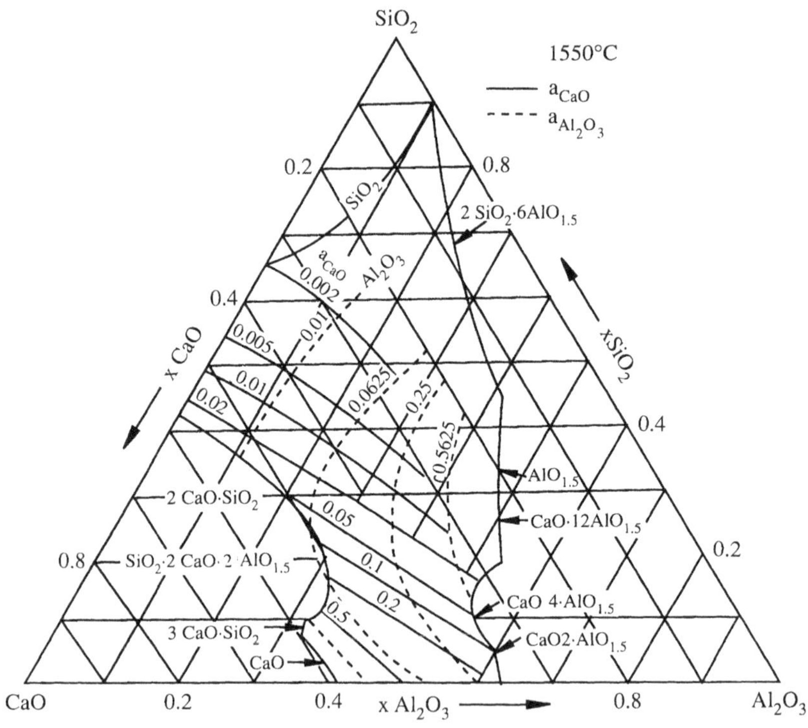

FIG. 2.15 CaO and Al_2O_3 activities in a CaO–Al_2O_3–SiO_2 slag at 1550°C.[27] Concentrations are given in mole fractions. Activity is referred to the pure solid oxide.

2.2 PHYSICAL AND TRANSPORT PROPERTIES OF MOLTEN METALS, GASES, SALTS, AND SLAGS

To describe the various processes that take place in refining reactors it is necessary to possess physical data concerning:

(1) viscosity

(2) diffusion coefficients

(3) surface tension (interfacial tension)

(4) thermal conductivity

(5) electrical conductivity.

The order given above reflects the importance of the various physical quantities. Viscosity is important in the study of flow and mixing, while the diffusion coefficients enter into the relations for the mass transfer coefficients. Surface tension is very important in the study of wetting and contacting problems. Interfacial tension and thermal conductivity must be known in order to describe nucleation and solidification phenomena. Other key subjects in connection with the use of such data are grain refining, recrystallization, grain growth, and fracture of metals and alloys. In the processing field one may mention gas purging, removal of inclusions to refractories, soldering and welding, and composites produced, for instance, by impregnation with a liquid metal binder of a porous high-melting framework.

The thermal conductivity of liquid metals is roughly proportional to the electrical conductivity.[68] Reliable data on electrical conductivity of various liquid metals is often available.

Data concerning the properties of gases is of interest in the study of various melt/gas operations such as gas purging and vacuum treatment. There are a number of analogies between the behaviour of molten metals and gases. Gases are discussed here also to throw light on some analogous problems concerning diffusion of the solute species in melts.

Salts and slags are only included when relations of a more general nature are available. Some references are given to assist the reader to find specific data.

2.2.1 Viscosity of gases

Consider a fluid (liquid or gas) flowing past a stationary wall. At a certain distance from the wall, the fluid has a velocity v_x, where x represents the direction of flow. Closer to the wall, the velocity will be less, that is there is a gradient in velocity dv_x/dz, where z represents the direction perpendicular to the flow. This gradient results in a shear force within the liquid

$$\tau_{zx} = -\mu \frac{dv_x}{dz}.$$

This is Newton's law for viscous flow in its simplest form, and the proportionality constant is called the viscosity of the fluid. The equation may alternatively be interpreted in terms of momentum transfer. A fluid moving in the x-direction imparts some of its momentum to an adjacent 'layer' of liquid causing it also to move in the x-direction. Hence, x-momentum is transmitted through the fluid in the z-direction. Consequently, τ_{zx} may be interpreted as the viscous flux of x-momentum in the z-direction. This interpretation ties in better with the molecular (atomic) nature of the momentum transport process. Viscous momentum flux is in the direction of the negative velocity gradient. A velocity gradient can thus be thought of as a 'driving force' for momentum transport.

In a gas at low density, momentum is transferred by collisions between the molecules or atoms in the gas. Since each molecule has an average kinetic energy of $\frac{1}{2}mv^2 = 3kT/2$ the velocity is proportional to $(kT/m)^{\frac{1}{2}}$. Now the number of collisions is proportional to the number of molecules per unit volume, N, and to the velocity of the molecules. The distance travelled between two collisions, λ, is inversely proportional to N and inversely proportional to the cross-sectional area for collision of two molecules πd_c^2 (see Fig. 2.16). It is assumed that all molecules have velocities representative of the region in which they last collided. The velocity difference between two consecutive collisions will then be proportional to $\lambda(\mathrm{d}v_x/\mathrm{d}z)$ and the momentum flux τ_{zx} will be proportional to:

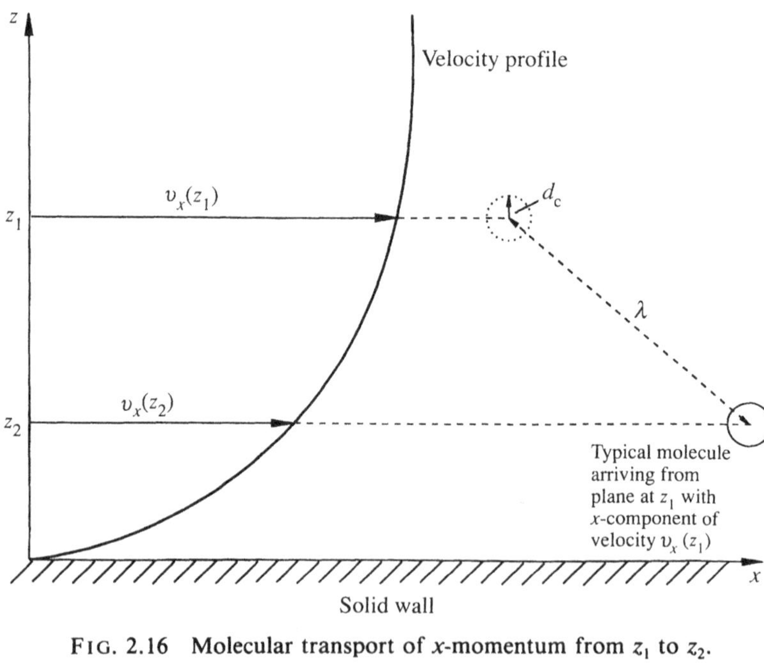

FIG. 2.16 Molecular transport of x-momentum from z_1 to z_2.

$$-Nm\left(\frac{kT}{m}\right)^{\frac{1}{2}}\lambda\frac{dv_x}{dz} \propto -\frac{(kTm)^{\frac{1}{2}}}{d_c^2}\frac{dv_x}{dz}.$$

According to kinetic gas theory[29] the proportionality constant is $5/(16\pi^{\frac{1}{2}})$ so that

$$\tau_{zx} = -\frac{5(mkT)^{\frac{1}{2}}}{16\pi^{\frac{1}{2}}d_c^2}\frac{dv_x}{dz}. \tag{2.53}$$

This gives a viscosity of

$$\mu = \frac{5(mkT)^{\frac{1}{2}}}{16\pi^{\frac{1}{2}}d_c^2} \tag{2.54}$$

or, since Boltzmann's constant $k = R/N_A$ and the atomic mass $m = M/N_A$, where N_A is Avogadro's number $N_A = 6.022 \times 10^{23}$ atoms/mol and M is the mass of one mole,

$$\mu = \frac{5(MRT)^{\frac{1}{2}}}{16\pi^{\frac{1}{2}}N_A d_c^2} \tag{2.55}$$

which represents the viscosity of a gas composed of hard spheres at low density. Note that measurements are needed to determine the collision diameter d_c. Equation (2.55) indicates that μ is proportional to $T^{\frac{1}{2}}$. However, data for various gases show that μ varies slightly more rapidly than $T^{\frac{1}{2}}$. To predict the temperature dependence of μ accurately, one has to replace the rigid sphere model by a more realistic molecular force field. Such a rigorous kinetic theory of gases was developed by Chapman and Enskog.[29] This theory gives expressions for the transport coefficients in terms of the interaction between a pair of molecules in the gas. This potential energy ϕ is related to the force of interaction F by the relation $F = -d\phi/dr$ in which r is the distance between the molecules. If one knew exactly how the forces depend on the distance between molecules, then one could substitute these relations into the Chapman–Enskog formulas and calculate the transport coefficients.

The Lennard–Jones potential

$$\phi(r) = 4\varepsilon\left\{\left(\frac{\delta}{r}\right)^{12} - \left(\frac{\delta}{r}\right)^{6}\right\} \tag{2.56}$$

has proved to be very useful in giving such a description of mono- and polyatomic gases. Note that it displays the characteristic features of molecular interactions: strong repulsion at small separations (roughly proportional to r^{-12}) and weak attraction at large separations (proportional to r^{-6}). δ is a characteristic diameter of the molecule called the collision diameter. Values of δ and ε are known for many substances; a partial list is given in Table 2.11. When values of δ and ε are not known for monoatomic gases, they may be estimated from the empirical relations:

TABLE 2.11 *Intermolecular force parameters.*[61]

Elements	δ (10^{-10}m)	$\frac{\varepsilon N_A}{R}$ K
H_2	2.915	38.0
Ar	3.418	124
N_2	3.681	91.5
Air	3.617	97.0
O_2	3.433	113
CO	3.590	110
CO_2	3.996	190
SO_2	4.290	252
Cl_2	4.115	357

$$\frac{\varepsilon N_A}{R} \approx 1.92\,T_m \qquad \text{and} \qquad \delta \approx 1.2 \times 10^{-8}\,V_m^{\frac{1}{3}} \tag{2.57}$$

where V_m is the molar volume of the liquid at the melting point.

The viscosity becomes

$$\mu = \frac{0.844 \times 10^{-24}}{\Omega_\mu \delta^2}(MT)^{\frac{1}{2}} \tag{2.58}$$

where Ω_μ is a slowly varying function of temperature $RT/(\varepsilon N_A)$ given in Table 2.12. If the gas were made up of rigid spheres of constant diameter δ, then Ω_μ would be unity. Ω_μ may therefore be interpreted as giving the deviation from rigid sphere behaviour. Notice that δ replaces d_c used in eqn (2.55). If $\Omega_\mu = 1$, then $d_c = \delta$ and eqn (2.58) becomes identical to eqn (2.55).

Equation (2.57) for δ inserted in eqn (2.58) gives

$$\mu = \frac{5.9 \times 10^{-9}(MT)^{\frac{1}{2}}}{V_m^{\frac{2}{3}}\Omega_\mu}. \tag{2.59}$$

To illustrate the use of eqn (2.58), the viscosity of CO_2 is calculated at 300 and 1000 K. From Table 2.11 $\varepsilon N_A/R = 190$, $\delta = 3.996 \times 10^{-10}$ m. The molecular weight of CO_2 is 44.01 g/mol = 44.01×10^{-3} kg/mol. At 300 K and 1000 K, $RT/\varepsilon N_A = T/190 = 1.58$ and 5.26, respectively. This gives from Table 2.12, $\Omega_\mu = 1.286$ and $\Omega_\mu = 0.9196$, respectively. Then at 300 K

$$\mu = \frac{0.844 \times 10^{-24}(44.01 \times 10^{-3} \times 300)^{\frac{1}{2}}}{3.996^2 \times 10^{-20} \times 1.286} = 1.494 \times 10^{-5}\,\text{Ns/m}^2$$

or

$$\mu = 1.494 \times 10^{-2}\ \text{centipoise.}$$

TABLE 2.12 *Values of Ω_μ for viscosity based on the Lennard-Jones potential*[61]

$\frac{RT}{\varepsilon N_A}$	Ω_μ
0.3	2.785
0.4	2.492
0.5	2.257
0.6	2.065
0.7	1.908
0.8	1.780
0.9	1.675
1.0	1.587
1.2	1.452
1.4	1.353
1.6	1.279
1.8	1.221
2.0	1.175
4.0	0.970
6.0	0.8963
8.0	0.8538
10	0.8242
20	0.7432
40	0.6718
60	0.6335
80	0.6076
100	0.5882
200	0.5320
400	0.4811

This may be compared with the experimental value of 1.495×10^{-5} Ns/m^2. At 1000 K the viscosity becomes

$$\mu = \frac{0.844 \times 10^{-24} \times 44.01^{\frac{1}{2}}}{3.996^2 \times 10^{-20} \times 0.9196} = 3.814 \times 10^{-5}\text{Ns/m}^2.$$

The change of Ω_μ from 1.2866 to 0.9196 takes into account the fact that the viscosity increases slightly more strongly than $T^{\frac{1}{2}}$.

The following correlation has been found to represent, with a reasonable degree of accuracy, the viscosity of multicomponent gas systems, with the exception of hydrogen-rich mixtures:[30]

$$\mu = \frac{\sum_{i=1}^{k} x_i M_i^{\frac{1}{2}} \mu_i}{\sum_{i=1}^{k} x_i M_i^{\frac{1}{2}}} \tag{2.60}$$

where x_i and M_i are the mole fraction and the molecular weight of fraction i, respectively.

2.2.2 The pair distribution function

In order to describe the transport behaviour of liquids (liquid metals) it is necessary to describe the structure of liquids. However, a liquid lacks a regular arrangement of its molecules, that is, it lacks the regular crystalline order of a solid. The molecular structure of a liquid, therefore, can be described and measured only on the basis of probability. The pair distribution function $g(r)$ is of central importance in the modern theory of liquids. Consider a liquid in equilibrium and then any atom with its centre at $r = 0$. The number of atoms, $\mathrm{d}N$, in the spherical shell between radial distances r and $r + \mathrm{d}r$ from the origin atom is

$$\mathrm{d}N = 4\pi r^2 \mathrm{d}r \frac{N}{V} g(r) \tag{2.61}$$

where N is the total number of atoms in the volume V. The pair distribution function $g(r)$ is proportional to the probability that an atom is found at distance r. If there is no correlation between atoms, as in the case of an ideal gas that consists of point atoms, $g(r) = 1$. Then the probability of finding another atom is the same for all distances r.

For real liquids the pair distribution function becomes complicated due to the forces of attraction and repulsion. When r is less than the collision diameter, $g(r)$ is essentially zero, since two molecules cannot overlap. At slightly larger separations there is a more than average chance of finding a molecule; then $g(r)$ is greater than one. This enhancement arises mainly from the geometric necessity of there being many pairs at such separations since the liquid density is only slightly less than that of a solid. Minor maxima and minima at larger separations are the remnants in the liquid of the much stronger extrema found in the solid (for which the pair distribution function $g(r)$ along a crystal axis consists of a series of narrow peaks separated by regions in which it is zero). Thus a liquid lacks the long-range order of a solid but shows appreciable order at short separations. Figure 2.17 gives the pair distribution function of iron near the melting point[31] as an example. All pair distribution functions for liquid metals have the same general shape.

A knowledge of the pair distribution function is the minimum information needed for the discussion of the thermodynamic properties of a liquid since molecules interact predominantly in pairs through the potential energy $\phi(r)$ and since $g(r)$ describes the statistics of such pairs. Therefore knowledge of $\phi(r)$ and $g(r)$ is necessary in order to link the microscopic energies and forces and the macroscopic properties. There are two methods of measuring the pair distribution function: by X-ray or neutron diffraction[32] and by computer simulation of the molecular structure and motions in a liquid. In the first, the

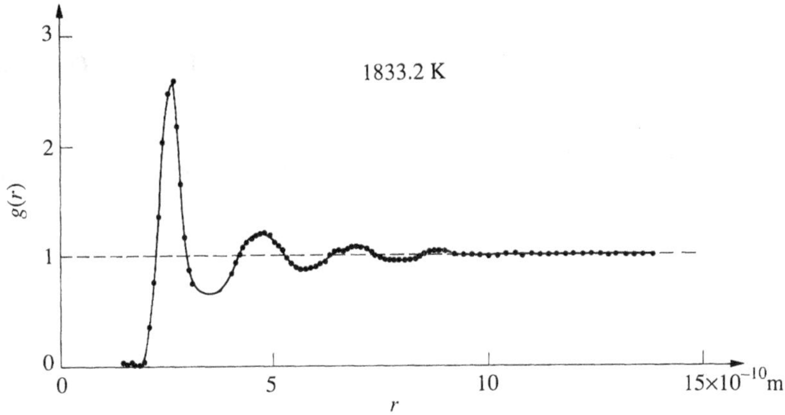

FIG. 2.17 The pair distribution function, $g(r)$, of iron near the melting-point.[31]

liquid is exposed to a specific monochromatic radiation. The intensity of the scattered radiation is measured for various angles. Then a Fourier transform treatment gives $g(r)$.[33] The second method of obtaining the pair distribution function $g(r)$ supposes that $\phi(r)$ is known. A computer model of the liquid is set up in which between 100 and 1000 molecules are contained within a cube. There are now two methods of proceeding: by Monte Carlo calculation[34] or by molecular dynamics. In the latter each molecule is initially assigned a random velocity and position, and Newton's equations of motion are solved to calculate the path of each molecule in the changing field of all the others. After a time the mean positions are those appropriate to the density and to the mean kinetic energy of the liquid under equilibrium conditions. Functions such as the pair distribution function can now be evaluated by taking suitable averages as the system evolves in time.

As already mentioned the potential energy $\phi(r)$ and the pair distribution function, $g(r)$, can be employed to derive various thermodynamic and macroscopic physical properties. The use of $g(r)$ and $\phi(r)$ to describe properties of a liquid is illustrated by deriving relations for the internal energy U (in J/mol) and for the enthalpy of vaporization. For a monoatomic liquid the internal energy per mole is given by

$$U = \tfrac{3}{2}RT + \langle\phi\rangle \tag{2.62}$$

where $\frac{3}{2}RT$ is the kinetic energy and $\langle\phi\rangle$ is the mean potential energy per mole obtained by integrating from $r = 0$ to $r = \infty$:

$$\langle\phi\rangle = \frac{N_A}{2}\int_0^\infty \phi(r)\mathrm{d}N = \frac{N_A}{2}\int_0^\infty \phi(r)g(r)4\pi r^2\mathrm{d}r. \tag{2.63}$$

N_A is the number of molecules per mole (Avogadro's number). The factor $\frac{1}{2}$

prevents us from counting each molecule twice in the summation of the potential energies.

The molar enthalpy of vaporization ΔH_v is obtained from eqn (2.62) and

$$\Delta H_V = U_g - U + RT \tag{2.64}$$

where RT is the work performed against the external pressure during vaporization. Since $U_g = \frac{3}{2} RT$ (for an ideal monoatomic gas)

$$\Delta H_V = RT - \langle\phi\rangle = RT - \frac{N_A^2}{V_m} 2\pi \int_0^\infty \phi(r) g(r) r^2 \mathrm{d}r \tag{2.65}$$

where finally eqns (2.61) and (2.63) have been inserted in eqn (2.62). V_m is the molar volume.

2.2.3 The viscosity of liquids

Born and Green[35] using kinetic theory have derived an equation for the viscosity in terms of the pair distribution function $g(r)$ and the pair interatomic potential ϕ(r):

$$\mu = \frac{2\pi}{15} \left(\frac{M}{RT}\right)^{\frac{1}{2}} \left(\frac{N_A}{V}\right)^2 \int_0^\infty g(r) \frac{\mathrm{d}\phi}{\mathrm{d}r} r^4 \mathrm{d}r \tag{2.66}$$

where V is the volume per mole. It is not a simple matter to derive eqn (2.66). However, it is seen that, reasonably enough, the interatomic force $\mathrm{d}\phi/\mathrm{d}r$ is multiplied by a probability density $g(r)\ 4\pi r^2\ \mathrm{d}r/V$ and by an area per volume of an atom $r^2\ N_A/V$.

There has been some discussion as to the validity of this equation. It gives results that coincide roughly with experimental data for metals, but it is difficult to determine whether the discrepancies are due to errors in the experimental measurements or errors in $g(r)$ or $\phi(r)$ employed in the calculations. The theory assumes that the atoms interact only in pairs so that the use of $g(r)$ and $\phi(r)$ suffice in order to describe the system.

Here it is also assumed that $\phi(r) = \varepsilon\phi^*(r/\delta, T^*)$ and $g(r) = g^*(r/\delta, T^*)$ where $\phi^*(X, T^*)$ and $g^*(X, T^*)$ are regarded as universal functions of $X = r/\delta$ and $T^* = T/5.2\ T_m$. It is also found that[36]

$$\varepsilon = \frac{5.2RT_m}{N_A}.$$

The factor 5.2 replaces 1.92 for gases in eqn (2.57). Then

$$\mu^* V^{*2} = f\left(\frac{1}{T^*}\right)$$

where

$$f\left(\frac{1}{T^*}\right) = \frac{2\pi}{15}\frac{1}{T^*}\int_0^\infty g^*(X, T^*)\frac{\partial\phi^*(X, T^*)}{\partial X}X^4 \mathrm{d}X$$

is an empirical function of temperature, T^*, and

$$\mu^* = \frac{\mu\delta^2 N_A}{(MRT)^{\frac{1}{2}}}$$

$$V^* = \frac{V}{N_A\delta^3}.$$

$f(1/T^*)$ is given in Fig. 2.18.[36] Furthermore it is found empirically that $N_A\delta^3 = 1.25\ V_m$ or $\delta = 1.28 \times 10^{-8}\ V_m^{\frac{1}{3}}$. Then

$$V^* = \frac{V}{V_m}0.8.$$

At the melting-point the theory gives that

$$\mu_m = \frac{1.8 \times 10^{-7}(MT_m)^{\frac{1}{2}}}{V_m^{\frac{2}{3}}}. \tag{2.67}$$

The similarity with eqn (2.59) for gases should be noted.

At higher temperatures

$$\mu\left(\frac{V}{V_m}\right)^2 = \mu_m\left(\frac{T}{T_m}\right)^{\frac{1}{2}}\frac{f(1/T^*)}{f(5.2)}. \tag{2.68}$$

Equation (2.67) is compared with experimental results in Fig. 2.19. It is seen that the correspondence is acceptable except for Al, Bi, Ge, Ti, Zn, Sb, and Zr.

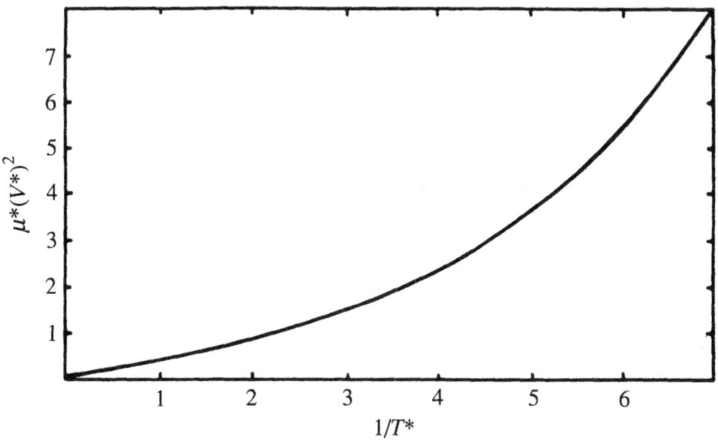

FIG. 2.18 The function $f(1/T^*)$ where $1/T^* = 5.2T_m/T$. T_m is the melting-point temperature.[29]

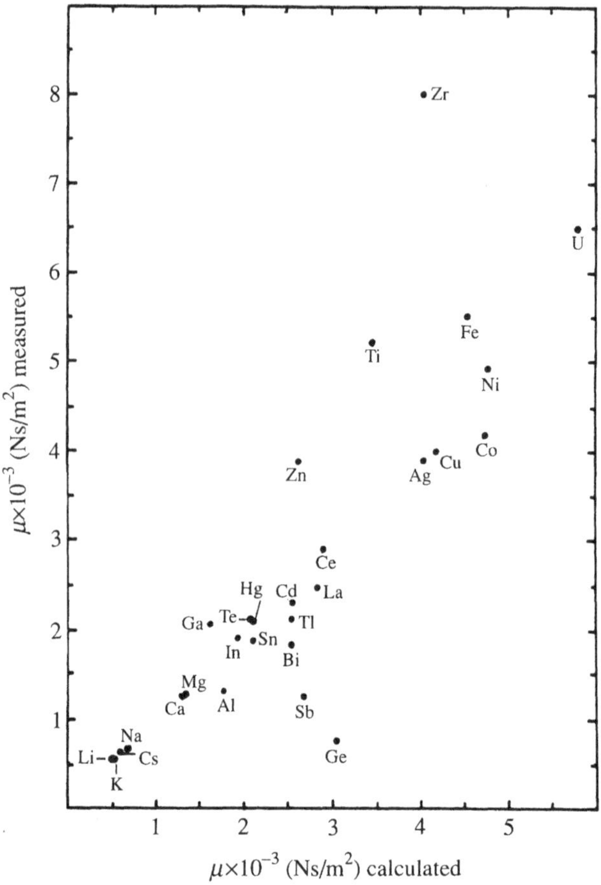

FIG. 2.19 Comparison of viscosity calculated from eqn (2.67) with experimental values[63] at the melting-points.

Of these Ge and Sb are metalloids.

To illustrate the use of eqn (2.67), the viscosity of iron at the melting-point is calculated. $M = 55.85 \times 10^{-3}$ kg/mol, $T_m = 1808$ K, and the density of iron at the melting point is 7030 kg/m^3. Then

$$V_m = \frac{55.85 \times 10^{-3}}{7030} = 7.94 \times 10^{-6}\ \mathrm{m^3/mol}.$$

This gives

$$\mu_m = \frac{1.8 \times 10^{-7}(55.85 \times 10^{-3} \times 1808)^{\frac{1}{2}}}{(10^{-6} \times 7.94)^{\frac{2}{3}}} = 4.54 \times 10^{-3}\ \mathrm{Ns/m_2}.$$

Similarly for Ga

$$\mu_m = \frac{1.8 \times 10^{-7}(69.7 \times 10^{-3} \times 303)^{\frac{1}{2}}}{(10^{-6} \times 11.4)^{\frac{2}{3}}} = 1.63 \times 10^{-3}\,\mathrm{Ns/m^2}.$$

These values probably lie within the errors in the experimental values.[31]

For alloys it is difficult to describe the dependence of viscosity on the content of solute elements and on the temperature. One approach is to use Hildebrand's 'free volume' theory.[37] For dilute systems where the solvent dominates one may as a rough approximation take viscosity values valid for the pure solvent.

It is not possible in this book to dwell on the subject of viscosity of salt and slag systems. For salts the viscosity is surprisingly low;[31,38] not much higher than in water or molten metals. The viscosity of slags[27,39,40] increases strongly with the content of network formers such as SiO_2 and is in the range 1–100 Pa s. As an illustration isoviscosity lines for the system CaO–Al_2O_3–SiO_2 at 1500°C are given in Fig. 2.20.

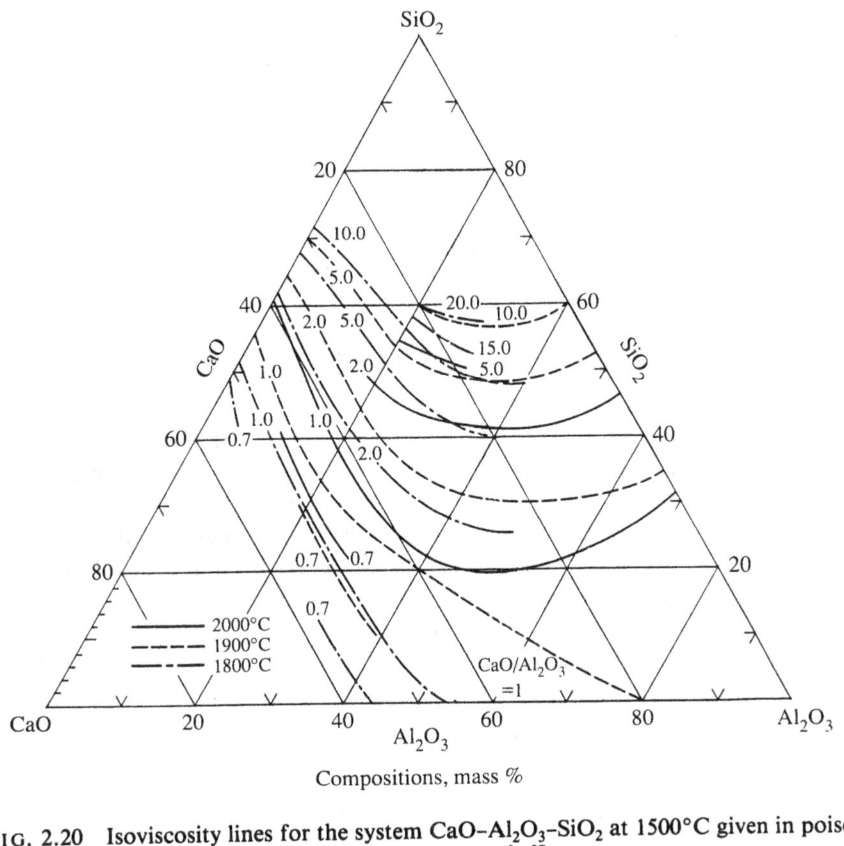

FIG. 2.20 Isoviscosity lines for the system CaO–Al_2O_3–SiO_2 at 1500°C given in poise (1 poise = 0.1 $\mathrm{Ns/m^2}$).[27]

2.2.4 Surface tension of pure molten metals

The surface tension, σ, may be defined in terms of the work $\sigma \mathrm{d}A$ required to create a surface area $\mathrm{d}A$. Alternatively σ may be thought of as the force per unit length needed to stretch the surface.

From statistical mechanics Fowler[41] has derived the following equation for the surface tension;

$$\sigma = \frac{\pi}{8}\left(\frac{N_\mathrm{A}}{V}\right)^2 \int_0^\infty g(r)\,\frac{\partial \phi}{\partial r}\,r^4\mathrm{d}r. \tag{2.69}$$

In the development of the equation it was assumed that density changes discontinously at a mathematical surface that separates the liquid and gas phase. This approximation has also been successfully employed in the study of the surface tension of alloys. This point will be discussed later.

The similarity of eqn (2.69) to the equation for viscosity is striking; the integrals in the two expressions are identical. This may be explained in part by pointing out that in viscous flow work is required to overcome the interatomic forces and move the atoms relative to each other. Similarly, it will be seen that work is needed to displace atoms from the bulk melt to the surface. Surface tension, however, is a purely thermodynamic quantity and should not depend on atomic mass m or on M. In fact eqn (2.69) does not contain M. Comparison of eqns (2.66) and (2.69) gives:

$$\frac{\sigma}{\mu} = \frac{15}{16}\left(\frac{RT}{M}\right)^{\frac{1}{2}}. \tag{2.70}$$

At the melting-point it is found from eqns (2.66), (2.67), and (2.69) that

$$\sigma_\mathrm{m} = 5.9 \times 10^{-8}\,\frac{RT_\mathrm{m}}{V_\mathrm{m}^{\frac{2}{3}}}. \tag{2.71}$$

Comparison with experiments indicates that agreement is better with a coefficient of 4.8 instead of 5.9. The discrepancy may be due partly to the discontinuity assumption in the derivation of eqn (2.69). However, the effect of impurities on the surface tension of metals should also be emphasized. Surface active elements all reduce the surface tension as will be seen in a later section.

From eqns (2.68) and (2.70) it is found that

$$\frac{\sigma}{\sigma_\mathrm{m}} = \frac{T}{T_\mathrm{m}}\left(\frac{V_\mathrm{m}}{V}\right)^2 \frac{f(1/T^*)}{f(5.2)}. \tag{2.72}$$

To illustrate the use of these equations the surface tension of Cu is calculated at the melting-point and at 1600 K. From eqn (2.71)

$$\sigma_\mathrm{m} = \frac{4.8 \times 10^{-8} \times 8.3145 \times (1083 + 273)}{(7.94 \times 10^{-6})^{\frac{2}{3}}} = 1.36\,\mathrm{N/m}.$$

From eqn (2.72) and Fig. 2.18

$$\frac{\sigma}{\sigma_m} = \left(\frac{1600}{1356}\right)\left(\frac{7.94 \times 10^{-6}}{8.1334 \times 10^{-6}}\right)^2 \frac{f\left(\dfrac{5.2 \times 1356}{1600}\right)}{f(5.2)}$$

$$= 1.18 \times 0.953 \times 0.77 = 0.87$$

or

$$\sigma = 1.18\,\mathrm{N/m}.$$

This may be compared with the published results of 1.35 N/m and 1.28 N/m,[42] at 1083°C and 1327°C, respectively, and 1.30 N/m and 1.24 N/m[31] at 1083°C and 1327°C.

For various salts it is found[31] that σ_m is proportional to $T_m/V_m^{\frac{2}{3}}$. Thus for the molten alkali halides

$$\sigma_m = \frac{1.3 \times 10^{-7} T_m}{V_m^{\frac{2}{3}}}.$$

Similarly for other molten halides

$$\sigma_m = \frac{2 \times 10^{-7} T_m}{V_m^{\frac{2}{3}}}$$

while for molten salts of sulphates, carbonates, nitrates, and so on, the proportionality factor in front of $T_m/V_m^{\frac{2}{3}}$ is 2.7×10^{-7}.

For non-ideal binary systems the surface tension may vary in a complex manner.[43] The surface tension of dilute multicomponent systems is discussed in the next section. Also the problem of the interfacial tension between solid and melt of a pure metal is treated. Furthermore it is indicated how one can estimate the surface tension of an ionic solid.

2.2.5 Thermodynamics of interfaces

The plane of separation of two phases is known as a surface or interface. From a mechanical point of view it may be regarded as having infinitesimal thickness. With such a surface we associate an interfacial tension, σ. σ is equal to the Gibbs energy required to stretch a unit area of the interface at constant pressure p and constant temperature with no exchange of atoms or molecules with the surroundings:

$$\sigma = \left(\frac{\partial G}{\partial A}\right)_{T,p,n_i}. \tag{2.73}$$

Often it is more practical in theoretical studies to use the Helmholtz energy F. Then

$$\sigma = \left(\frac{\partial F}{\partial A}\right)_{T,V,n_i}. \tag{2.74}$$

Strictly speaking a system should be regarded as composed of two bulk phases and a region of space between them whose properties are different from those of the bulk phases. At the extremeties the properties in the interphase layer should merge with the properties in the bulk phases. The interphase is composed of a multilayer of atoms. However, such a treatment is beyond the scope of this book. Here it is only assumed that the interphasial layer is only one atom or molecule thick or that there is a discontinuity of properties at the interface. This is the approximation made by Fowler in the derivation of eqn (2.69). The Gibbs energy, G^σ for the interphasial layer is given by

$$G^\sigma = G - G_1 - G_2 \tag{2.75}$$

where G is the Gibbs energy for the total system and G_1 and G_2 refer to the two bulk phases. If one of the phases (index 2) is a vacuum or an inert gas we speak of surface tension instead of interfacial tension. Then

$$G^\sigma = G - G_1. \tag{2.76}$$

Here G takes into account conditions on the surface while G_1 assumes constant properties of the bulk all the way to the surface.

One mole of atoms in a monolayer approximately covers a surface

$$A \approx N_A\left(\frac{V_m}{N_A}\right)^{\frac{2}{3}} = N_A^{\frac{1}{3}} V_m^{\frac{2}{3}} \tag{2.77}$$

where V_m is the volume of one mole of the substrate and $(V_m/N_A)^{\frac{2}{3}}$ is roughly the area covered by an atom. For a closest packed configuration[11]

$$A = 1.09 N_A^{\frac{1}{3}} V_m^{\frac{2}{3}}. \tag{2.78}$$

The surface tension for a one-component system is

$$\sigma = \frac{G^\sigma}{A}.$$

Now

$$G^\sigma = H^\sigma - TS^\sigma$$

and

$$\sigma = \frac{H^\sigma}{\mathrm{A}} + T\frac{\mathrm{d}\sigma}{\mathrm{d}T}$$

since

$$S^\sigma = -\left(\frac{\partial G^\sigma}{\partial T}\right)_{p,n_i}.$$

If the temperature dependence of σ is disregarded

$$\sigma \approx H^\sigma/A. \tag{2.79}$$

H^σ is the difference between the molar enthalpy on the surface and in the bulk. To estimate H^σ for a metal, one can compare the number and type of bonds in the bulk phase and on the interface. The surface tension is then obtained by counting the number of broken bonds, m, when an atom or molecule is brought to the surface. Thus, if Z is the number of neighbours to the atom in the bulk,

$$\sigma \approx \frac{\Delta H_V m}{1.09 N_A^{\frac{1}{3}} V_m^{\frac{2}{3}} Z}$$

where ΔH_V is the molar enthalpy of evaporation.

This procedure bypasses the theoretical calculation of the exact cohesion

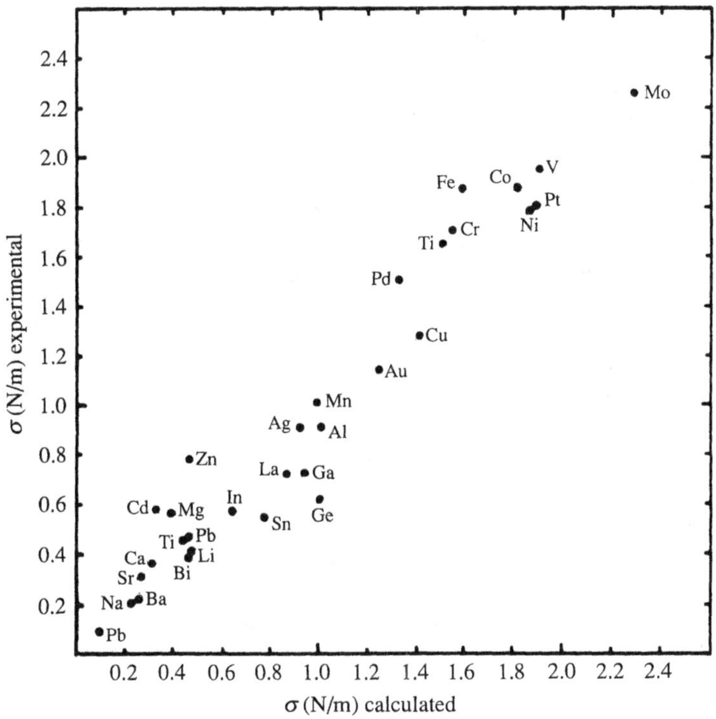

FIG. 2.21 Comparison of surface tension calculated from eqn. (2.80) with experimental values[63] at the melting-points.

forces and ignores the matter of surface distortion. The surface atoms, being subject only to a one-sided attraction, lie closer to their underlying neighbours than they do in the bulk. Consequently, surface atoms should be bound to the atoms immediately beneath the surface by greater binding energies. The energies of the bonds at the surface are approximately 13 per cent greater.[11]

$$\sigma = \frac{\Delta H_V \, 1.09 \times 10^{-8}}{V_m^{\frac{2}{3}}} \left(\frac{Z - (Z - m)1.13}{Z} \right).$$

For a face-centered cubic structure (fcc structure) $m/Z = 0.25$.[11] Then

$$\sigma = \frac{\Delta H_V \, 1.7 \times 10^{-9}}{V_m^{\frac{2}{3}}}. \tag{2.80}$$

Figure 2.21 compares eqn (2.80) using ΔH_V values from[3] with surface tensions of the various metals at the melting-point.[63] For metals crystallizing in hexagonal or rhombohedral forms such as Cd, Mg, Zn, $m/Z = 0.36$. Then the factor 1.7 in eqn (2.80) should have been replaced by 3.0.

2.2.6 Surface energy of compounds

In order to study wetting it is important to know the surface energies. The surface energy of compounds may be obtained by a procedure similar to that applied to metals in the previous section. The surface energy H^σ is the difference between the energy in the surface layer of atoms or molecules and in the bulk. The bonding energy between atoms in compounds, based on the nature of the interactions of the electrons in the outer shells or their electronic structure, can be divided into the general categories of (a) ionic, (b) covalent, and (c) metallic. In addition there are weak forces of attraction between atoms which are referred to as van der Waals forces. Actual bonds are mixtures of several types. Usually one of the types is dominant.

Ionic bonds are easier to understand than other types of bonding. Typically they form between strongly electropositive elements (see Chapter 3) and electronegative ones (non-metals, F, O, S). In sodium one electron is held in the outer shell of the atom and is held by an effective potential of a single charge. It loses this outer electron comparatively easily to become positively charged Na^+. To remove an electron from each atom requires 5×10^5 J/mol.[44] For fluorine the situation is quite different. Each of the seven outer electrons is attracted to a kernel with a charge +7 so that the ionization energy of F is very high, 17×10^5 J/mol. On the other hand there is room for an eighth electron in the outer shell of F. Adding an electron yields 3.5×10^5 J/mol. To convert a mole of Na atoms and a mole of F atoms to Na^+ and F^- requires an energy $(5 - 3.5) \times 10^5$ J/mol. The resulting ions are capable of forming ionic bonds whose energy yield comes from the electrostatic forces of attraction. From this energy of attraction must be subtracted 1.5×10^5 J/mol necessary to form the ions.

The energy of an ionic bond is calculated from simple electrostatic forces. Consider an ion with charge $Z_1 e$ attracting an ion with a charge $-Z_2 e$. The force of attraction is $Z_1 Z_2 e^2/r^2$ where r is the distance between the ions. The attractive energy as the ions approach each other from infinity is

$$U = \frac{e^2 Z_1 Z_2}{4\pi\varepsilon_0} \int_\infty^{r_0} \frac{dr}{r^2} = -\frac{e^2 Z_1 Z_2}{4\pi\varepsilon_0 r_0}. \tag{2.81}$$

ε_0 is the dielectric constant 8.854×10^{-12} C^2/Nm^2. As the ions approach each other their electron clouds overlap, building up a very strong repulsive potential which may be taken to be proportional to exp $(-r/\rho)$ and called the central field repulsive potential.[45] ρ is called the screening length. At the equilibrium distance, r_0, the forces of repulsion balance the forces of attraction. At this point the energy is at a minimum as seen in Fig. 2.22 which gives a schematic plot of energy, U, versus interionic distance r.

In calculating the energy of an ionic crystal it is necessary to sum up all the attractive and repulsive forces between all the ions. Various elegant methods[46]

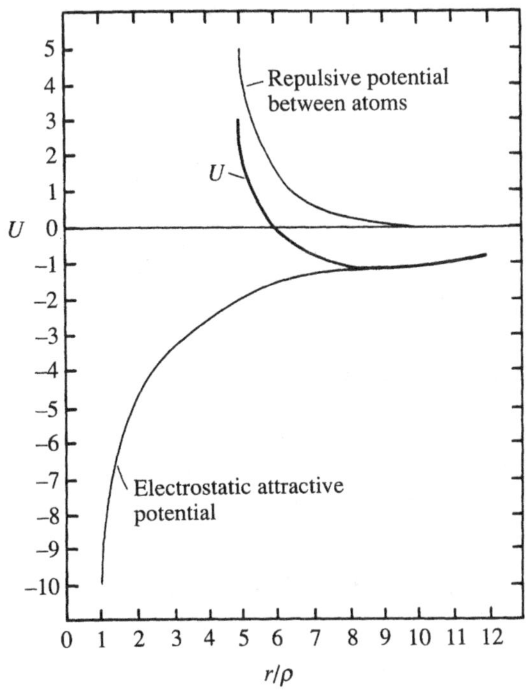

FIG. 2.22 The energy, U, between atoms is determined by the repulsive potential and electrostatic attractive potential.

exist. The original approach according to Madelung[47] considered, for an NaCl structure, an infinite number of charges with periodic repetition after a length $2r_0$ in all three normal directions. The potential at a charge in a neutral row of alternating charges e and $-e$ is

$$U = -\frac{2e^2}{4\pi\varepsilon_0 r_0}\left(\frac{1}{1} - \frac{1}{2} + \frac{1}{3} - \frac{1}{4} + \cdots\right) = -\frac{2e^2}{4\pi\varepsilon_0 r_0}\ln 2$$

since for $x \leqslant 1$

$$\ln(1 + x) = x - \frac{x^2}{2} + \frac{x^3}{3} - \frac{x^4}{4} + \cdots$$

If all the rows on the two sides and above and below are taken into account

$$U = -\frac{e^2}{4\pi\varepsilon_0 r_0} \times 3.495.$$

The contribution to 3.495 from the rows above or below is 0.132. The main contribution to 0.132 is from the neighbouring plane. The number 0.132 will be used in the following to obtain H^o.

If all the N_A positive and negative ions are taken into account the electrostatic energy per mole becomes

$$H = \frac{N_A U}{2}. \tag{2.82}$$

The electrostatic factor $\frac{1}{2}$ arises from the fact that in the double summations each pair of ions is counted twice. Then

$$H_{\text{electrostatic}} = -N_A\frac{e^2 1.747\,56}{r_0 4\pi\varepsilon_0}$$

where the factor $A_M = 1.747\,56$ is called the Madelung constant for NaCl. Values of the Madelung constant for various ionic structures are given in the literature.[46]

From the electrostatic energy must be substracted the coulomb energy of repulsion from the overlapping of electron clouds around the ions. This energy which usually amounts to only about 10 per cent of the electrostatic energy is here obtained from experimental data. The crystal energy is

$$H = -\frac{A_M N_A e^2}{4\pi\varepsilon_0 r_0} + B\exp(-r_0/\rho). \tag{2.83}$$

The values of B (the repulsive coefficient) and ρ which takes screening into account must be determined. ρ can be obtained from compressibility data. The values obtained for various crystals cluster around $0.33\,\text{Å} = 0.33 \times 10^{-10}\,\text{m}$. For NaCl $\rho = 0.321\,\text{Å}$.[45] B is proportional to the number of nearest neighbours, Z. One relation is fixed from the condition that r_0 is the equilibrium distance so that

$$\frac{\mathrm{d}H}{\mathrm{d}r} = 0 \qquad \text{at } r = r_0. \tag{2.84}$$

This gives

$$\frac{A_\mathrm{M} N_\mathrm{A} e^2}{4\pi\varepsilon_0 r_0^2} - \frac{B}{\rho} \exp(-r_0/\rho) = 0. \tag{2.85}$$

Equations (2.83) and (2.85) give

$$H = -\frac{A_\mathrm{M} N_\mathrm{A} e^2}{4\pi\varepsilon_0 r_0}\left(1 - \frac{\rho}{r_0}\right). \tag{2.86}$$

This equation agrees with experimental data within 2 per cent for essentially ionic crystals. For instance, for NaCl the crystal energy is 0.75×10^6 J/mol. With $Z_1 = 1, r_0 = 2.82 \times 10^{-10}$ m,[45] eqn (2.85) gives

$$\begin{aligned} H &= \frac{1.748 \times 6.022 \times 10^{23} \times 1.6022^2 \times 10^{-38}}{4\pi \times 8.854 \times 10^{-12} \times 2.82 \times 10^{-10}}\left(1 - \frac{0.321}{2.82}\right) \\ &= -0.763 \times 10^6 \text{ J/mol.} \end{aligned}$$

In the calculation, 1.6022×10^{-19} is the electron charge in coulombs.

To calculate H^σ one may note that H^σ is the difference between the molar enthalpy on the surface and in the bulk. On the surface the electrostatic energy is less than in the bulk since there are no charges above the surface layer. Therefore a Madelung constant A_M^S is introduced for the surface layer, $A_\mathrm{M}^\mathrm{S} < A_\mathrm{M}$. For instance for NaCl

$$\frac{A_\mathrm{M} - A_\mathrm{M}^\mathrm{S}}{A_\mathrm{M}} = \frac{0.132}{3.495} = 0.0378.$$

On the other hand the surface layer is drawn closer to the sublayer since the atoms in the surface layer are not attracted from above. Thus the distance between the two layers, r_0^S, becomes less than r_0. This increases the electrostatic energy stored in the surface. To calculate r_0^S eqn (2.84) is applied to the two surface layers:

$$\frac{(A_\mathrm{M}^\mathrm{S} + A_\mathrm{M})\mathrm{N}_\mathrm{A} e^2}{4\pi\varepsilon_0 (r_0^\mathrm{S})^2} - \frac{B}{\rho}\left(\frac{Z - m}{Z} + 1\right)\exp\left(-\frac{r_0^\mathrm{S}}{\rho}\right) = 0 \tag{2.87}$$

where Z is the number of nearest neighbours and m is the number of missing nearest neighbours for the top layer. For NaCl, $Z = 6$. If the surface lies along a crystal plane, $m = 1$.

Equations (2.85) and (2.87) give

$$\left(1+\frac{A_{\mathrm{M}}^{\mathrm{S}}}{A_{\mathrm{M}}}\right)=\left(1+\frac{Z-m}{Z}\right)\exp\left(\frac{r_0-r_0^{\mathrm{S}}}{\rho}\right)\left(\frac{r_0^{\mathrm{S}}}{r_0}\right)^2. \tag{2.88}$$

For NaCl one obtains $r_0^{\mathrm{S}} = 2.80$ Å from eqn (2.88).

For the surface it is assumed that the screening term is proportional to the number of remaining bonds:

$$\begin{aligned} H^{\sigma} &= H^{\mathrm{S}} - H \\ &= -\frac{A_{\mathrm{M}}^{\mathrm{S}} N_{\mathrm{A}} e^2}{4\pi\varepsilon_0 r_0^{\mathrm{S}}}\left(1-\frac{(Z-m)\,\rho}{Z r_0^{\mathrm{S}}}\right)+\frac{A_{\mathrm{M}} N_{\mathrm{A}} e^2}{4\pi\varepsilon_0 r_0}\left(1-\frac{\rho}{r_0}\right) \end{aligned}$$

or

$$\frac{H^{\sigma}}{H}=1-\frac{A_{\mathrm{M}}^{\mathrm{S}} r_0}{A_{\mathrm{M}} r_0^{\mathrm{S}}}\left(\frac{1-\dfrac{(Z-m)\rho}{Z r_0^{\mathrm{S}}}}{1-\dfrac{\rho}{r_0}}\right). \tag{2.89}$$

For NaCl

$$\begin{aligned} \frac{H^{\sigma}}{H} &= 1-\frac{3.363\times 2.82}{3.495\times 2.80}\left(\frac{1-\dfrac{5}{6}\times\dfrac{0.321}{2.80}}{1-\dfrac{0.321}{2.82}}\right) \\ &= 0.011 \end{aligned}$$

or

$$H^{\sigma} = 0.011 \times 0.765 \times 10^6 = 8415 \text{ J/mol}.$$

One mole occupies an area

$$\begin{aligned} A &\approx N_{\mathrm{A}}^{\frac{1}{3}} V_{\mathrm{m}}^{\frac{2}{3}} = (6.022\times 10^{23})^{\frac{1}{3}}\left(\frac{58.45}{2.17\times 10^6}\right)^{\frac{2}{3}} \\ &= 76\,000 \text{ m}^2/\text{mol}. \end{aligned}$$

Then the surface tension becomes

$$\sigma = \frac{H^{\sigma}}{A} = 0.11 \text{ N/m}.$$

This value may be compared with $\sigma = 0.11$ J/m^2 obtained by fracture of NaCl.[48] The excellent correspondence between the computed value and the experimental value is probably fortuitous. It should be recalled that the calculation is only approximate. Several corrections have been omitted deliberately to obtain equations in simple forms. These corrections include the van der Waals attraction and the repulsive forces between non-nearest neighbours.

In this book we do not pursue the theory further. The main point is to illustrate that the surface energy is given by the difference between the energy associated with the surface and with the bulk. For ionic crystals it is given by the difference between the electrostatic energy associated with the surface and with the bulk. From this difference we subtract the repulsion energy difference between bulk and surface. It is recognized that the numerical problems are considerable: the differences between the various energies are small. Furthermore the electrostatic terms and the repulsion energy terms are of the same order of magnitude. The net result is that the surface energy is only around 1 per cent of the crystal energy. From these facts it is recognized that the surface energy (surface tension) is sensitive to minor changes in structure and crystal radii caused by the presence of small amounts of impurities in the surface layer.

2.2.7 Interfacial tension of liquids with several components

Only for one-component systems is the interfacial tension σ equal to the Gibbs energy per unit area $g^\sigma = G^\sigma/A$. For systems with several components σ is in general different from g^σ.[49] Then

$$\sigma = \left(\frac{\partial G^\sigma}{\partial A}\right)_{n_i} = \left(\frac{\partial (A g^\sigma)}{\partial A}\right)_{n_i} = g^\sigma + A\left(\frac{\partial g^\sigma}{\partial A}\right)_{n_i}. \qquad (2.90)$$

σ is equal to g^σ only if $(\partial g^\sigma/\partial A)_{n_i} = 0$, that is, only if g^σ does not change when the surface area is changed. In general this is the case only for a one-component system. If the concentration of the various components at the surface is given by

$$\Gamma_i = n_i^\sigma/A$$

then

$$\left(\frac{\partial g^\sigma}{\partial A}\right)_{n_i} = \sum_{i=1}^{k} \frac{\partial g^\sigma}{\partial \Gamma_i}\left(\frac{\partial \Gamma_i}{\partial A}\right)_{n_i}$$

where $\partial g^\sigma/\partial\Gamma_i = \mu_i^\sigma$ is the chemical potential of component i at the interface. Under equilibrium conditions $\mu_i^\sigma = \mu_i$, that is, μ_i^σ is equal to the bulk chemical potential.

$$\left(\frac{\partial \Gamma_i}{\partial A}\right)_{n_i} = \left(\frac{\partial (n_i^\sigma/A)}{\partial A}\right)_{n_i} = n_i^\sigma \frac{\partial 1/A}{\partial A} = \frac{-n_i^\sigma}{A^2} = -\frac{\Gamma_i}{A}.$$

Then eqn (2.90) becomes

$$\sigma = g^\sigma - \sum_{i=1}^{k} \mu_i^\sigma \Gamma_i. \qquad (2.91)$$

This shows that in general the interfacial tension σ cannot be identified with the interfacial free energy per unit area g^σ. σ depends also on the chemical potential at the surface μ_i^σ and on the surface coverage $\Gamma_i = n_i^\sigma/A$. It will seen that the ratio between the n_i^σ at the surface in general is different from the molar ratios in the bulk. Electronegative elements in metals such as F, O, S, and so on tend to accumulate at the surface. These elements are therefore called surface active. We have seen that the interfacial tension is given by the difference between the potential for the surface and the potential for the bulk. The electronegative elements give a deeper surface potential. This reduces the surface tension.

The interfacial (surface) tension of the pure metal is often known and one would like to determine the effect of various alloy additions and impurity levels. Then differentiation of eqn (2.91) gives

$$d\sigma = dg^\sigma - \sum_{i=1}^{k} \mu_i^\sigma d\Gamma_i - \sum_{i=1}^{k} d\mu_i^\sigma \Gamma_i. \tag{2.92}$$

g^σ depends on T, P, and the Γ_i. It has a total differential

$$dg^\sigma = \left(\frac{\partial g^\sigma}{\partial T}\right)_p dT + \left(\frac{\partial g^\sigma}{\partial p}\right)_T dp + \sum_{i=1}^{k} \mu_i^\sigma d\Gamma_i.$$

If the two equations are added, the following is obtained;

$$d\sigma = \left(\frac{\partial g^\sigma}{\partial T}\right)_p dT + \left(\frac{\partial g^\sigma}{\partial p}\right)_T dp - \sum_{i=1}^{k} \Gamma_i d\mu_i^\sigma$$

where $(\partial g^\sigma/\partial T)_p = -s^\sigma$. Usually the effect of pressure can be neglected (compare problem 2.2).

Then

$$d\sigma = -s^\sigma dT - \sum_{i=1}^{k} \Gamma_i d\mu_i^\sigma.$$

At equilibrium, where $\mu_i^\sigma = \mu_i$, and at constant temperature[55]

$$d\sigma = -\sum_{i=1}^{k} \Gamma_i d\mu_i. \tag{2.93}$$

If there is only one component so that $k = 1$, $d\mu_1 = RT d(\ln a_1) = 0$ since $a_1 = 1$ and $\sigma = g = G/A$. This equation has been used previously in the discussion of one-component systems.

To obtain the interfacial tension σ at equilibrium when there are several components in addition to the solvent metal, the area occupied by one mole of species i in the monolayer, A_i^σ is introduced. Then

$$\Gamma_i = \frac{n_i^\sigma}{\sum_{j=1}^{k} n_j^\sigma A_j^\sigma}$$

or in terms of mass per cent at the surface

$$\Gamma_i = \frac{[\%\, i]^\sigma/M_i}{\sum_{j=1}^{k} [\%\, j]^\sigma A_j^\sigma/M_j}.$$

An *interfacial activity coefficient* f_i^σ is defined as follows

$$f_i^\sigma = \frac{a_i}{[\%\, i]^\sigma} \tag{2.94}$$

where the activity in the bulk a_i is given in mass per cent so that f_i^σ is dimensionless. Notice that f_i^σ compares activity in the bulk liquid with the concentration in the interface. It will be seen that f_i^σ is a useful tool in the derivation that follows. Finally f_i^σ can be eliminated.

If the liquid is in contact with a vacuum or an inert gas, the mass percentages on the surface sum up to 100 per cent:

$$\sum_{i=1}^{k} [\%\, i]^\sigma = 100. \tag{2.95}$$

For a liquid in contact with a solid or another liquid the percentage on the right-hand side is an unknown quantity, A, less than 100. In terms of activities and surface activity coefficients eqn (2.95) becomes

$$\sum_{i=1}^{k} \frac{a_i}{f_i^\sigma} = 100. \tag{2.96}$$

Similarly for the bulk phase

$$\sum_{i=1}^{k} \frac{a_i}{f_i} = 100.$$

According to the Gibbs–Duhem equation (eqn (2.26))

$$\sum_{i=1}^{k} x_i \frac{\mathrm{d}a_i}{a_i} = 0.$$

Now for the mole fraction

$$x_i = \frac{[\%\, i]/M_i}{\sum_{i=1}^{k} [\%\, j]/M_j}. \tag{2.97}$$

Inserted in the Gibbs–Duhem equation this gives

$$\sum_{i=1}^{k} \frac{[\%\, i]\,\mathrm{d}a_i}{M_i a_i} = 0$$

or

$$\sum_{i=1}^{k} \frac{\mathrm{d}a_i}{M_i f_i} = 0. \tag{2.98}$$

To obtain an equation for the surface tension, the expressions for Γ_i and $\mathrm{d}\mu_i = RT\mathrm{d}\ln a_i$ are introduced in eqn (2.93). Furthermore, concentrations are replaced by activities. This gives

$$\frac{\mathrm{d}\sigma}{RT} = \frac{-\sum_{i=1}^{k} \frac{a_i}{M_i f_i^{\sigma}} \mathrm{d}(\ln a_i)}{\sum_{i=1}^{k} \frac{a_i A_i^{\sigma}}{M_i f_i^{\sigma}}}$$

or, since $a_i \mathrm{d}(\ln a_i) = \mathrm{d}a_i$,

$$\frac{\mathrm{d}\sigma}{RT} = \frac{-\sum_{i=1}^{k} \mathrm{d}a_i/(M_i f_i^{\sigma})}{\sum_{i=1}^{k} a_i A_i^{\sigma}/(M_i f_i^{\sigma})}. \tag{2.99}$$

We notice that if $f_i^{\sigma} = f_i$ then according to the Gibbs–Duhem eqn (2.98), $\mathrm{d}\sigma = 0$. This means that if the activity coefficients were the same in the bulk as on the surface, surface tension would not change with composition. In general $f_i^{\sigma} \neq f_i$. In this case eqn (2.99) may be written

$$\sum_{i=1}^{k} \frac{a_i A_i^{\sigma}}{M_i f_i^{\sigma}} + \sum_{i=1}^{k} \frac{\mathrm{d}a_i}{\mathrm{d}\sigma} \frac{RT}{M_i f_i^{\sigma}} = 0. \tag{2.100}$$

We wish to study the effect of the solute additions. Therefore, the solvent component, $i = 1$, is eliminated from eqn (2.100). To accomplish this, eqn (2.98) is multiplied by $f_1 RT/(f_1^{\sigma}\mathrm{d}\sigma)$ to give

$$\sum_{i=1}^{k} \frac{\mathrm{d}a_i}{\mathrm{d}\sigma} \frac{RTf_1}{f_1^{\sigma} M_i f_i} = 0.$$

Furthermore, eqn (2.96) is multiplied by A_1^{σ}/m_1 to give

$$\sum_{i=1}^{k} \frac{A_1^{\sigma} a_i}{M_1 f_i^{\sigma}} = 100 \frac{A_1^{\sigma}}{M_1}.$$

If these two equations are subtracted from eqn (2.100), one obtains

$$\frac{100A_1^{\sigma}}{M_1} + \sum_{i=2}^{k} \left\{ \left(A_i^{\sigma} - \frac{A_1^{\sigma} M_i}{M_1} \right) a_i + RT \left(1 - \frac{f_1 f_i^{\sigma}}{f_1^{\sigma} f_i} \right) \frac{\mathrm{d}a_i}{\mathrm{d}\sigma} \right\} \frac{1}{M_i f_i^{\sigma}} = 0. \tag{2.101}$$

In the following only a binary system is studied. For a dilute binary system f_2 and f_1 may as an approximation be regarded as constant with respect to a_2 according to Henry's and Raoult's laws. However, the concentration of component 2 may be high on the surface. Therefore, f_2^σ and f_1^σ depend on a_2. This relationship must be determined before eqn (2.101) can be integrated. A Langmuir type model is employed for this purpose. According to the model adsorption is limited to a monolayer on the surface and gives an 'exchange reaction':

$$\begin{aligned}&(\text{component 2 in bulk}) + (\text{component 1 on surface})\\&= (\text{component 2 on surface}) + (\text{component 1 in bulk})\end{aligned} \tag{2.102}$$

The equilibrium constant for the exchange reaction is expressed by ΔG^0, the Gibbs energy gained by moving one mole of component 2 from the bulk to the surface, the *Gibbs energy of adsorption*.

$$\frac{a_2[\%\,1]^\sigma}{[\%\,2]^\sigma a_1} = \exp\left(\frac{\Delta G^0}{RT}\right) \tag{2.103}$$

where a_1 and a_2 are the solvent and solute activities in solution and the activities in the adsorbed layer are approximated by the respective mass percentages. Notice that eqns (2.102) and (2.103) differ from the Langmuir model in that the size of the species 1 and 2 is only taken into account in ΔG^0. ΔG^0 must be determined experimentally.

Equation (2.103) combined with the definition of the surface activity coefficients, eqn (2.94), gives, when $[\%\,1]^\sigma$ and a_2 are eliminated using eqn (2.95) and $f_1 = a_1/[\%\,1]$ and $f_2 = a_2/[\%\,2]$:

$$f_2^\sigma = \frac{a_2}{100}\left(1 - \frac{f_1}{f_2}\exp\left(\frac{\Delta G^0}{RT}\right)\right) + f_1\exp\left(\frac{\Delta G^0}{RT}\right). \tag{2.104}$$

If eqn (2.103) is written in activities only, we find that

$$\frac{f_2^\sigma}{f_1^\sigma} = \exp\left(\frac{\Delta G^0}{RT}\right). \tag{2.105}$$

Equation (2.101) for a two-component system ($k = 2$) can then be integrated to give

$$\sigma = \sigma_1 - \frac{M_1}{A_1^\sigma M_2}\frac{RT}{b_2}\left(1 - \frac{f_1}{f_2}\exp\left(\frac{\Delta G^0}{RT}\right)\right)\ln\left(b_2\frac{f_2}{f_1}\exp\left(-\frac{\Delta G^0}{RT}\right)\frac{\%[2]}{100} + 1\right) \tag{2.106}$$

where

$$b_2 = \frac{A_2^\sigma M_1}{A_1^\sigma M_2} - \frac{f_1}{f_2}\exp\left(\frac{\Delta G^0}{RT}\right). \tag{2.107}$$

σ is now given in terms of [% 2] and the surface tension of the pure solvent σ_1. This may be regarded as a further development of an equation given by Belton.[50] In the following we set both f_2 and f_1 equal to one.

So far we have considered a melt surface in contact with a vacuum. In principle the same argument can be applied for a melt surface in contact with a solid or with a slag.[56] Then

$$\sigma_{sl} = \sigma_{sl}^{p} - \frac{M_1}{A_1^{\circ} M_2}\frac{RT}{b_2}\left(1 - \frac{f_1}{f_2}\exp\left(\frac{\Delta G^0}{RT}\right)\right)\ln\left(b_2\frac{f_2}{f_1}\exp\left(\frac{\Delta G^0}{RT}\right)\frac{[\%\,2]}{A} + 1\right) \tag{2.108}$$

where σ_{sl}^{p} is the interfacial tension between the pure melt and the solid and σ_{sl} is the interfacial tension when the melt contains [% 2] mass per cent component 2. b_2 is defined by analogy with eqn (2.107). ΔG^0 must of course be determined experimentally for each solid–melt system. A is the percentage of the interface occupied by the liquid. For pure metals at the melting-point it is possible to determine A as indicated in the next section.

For metallic additions the metallic radius is used to calculate A_2°, (see paper by Frisvold and the author, to be published).

For surface active elements such as oxygen or sulphur in iron, the ionic radius

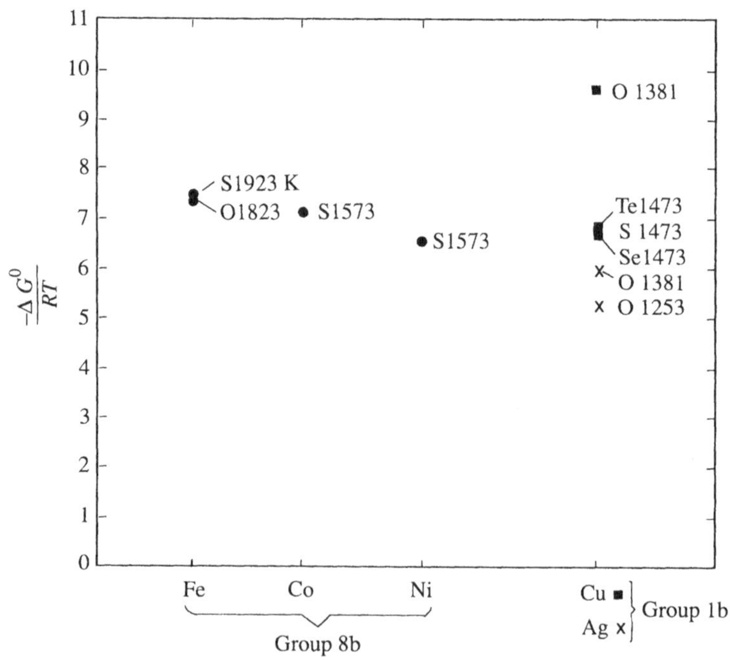

FIG. 2.23 Gibbs energy of adsorption from bulk to surface, ΔG^{O}, of O, S, Se, and Te in solvent metals: Fe, Cu, and Ag. R is the gas constant. The absolute temperatures T are indicated.

is used to calculate A_2^σ. For oxygen in iron the ionic radius, r_i, is 1.40 Å.[69] Then an estimate for the area occupied by one oxygen atom is

$$\frac{A_2^\sigma}{N_A} = \pi r_i^2 = \pi \times 1.4^2 = 6.16\ \text{Å}^2. \tag{2.109}$$

The metallic radius of iron is 1.24 Å. The area occupied by a neutral iron atom is (see Table 2.13).

$$\frac{A_2^\sigma}{N_A} = \pi r^2 = \times 1.24^2 = 4.83\ \text{Å}^2.$$

For various solvent metals with impurities of oxygen, sulphur, selenium, and tellurium eqn (2.106) has been compared with experimental data[50, 52] and the ΔG^O that give the best fit for eqn (2.105) have been determined. The ΔG^O values are shown in Fig. 2.23. Figures 2.24 and 2.25 compare eqn (2.106) and experimental data for the two important systems Fe–O and Cu–S, respectively. For the iron–oxygen system at 1550°C

$$\sigma_1 = 1.8\,\text{N/m}, \qquad b_2 = \frac{6.16 \times 55.85}{4.83 \times 16} - \exp\left(\frac{-111156}{8.3145 \times 1823}\right) = 4.44$$

$$A_1^\sigma = 4.83 \times 10^{-20} \times 6.022 \times 10^{23}.$$

Equation (2.108) gives

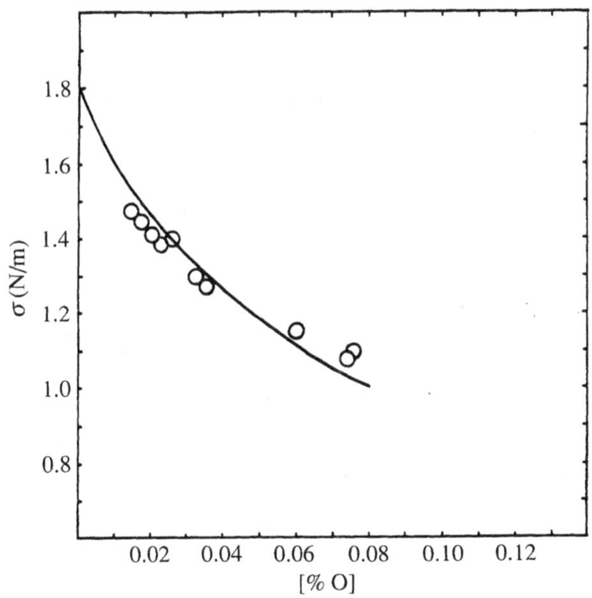

FIG. 2.24 The depression of the surface tension of iron by oxygen at 1550°C[50] compared with eqn (2.110).

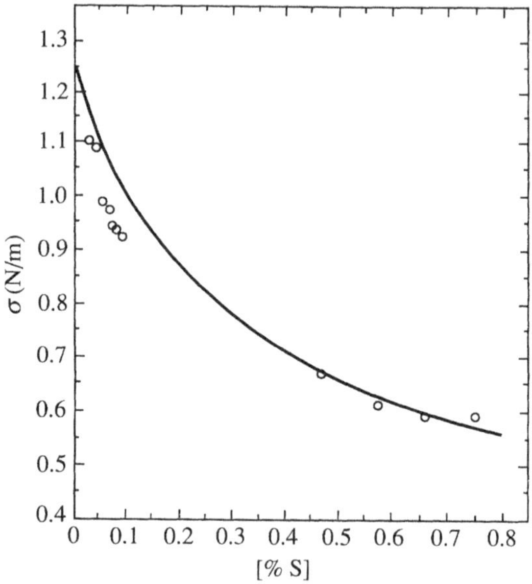

FIG. 2.25 The depression of the surface tension of copper by sulphur at 1200°C[51] compared with eqn (2.111).

$$\sigma = 1.8 - 0.409\,(\ln(68[\%\,\mathrm{O}] + 1). \tag{2.110}$$

Similarly for sulphur in copper an ionic radius for sulphur[53] of 1.84 Å and a metallic radius of copper of 1.28 Å[68,54] gives, using $-\Delta G^{\sigma} = 84\,100$ J/mol (see Fig. 2.23),

$$\sigma = 1.244 - 0.191\,(\ln(39.3[\%\,\mathrm{S}] + 1). \tag{2.111}$$

Figures 2.24 and 2.25 are typical examples that show good correspondence between equation and experiments when employing the 'fitted' values of ΔG^0 as the one adjustable parameter.

2.2.8 Solid–liquid interfacial tension for pure metals

Turnbull[57,58] has found experimentally for many pure metals that the interfacial tension, σ_{sl}, at the melting point is given by

$$\sigma_{\mathrm{sl}} = \frac{0.45\Delta H_{\mathrm{m}}}{N_{\mathrm{A}}^{\frac{1}{3}} V_{\mathrm{m}}^{\frac{2}{3}}} \tag{2.112}$$

where ΔH_{m} is the molar enthalpy of melting and $N_{\mathrm{A}}^{\frac{1}{3}}\, V_{\mathrm{m}}^{\frac{2}{3}}$ is roughly equal to the area of one mole of atoms in a monolayer.

Equation (2.112) can be derived as follows. For the one-component system $\sigma_{sl} = G^{\sigma}/A$. It may be assumed that no bonds are broken in passing from the bulk phases to the interface if the structure in the solid and liquid state is similar. Then, for the transfer to the interface, $H^{\sigma} = 0$ on both sides of the interface so that

$$A\sigma_{sl} = -T_m S^{\sigma} \tag{2.113}$$

where S^{σ} is the difference between the molar entropy of the interface and the two bulk phases. A is the area of a mole of atoms in a monolayer. We introduce the monolayer model and assume that this layer is a mixture (configuration) of atoms in the liquid and solid state, with mole fractions x_l and x_s of the layer belonging to the liquid and solid phases, respectively. The enthalpy content of the liquid atoms is ΔH_m (heat of fusion per mole) higher than for the solid atoms while the partial molar entropy difference between liquid and solid at the interface, assuming an 'ideal solution', is[2]

$$\Delta S_m = -R \ln x_l + R \ln x_s. \tag{2.114}$$

The reduction of entropy $Rx_l \ln x_l$ on the liquid side of the interface compared with the bulk liquid phase and the corresponding reduction $Rx_s \ln x_s$ on the solid side gives for eqn (2.113) on employing eqn (2.4) for an ideal solution:

$$A\sigma_{sl} = -T_m R(x_l \ln x_l + x_s \ln x_s). \tag{2.115}$$

Since

$$x_l + x_s = 1 \tag{2.116}$$

eqns (2.114) and (2.116) give

$$x_l = \frac{\exp(-\Delta S_m/R)}{1 + \exp(-\Delta S_m/R)} \tag{2.117}$$

$$x_s = \frac{1}{1 + \exp(-\Delta S_m/R)}. \tag{2.118}$$

Then eqn (2.115) becomes

$$A\sigma_{sl} = T_m R\left(\ln(1 + b) - \frac{b}{1 + b}\ln b\right) \tag{2.119}$$

where b is a constant:

$$b = \exp(-\Delta S_m/R). \tag{2.120}$$

According to Richard's rule[3] that $\Delta H_m/T_m \approx 9.2$ J/mol

$$\frac{\Delta S_m}{R} = \frac{\Delta H_m}{RT_m} \approx \frac{9.2}{8.3145} = 1.1$$

so that

$$b = 0.33$$

and

$$A\sigma_{sl} = T_m R\, 0.56. \tag{2.121}$$

If the Richard's rule approximation is employed once more

$$\sigma_{sl} = 0.51\Delta H_m / A. \tag{2.122}$$

Equation (2.78) now gives

$$\sigma_{sl} = \frac{0.46\Delta H_m}{N_A^{\frac{1}{3}} V_m^{\frac{2}{3}}}. \tag{2.123}$$

This equation may be compared with experimental data[59] where the proportionality factor is 0.436, 0.490, 0.439, 0.445, 0.436, 0.444, and 0.450 for Au, Co, Cu, Fe, Ga, Ni, and Pd, respectively.

For Si, Ge, and Bi which are non-metals in the solid state and acquire metallic properties on melting, the above theory cannot be applied. The number of nearest neighbours is about four in the solid state and increases to about eight on melting. A rough estimate of the proportionality constant may be carried out as follows. For the two phases the 'average' number of bonds broken between bulk phase and the interface is $(8 - 4)/2 = 2$. The enthalpy contribution of the surface energy is then $\Delta H_m 2/8 = 0.25\Delta H_m$ while the entropy contribution is given by the right-hand side of eqn (2.119). Richard's rule cannot be applied to the metalloids. For instance for Si $\Delta H_m/(RT_m) = 3.62$ so the right-hand side of eqn (2.119) becomes $0.121RT_m$. Then for Si

$$\sigma_{sl} = \left(0.25 + \frac{0.121}{3.62}\right)\frac{\Delta H_m}{A}$$

and

$$\sigma_{sl} = \frac{0.26\,\Delta H_m}{N_A^{\frac{1}{3}} V_m^{\frac{2}{3}}}.$$

For Ge and Bi a similar calculation gives 0.26 and 0.34, respectively, for the proportionality constant. For these elements it is found experimentally that the proportionality constant in eqn (2.123) is 0.32–0.35.[59,60]

2.2.9 Diffusion in molten metals

It is recalled that the diffusion coefficient D enters into Fick's law of diffusion:

$$\dot{n} = -D\frac{\partial c_2}{\partial z} \tag{2.124}$$

where $\dot{n}$ is the flux of solute particles and c_2 is their concentration. Equation (2.124) applies when there is no convection.

The remarkable characteristic of diffusion in molten metals, and liquids in general, is the fact that the values of the diffusion coefficient D are almost all of the same order of magnitude, 10^{-8}–10^{-9} m^2/s, even though their solid state properties differ widely.[31,53-55] However, D does in fact increase with decreasing mass of the diffusion species. This may partly be explained in terms of interstitial movement between the larger atoms. In the following a rough relation for D is derived that takes mass into account. It will be seen that such a simple equation for estimating D is very convenient.

The Stokes–Einstein equation[45] is a very useful starting point in the description of diffusion in molten metals and other liquids. This is the case in spite of the fact that the equation is derived for particles moving randomly in a continuum (called Brownian motion). The assumption is that the particles are much larger than the particles in the surrounding continuous liquid. However, the diffusing atoms in a molten metal are about the same size as the solvent atoms; they may even be smaller as is the case with carbon in iron.

According to Einstein the mean squared displacement $\langle z^2 \rangle$ in a random walk[45] is

$$\langle z^2 \rangle = \frac{2kT}{\zeta} t \qquad (2.125)$$

where t is the time elapsed from position $z = 0$. Equation (2.125) is valid only after a large number of collisions has taken place between the Brownian particle and the surrounding particles (continuum). $\langle z^2 \rangle$ is the square of the distance that the particle has drifted in the direction z. The total distance travelled is of course much greater. ζ is the factor given by the ratio between the force F acting on the Brownian particle that moves through a continuum with velocity u:

$$\zeta = \frac{F}{u}. \qquad (2.126)$$

For a large (Brownian) particle moving in a continuum we have the relation first obtained by Stokes:[61] $F = 3\pi\,\mu u d$ so that

$$\zeta = 3\pi\mu d \qquad (2.127)$$

where μ is the viscosity of the fluid and d is the particle diameter. Stokes' equation is based on the boundary condition (no-slip condition) that there is no relative motion between the particle and the fluid (continuum) immediately in contact with it. This condition cannot be applied to the case of collisions between a particle and neighbouring particles of greater or equal mass. In this case we assume—somewhat arbitrarily—that the tangential forces are negligible when the masses are equal. If there are no tangential forces the factor 2 replaces 3 in eqn (2.127).[62] Then eqn (2.125) becomes

$$\langle z^2 \rangle = \frac{kTt}{\pi\mu d}. \tag{2.128}$$

The diffusion coefficient D may be obtained by first using eqn (2.124) to derive an equation for c_2, then calculating $\langle z^2 \rangle$, and finally comparing the result with eqns (2.125) and (2.127) or with eqn (2.128). The accumulation of component c_2 per unit time is equal to the difference between the flux in and out of the control volume element. Insertion of eqn (2.124) in this equation gives the diffusion equation

$$\frac{\partial c_2}{\partial t} = D\frac{\partial^2 c_2}{\partial z^2} \tag{2.129}$$

(see also Chapters 4, 7, and 8). If n_2 moles of component 2 are added at position $z = 0$ at time $t = 0$ it is found[45] that the solution of eqn (2.129) is

$$c_2 = \frac{n_2}{2}\exp\left(-\frac{z^2}{4Dt}\right)(\pi Dt)^{-\frac{1}{2}}.$$

Then

$$\langle z^2 \rangle = \frac{\int_{-\infty}^{\infty} x^2 c_2(z, t)\,\mathrm{d}x}{n_2} = 2Dt. \tag{2.130}$$

Comparison of eqn (2.130) with eqns (2.125) and (2.127) gives that for Brownian motion

$$D = \frac{kT}{3\pi\mu d}.$$

However, for diffusion of atoms, D should depend on mass, m_2; in fact D depends on a dimensionless ratio such as m_2/m_1. We assume that

$$D = \frac{kT}{2\pi\mu d}\left(\frac{1}{2} + \frac{m_1}{2m_2}\right)^a \tag{2.131}$$

where the exponent a is not as yet determined. Equation (2.131) is in accordance with eqns (2.128) and (2.130) when $m_1 = m_2$.

If $m_2 \gg m_1$, we should recover Einstein's equation for Brownian motion $D = kT/3\pi\mu d$. This requires that $a = 0.585$. However, we take $a = 0.5$ since the procedure only gives a rough estimate of D. Then

$$D = \frac{kT}{2\pi\mu d}\left(\frac{m_1 + m_2}{2m_2}\right)^{\frac{1}{2}}. \tag{2.132}$$

It is seen that D increases with decreasing mass ratio between solute and solvent

m_2/m_1. Within the experimental errors eqn (2.132) seems to be in accordance with experimental results. For instance, experimental values of the diffusion coefficients for hydrogen in molten metals agree reasonably well with the equation.

There is some evidence that the size d of a solute atom should vary from solvent to solvent depending upon the chemical affinity ΔG^{O}, between the two species. However, as a first approximation it is only assumed that $d/2$ is proportional to the metallic radius. Suggested radii are given in Table 2.13. For the non-metals it is proposed that the Goldschmidt atomic radii be employed:[63] $r = 0.46, 0.77, 0.71, 0.60$ Å for H, C, N, and O, respectively. $r = 1.17, 1.09, 1.04$ Å for Si, P, and S.

To illustrate the use of eqn (2.132) for the extreme case of diffusion of hydrogen, the diffusion coefficients in iron at 1823 K and aluminium at 1000 K are calculated. For H in iron

$$D = \frac{1.38 \times 10^{-23} \times 1823}{2\pi \times 5 \times 10^{-3} \times 0.92 \times 10^{-10}} \left(\frac{55.85 + 1}{2}\right)^{\frac{1}{2}}$$

$$= 0.5 \times 10^{-7}\,\mathrm{m^2/s}.$$

Similarly for H in Al at 1000 K, $D = 0.9 \times 10^{-7}\,\mathrm{m^2/s}$. These values may be compared with the experimental values $D = 10^{-7}\,\mathrm{m^2/s}$ for H in Fe[31] and $D = 10^{-7} - 10^{-8}\,\mathrm{m^2/s}$ for H in Al.[63,64] Experimental values given in the literature vary widely; even by a factor of ten. In calculations of mass transfer D usually enters as the square root. Thus the factor of ten can lead to an error of three in mass transfer calculations. Probably this is the most serious problem we face in the development of quantitative models of refining operations.

2.2.10 Thermal and electrical conductivity

At low temperatures the thermal conductivity of metals may drop with increasing temperatures. The reason is that for impure metals or disordered alloys the contribution from phonons (lattice waves) can be comparable to the electronic contribution. The contribution of the lattice waves to the thermal conductivity drops with temperature. In normal pure metals even at room temperature the electrons carry almost all the heat current. The contributions from electrons increase with temperature, therefore at higher temperatures, both for pure and impure metals, the electronic contribution overwhelms the phonon contribution and then the thermal conductivity finally increases with temperature.

Although there are several compilations of transport properties of elements, no comprehensive survey has been made for metals and alloys. Only a few studies for predicting metal thermal conductivities have been carried out.[65,66] At high temperatures the Wiedemann–Franz law which gives a relation between the thermal and electrical conductivity of metals applies. On combining this law and measurements of electrical conductivities determined experimentally, the

TABLE 2.13 *Valence, relative atomic mass, metallic radius, electronegativity, and reduction potential in aqueous solution (at 25°C) of the elements.*[69] *The reduction potential is for the lowest valence in the table. For instance for iron $Fe^{2+} + 2e^{-} = Fe(s)$*

	1a	2a	3b	4b	5b	6b	7b		8b		1b	2b	3a	4a	5a	6a	7a
	Li	Be											B	C	N	O	F
Valence	1	2											3	4	3	2	1
atomic mass	6.94	9.01											10.82	12.01	14.01	16.00	19.00
radius, Å	1.52	1.12											–	–	–	–	–
electronegativity	0.98	1.57											2.04	2.55	3.04	3.44	3.98
potential	−3.0	−1.97															2.87
	Na	Mg											Al	Si	P	S	Cl
Valence	1	2											3	4	3	2	1
atomic mass	22.99	24.31											26.98	28.09	30.97	32.06	35.45
radius, Å	1.86	1.60											1.43	–	(1.10)		
electronegativity	0.93	1.31											1.61	1.90	2.19	2.58	3.16
potential	−2.71	−2.36											−1.66			−0.48	1.36
	K	Ca	Sc	Ti	V	Cr	Mn	Fe	Co	Ni	Cu	Zn	Ga	Ge	As	Se	Br
Valence	1	2	3	4	5	3, 6	4, 6	2, 3	2, 3	2	1, 2	2	3	4	3	2	1
atomic mass	39.10	40.08	44.96	47.90	50.94	52.00	54.94	55.85	58.93	58.70	63.55	65.38	69.72	72.59	74.92	78.96	79.90
radius, Å	2.27	1.97	1.61	1.45	1.31	1.25	1.37	1.24	1.25	1.25	1.28	1.33	1.22	1.23	(1.22)	(1.16)	–
electronegativity	0.82	1.0	1.36	1.54	1.63	1.66	1.55	1.83	1.88	1.91	1.90	1.65	1.81	2.01	2.18	2.55	2.96
potential	−2.92	−2.87						−0.41	−0.28	−0.23	0.52	−0.76	−0.56			−0.92	1.07
	Rb	Sr	Y	Zr	Nb	Mo	Tc	Ru	Rh	Pd	Ag	Cd	In	Sn	Sb	Te	I
Valence	1	2	3	4	5	6		4	3	2	2	2	3	4	3	2	1
atomic mass	85.47	87.62	88.91	91.22	92.91	95.44	(97)	101.07	102.91	106.4	107.87	112.41	114.82	118.69	121.75	127.60	126.90
radius, Å	2.48	2.15	1.78	1.59	1.43	1.36	1.35	1.33	1.34	1.38	1.44	1.49	1.36	1.51	1.45	1.43	–
electronegativity	0.82	0.95	1.22	1.33	1.6	2.16	1.9	2.2	2.28	2.20	1.93	1.69	1.78	1.96	2.05	2.1	2.66
potential	−2.92	−2.89								0.92	0.80	−0.40	−0.34			−0.95	0.54
	Cs	Ba	La	Hf	Ta	W	Re	Os	Ir	Pt	Au	Hg	Tl	Pb	Bi		
Valence	1	2	3	4	5	6		4	2	2		2	3	2, 4	3		
atomic mass	132.91	137.33	138.91	178.49	180.95	183.85	186.2	190.2	192.2	195.1	196.97	200.59	204.37	207.2	208.98		
radius, Å	2.65	2.17	1.87	1.56	1.43	1.37	1.37	1.34	1.36	1.39	1.44	1.50	1.70	1.75	1.55		
electronegativity	0.79	0.89	1.10	1.3	1.5	2.36	1.9	2.2	2.20	2.28	2.54	2.00	2.04	2.33	2.02		
potential	−2.92	−2.90	−2.52							1.2	1.4	0.85		−0.13			

prediction of thermal conductivities of solid or liquid metals and alloys is possible since reliable data on electrical conductivities is often available. Various data on electrical conductivities (or on the reciprocal quantity, electrical resistivity) is found in Smithells' Metals Reference Book[63] and in a textbook by Meaden.[67]

The Wiedemann–Franz law states that for metals at temperatures which are not too low the ratio of the thermal conductivity to the electrical conductivity is directly proportional to the absolute temperature. The constant of proportionality is independent of the particular metal. The basis for the law is that the electrons transport charge and energy by the same mechanism. The ratio between the thermal conductivity k and electrical conductivity σ is[68]

$$\frac{k}{\sigma} = \frac{\pi^2}{3}\left(\frac{k}{e}\right)^2 T = LT. \tag{2.133}$$

L, called the Lorenz number, is

$$L = \frac{\pi^2}{3} \times \left(\frac{1.38 \times 10^{-23}}{1.6 \times 10^{-19}}\right)^2 = 2.45 \times 10^{-8}\ \text{watt-ohm/K}^2.$$

TABLE 2.14 *Melting-point, T_m, electrical resistivity, thermal conductivity, and Lorenz number at the melting-point for various molten elements*[63]

Metal	Melting-point (K)	$1/\sigma$ (10^{-8} Ωm)	k (W/mK)	L (10^{-8}W Ω/K^2)
Ag	1234	17.25	175	2.45
Al	933	24.25	94	2.44
Au	1337	31.25	104.4	2.44
Bi	544	129	17.1	4.05
Cd	594	33.7	42	2.38
Cs	301.5	37.0	19.7	2.42
Cu	1358	20.0	165.6	2.44
Ga	302.8	26	25.5	2.19
Hg	234.1	90.5	6.78	2.62
In	429.6	32.3	(42)	3.16
K	336.7	13.65	53.0	2.15
La	1193	138	(21.0)	2.43
Li	454	24.0	46.4	2.45
Mg	923	27.4	78	2.31
Na	370.8	9.64	89.7	2.33
Pb	600.5	94.85	15.4	2.43
Rb	311.9	22.83	33.4	2.44
Sb	903.5	113.5	21.8	2.74
Sn	504.9	47.2	30	2.80
Te	723	550	2.5	1.90
Tl	576.5	731	(24.6)	3.12
Zn	692.4	37.4	49.5	2.67

Table 2.14 compares the thermal conductivity σ and electrical resistivity at the melting-point of various metals. It is seen that, apart from the metalloids, the ratios between thermal and electrical conductivities are not far from $2.45 \times 10^{-8} W\Omega/K^2$.

The thermal conductivity of steel alloys becomes close to that for iron above 850°C. Above 1000°C the Lorenz ratio for steel alloys is around $2.55 \times 10^{-8} W\Omega/K^2$.

REFERENCES

1. Denbigh, K. (1981). *The principles of chemical equilibrium*, (4th edn). Cambridge University Press.
2. Gaskell, D.R. (1981). *Introduction to metallurgical thermodynamics*, (2nd edn). Hemisphere Publ. Co./McGraw-Hill, New York.
3. Kubaschewski, O. and Alcock, C.B. (1983). *Metallurgical thermochemistry*, (5th edn). Pergamon Press, Oxford.
4. Sigworth, G.K. and Engh, T.A. (1982). Refining of liquid aluminium—a review of important chemical factors. *Scandinavian Journal of Metallurgy*, **11**, 143–9.
5. Sigworth, G.K. and Elliott, J.F. (1974). The thermodynamics of liquid iron alloys. *Met. Sci.*, **8**, 298.
6. Sigworth, G.K. and Elliott, J.F. (1974). The thermodynamics of dilute liquid copper alloys. *Can. Metall. Q.*, **13**, 455–61.
7. Sigworth, G.K. and Elliott, J.F. (1976). The thermodynamics of liquid cobalt alloys. *Can. Metall. Q.*, **15**, 123–7.
8. JANAF Thermochemical Tables, (3rd edn) (1985), (ed. M.W. Chase *et al.*). American Chem. Soc. and The American Institute of Physics for The National Bureau of Standards. J. Phys. Chem. Ref. Data Vol 14 (1985) Supplement No. 1, Michigan 48674, USA.
9. Barin, I (1989). *Thermochemical Data of Pure Substances*. Parts I and II VCH Verlagsgesellschaft, Weinheim, Germany.
10. Hultgren, H., Desai, P.D., Hawkins, D.T., Gleiser, M., Kelley, K.K., and Wagman, D.D. (1973). *Selected values of the thermodynamic properties of the elements*. Am. Soc. Met., Metals Park, Ohio.
11. Lupis, C.H.P. (1983). *Chemical thermodynamics of materials*. North-Holland, New York.
12. Ohtani, M. and Gokcen, N.A. (1960). Thermodynamic interaction parameters of elements in liquid iron. *Trans. Am. Inst. Min. Metall. Engrs.*, **218**, 533–40.
13. Bodsworth, C. and Bell, H.B. (1972). *Physical chemistry of iron and steel manufacture*. Longman London.
14. Wilson, F.G., Heesom, M.J., Nicholson, A., and Hills, A.W.D. (1987). Effect of fluid flow characteristics on nozzle blockage on aluminium-killed steels. *Ironmaking and Steelmaking*, **14**, 296–309.
15. Saxena, S.K., Sandberg, H., Waldenstrøm, T., Person, A., and Steensen, S. (1978). Mechanism of clogging of tundish nozzles during continuous casting of aluminium killed steel. *Scandinavian Journal of Metallurgy*, **7**, 126–33.
16. Hilty, D.C. and Farrell, J.W. (1975). Steel flow through nozzles: Influence of deoxidizers. *Iron and Steelmaker*, **2** (4), 17–22.

17. Väinölä, R., Karvonen, P., and Helle, L.W. (1986). Establishment of calcium treatment practices at Ovako—historical development and evaluation of alternative methods. *Proceedings 4th International Conference on Injection Metallurgy*, **23**, 1-19, MEFOS, Luleå, Sweden.
18. Sharma, R.A. and Richardson, F.D. (1961). Activities in lime-alumina melts. *J. Iron and Steel Inst., London*, **198**, 386-90.
19. Dewing, E.W. (1972). Thermodynamics of the system NaF-AlF_3. Part 3: Activities in liquid mixtures. *Metall. Trans.*, **3**, 495-501.
20. Warren, B.E. (1937). X-ray determination of the structure of liquids and glass. *J. Appl. Phys.* **8**, 645.
21. Wong, J. and Angell, C.A. (1976). *Glass structure by spectroscopy*. Marcel Dekker, New York.
22. Flood, H. and Førland, T. (1947). The acidic and basic properties of oxides. *Acta Chem. Scand.*, **1**, 592-604, 781-9.
23. Richardson, F.D. (1955). Thermodynamic aspects of slags and glasses. In *Vitreous State*. The Glass Delegacy of the University of Sheffield, Sheffield.
24. Turkdogan, E.T. (1980). *Physical chemistry of high temperature technology*. Academic Press, New York.
25. Shirota, Y., Katohgi, K., Klein, K., Engell, H.-J., and Janke, D. (1985). Phospate capacity of FeO-Fe_2O_3-CaO-P_2O_5 and FeO-Fe_2O_3-CaO-CaF_2-P_2O_5 slags by levitation melting. *Trans ISIJ*, **28**, 1132-40.
26. Selin, R. (1987). The role of phosphorus, vanadium and slag forming oxides in direct reduction based steelmaking. Ph.D. Thesis, Royal Inst. of Technology, Stockholm, Sweden.
27. Elliott, J.F., Gleiser, M., and Ramakrishna, V. (1963). *Thermochemistry for steelmaking I, II*. Addison-Wesley, Reading, Massachusetts.
28. Muan, A. and Osborn, E.F. (1965). *Phase equilibria among oxides in steelmaking*. Addison-Wesley, Reading, Massachusetts.
29. Chapman, S. and Cowling, T.G. (1970). *The mathematical theory of nonuniform gases*, (3rd edn). Cambridge University Press.
30. Herning, F. and Zipperer, L. (1936). Beitrag zur Berechnung der Zähigkeit technischer Gasgemische aus den Zähigkeitswerten der Einzelbestandteile. *Gas- und Wasserfach*, **79**, 49-54, 69-73.
31. Iida, T. and Guthrie, R.I.L. (1988). *The Physical properties of liquid metals*. Clarendon Press, Oxford.
32. Eisenstein, H. and Gingrich, N.S. (1942). The diffraction of x-rays by argon in the liquid, vapor and critical regions. *Phys. Rev.*, **62**, 261-9.
33. Ocken, H. and Wagner, C.N.J. (1966). Temperature dependence of the structure of liquid indium. *Phys. Rev.*, **149**, 122-30.
34. Longuet-Higgins, H.C. and Pople, J.A. (1956). Transport properties of a dense fluid of hard spheres. *J. Chem. Phys.*, **25**, 884-9.
35. Born, M. and Green, H.S. (1947). A general kinetic theory of liquids III. Dynamical properties. *Proc. Roy. Soc.*, **A190**, 455-74.
36. Chapman, T.W. (1966). The viscosity of liquid metals. *AICHE*, **12**, 395-400.
37. Chhabra, R.P. (1989). Application of Hildebrands fluidity model to molten metallic alloys. *Zeitschrift für metallkunde*, **80**, H9, 658-62.
38. Janz, G.J. (1967). *Molten salts handbook*. Academic Press, New York.
39. Turkdogan, E.T. and Bills, P.M. (1960). A critical review of viscosity of CaO-MgO-Al_2O_3-SiO_2 melts. *Am. Ceram. Soc. Bull.*, **39**, 682.
40. Kozakevitch, P. (1961). In *Physical chemistry of process metallurgy* (ed.

G.R. St. Pierre), p. 97. Wiley Interscience, New York.
41. Fowler, R.H. (1937). A tentative statistical theory of Macleods equation for surface tension and the parachor. *Proc. Roy. Soc.*, **A159**, 229–46.
42. Nogi, K., McLean, A., and Miller, W.A. (1986). The temperature coefficient of the surface tension of pure liquid metals. *Met. Trans.*, **17B**, 163–70.
43. Yeum, K.S., Speiser, R., and Poirier, D.R. (1989). Estimation of the surface tensions of binary liquid alloys. *Met. Trans.*, **20B**, 693–703.
44. Hultgren, R.R. (1967). Interatomic forces and chemical bonding energies. In *An atomistic approach to the nature and properties of materials.* (ed. J.A. Pask), Chapter 3. Wiley, New York.
45. Berry, R.S., Rice, S.A., and Ross, J. (1980). *Physical chemistry.* Wiley, New York.
46. Sherman, J. (1932). Crystal energies of ionic compounds and thermochemical applications. *Chem. Rev.*, **11**, 93–170.
47. Madelung, E. (1918). Das elektrische Feld in systemen von regelmässig angeordneten punktladungen. *Physik Z.*, **19**, 524.
48. Adamson, A.W. (1982). *Physical chemistry of surfaces.* Wiley, New York.
49. Defay, R., Prigogine, I., and Bellemans, A., (1966). Surface tension and adsorption (trans. D.H. Everett). Longmans, Green and Co., London.
50. Belton, G.R. (1976). Langmuir adsortion, the Gibbs adsorption isotherm and interfacial kinetics in liquid metal systems. *Met. Trans.*, **7B**, 35–42.
51. Frisvold, F. (1990). Filtration of aluminium. Theory, mechanisms, and experiments. Dr. Ing. Thesis, The Norwegian Institute of Technology, Trondheim, Norway.
52. Sahao, P., Debroy, T., and McNallan, M.J. (1988). Surface tension of binary metal-surface active solute systems under conditions relevant to welding metallurgy. *Met. Trans.*, **19B**, 483–90.
53. Lide, D.R. (1990–1991). *CRC handbook of chemistry and physics*, (71th edn). CRC Press, Ann Arbor, Boston.
54. Darken, L.S., Gurry, R.W., and Bever, M.B. (1953). *Physical chemistry of metals.* McGraw-Hill, New York.
55. Elliott, J.F. (1981). *Physical chemistry of liquid metal solutions.* Metallurgical Treatises (ed. J.K. Tien and J.F. Elliott). Metallurgical Soc. of AIME, Warrendale, Pa.
56. Ogino, K., Hara, S., Miwa, T., and Kimoto, S. (1984). The effect of oxygen content in molten iron on the interfacial tension between molten iron and slag. *Trans. ISIJ*, **24**, 522–31.
57. Turnbull, D. (1950). Formation of crystal nuclei in liquid metals. *J. Appl. Phys.*, **21**, 1022–8.
58. Turnbull, D. (1950). Correlation of liquid-solid interfacial energies calculated from supercooling of small droplets. *J. Chem. Phys.*, **18**, 769.
59. Nishi, Y., Mikagi, K., and Igarashi, A. (1988). Solid-liquid interfacial energy of non-metallic bonded elements. *J. of Materials Sci.*, **23**, 400–4.
60. Nishi, Y., Igarashi, A., Kubo, Y., Ninomiya, N., and Mikagi, K. (1988). Solid-liquid interfacial energy of Pd-16.5 at.% Si glasses. *Materials Sci. and Engng.*, **97**, 199–201.
61. Geiger, G.H. and Poirier, D.R. (1973). *Transport phenomena in metallurgy.* Addison-Wesley, Reading, Mass.
62. Lamb, H. (1932). *Hydrodynamics*, (6th edn). Dover, New York.
63. Smithells, C.J. and Brandes, E.A. (1983). *Smithells metals reference book* (6th edn). Butterworth, London.

64. Sigworth, G.K., and Engh, T.A. (1982). Chemical and kinetic factors related to hydrogen removal from aluminium. *Met. Trans.*, **13B**, 447–60.
65. Viswanath, D.S. and Mathur, B.C., (1972). Thermal conductivity of liquid metals and alloys. *Met. Trans.*, **3**, 1769–72.
66. Ho, C.Y., Powell, R.W., and Liley, P.E. (1974). *Thermal conductivity of elements: a comprehensive review*, Vol 3, Sup. No. 1. American Chemical Society and the American Institute of Physics for the National Bureau of Standards.
67. Meaden, G.T. (1965). *Electrical resistance of metals*. Plenum, New York.
68. Kittel, C. (1976). *Solid state physics* (6th edn). Wiley, New York.
69. Aylward, G.H. and Findlay, T.J.V. (1974). *SI chemical data*. Wiley, Hong Kong.
70. Barin, I. and Knacke, O. (1973). *Thermodynamic properties of inorganic substances*. Springer Verlag, Berlin.
71. Barin, I., Knacke, O., and Kubaschewski, O. (1977). *Thermochemical properties of inorganic substances, supplement*. Springer Verlag, Berlin.
72. Baker, E.H. (1964). The barium oxide-carbon dioxide system in the pressure range 0.01–450 atmospheres. *J. Chem. Soc.*, Part I, 699–704.
73. Schurmann, E. and Schmidt, R. (1975). Vapor pressure equations and caloric variables of pure liquid and solid (β) calcium. *Arch. für Eisenhüttenwesen*, **46**, 773–5.
74. Kubaschewski, O. (1972). The thermodynamic properties of double oxides. *High temperatures—High pressures*, **4**, 1–12.
75. Hills, A.W.D. (1967). Equilibrium decomposition pressure of calcium carbonate between 700 and 900°C. *Inst. Min. Metall Trans.*, Sect. C76, 241–5.
76. Turkdogan, E.T., Rice, B.B., and Vinters, J.V. (1974). Sulfide and sulfate solid solubility in lime. *Metall. Trans.*, **5**, 1527–35.
77. Gschneider, K.A. and Kippenhan, N. (1971). Thermochemistry of the rare earth carbides, nitrides and sulfides for steelmaking. IS-RIC-5, Rare Earth Inf. Cent., Iowa State Univ., Ames.
78. O'Dell, K.D. and Hensley, E.B. (1972). The decomposition pressure, congruent melting point, and electrical resistivity of cerium nitride. *J. Phys. Chem. Solids*, **33(2)**, 443–9.
79. Baker, F.B. and Holley, C.E. (1968). Enthalpy of formation of cerium sesquioxide. *J. Chem. Eng. Data*, **13**, 405–8.
80. Fruehan, R.J. (1979). The free energy of formation of Ce_2O_2S and the non-stoichiometry of cerium oxides. *Metall. Trans.*, **B10**, 143–8.
81. Rosenqvist, T. (1954). A thermodynamic study of the iron, cobalt and nickel sulphides. *J. Iron and Steel Inst., London*, **176**, 37–57.
82. Schwerdtfeger, K. (1967). The nitriding of chromium in N_2–H_2 gas mixtures at elevated temperatures. *Trans Metall. Soc. AIME*, **239**, 1432–8.
83. Mazandarany, F.N. and Pehlke, R.D. (1974). Standard free energy of formation of Cr_2O_3. *J. Electrochemical Soc.*, **121**, 711–4.
84. Toker, N. (1977). Ph.D. Thesis, Pennsylvania State University.
85. Hager, J.P. and Elliott, J.F. (1967). The free energy of formation of $\overline{CrS}$, $\overline{MoS}_2$ and $\overline{W}S_2$. *Trans. Metall. Soc. AIME*, **239**, 513–28.
86. Jacob, K.T., Rao, D.B., and Nelson, H.G. (1979). Stability of chromium (III) sulfate in atmospheres containing oxygen and sulfur. *Metall. Trans.*, **A10**, 327–31.
87. Kellogg, H.H. (1969). Thermochemical properties of the system Cu-S at elevated temperatures. *Can. Metall. Q.*, **8**, 3–23.
88. King, E.G., Mah, A.D., and Pankratz, L.B. (1973). *The metallurgy of copper*. INCRA Monograph II, US Bureau of Mines, Washington, DC.

89. Pemsler, J.P. and Wagner, C. (1975). Thermodynamic investigations on chalcopyrite. *Metall. Trans.*, **B6**, 311–20.
90. Kellogg, H.H. (1964). A critical review of sulfation equilibria. *Trans. Metall. Soc. AIME*, **230**, 1622–34.
91. Chipman, J. (1972). Thermodynamics and phase diagram of the Fe–C system. *Trans. Soc. AIME*, **3**, 55–64.
92. Hillert, M. and Jarl, M. (1975). A thermodynamic analysis of the iron-nitrogen system. *Metall. Trans.*, **A6**, 553–9.
93. Hillert, M. and Staffanson, L.-I. (1975). An analysis of the phase equilibria in the Fe-FeS system. *Metall. Trans.*, **B6**, 37–41.
94. Jacob, K.T. and Alcock, C.B. (1975). The oxygen potential of the systems Fe-$FeCr_2O_4$ + Cr_2O_3 and Fe + FeV_2O_4 + V_2O_3 in the temperature range 760-1600°C. *Metall. Trans.*, **B6**, 215–21.
95. Skeaff, J.M. and Espelund, A.W. (1973). An E.M.F. method for sulphate-oxide equilibria results for the Mg, Mn, Fe, Ni, Cu and Zn systems. *Can. Metall. Q.*, **12**, 445–58.
96. Alcock, C.B. and Zador, S. (1967). Thermodynamic study of the manganese/manganese-oxide system by the use of solid oxide electrolytes. *Electrochem. Acta*, **12**, 673–7.
97. Schwerdtfeger, K. and Muan, A. (1967). Equilibria in the system Fe–Mn–O involving '(Fe, Mn) O' and $(Fe, Mn)_3O_4$ solid solutions. *Trans. Metall. Soc. AIME*, **239**, 1114–9.
98. Ingraham, T.R. (1966). Thermodynamics of the Mn–S–O system between 1000 K and 1250 K. *Can. Metall. Q.*, **5**, 109–22.
99. Turkdogan, E.T., Olsson, R.G., and Vinters, J.V. (1977). Sulfate and sulfide reactions in the Mn–S–O system. *Metall. Trans.*, **B8**, 59–65.
100. Kellogg, H.H. (1969). Thermodynamic properties of the oxides of copper and nickel. *J. Chem. Eng. Data*, **14**, 41–4.
101. Pehlke, R.D. and Elliott, J.F. (1959). High-temperature thermodynamics of the silicon, nitrogen, silicon-nitride system. *Trans. Metall. Soc. AIME*, **215**, 781–94.
102. Rosenqvist, T. and Tungesvik, K. (1971). Thermodynamics of silicon sulphides and zinc sulphide. *Trans. Faraday Soc.*, **67**, 2945–51.
103. Baker, E.M. (1963). The strontium oxide-carbon dioxide system in the pressure range 0.3-400 astmospheres. *J. Chem. Soc.*, Part I, 339–46.
104. Ackermann, R.J. and Rauh, E.G. (1974). Thermodynamic properties of ZrO(g) and HfO(g). *J. Chem. Phys.*, **60**, 2266–71.
105. Gupta, D.K. and Seigle, L.L. (1975). Free energies of formation of WC and W_2C, and the thermodynamic properties of carbon in solid tungsten. *Metall. Trans.*, **A6**, 1939–44.
106. Chang, Y.A. and Ahmad, N. (1982). *Thermodynamic data on metal carbonates and related oxides.* AIME, Warrendale, Pa.
107. Rosenqvist, T. (1988). *Principles of extractive metallurgy*, (2nd edn). McGraw-Hill, New York.
108. Flynn, H., Morris, A.E., and Carter, D. (1986). An iterative gas-phase removal version of SOLGASMIX. Proceedings of the 25th CIM conference of metallurgists, TMS-CIM, Toronto, Canada.
109. Tipp, J.-E. (1984). The development of a thermochemical database and its application in metallurgical process modelling. Teknisk doktor Thesis, Institutionen för Teoretisk Metallurgi, KTH, Stockholm, Sweden.
110. Bale, C.W. and Eriksson, G. (1990). Metallurgical thermochemical databases—a review. *Canadian Metall. Q.*, **29**, 105–32.

111. Sundman, B. (1974) *Thermo-calc system*. Division of Physical Metallurgy, KTH, Stockholm, Sweden.
112. Iguchi, Y. (1990) The thermochemistry of ferrous melts. The Elliott Symposium, June 10–13, Cambridge, Mass., pp 129–47. AIME, Warrendale, Pa.
113. JANAF Thermochemical Tables (2nd edn) (1971) (ed. D. R. Stull and H. Prophet) US. Department of Commerce, National Bureau of Standards, Washington, DC.
114. Barin, I. (1989). *Thermochemical data of pure substances*, Parts I and II. VCH Verlagsgesellschaft, Weinheim, Germany.

3

PROPERTIES OF THE VARIOUS METALS, IMPURITY CONTENT, AND REFINING

3.1 INTRODUCTION

Partly as a background and a framework for the other chapters a very brief and introductory presentation is given in this chapter of the various metals, their physical properties, how they are extracted, and how this affects the content of impurities and inclusions. The aim has certainly not been to produce a handbook but rather to outline the problems, point out properties of interest to workers in refining, and to encourage a general systematic approach that applies to any metal. The review is based on the periodic table of the elements (see Table 2.13). Metals are included because of their commercial importance or, as in the case of the alkali metals, because they can be used to illustrate some important problems or principles in refining. When an element is not dealt with, it is hoped that the position in the periodic table will give an indication of the impurities and problems that will be encountered in purification.

3.2 ALKALI METALS

The alkali metals are the elements in group 1a of the periodic table (see Table 2.13 and 3.1). Metals are ordinarily characterized by their high thermal and electrical conductivity and by their lustre, ductility, and malleability. The alkali metals exhibit these properties to a high degree. However, they bear little resemblance to the more familiar metals such as iron and copper. They are silver-white in colour and soft enough to cut with a knife. The alkali metals are the most electropositive (the least electronegative) of the elements and they react with a great variety of non-metals. All of the alkali metals react with oxygen. The rate of reaction with oxygen, or with air, depends on whether the metals are in the solid or liquid state, as well as upon the degree of mixing with oxygen or air. In the liquid state alkali metals can ignite in air. The corrosion

TABLE 3.1 *Physical and themodynamic properties of the alkali metals*[68,2] *(see also Appendix 1)*

	Li	Na	K	Rb	Cs
Melting point (°C)	181	97.8	63.2	39.5	28.5
Boiling point (°C)	1347	897	766	697	675
Density at 20°C (kg/m^3)	534	970	862	1530	1870
Heat of fusion (joule/mol)	3000	2600	2334	2190	2090

of solid sodium by air is accelerated by the presence of small amounts of impurities in the sodium. Lithium also reacts with nitrogen to form a nitride, Li_3N, and with hydrogen and with carbon to form LiH, Li_2C_2. The alkali metals burn in an atmosphere of the various halogens to form the corresponding halides. The reactions are highly exothermic. The alkali metals react with non-metals in groups 5a and 6a such as phosphorus and sulphur. Further details can be found in the *Handbook of thermodynamic and transport properties of alkali metals.*[1]

The reactivity of the alkali metals may be described in terms of their electronegativity. They have a valence of one, that is, they have only a single electron in the outermost shell of their atoms. In passing from lithium to caesium, this single electron tends to be less strongly held. This can be understood from Table 2.13 which shows that the electronegativity of Li is 0.98, and that of Cs is 0.79. The mobility of the electron gives a high electrical and thermal conductivity to the alkali metals (see Table 2.14). Thus Na is used as a heat exchange fluid in atomic reactors. Generally, the energy necessary to remove the outermost electron from the atoms of an element, the ionization energy, decreases in the periodic table toward the left and downwards in each vertical column, with the result that the most easily ionizable element is francium below caesium, followed closely by caesium. The electronegativity scale (see Table 2.13) compares the ability of the atoms of the various elements to attract electrons to themselves. In the periodic table the electronegativities range from 0.79 fro caesium, the least electronegative of the elements, to 3.98 for fluorine, the most electronegative. Metals are ordinarily considered to be those elements having values less than 2.0 on the electronegativity scale. The electronegativity of the alkali metals ranges from 0.8 to 1.0.

The relatively simple electronic structure of the atoms (a single electron in the outer shell) results in a considerable degree of regularity and predictability of their physical and chemical properties. Bonds of the alkali metals with all non-metals are essentially ionic resulting from attractive forces between the excess positive charges on the alkali ions and the corresponding negative charges on the nonmental ions. Table 3.1 summarizes the important physical and thermodynamic properties of the alkali metals.

The closest distance between atoms, a characteristic property of crystals, is large and increases with increasing atomic weight of the atoms. As is seen from the table lithium, sodium and potassium are lighter than water. The melting points of the alkalis range from 181°C for lithium down to 28.5 for caesium. The low melting points and high reactivities seem to preclude the direct use of the alkali metals for light-weight structural applications. However, lithium is used as an alloy addition to aluminium and magnesium. Addition of lithium to aluminium significantly reduces density. Therefore, aluminium–lithium alloys offer interesting low density properties which are especially attractive for aerospace applications.[3] The problems with such alloys reflect some of the properties found in the alkali metals. The lithium additions exacerbate high temperature oxidation at all stages of melt treatment, processing and fabrication.[4] As for aluminium hydrogen contamination is primarily caused by moisture in air and in refractories. Aluminium-lithium alloys are reported to contain many times the amount of hydrogen normally observed in aluminium base alloys. To attain acceptable ductility and fracture toughness values hydrogen must be suppressed. Therefore production of Al–Li alloys requires radical improvements in melting and casting procedures.

The alkalis illustrate very well that elements with similar atomic volumes and ionization potentials mix completely in all proportions (they form solid solutions); some dissimilarity results in eutectic-type systems and further dissimilarity gives totally immiscible systems. The metallic radius increases downwards in the periodic table. The elements potassium, rubidium, and caesium are not very different in size and form complete solid solutions and mixed crystals. Sodium, which is a significantly smaller atom than potassium, tends to form eutectic systems with K, Rb, and Cs. The dissimilarity in the atomic volumes of Li and Na results in insolubility of the liquid phases. The solubility of K in Li is less than for Na in Li. With purification and separation of elements in mind insolubility is a very useful property. This point is discussed in Chapter 9. For instance, lithium metal is produced by electrolysis of lithium and potassium chloride at about 400°C. (The addition of potassium chloride allows the temperatures to be significantly lowered compared to the melting-point of LiCl, 610°C.) Since the voltage at which decomposition of LiCl takes place is lower than that of KCl, lithium is deposited in purities greater than 97 per cent. The lithium is remelted after casting and materials insoluble in the melt, such as potassium, either float or sink to the bottom of the melt pot. The remelting step reduces the potassium content to less than 100 p.p.m. by mass.

3.3 THE ALKALINE-EARTH METALS

The alkaline-earth metals are the elements in group 2a of the periodic table (see table 2.13).[5] They are extremely electropositive; that is, like the alkali metals

TABLE 3.2 *Physical and thermodynamic properties of the alkali metals*[2,67,68] *(see also Appendix 1)*

	Be	Mg	Ca	Sr	Ba	Ra
Melting point (°C)	1287	650	842	777	727	700
Boiling point (°C)	2468	1093	1500	1412	1894	1530
Density at 20°C (kg/m^3)	1850	1740	1540	2600	3510	
Heat of fusion (joule/mol)	8000	8500	8500	7400	8000	
Youngs modulus E (G Pa)	318	44.7				
Poisson's ratio	0.025	0.291				

of group 1a, their atoms easily lose electrons to become positive ions (cations). The elements are highly metallic and are good conductors of electricity. They have a grey-white lustre when freshly cut but tarnish readily in air. Beryllium is sufficiently hard to scratch glass, but barium is only slightly harder than lead. The melting and boiling-points of the group are higher than for the corresponding alkali metals, beryllium having the highest melting-point, 1287°C, and magnesium the lowest, 650°C (see Table 3.2).

3.3.1 Beryllium

Metallic beryllium is greyish and its chemical properties resemble these of aluminium.[6] Beryllium is resistant to air oxidation under normal conditions due to the formation of a thin protective layer of oxide. However, it is readily attacked by acids and alkalis. The applications of beryllium depend upon its unusual combination of physical properties; it is a very light metal with a high mechanical strength[6] and a high melting-point. The modulus of elasticity is about half as much again as that of steel. The brittleness and ductility of commercially pure beryllium differ from those of most common metals in that beryllium's ductility is limited and behaves more erratically. The mechanical properties are highly dependent on grain size and orientation.[6] Also the ductility behaviour depends on the impurity or 'alloy' content.

Much beryllium is used as a low percentage component of hard alloys, especially with copper as the main constituent.

Two industrial routes are employed at the present time to produce beryllium. Both have a beryllium halide as an intermediate step from the oxide. Considerable effort is expended hydrometallurgically in producing a pure halide. In the Brush reduction process[6] BeF_2 is reduced with Mg:

$$BeF_2 + Mg = Be + MgF_2. \tag{3.1}$$

Another attractive process is the reduction of BeF_2 or $BeCl_2$ by electrolysis.

A major problem with both processes is the high content of salts in the raw metal produced. These problems seem to be similar to those that have been experienced in magnesium production. Purification of Be is often carried out by vacuum melting. Probably due to the effect of stirring of the melt (see Chapter 5), BeO and Be_2C inclusions can be removed to the ladle walls or top surface. Dissolved hydrogen, together with other elements that have a high vapour pressure, are partly extracted by the vacuum treatment. Due to the high reactivity of the Be, all other undesirable elements must be eliminated at an earlier stage in the process. Alternatively, if very high purity is desired, re-electrolysis can be carried out.

Soluble beryllium compounds in the form of solutions, dry dust, or fumes are toxic. When inhaled they may produce acute effects similar to those caused by phosgene.

3.3.2 Magnesium

Magnesium is the lightest metal that can be commonly used for structural purposes.[8] Its density is 65 per cent that of aluminium and 22% that of iron, a property widely exploited. Its thermal and electrical conductivity and its melting point are very similar to these of aluminium. Whereas aluminium is attacked by alkalis but is resistant to most acids, magnesium is resistant to most alkalis but is readily attacked by most acids. Due to the formation of a thin but dense layer of oxide, it is stable in air and water at normal temperatures, but burns when heated in air and is attacked by steam. Magnesium is a powerful reducing agent and is used to produce metals such as titanium, hafnium, and zirconium from their compounds.

Magnesium is mainly used in the form of alloys, which usually contain 10 per cent or less of other elements, generally added to increase the strength of the metal. The most important alloys are those of aluminium and zinc. Casting, rolling, extruding, and forging operations are all employed with the alloys. Sheet, plate, or extruded profiles are produced by normal forming, joining, or machining operations. Magnesium is the easiest structural metal to machine.

Magnesium is commercially produced chiefly by the electrolysis of molten magnesium chloride and by the thermal reduction of magnesium oxide by ferrosilicon.

Impurities are brought into the magnesium from the raw materials. In fact all the elements of the periodic table are found in small or trace amounts.[7]

A continuous pick-up of soluble elements from the equipment or tools may occur if the melt is not saturated with the element. Iron and carbon especially may enter the melt in this manner.[8] Iron has a solubility in magnesium of ~ 0.03% at 700°C. Iron or iron-containing particles are found in commercially pure Mg in the range 0.005–0.003%. During the electrolysis process finely dispersed $MgCl_2$ may be formed. Treatment with salts (flux treatment) such as NaCl, $CaCl_2$, and KCl may result in further salt contamination. Traces of salts can lead to very serious corrosion problems.

Oxides, MgO, and nitrides, Mg_3N_2, are formed due to the reaction between magnesium and air. These are found to exist as thick films or flakes of loosely connected particles.

To prevent oxidation or burning of molten metal, SF_6 is used at the present time as a protective gas cover. The SF_6 blend reacts with magnesium forming a dense oxide layer that contains small amounts of MgF_2. The MgF_2 crust may enter the melt.

Hydrogen is readily soluble in liquid magnesium (see Fig. 1.1). As for other metals hydrogen solubility is reduced during solidification. However, for Mg this reduction is relatively small. As is shown in some detail in Chapter 2 for aluminium, solubility depends not only on temperature but also on alloy composition. The sources of hydrogen contamination are the same as for aluminium: mainly moisture in the atmosphere.

Ni, Co, and Cu create corrosion problems already in the part-per-million range (see Chapter 1). Since they are more noble than the solvent metals, and have low vapour pressures it is not a simple matter to remove them.

3.3.3 Calcium

Calcium reacts with water, and on heating with all the electronegative elements: oxygen, nitrogen, halogens, boron, sulphur, carbon, and phosphorous. Commercial applications depend largely on these reactions. Calcium compares favourably with sodium as a reducing agent. In Chapter 2 the use of Ca for deoxidation, desulphurization and inclusion modification of steel was treated.

Alkaline-earth metals can be sublimed or distilled to give metal of very high purity. The metallic impurities are the most easily separated although removal of the other alkaline-earth metals requires a fractional sublimation. The interstitial elements O, N, and C can be reduced to low levels by subliming or distilling under a small pressure (less than 0.01 bar) of an inert gas. Typically calcium metal purified by batch sublimation contains:[5] 45–90 p.p.m C, 20 p.p.m N, 20–80 p.p.m O, and 75 p.p.m H. Hydrogen is the most troublesome to remove and may be reintroduced by exposure to H_2O.

3.4 ALUMINIUM

Aluminium is the second element in the 3a (boron) group of the periodic table. It has a value of 1.43 on the electronegative scale, so it is not as strongly attracted to the group 7a elements, F_2, Cl_2, and so on as the group 1a and 2a elements. This property, as we shall see, makes it possible to remove the group 1a and 2a elements from aluminium. Aluminium has a density of about 2700 kg/m^3, less than one-third that of steel. The melting-point is 660°C and pure aluminium (99.996 per cent) is quite soft and weak. Aluminium is an excellent conductor of heat and electricity. Its thermal conductivity is about one-half that of copper, its electrical conductivity about two-thirds. Ductile and malleable, aluminium can be drawn into wire or rolled into thin foil. Aluminium is protected by a thin, dense, transparent film that prevents further oxidation in air (or even by concentrated nitric acid). However, aluminium is attacked by other acids and by alkalis. Due to its strong attraction to oxygen, Al is often used in the production of various metals by reduction of the oxides (aluminothermal reduction).

The strength properties of aluminium may be greatly improved by alloying additions of Mg, Zn, Si, Cu, and Mn. In the wrought or cast form the metal and its alloys are used extensively for aircraft, building materials, consumer durables, electrical conductors, chemical and food processing equipment, beverage cans, and foil of thickness down to 5 μm.

TABLE 3.3 *Inclusion sources[10] in aluminium*

Particle type	Particle source	Typical size (microns)	Typical concentration p.p.m.
Carbides	Al_4C_3 from Hall-Heroult cell	1–20	5
Borides	TiB_2 from grain refiner	1–3	10–50
	VB from Hall-Heroult cell	1	0.5
Oxides	Al + Mg oxides from melt surface	1–100	0.01–2
	Al_2O_3 from undissolved alumina in Hall-Heroult cell	10–20	0.01
	Various oxides from several alloy additions	100	0.01
	Refractories	100–500	0.01
Salts	Cryolite from	10–20	0.01
	Mg–Al chlorides from Cl_2 fluxing		
	Na–K chlorides from dross fluxing		

The aluminium from the Héroult–Hall electrolysis cell[9,10] contains about 0.01% Na, 0.01% V, and 0.5 p.p.m. H, in addition to 0.1% Fe and 0.04% Si (see Table 1.1). Sodium may lead to edge cracks during rolling, while hydrogen can give porosity during solidification. Table 3.3 reviews some of the more important inclusions found,[10] as well as their origin. Notice that various chlorides may contaminate the metal. These chlorides stem partly from the cryolite in the reduction cell and partly from the salf fluxes employed to protect the bath surfaces; the fluxes also serve to release aluminium from the dross layer (that contains 60–90% Al droplets). The term dross refers to any solid or liquid scum floating on top of a metal bath that comprises metal oxides, sulphides, halides, metallic compounds, and so on. Salts may also arise from the use of Cl_2 to remove group 1a and 2a elements from aluminium.

Figure 3.1 shows the stability of alkali and magnesium chloride salts relative to $AlCl_3$ (gas).[11] In essence, the lower the Cl_2 pressure required to form a salt, MCl, the more stable the salt is, and the easier it is to remove, $\underline{M}$, by fluxing. It can be seen that calcium chloride is the most stable, followed by sodium, lithium, magnesium, and finally aluminium chlorides. Thus, when using chlorine as a fluxing agent, calcium will be the element first removed. If chlorine is used in amounts exceeding that which is required for lithium removal, magnesium levels will start to be reduced. Finally, if the magnesium level has been reduced to its thermodynamic minimum concentration, aluminium will be consumed. Figure 3.2 shows the same type of information for

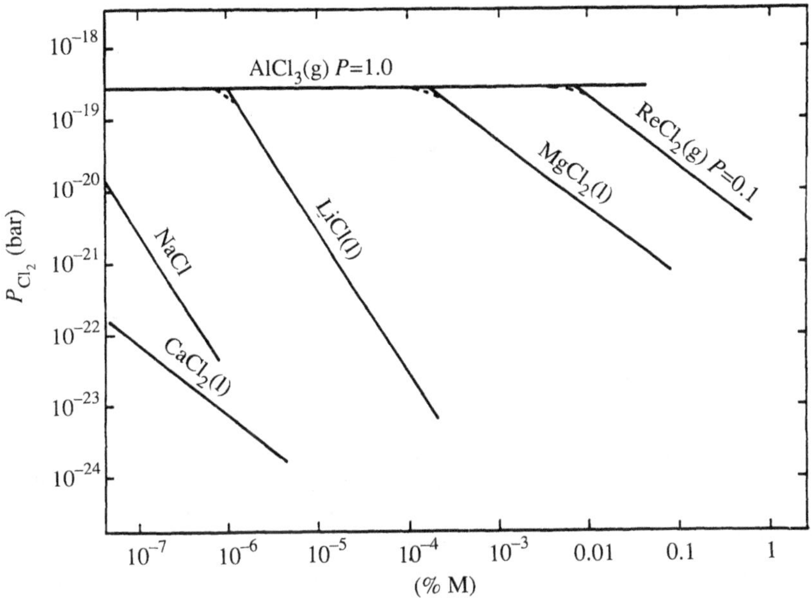

FIG. 3.1 Calculated equilibria for ternary Al–M–Cl systems at 1000 K.[11]

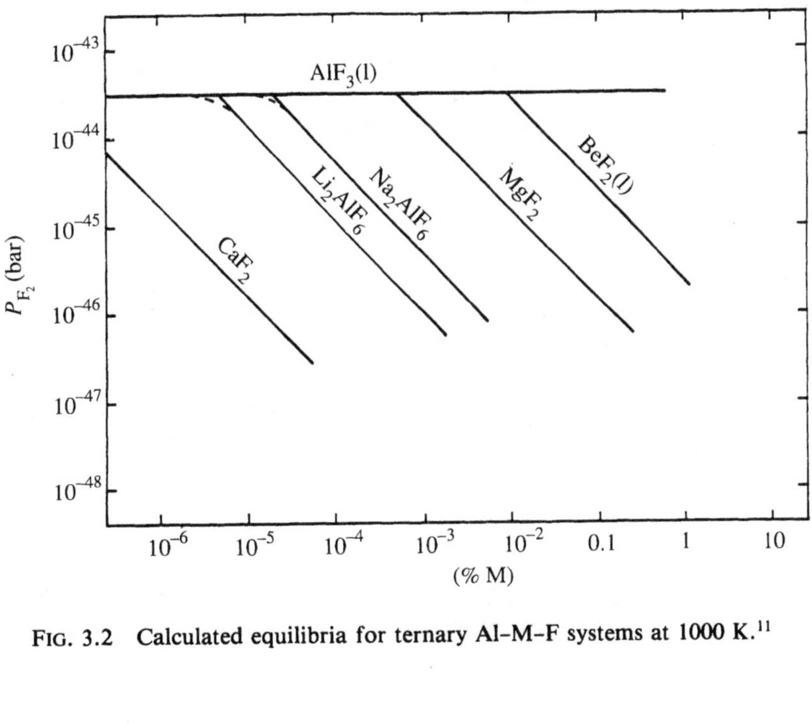

FIG. 3.2 Calculated equilibria for ternary Al–M–F systems at 1000 K.[11]

TABLE 3.4 *Approximate minimum impurity levels attainable in aluminium by reactive gases[11] (T = 1000 K)*

Element	Content remaining in p.p.m.	
	chlorine	fluorine
Be	65	85
Ca	3×10^{-5}	5.7×10^{-4}
Li	0.01	0.05
Mg	1.4	4.9
Na	9×10^{-5}	0.19

the fluoride salt stability in liquid aluminium.[11] Table 3.4 gives the minimum realizable concentrations of solute elements when fluxing with reactive gases at 1000 K. The values given for Cl_2 are the intersection between the $AlCl_3$ and the various MCl lines. Lower levels cannot be obtained, since the chloride will then be used to oxidize the parent metal. High-purity aluminium is produced by electrolytic refining using the so-called *three-layer method*. The electrolyte consists of a mixture of aluminium, sodium, and barium halides. The pure aluminium cathode floats on top of the electrolyte. The anode consists of aluminium scrap with enough copper added to make it heavier than the electrolyte. To prevent contamination a large anode-to-cathode distance (15–18 cm) is used. The electrolysis is carried out in cells which resemble production cells except that the anode is at the bottom and the cathode at the top. As in other electrolytic refining processes where a metal is first dissolved in an electrolyte and then deposited, the theoretical voltage is close to zero. However, in order to generate enough heat in the relatively small cells, energy consumption per kilogram of metal is as high as in the Hall–Héroult cell.[12]

3.5 GROUP 4A METALS

Carbon and silicon are the first two of the group 4a elements. They play a very important role in reducing oxides to their metals and as alloying elements. In Chapter 6 carbon removal (decarburization) is treated. Chapter 9 includes some material concerning silicon.

The element below silicon, germanium, is a metalloid with properties intermediate between metals and non-metals. Germanium is followed by the two metals tin and lead.

3.5.1 Germanium

Germanium is a rare element in nature. In refining germanium, the low-grade residues obtained from its ores are treated with strong hydrochloric acid and the resulting germanium tetrachloride is distilled, purified by repeated redistillation, and hydrolyzed to form germanium dioxide.[13-15] This is then reduced by hydrogen to a powdery form of the metal which is melted in an inert atmosphere and cast into ingots or billets.

For use in electronic devices, germanium requires further purification, usually by zone refining (see Chapter 9). The highly pure germanium is then melted and 'doped' by adding minute amounts of arsenic, gallium, or other elements. Finally, single crystals are generated from the melt at a carefully controlled temperature, using a seed crystal as a nucleus.

3.5.2 Tin

Tin (Sn) is a soft, silvery-white metal with a bluish tinge and known to the ancients as a constituent of bronze, an alloy with copper.[2] Tin is most widely used today for plating steel cans used as food containers, in metals for bearings, and in solder. Tin is non-toxic, ductile, malleable, and adapted to all kinds of cold working, such as rolling, spinning, and extrusion. The colour of pure tin is retained during exposure because a thin invisible film of tin (IV) oxide is formed spontaneously by reaction with oxygen in air. The low melting-point of tin (232°C) and its firm adhesion to clean surfaces of iron, steel, and copper facilitate its use as an oxidation-resistant coating material. Tin exists in two forms or allotropes: white (or beta) tin which is thermodynamically stable above 13°C, and a grey (or alpha) form which is powdery and of little use. This form is thermodynamically stable below 13°C. However, small amounts of impurities normally present in commercial grades of tin prevent the transformation from β to α-Sn.

The principal mineral of tin is cassiterite, SnO_2, found in association with Fe_3O_4, $FeWO_4$ and FeS_2 as well as sulphides of Cu, Ni, Bi, As, Sb, and Pb. Carbon and CO are used to reduce SnO_2 to Sn. However, SnO_2 and FeO are very similar in reducibility by C and CO and tin ores always contain small amounts of iron oxide. Therefore both Sn and Fe are found in molten metal in equilibrium with SnO and FeO slags. Hence, all the tin can be reduced only if all the iron is reduced. To avoid this a two- or even three-stage 'counter-flow' process is necessary. In the first stage, a fairly pure tin with a low iron content is produced, but with poor recovery, leaving about 20–25% Sn in the slag.[16,17]

The temperature is kept as low as possible to avoid excessive fuming of tin as the volatile SnO. Thus, the first stage is run at temperatures of around 1200°C, merely sufficient to melt the calcium–iron–tin silicate slag. A higher temperature (above 1400°C) is required in the second stage for two reasons. The tin–iron alloy must be molten, and a higher temperature displaces the equilibrium

$$Fe + SnO = Sn + FeO \tag{3.2}$$

to the right, that is, the partition coefficient

$$K(T) = \frac{[\%\,\mathrm{Sn}]\,(\%\mathrm{FeO})}{[\%\,\mathrm{Fe}]\,(\%\,\mathrm{SnO})} \tag{3.3}$$

increases. The tin is recovered fairly completely from the slag leaving only about 1 per cent. This slag is usually discarded. In some plants, however, Sn is fumed off from slag as $SnCl_2$, SnS, or SnO and recovered using a gas cooler, baghouse system.[18] The tin alloy produced (known as *hard head*) in the second stage, contains at least 20% Fe. This is recycled back to the first stage, utilizing the iron content as a reducing agent to replace some carbon. This batch counter-flow process, shown schematically in Fig. 3.3, was described by Agricola in *De re metallica*, and is typical of many extractive and refining processes. The first unit may be a fuel-fired shaft, reverberatory or rotary furnace, or a submerged-arc electric furnace. In the second stage a converter for slag may be used.

Figure 3.3 is easy to understand if the equilibrium given by eqn (3.2) is kept in mind. If the discarded slag in stage 2 is to be low in SnO, and high in FeO, the metal in equilibrium with this slag must contain considerable amounts of Fe. Since the slag in stage 1 is relatively high in SnO, the metal in this unit is a fairly pure tin.

The best examples of counter-flow processes are perhaps continuous distillation units. These will be discussed later in this chapter and in Chapter 4.

The remaining 1% per cent iron in tin may be removed by making use of part of the tin–iron phase diagram[19] as shown in Fig. 3.4. If tin is cooled FeSn will

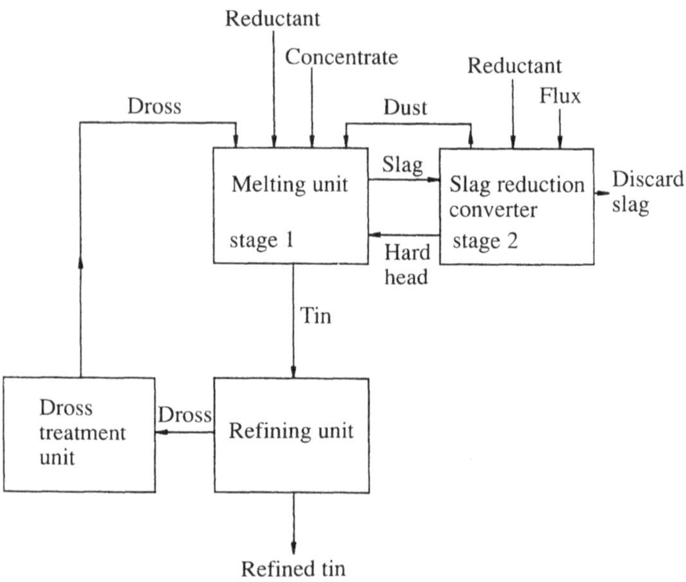

FIG. 3.3 Counter-flow process for the production of refined tin.

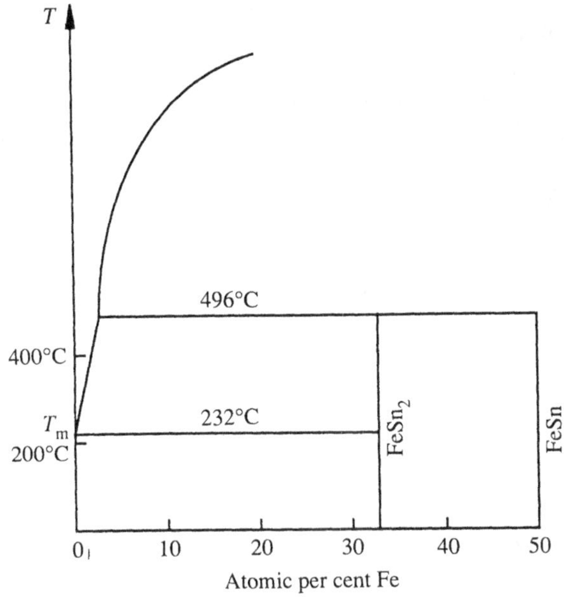

FIG. 3.4 Part of the tin–iron phase diagram.[19]

separate out and finally, between 496°C and the melting-point at 232°C, the compound $FeSn_2$ will appear as a solid phase (a dross): the FeSn + Sn reacts to $FeSn_2$. Copper, As, and Sb are removed as dross in various precipitation reactions by adding S and Al. Tin has a low vapour pressure, so Pb, Bi, As, and Sb can also be removed by vacuum distillation.[16]

Significant quantities of tin are recovered by the treatment of scrap from the manufacturing of tinplate and tin cans, and, increasingly, by the treatment of tin cans recovered from domestic waste. The tin is dissolved in hot caustic soda solution. Tin may then be recovered, for instance, by electrolysis.[20]

3.5.3 Lead

Like tin, lead (Pb) is soft and silvery white. Its melting-point is low, 327°C, and it is very malleable, ductile, and dense. When freshly cut, a dull grey coating is formed which protects the metal from further corrosion. Lead is highly durable and resistant to corrosion. Because of this general chemical resistance, considerable amounts of lead are used in roofing, as coverings for electric cables placed in the ground or under water, and as linings for water pipes and conduits. Lead has many other applications; the largest is in the manufacture of storage batteries. It is used as a constituent of various low-melting alloys such as solder, type metal, and pewter. Because lead effectively absorbs high-energy electromagnetic radiation and neutrons it is used as a protective shielding around nuclear reactors, X-ray equipment, radioactive materials, and so on.

Lead sulphide (PbS) is the major source of lead production throughout the world. Lead may be extracted by roasting the ore (heating it in air or oxygen-enriched air to remove S as SO_2) in a reverberatory, or open-hearth furnace, or in a blast furnace. The output is lead bullion, which contains dissolved

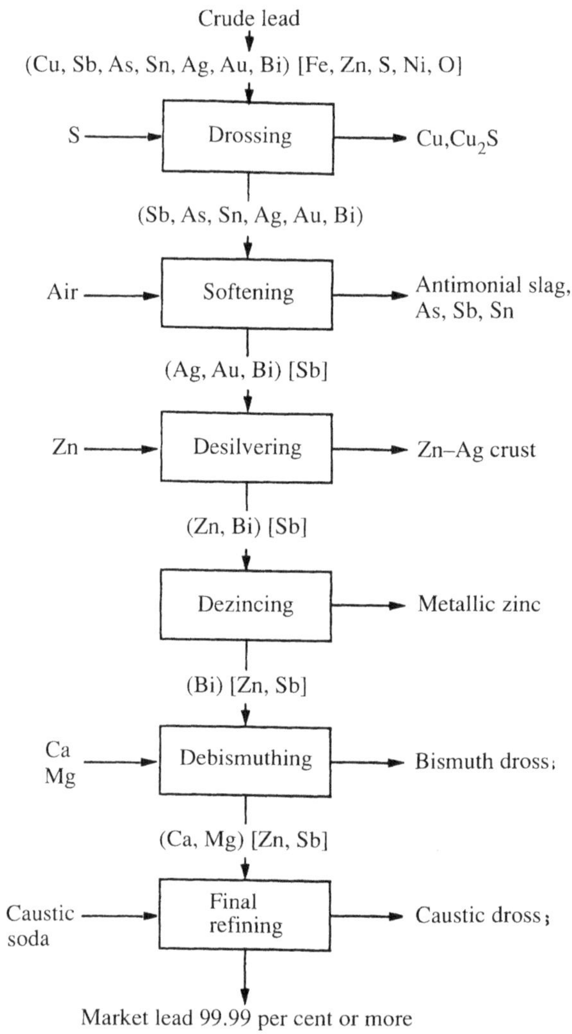

FIG. 3.5 Lead refining; round brackets signify major impurities, square brackets minor impurities.[21] Additions made are indicated to the left.

metallic and non-metallic (oxide and sulphide) impurities. Figure 3.5 shows the various steps in the refining of lead. As lead is an almost universal solvent at temperatures around 1000°C, it contains dissolved impurities not only from the concentrates but also from the coke used for reduction. These impurities are iron, zinc, copper, nickel, tin, and various group 1b and 5a elements. Usually all impurities are removed in a series of pyrometallurgical operations (normally performed batchwise, although continuous versions are also in use). Refining involves the following stages.[21] First, as many impurities as possible are brought out of solution by cooling the molten lead down to its freezing-point. In this stage termed drossing, solid particles are skimmed off. Drossing is performed in large kettles of mild steel of up to 350 tonne capacity. Impurities are brought to the surface by stirring, by the addition of small amounts of sulphur, and by agitation with submerged air lances. For instance, Cu and Cu_2S are removed in this manner. (The lead–copper phase diagram is somewhat similar to the tin–iron diagram.)

In the second stage more oxidizable impurities are removed by blowing air or oxygen through the molten lead, or by stirring in sodium nitrate. This stage is termed softening as the impurities removed (arsenic, antimony, and tin) tend to harden lead. Air oxidation removes the impurities as a slag phase.

In the third stage other impurities are precipitated from solutions by addition of specific reagents. Silver is extracted (Parkes process) by precipitation with zinc.[12] Bismuth is removed by precipitation with alkaline-earth metals (calcium and magnesium).

Finally, all remaining impurities or reagents are removed by stirring the lead with caustic soda.

The reagents introduced into the lead are expelled to the gas phase by a vacuum distillation process (zinc) or by chlorination (zinc, calcium, and magnesium).

Electrolytic refining, using a fluosilicic acid electrolyte, may replace the desilvering and debismuthing operations. An electrolytic method is also used to recover lead from scrap batteries.[22,23] Lead scrap which is largely metallic in nature may simply be melted in a steel kettle and refined if necessary to remove any impurities present.

Lead and its compounds are toxic and retained by the body, accumulating over a long period of time. The poison affects the entire body, especially the nervous system. A lengthy medical treatment with substances that extract lead from body tissues is possible.

3.6 GROUP 5A METALS; ANTIMONY AND BISMUTH

Both antimony (Sb) and bismuth (Bi) have a hard, brittle crystalline structure. Bismuth is tin-white in colour with a reddish tinge that distinguishes it from all other metals. Antimony and bismuth occur in nature often as the sulphides

Sb_2S_3 and, Bi_2S_3. Bismuth is also found as the oxide Bi_2O_3 or in metallic form. Antimony or bismuth can be obtained by heating the sulphide in air (roasting) and then reducing the oxide with carbon. Alternatively the sulphide ore can be melted with scrap iron. Then sulphur is removed by the iron as an FeS (matte) phase.

Because antimony and bismuth expand on solidifying (a rare characteristic that they share with water), they are valuable ingredients in casting alloys. The expansion forces the alloy to fill the small crevices of casting moulds.

Combined with tin, lead, and copper, antimony forms antifriction alloys called Babbitt's metal that are used as components of machine bearings.

Bismuth has a melting-point of 271 °C and alloys that contain bismuth may have very low melting-points. For this reason they are employed in special solders as well as in automatic sprinkler heads, fuses, and fire detection devices.

Ingesting or inhaling certain compounds of antimony may have harmful effects on body tissues and functions. However, the industrial use of antimony has not appeared to be associated with serious occupational poisoning.

3.7 GROUP 1B METALS

Group 1b is composed of the metals copper, silver, and gold. The characteristics of these metals are that they are chemically inert, have a high electrical conductivity, and are malleable and ductile. Thus a gram of gold (Au) can be hammered out to more than one square metre. Some of the physical properties of the three metals are given in Table 3.5.

Gold and silver are widely used in jewellery and ornaments. All three metals have played an important role in coinage. An important industrial use of the three metals is in the electric and electronic industry for plating contacts, terminals, printed circuits, and semiconductor systems.

Like many other elements gold is widespread in small amounts in all igneous rocks. It occurs mostly in the free state, but can combine with tellurium and possibly selenium and bismuth. Two types of deposits containing significant amounts of gold are known: hydrothermal veins, where it is associated with quartz and pyrite, and placer deposits.

TABLE 3.5 *Physical properties of the group 1b metals*[2,67,68] *(see also Appendix 1)*

	Cu	Ag	Au
Melting point (°C)	1085	961	1064
Boiling point (°C)	2570	2 160	2854
Density at 20°C (kg/m^3)	8960	10 500	19 300
Youngs modulus E (G Pa)	129.8	82.7	78.5
Poisson's ratio	0.343	0.367	0.42

Unlike gold, silver is present in many naturally occurring minerals. Deposits of free silver are also commercially important. Because the majority of silver ores contain the important metals lead, copper, or zinc or a combination of the three, the extractive metallurgy of silver is essentially that of these three metals, the silver being removed as a by-product. Depending on the ores, pure silver is extracted from the crude by-product either by electrolysis, smelting, or hydrometallurgical methods.

Silver occurring in copper is first extracted with the copper and then recovered as a silver–gold anode mud after electrolysis. It has already been mentioned that silver and gold occurring with lead ores are extracted with a zinc addition from the lead bullion as a dross that also contains some lead. First the zinc is distilled off by heating in a retort to above the boiling-point of zinc. Then the remaining lead–silver alloy is oxidized in a hearth furnace (cupellation) until all the lead in it has been oxidized. The remaining noble metal is refined by electrolysis to produce pure silver and to recover its gold content.

Silver occurring with zinc ores which also may contain lead (see Section 3.8.1 on zinc) is recovered with the lead bullion.

For ores that contain only silver and gold as the valuable constituent the cyanide process[24] is usually employed. The ore is pulverized and added to the sodium cyanide solution. The silver and gold form water-soluble Na–Ag–Au complexes. The resulting mixture is filtered and the solids are discarded. Then the silver and gold is precipitated with the less noble metal zinc which thereby goes into solution. (This process where metal ions are precipitated out of solution by a less noble metal is important in hydrometallurgy and is called *cemention*.) The precipitated Ag and Au is filtered off and cast into bullion bars.

The two most important methods for separating silver and gold in bullion are electrolysis and parting. In electrolysis, a silver nitrate water solution is used, with the bullion as the anode. Pure silver is deposited at the cathode. Gold is recovered from the anode mud.

In the parting method the bullion is placed in a bath of hot concentrated sulphuric acid or strong nitric acid. Then the silver dissolves and gold is recovered from the residues.

3.7.1 Copper

Until the 1960s copper remained the metal second in use and importance to iron. Now cheaper and more plentiful aluminium has moved into second place in world production. Most of the copper produced is used by the electrical industries. Most of the remainder is combined with other metals to form alloys. Important series of alloys in which copper is the chief constituent are brasses (Cu–Zn), bronzes (Cu–Sn), and nickel silvers (Cu–Zn–Ni, no Ag). There are many useful alloys of Cu and Ni, including Monel. The two metals are completely miscible. Copper also forms an important series of alloys with aluminium, called aluminium bronzes.

Copper is one of the most ductile metals, not especially strong or hard. Strength and hardness are appreciably increased by cold working due to the formation of crystals of the same face-centred cubic (fcc) structure that is present in the softer annealed copper. Common gases, such as O_2, N_2, CO_2, H_2O, and SO_2 are soluble in molten copper and greatly affect the mechanical and electrical properties of the solidified metal. As is seen from Table 2.13, copper has a high positive reduction potential in water; it is therefore not soluble in acids with the evolution of hydrogen. Copper resists the atmosphere and sea water. Exposure for long periods to air, however, results in the formation of a thin protective coating of green, basic copper carbonate (patina).

As for Pb, Cu is mainly obtained directly from the sulphides (chalcocite, chalcopyrite, bornite) in the pyrometallurgical extraction process. The first step involves the physical separation of sulphides from gangue. The copper concentrate is charged into the reverberatory, flash, or blast furnace with fluxes,[25,26] and the charge is melted at 1500–1650 K. The copper and iron sulphides Cu_2S and FeS, which are stable in the absence of oxygen, are insoluble in the slag formed from the gangue and flux. Due to their greater specific gravity, they form a separate lower liquid layer, called matte, which contains 25–70% Cu. In the second stage, molten matte from the smelting furnace is transferred to a converter where air or oxygen-enriched air is blown by way of submerged tuyeres in the side walls. Due to its greater affinity for oxygen, the iron is oxidized before the copper. The FeO is fluxed with SiO_2 and tapped from the converter. In a second oxygen blowing period the remaining sulphur is oxidized until the first appearance of Cu_2O. The metal produced at this point, called blister copper, contains about 98% Cu. The name derives from the appearance of blisters on the upper surface of solidifying slabs, due to the deformation of the thin solidified crust during formation of SO_2 by the reaction between $\underline{O}$ and $\underline{S}$. Blister copper contains about 0.6% O and 0.01–0.03% S. In addition, it contains metallic and non-metallic impurities such as Ag, Au, Se, Te, and As that have not reacted with oxygen. Metallic impurities less noble than Cu, such as Zn and Pb, were essentially removed in the first blowing period. In problem 3.1, standard free energy data from Appendix 1 is used to calculate the activity of oxygen in equilibrium with Cu_2O at 1100°C and the partial pressure of SO_2 in equilibrium with this oxygen and 0.03% S slightly above the solidification temperature of 1358 K.

Blister copper is clearly too impure to cast into slabs for working or even to cast into anodes for purification by electrolysis. The content of oxygen, sulphur, metallic impurities, and group 6a and 5a impurities must be reduced further. This can be performed by further oxidation and/or inert gas purging. The blister copper is fed into cylindrical furnaces capable of holding up to 300 tonnes and air is blown through until there is a separate layer of Cu_2O with other oxides in solution. At this stage the metal will contain about 0.001% $\underline{S}$ and 1% $\underline{O}$. Especially at the lower impurity levels, removal of impurities will be limited by mass transfer (diffusion) of the impurity element to the melt–gas interface. Therefore it is important that this contact area is large. As will be shown in Chapter 4, the contact area is inversely proportional to the bubble size and proportional to the flow of gas into the melt. Now the melt already contains about 1% O̲ so that oxidizing refining can be accomplished simply by blowing in an inert gas such as nitrogen. Figure 3.6 shows experimental results in the laboratory on the removal of S̲ (and (O̲)) by the bottom injection of N_2.[27] The experiment is the object of analysis in problem 3.2. By injection of lime and soda, removal of As and Sb can be speeded up.[28]

When refining using oxidizing gas, inert gas, or powder injection is completed and the top slag has been removed, a reducing gas such as propane or natural gas is employed in order to deoxidize. The progress of oxygen removal can be followed by an oxygen probe, or by casting small samples and noting the development of a flat upper surface. The product can be cast into anodes, or if the precious metal content is low and certain impurities are absent, into slabs for rolling and other working processes. The final oxygen content is 0.05–0.15%.

Tough-pitch copper contains 0.04%–0.02% oxygen by mass. At these oxygen concentrations the formation of internal H_2O pores can compensate for the 4.7 per cent shrinkage at the melting-point during solidification (compare problem 2.7), so that the surface remains level and flat.

Electrorefining[29] is carried out by the electrochemical solution of anodes that are about 350 kg, 1 m square and 5 cm thick. They hang from connecting lugs in a circulating acid copper sulphate solution, contain about 50 kg/m^3

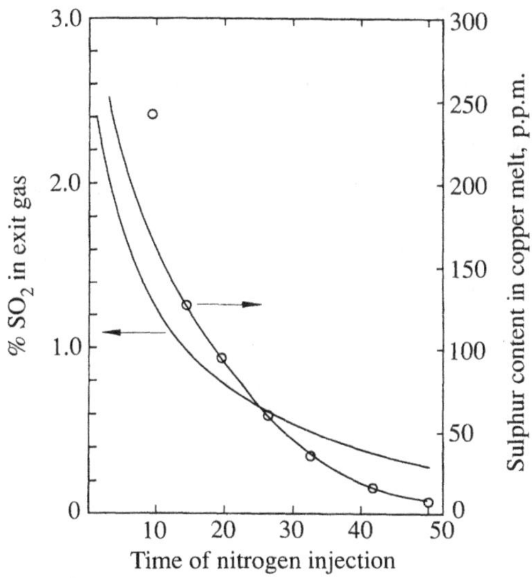

FIG. 3.6 Simultaneous removal of sulphur and oxygen from a laboratory copper melt by gas purging with N_2.[27] Reprinted with permission from JOM (formerly Journal of Metals).

of Cu and 150 kg/m^3 of H_2SO_4. Copper is deposited on to cathodes of pure copper 1 m square and 2–5 mm thick, weighing 5–15 kg when new. The behaviour of the impurities in the anodes depends on their solubility in the electrolyte and the solubility of some of their compounds as well as their reduction potential (see Table 2.13). Gold and platina are insoluble and fall to the slime. Any Ag that dissolves in the electrolyte is precipitated as AgCl by reaction with NaCl added for this purpose. Sulphur, Se, and Te enter the system as Cu_2S, Cu_2Se, Cu_2Te, Ag_2S, Ag_2Se, and Ag_2Te inclusions in the anodes. All are insoluble and normally pass to the slimes. Lead and tin form insoluble $PbSO_4$ and $Sn(OH)_2SO_4$. Nickel, Co, and Fe dissolve but are so electronegative that they do not deposit at the cathode. Copper, Ni, Co, and Fe are prevented from accumulating beyond prescribed limits by bleeding off part of the circulating electrolyte to a purification circuit. When the cathodes are taken out they are washed free of electrolyte, melted down, and cast into slabs for rolling or extrusion.

Electrorefined Cu is the purest form available. The Cu content is given as 99.99 per cent. A series of examples give the following figures[29] for the impurity content in p.p.m. by mass: Ag 4–10, As 0.4–4, Sb 0.5–3.5, Bi 0–0.3, Fe 0.5–2.5, Ni 0.7–1, Pb 0.6–1.3, Te 0.1–0.4, Zn 0.1–0.3, and S 4–9 p.p.m.

3.8 GROUP 2B METALS; THE PROBLEM OF HIGH VAPOUR PRESSURES

Physical and thermodynamic properties of the elements in the group 2b:zinc (Zn), cadmium (Cd), and mercury (Hg), also called the zinc group, are summarized in Table 3.6. None of the elements are found in nature in the free state. They have comparatively low melting and boiling-points and correspondingly high vapour pressures. The boiling-point of mercury—the only common metal that is liquid at room temperature—is lower than that of any other metal.

The pyrometallurgical extraction and refining of metals with high vapour pressures present special problems in pyrometallurgy. It will be seen that these problems can be overcome in the case of mercury. For the production of highly pure zinc it is in general more economic to use an electrolytic process.

TABLE 3.6 *Physical properties of the group 2b metals*[2,67,68]

	Zn	Cd	Hg
Melting point (°C)	420	321	−38.9
Boiling point (°C)	907	766	356
Density at 20°C (kg/m^3	7140	8640	13 546
Youngs modulus *E* (G Pa)	104.5	62.6	
Poisson's ratio	0.249	0.30	

3.8.1 Zinc

The major uses of Zn are as a protective covering for iron and steel (galvanizing). When iron and zinc together are exposed to a corrosive medium, they constitute an electrolytic cell, and the zinc is attacked (oxidized to the Zn^{2+} ion) preferentially and sacrificially since its reduction potential (see Table 2.13) of −0.76 is lower than that of iron, which is −0.41. This reaction, coupled with the much greater corrosion resistance of zinc under atmospheric conditions is the basis for galvanizing.

Freshly cast zinc has a bluish-silver surface, but slowly oxidizes in air to form a greyish protective oxide film. Highly pure zinc (99.99 per cent) is ductile; the so-called prime western grade (99.98 per cent) is brittle when cold but above 150°C can be rolled into sheets that remain flexible at lower temperatures.

In addition to galvanizing, zinc is a component in brasses and alloys for die casting. Zinc also serves as the cathode (outside can) in one common type of electric dry cell.

The chief mineral for zinc is the sulphide sphalerite which, together with its oxidation products, constitute nearly all the world's zinc ore. Metallic zinc appeared much later in history than the other common metals Cu, Pb, Sn, and Fe. The reason is that ZnO cannot be reduced to Zn by carbon until temperatures are reached well beyond the relatively low boiling-point of the metal (907°C). In furnaces developed to smelt other metals, zinc would go off as fumes, then oxidize in the air and be lost as a fine ZnO powder. Brass, an alloy of Zn and Cu, however, was produced by the early Romans. They heated a mixture of zinc oxide and charcoal covered with lumps of metallic copper. These facts are easy to explain by making use of the thermodynamic data given in Appendix 1 to calculate the Gibbs free energy for the equilibrium:

$$\mathrm{ZnO + C = Zn + CO.} \tag{3.4}$$

For

$$\mathrm{ZnO} = \tfrac{1}{2}\mathrm{O_2} + \mathrm{Zn(g)} \qquad \Delta G^0 = 460\,240 - 190.4T;$$

also for

$$\mathrm{C} + \tfrac{1}{2}\mathrm{O_2} = \mathrm{CO} \qquad \Delta G^0 = -114\,400 - 85.77\,T.$$

These equations give in sum

$$\mathrm{ZnO} + C = Zn(g) + \mathrm{CO} \qquad \Delta G^0 = 345\,840 - 276.17\,T.$$

Then

$$\frac{p_{\mathrm{Zn}}p_{\mathrm{CO}}}{a_{\mathrm{ZnO}}a_{\mathrm{C}}} = \exp\left(-\frac{\Delta G^0}{RT}\right) = \exp\left(-\frac{41\,595}{T} + 33.22\right). \tag{3.5}$$

The lowest temperature that will give Zn (vapour) at a total pressure of one atmosphere is found using the stoichiometry in eqn (3.4) by setting

$p_{Zn} = p_{CO} = \frac{1}{2}$. Also $a_{ZnO} = 1$ and $a_C = 1$. This gives a temperature of 929°C. Thus by the reduction zinc is formed as an unsaturated vapour. If this vapour composed of Zn and CO is cooled, eqn (3.4) reverses.

If the reduction of ZnO is carried out in the presence of another metal which dissolves Zn, such as copper, the reduction temperature can be lower that 929°C for one atmosphere total pressure. For instance, at 907°C and $p_{CO} = 1$ bar, eqn (3.5) gives $p_{Zn} = 0.13$ bar. From the activity diagram of Cu–Zn alloys[30] it is found that the Zn content is about 40 mass% in equilibrium with $p_{Zn} = 0.13$ bar. If the Zn can be absorbed by the copper, losses as zinc vapour can be reduced. This is the basis for the direct production of brass from mixed Cu and Zn ores.

Production of Zn is based on being able to condense Zn vapour at temperatures much lower than the reduction temperature. During condensation Zn and CO must be separated in a very short time so that reaction (3.4) does not proceed back to ZnO. A problem is that in addition to CO, the gas will always contain a small amount of CO_2(0.1–1%).[12] The CO_2 will, on cooling, invariably react with an equivalent amount of Zn vapour:

$$Zn(g) + CO_2 = ZnO + CO. \tag{3.6}$$

This gives a so-called *blue powder* which is Zn droplets covered with oxide.

In the horizontal and vertical retort processes[20] the condenser is a molten zinc bath operated at about 500°C. In the Imperial smelting blast furnace, the hot gas mixture is withdrawn at 1000°C and shock-cooled. In the condenser the gas is met by a shower of liquid lead which causes the zinc to condense and dissolve in the lead without appreciable oxidation.

The raw zinc from the retort and blast furnace processes contains as major impurities, lead, cadmium, and iron. Of these Cd is more volatile than Zn, Pb and Fe less volatile than Zn. The zinc can be purified by fractional distillation as shown in Fig. 3.7 (see also Section 4.12). The columns are built of superposed trays of silicon carbide. The lower half of the columns are heated externally through the SiC walls which have a high thermal conductivity. The feed Zn is introduced in the middle of the first column. During the descent down the column, zinc and cadmium evaporate, while the lead and iron remain in the downflow. In the upper part further condensation of lead and iron vapour take place. In the second column Zn tends to condense in the downflow and Cd vapour leaves at the top.

The refining of zinc by distillation is a heat-consuming process, but it may be economic in localities with cheap fuel. On the whole, however, the demand for high-purity zinc is met by the electrolytic process.

Zinc is the metal with the lowest reduction potential (see Table 2.13) capable of being electrodeposited from an aqueous solution. Any other depositable metal will therefore be deposited preferentially if it is in the electrolyte and will tend to cause solution of cathode zinc by galvanic action. The other metals and zinc change places at the cathode. Also, at first sight it appears impossible to

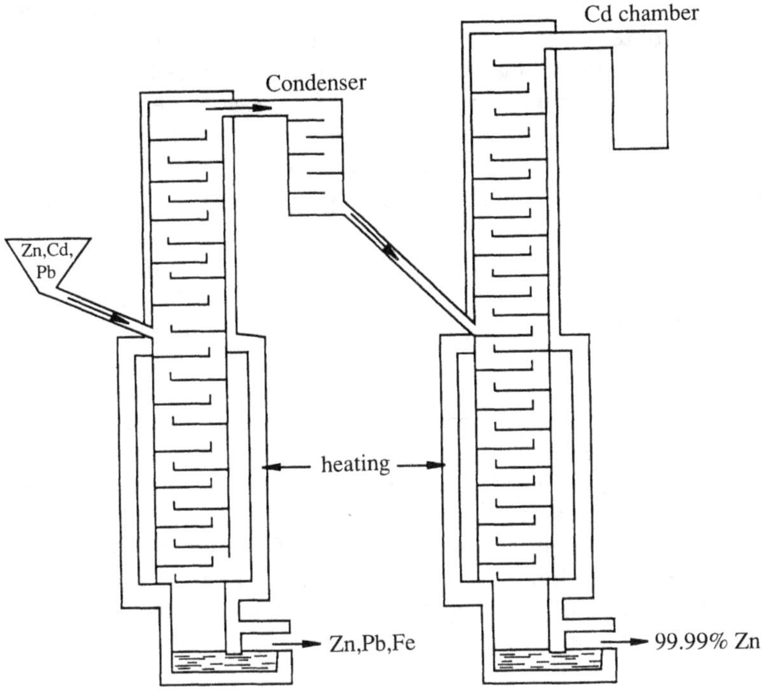

FIG. 3.7 Double distillation column for purification of Zn.

electrodeposit zinc from acid solution because then H_2 should be formed in preference to Zn. However, the overvoltage for H_2 formation on pure Zn increases so rapidly with the current density that Zn is deposited instead of H_2 at high current densities.[20]

Zinc production by electrolysis was established in about 1920, with the immediate advantage of producing 99.99 per cent pure metal. The initial problem was the poor yield from concentrates containing Fe and electrochemical losses from trace impurities. After the removal of iron there are three or four stages of addition of zinc dust to remove ions of Cu, Ni, Cd, Sb, Co, and Ge from the electrolyte.[31] The Cu, Co, Cd, and other metals can be recovered as by-products. The zinc deposits are stripped from the cathodes every twenty-four hours and remelted in an induction furnace before casting into ingots or pigs.

3.8.2 Cadmium

Silver-white and capable of taking a high polish, cadmium (Cd) is nearly as soft as tin. Cadmium melts and boils at low temperatures. The metal is permanent in dry air but becomes coated with oxide in moist air. Most Cd produced is electroplated on to steel, iron, copper, brass, and other alloys to protect them from corrosion. Cadmium plating is especially resistant to attack by alkalis. Some Cd is used as the anode material in rechargeable storage batteries. Cadmium is a component in bearing alloys and low melting alloys used for solder. Due to its ability to absorb neutrons, Cd is used in control rods for some nuclear reactors.

Most Cd is produced from the fumes emitted during the sintering of zinc concentrates, from the dust collected from the gases leaving lead blast furnaces, and from the various residues produced during the electrolytic refining of zinc. If Cd is present in the residues in metallic particle form or slightly oxidized on the surface of the particles, purification may be accomplished by distillation. The Cd-bearing dusts are heated with a carbon reducing agent in the absence of air. The method is based on the lower reduction temperatures and higher vapour pressures of Cd compared with Zn and other metals. As is seen from Table 3.6, the boiling-point of Cd is 766°C and that of Zn is 907°C. Cadmium oxide is reduced by carbon between 775°C and 815°C, approximately 250°C lower than the temperature where reduction of ZnO takes place.[32]

Poisoning results from the inhalation of vapour or dust of cadmium. Inhalation of fumes under industrial conditions can produce an acute, extremely severe inflammation of the lungs that may be fatal. Chronic poisoning from inhalation may injure the liver and kidneys.

3.8.3 Mercury

Mercury is the only common metal that is liquid at room temperature. It is a dense mobile liquid having a lustrous silvery-white colour. It is not attacked by dry air, oxygen, or carbon dioxide at ordinary temperatures and tarnishes only slowly in moist air. Mercury (Hg) does not wet glass or cling to it. This property coupled with its uniform volume expansion throughout the liquid range, makes it ideal for thermometers. Barometers and thermometers utilize its high density and low vapour pressure. The good electrical conductivity of Hg (see Table 2.14) makes it useful in sealed electrical switches and relays. An electrical discharge through mercury vapour produces a bluish glow rich in ultraviolet radiation. This is exploited in ultraviolet, fluorescent, and high-pressure mercury vapour lamps. Mercury alloys with most metals to form amalgams (iron is an exception). Gold and silver can be extracted by amalgamation. An important application for Hg is as the cathode in a cell for the production of chlorine and caustic soda by electrolysis of brine. Mercury–silver amalgams are used in dentistry for filling cavities.

The principal ore mineral is cinnabar, a soft red to reddish-brown HgS. Cinnabar is usually found alone, but sometimes in association with Sb, Au, Fe, and Zn. Secondary mercury, that is mercury that has been recovered after primary uses, is a significant source in most developed countries.

The recovery of the metal from the ore is one of the simpler extraction processes. Most mercury is recovered by furnace treatment of ore that has come directly from the mine without processing other than crushing. If the ore is low grade or contains certain impurities that create problems in the furnace, the mercury content may be upgraded by an ore flotation process. The process makes use of mercury's non-wetting properties to separate it from the other substances. Mercury is removed from the ore or concentrate by roasting it at temperatures above 587°C (the temperature where HgS sublimes).[33] The sulphur vapours react with oxygen in the air to form sulphur dioxide:

$$HgS(s) = Hg(g) + \tfrac{1}{2}S_2(g) \tag{3.7}$$

$$\tfrac{1}{2}S_2(g) + O_2 = SO_2. \tag{3.8}$$

The vapour is passed through a dust collector and into a series of condenser pipes. Both reactions (3.7) and (3.8) shift to the left on cooling, and if equilibrium were established the reactions would reverse completely during condensation of the product. Fortunately, the back-reactions are slow. Nevertheless, if the ores or concentrates contain relatively large quantities of sulphur, iron fillings, scrap iron, or brine are added to tie up the sulphur:

$$HgS + Fe = Hg(g) + FeS \tag{3.9}$$

$$4\,HgS + 4\,CaO = 4\,Hg(g) + 3\,CaS + CaSO_4. \tag{3.10}$$

In this case the vapour contains only mercury, and there is no precipitation of synthetic cinnabar or free sulphur in the condenser pipes.

Mercury of prime virgin grade averaging not less that 99.9 per cent recovered in this manner is pure enough for most commerical uses. To meet the special requirements of dental and medical instrument and mercury dry-cell battery applications, mercury must be further purified by redistilling or by treatment with nitric acid.

Mercury is toxic. Poisoning may result from inhalation of the vapour, ingestion of soluble compounds or absorption through the skin. Certain modern industrial and biological processes concentrate mercury compounds to dangerous levels. Ingested mercury is concentrated in the kidneys where it may destroy the blood-filtering structures. This can cause the accumulation of toxic substances in the blood and final death. Chronic mercury poisoning from inhalation or absorption through the skin may result in a loss of weight and appetite, trembling, and mental and personality changes marked by depression.[2]

3.9 GROUP 3B METALS; RARE-EARTH AND LANTHANIDE SERIES METALS

The rare-earth elements form a series of chemically similar metals. They are scandium (Sc), yttrium (Y), lanthanum (La), cerium (Ce), praseodymium (Pr), neodymium (Nd), promethium (Pm), samarium (Sm), europium (Eu), gadolinium (Gd), terbium (Tb), dysprosium (Dy), holmium (Ho), erbium (Er), thulium (Tm), ytterbium (Yb), and lutetium (Lu). The first three elements are in the 3b group, while the remaining fourteen members, called the lanthanide series, are written apart from the main periodic table (Table 2.13). The aqueous chemistry of all rare earths is very similar and changes only slightly in progressing along the lanthanide series. Therefore it is difficult to separate individual rare earths. All of the elements form trivalent compounds and in the crystal lattices of such compounds one rare-earth ion readily replaces another. The rare-earth elements when heated react strongly with non-metallic elements to form very stable compounds. They are never found as free metals in the earth's crust. The chemical properties of scandium differ sufficiently from the other rare-earth elements to have become segregated from them by the action of geological processes. Thus Sc is seldom associated with the rare earths in minerals. The fourteen lanthanide elements, from Ce to Lu, are much alike because the differences in their electronic structures chiefly involve the inner 4f electrons, whereas it is the outer s and p (sometimes d) electrons that determine the chemical behaviour. Physical properties, however, differ as much across the lanthanum series as they do for most other series in the periodic table. The melting-point of Lu, for example, is almost twice that of La and the vapour pressures of Yb and Eu at 1000°C are millions of times greater than those of La and Ce. Each of the rare-earth metals readily combines with almost any other metallic element to form alloys that can be hard or soft brittle or ductile, and they can have high or low melting-points. The lanthanides are invaluable in the study of the metallic state, properties of alloys, the existence of intermetallic compounds, and the effect of impurities.

The pure rare-earth metals are bright and silvery. Some physical properties of the first six rare earths are presented in Table 3.7. Lanthanum Ce, and Nd corrode readily in air. Metallic Y on the other hand remains bright and shiny for years. Properties are frequently sensitive to impurities. For example, the light lanthanide metals will corrode much more rapidly if small amounts

TABLE 3.7 *Physical properties of the group 3b metals and elements in the lanthanide series*[2,67,68] *(see also Appendix 1)*

	Sc	Y	La	Ce	Pr	Nd
Atomic mass	44.96	88.91	138.91	140.12	140.91	144.24
Metallic radius (A)	1.61	1.78	1.87	1.83	1.83	1.81
Electronegativity	1.36	1.22	1.10	1.12	1.13	1.14
Reduction potential			−2.52	−2.48		
Melting-point (°C)	1539	1526	920	798	931	1016
Boiling-point (°C)	2828	3334	3453	3422	3507	3064
Density at 20°C (kg/m^3)	2990	4478	6145	6750	6773	7004
Youngs modulus E (G Pa)		66.3	37.9	33.5		
Poisson's ratio		0.265	0.28	0.248		

of Ca or Mg or rare-earth oxides are present in the metal.

Until the mid-20th century, there was not much use for pure rare-earth elements or compounds except cerium or lanthanum. However, a mixture of La, Ce, Pr, and Nd *mischmetal* was used for deoxidation and desulphurization of steel[34,35] and for the removal of tramp elements such as P, Sn, Sb, and As from steel and other metals. By the 1970s Y, Gd, and Eu were being used in the red phosphors for colour television. Very few uses have developed for Sc.[36] However, its low density and high melting-point suggest application as an alloying agent for lightweight metals. Yttrium, Ce, Pr, and Nd are constituents of numerous ferrous and non-ferrous alloys. Cerium, Pr, and Nd give Mg alloys higher strength at all temperatures. Rare earths are added to Fe–Cr and Fe–Cr–Ni alloys to increase their resistance to corrosion. Rare earths are used in tonne quantities as catalysts for the cracking of crude petroleum. The oxides play an important role also in the ceramics and glass industries.

Chemists of the 19th and early 20th century resorted to laborious fractionating processes to isolate individual rare-earth elements from aqueous solution. Many different processes were used[37] such as fractional crystallization, fractional precipitation, fractional decomposition, and fractional extraction. All of these consisted of separating the mixed rare earths into approximately equal fractions, one enriched in the lighter elements, the other in the heavier elements. Both fractions would then be put back into solution and the process repeated on each of them. Based on such methods pure lanthanum and cerium compounds have been commercially available in tonne quantities for many years.

Separation of the rare earths may be carried out by ion exchange methods or liquid/liquid extraction. Space considerations allow only a reference to be made to these methods.[38]

It is relatively easy to reduce halides of the rare earths to metals. What is difficult, however, is to reduce them to high-purity metals in ingot form. As already pointed out, the rare earths have a great affinity for all the non-metallic elements, H_2, B, C, N_2, O_2, Si, S, P, Cl_2, and Br. The molten rare-earth metals come close to being a universal solvent and dissolve almost all other metals. The molten metal attacks any crucible in which it is melted, and the final product generally is a rare-earth-rich alloy of the crucible elements. Tungsten and tantalum make the best crucibles. They are only weakly attacked by the molten metals below 1000°C. Impurities from the atmosphere can be largely eliminated by carrying out all operations in an environment of helium or argon and by the use of modern high-vacuum ion pumps instead of oil pumps. Finally, if ultrapure metals are to be obtained, the raw materials from which they are made must be ultrapure or their impurities will end up in the ingot. Anhydrous rare-earth halides form oxyhalides upon contact with water vapour. Dry HF is used to keep the oxygen content of the rare-earth fluorides below 10 p.p.m. Then a metallothermic reduction can be employed to produce the metal:

$$3Ca + 2MF_3 = 2M + 3CaF_2. \tag{3.11}$$

For instance, from problem 3.4 it is seen, using the thermodynamic data in Appendix 1, that Ca displaces eqn (3.11) with M = Ce well over to the right.

3.10 GROUP 4B METALS; TITANIUM, ZIRCONIUM, AND HAFNIUM

One of the important characteristics of these three metals is their great affinity to oxygen and nitrogen. The equilibrium

$$\tfrac{1}{2}O_2 = \underline{O} \tag{3.12}$$

is displaced far over to the right. Even at a very low oxygen pressure p_{O_2}, the oxygen content in the metal may be high enough to render the metal brittle and hard, useless for most purposes. Figure 3.8 shows the Gibbs energy of solution of oxygen in titanium according to eqn (3.12) at 1000°C.[39] Also on the diagram the oxygen potentials for reduction with Al, Mg, and Ca are given. It is seen that only Ca is capable of reducing oxygen to low levels. The concentration of oxygen in equilibrium with Ca is 0.07 mass% at 1000°C.

Similar problems with oxygen and sulphur arise, not only form the group 4b but also for the 5b metals, vanadium, niobium, and tantalum. Each of these systems contains several compounds, and in addition the oxygen and sulphur have a tendency for solution in the metallic lattice.

In Table 3.8 some of the physical properties of the 4b elements are presented. The high values of the melting-points should be noted. The pure metals Ti and Zr are ductile. All three metals have a high corrosion resistance due to the formation of an inert oxide surface film. For titanium the density is low at

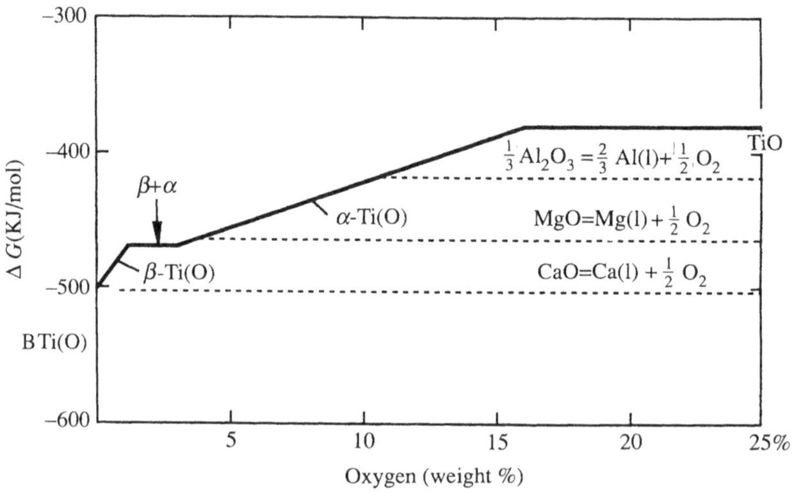

FIG. 3.8 The Gibbs energy of solution of $\tfrac{1}{2}$ O_2 in titanium.[39] The dashed lines indicate the oxygen potential for reduction with Ca, Mg, and Al.

TABLE 3.8 *Physical properties of the group 4b metals*[2,67,68] *(see also Appendix 1)*

	Ti	Zr	Hf
Melting point (°C)	1666	1852	2 227
Boiling point (°C)	3358	4429	398
Density at 20°C (kg/m^3)	4500	6490	13 100
Youngs modulus E (G Pa)	120.2	98	141
Poisson's ratio	0.361	0.38	0.26

4500 kg/m^3. Titanium is used structurally in aircraft due to its light weight and high strength. Perhaps the most important property of titanium is its ability to be alloyed with most of the other metals (and, alas, many of the non-metals!). Alloying additions increase the strength of titanium both at room temperature and at elevated temperatures. Titanium has long been used as a deoxidizer in steel (see Fig. 2.5), to which it is added as a ferrotitanium or ferrocarbon-titanium alloy. It has been added to low-alloy steels to reduce grain size and in stainless steel to control carbon content. It is also used in aluminium as a grain refiner (see Chapter 9). The most important use for zirconium is in nuclear reactors for cladding fuel rods and for reactor-core structures due to its unique combination of properties. It has good strength at elevated temperatures, resists corrosion from the rapidly circulating coolants, does not form highly reactive isotopes, and withstands mechanical damage from neutron bombardment. Hafnium, similar to zirconium in chemical properties, must be scrupulously removed from the metal intended for reactors. The reason is that it strongly absorbs neutrons. This last property makes Hf useful for fabricating nuclear reactor control rods.

Rutile (TiO_2) and ilmenite (FeO–TiO_2) are the two most important ores of titanium. In general in extractive metallurgy, oxide intermediates are preferred in metal production due to their low cost and ease of handling. The selection of oxide reduction routes, however, depends upon the possibilities of adequately refining the as-reduced metals by available deoxidation processes. As will be pointed out in Chapter 4, oxygen can be removed by transfer to a second phase insoluble in the metal. This second phase can be solid (inclusions), liquid (slag), or a gas. Calcium is obviously a candidate for removing oxygen from Ti as solid or liquid inclusions. However, Ca has a high vapour pressure. In Chapter 2 we discussed the problems of adding calcium at steelmaking temperatures. The melting point of Ti is even higher than that of Fe. These facts seem to preclude the use of calcium. Removal of oxygen over the gaseous phase can often take place by special mechanisms involving formation and evaporation of volatile suboxides of the parent metals, alloying components, and residual reducing agents (mainly carbon and aluminium). However, for the group 4b metals none of these mechanisms seem to be capable of removing a substantial amount of oxygen.[40] For the group 5b metals removal over the gaseous phase is possible; this will be touched on in the next section. For the group 4b metals the use of oxide as the starting material invariably results in considerable amounts of oxygen in the as-reduced metal.

3.10.1 The Kroll process

We conclude that group 4b metal production must be based on a different route from the oxide one. Halogens, in particular chlorine salts or halides, are obvious candidates. Due to the problems with oxygen, the halides must be anhydrous. Production of titanium by electrowinning presents serious problems,[41] partly due to the deposition of Ti as porous dendritic deposits. By elimination this leaves a metallothermic reduction of the halides. Then in order to obtain a favourable equilibrium for producing the halide the oxide is mixed with a suitable reducing agent, usually carbon:

$$TiO_2 + 2Cl_2 + C = TiCl_4\,(g) + CO_2. \qquad (3.13)$$

The $TiCl_4$ is then condensed in a fluidized bed reactor. The liquid $TiCl_4$ contains many impurities, principally other metal chlorides, and it must be purified before reduction to titanium metal. A sequence of distillations above or below the boiling-point is used to remove the undesired compounds. The high-purity $TiCl_4$ produced is reduced to metallic Ti by reaction with magnesium or sodium. The Kroll process uses magnesium, or modifications based on sodium, as the reducing agents. The batch-type processes are conducted in large sealed reactors where argon or helium have been introduced to exclude air or moisture. Both processes produce titanium in the form of a highly porous material, termed titanium sponge, with the chlorides entrapped in the pores. This sponge is crushed to a suitable size for handling and then leached with water and acid or heated in a vacuum to remove the chlorides.

3.10.2 Water-cooled moulds

The combination of high melting-points and the reactivity of molten group 4b metals with refractory mould materials initially posed a major problem for the production of commercially sized ingots. The first ingots were produced by side feeding the sponge and alloy additions into a tungsten electrode arc. The crucible was a water-cooled copper mould and the melting was carried out in an argon atmosphere. Although crude, this method had most of the basic elements of the newer methods[42] employed to produce ingots of reactive and refractory metals: protective environment, concentrated high-energy input, and water-cooled moulds (see Fig. 3.9). The newer techniques, termed consumable-electrode melting, may also involve the manufacture of a sponge electrode from pressed compacts that contain a mixture of sponge and alloy additions. Individual compacts are welded together to form a single electrode. The electrode is

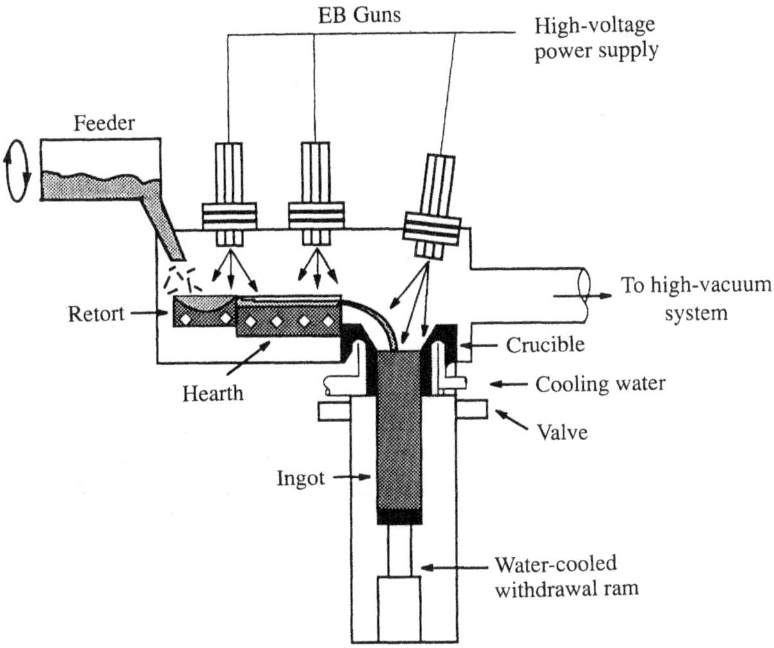

FIG. 3.9 A schematic drawing of an electron-beam melting furnace showing major components, electron beam guns and including the cold-wall copper refining hearth. The furnace can be used with bar or loose feed.[42]

placed in a consumable electrode furnace under a vacuum or inert atmosphere. The sponge electrode is lowered into an externally water-cooled cylindrical copper crucible and an arc struck on a starter charge, or a plate placed in the bottom of the crucible. The sponge electrode is then consumed as the molten metal drips down to the water-cooled mould. The transferred metal solidifies and the ingot is gradually built up in the crucible. The initial ingot is remelted either once or twice to achieve greater size and chemical homogenity. These consumable electrode melting furnace units are similar to the refining and remelting units for steel described in the section on group 8b metals (see Fig. 3.11).

3.11 GROUP 5B METALS; VANADIUM, NIOBIUM, AND TANTALUM

As seen from Table 3.9 all these three metals have high melting-points, niobium (Nb) and tantalum (Ta) being near 2500°C and 3000°C, respectively. In the previous section it was pointed out that the problems of extracting and refining the pure metals are similar for the 4b and 5b elements. They have a strong

TABLE 3.9 *Physical properties of the group 5b metals*[2,67,68] *(see also Appendix 1)*

	V	Nb	Ta
Melting-point (°C)	1902	2467	3014
Boiling point (°C)	3406	4740	5453
Specific density at 20°C (kg/m^3)	6100	8600	16 600
Youngs modulus			
E (G Pa)	127.6	104.9	185.7
Poisson's ratio	0.365	0.397	0.342

affinity to oxygen as may be seen from the Gibbs energy diagrams of the oxides [12] or from the data in Appendix 1. They also have a strong affinity to carbon and nitrogen. Tantalum especially can dissolve large amounts of hydrogen, oxygen, nitrogen, and carbon interstitially. If the solubility is exceeded, hydrides as well as very hard oxides, nitrides, carbides, and mixed phases are formed. In fact the hardmetals industry is a large user of Ta and Nb carbides. This industry consumes 30 per cent of the Ta production.

The group 5b metals are used as alloy additions to steel, superalloys, titanium, and aluminium. They strengthen the alloys by the precipitation of carbides and act as grain refiners (see Chapter 9). When added to steel the ferroalloys FeV, FeNb, and FeTa are used. In fact the bulk of V produced ultimately becomes FeV.

Vanadium metal, sheet, strip, foil, bar, and wire have found use in high-temperature service and in the chemical industry. The most important use for Ta is in electrolytic capacitors, special alloys, and corrosion-resistant equipment. The pure 5b metals are soft and ductile. Hardness is a very sensitive indicator for the content of metalloids and is therefore used as a criterion of metal purity. Hardness differences of 100–200 per cent exist between technically pure tantalum, which contains 99.5% Ta and cold ductile Ta which is purer than 99.95% by weight.[45] The influence of cold deformation also depends on the content of metalloids.

In order to remove nitrogen, hydrogen and carbon, a high temperature and a good vacuum or very pure inert gas is required.[43,20] Niobium or Ta may be deoxidized with C, Al, or sacrifically. Then the oxygen is removed over the gas phase as CO, Al_2O, NbO, or TaO.[40]

The main routes for the ferroalloys FeM go through the pentoxide M_2O_5 or MCl_5.[44] The cost of reducing the pentoxide to FeM is only a fraction of the total costs. The pentoxide is reduced to the metal by Al, Si, or C.

For the production of pure Nb and Ta another route is to start with the double fluoride:

$$K_2MF_7 + Na = 2KF + 5NaF + M \tag{3.14}$$

where M is Nb or Ta. This process and the fused salt electrolysis of K_2TaF_7 in

Ta_2O_5-containing baths have been successful on an economic scale.[44] The electrolysis is carried out under air at temperatures of 700–900°C in cast iron vessels connected to act as the cathode. Tantalum is deposited at the cathode in dendritic form. After cooling, the mixture of crystals and salts is taken out, crushed, and washed on a sorting and vibrating bed. Then the tantalum powder is washed again with mineral acids and classified. The metal grade is 99.8% Ta.

Tantalum and niobium are found together in high-grade tin slags. Usually they are separated as fluorides or oxides by liquid/liquid extraction.[44] Niobium metal is also produced by electrolysis. Kroll processes where $NbCl_5$ is reduced by Mg have been used.

Pure niobium and tantalum are produced in a similar manner to the 4b metals: by electron-beam melting and by vacuum electric arc melting. Vacuum sintering by direct current passage is also applied to make compacted Nb or Ta metal.

3.12 GROUP 6B METALS; CHROMIUM, MOLYBDENUM, AND TUNGSTEN

Chromium (Cr) is white, hard, lustrous, and brittle and is extremely resistant to ordinary corrosive reagents. This resistance accounts for its widespread use as an electroplated protective coating. At elevated temperatures chromium unites directly with oxygen, sulphur, silicon, boron, nitrogen, and carbon and with the halogens. Ferrochrome is an alloy extensively added to steel to increase corrosion and oxidation resistance, and strength. Up to 3% Cr is added to improve the physical properties and increase heat treatability. Steels that contain Cr in combination with other elements such as Mo, Ni, Mn, and V are used for springs, roller- and ball-bearings, dies, rails, and high-strength structures. Steels with 5–6% Cr have increased corrosion and oxidation resistance and are employed in the form of tubes in the oil industry. Stainless steels which have a high resistance to oxidation and atmospheric corrosion contain 10–20% Cr. The low carbon varieties (0.1% C or less) cannot be heat treated if Cr is more than 13 per cent. The higher Cr steels are used as automobile trim and for equipment to handle nitric acid. High-carbon (1–2 per cent) steels with about 13% Cr are used when hardeners and abrasion resistance are required. Steels that contain 25–30% Cr are found in parts subjected to high temperatures such as furnace and kiln linings and burner nozzles.[46] In Chapter 6 on decarburization, production of stainless steel is described in some detail.

Molybdenum (Mo) and tungsten (W) are very similar chemically. The metals are quite refractory with melting-points of 2624°C and 3407°C, respectively (see Table 3.10). The metals are inert to oxygen at normal temperatures, but combine with it readily at red heat. Like Cr, both metals react with B, C, N_2 and Si at high temperatures to form hard, refractory, and chemically inert compounds. Tungsten carbide is extensively used for high-speed cutting tools. An important use for W is in making high-speed cutting steels that retain their

TABLE 3.10 *Physical properties of the group 6b metals*[2,67,68] *(see also Appendix 1)*

	Cr	Mo	W
Melting point (°C)	1857	2624	3407
Boiling point (°C)	2669	4705	5658
Specific density at 20°C (kg/m^3)	7100	10 200	19 300
Youngs modulus			
E (G Pa)	279	324.8	411
Poisson's ratio	0.21	0.293	0.28

hardness at high temperatures. These steels contain 6–20% W with Mo, Cr, V, or Nb. Tungsten itself has wide applications as a lamp filament. Molybdenum is used in a number of small-scale applications that require high resistance to erosion at high temperatures: in jet engines, combustion liners, and afterburner parts. Molybdenum and W are added to ferrous and non-ferrous alloys to improve hot strength and corrosion resistance. Molybdenum is one of the most effective elements for increasing the hardenability of iron and steel, and it also contributes to the toughness of quenched and tempered steels.

High-carbon FeCr is produced carbothermically from chromite ore. Ore and coke is fed into the top of an open-top submerged arc furnace and the alloy is tapped from the furnace into chill moulds. After cooling it is broken into lumps and graded.[47]

FeMo and FeW may also be produced by carbothermic reduction. Due to the high melting-points of the alloys, the furnaces cannot be tapped. The metal collects and cools inside the furnace. After a sufficient amount of metal has been produced, the furnace is closed down and the bottom section is broken up to recover the lump of solidified alloy.

Production of low-carbon FeCr is carried out in two steps. In the first step a high-silicon FeCr is produced in the submerged arc furnace by feeding in a mixture of chrome ore, quartz, and coke. Reduction takes place analogously to the production of FeSi.[12] The solubility of carbon in Fe–Cr–Si decreases strongly with the content of Si. Above 40% Si it is low. The alloy in the first stage contains about 40% Si and 40% Cr and is low in carbon. In a second furnace this alloy is mixed with a slag rich in Cr_2O_3 so that the silicon reduces the chromite. An effort is made to obtain a final alloy low in Si and a waste slag low in Cr. A similar process is used for the production of FeMn.

Silicon and Al may be employed instead of C as reductants to supply metals of low carbon content.

Due to their high melting-points, the metals Mo and W are usually produced as powders. Molybdenum trioxide (MoO_3) or $(NH_4)_2MoO_4$ is reduced by hydrogen. Most W is produced by H_2 reduction of H_2WO_4, WO_3, or ammonium paratungstate crystals. The basic reduction process is to pass metal boats charged with the oxide through a heated tube with a counter-flow of H_2. The temperature of reduction varies from 700°C to 1100°C; the higher the temperature, the coarser the powder. The particle size of the powder is important in determining the success of the subsequent operations, and different types of powder are employed in various applications. For the production of massive metal (sheet, rod, and wire) a wide particle size range is required to give maximum compaction. A narrow size range is necessary when the particles are to be converted to carbides for hard metal production.[48] The composition of the tungsten carbides is usually W_2C or WC. For increased shock resistance Co is used as a binder.

When tungsten powder is converted to massive metal, the powder is presintered to a bar in a hydrogen atmosphere. A current of several thousand amperes is passed directly through it to raise the temperature to about 3000°C. This bar (with a porosity of about 5 per cent) is too brittle to be handled at ordinary temperatures, but can be forged and rolled at 1500°C. The consumable-electrode melting techniques discussed in the section on titanium can be applied to tungsten. However, the ingot has a coarse columnar grain structure unsuitable for working. Hot isostatic compaction is a new technique for consolidating tungsten powder. The powder is sealed in a metal container and held at temperatures of around 1500°C and pressures of 700–1400 bar. This gives the full density of tungsten.

3.13 GROUP 7B METALS; MANGANESE, TECHNETIUM, AND RHENIUM

The only important group 7b metal is manganese. Rhenium (Re) has only minor uses such as in an alloy with W in flash blub filaments. Technetium (Tc) has essentially no application. Manganese (Mn) has a melting-point of 1244 (±3)°C, a boiling-point of 2059°C, and a density of 7440 kg/m^3 20°C.

Ferromanganese and manganese are primarily used as alloy additions to steel, Al, and Mg. Manganese increases the wear resistance of steel. Like Cr, FeMn may be produced by smelting in the electric furnace. Ores of Mn and Cr always contain some iron oxides, and iron will enter the metallic phase. Ferromanganese usually contains about 80 mass% Mn and 7% C[47] and the rest is iron. The main impurities are P and S. The source of P is the ores, while S derives from the coke. As pointed out in the previous section, medium- and low-carbon FeMn (1% and 0.5% C) can be produced in two stages. The first stage is carbothermic and the second is silicothermic. The first-stage SiMn alloy is also an important commercial alloy. It contains 15–20% Si and more than 65% Mn. For high-purity applications electrolytic manganese is used. The deposition of Mn from an aqueous solution is even more difficult than for Zn.[12] The current efficiency is low and the energy consumption is high at about 8.5 kWh/kg.

3.14 GROUP 8B METALS

3.14.1 Introduction

There are nine elements in the group 8b metals; the iron group, Fe, Co, and Ni; the light platina group Ru, Rh, and Pd; and the heavy platina group Os, Ir, and Pt. (See Tables 3.11, 3.12, and 2.13). Iron and the iron group play the dominant role for a number of reasons. Iron is number four in relative abundance in the earth's solid crust and has some unique chemical and metallurgical properties. The iron group elements can be melted by standard metallurgical techniques, their densities are lower than for the two other subgroups, and their relative abundance is much higher (and the price much lower).

3.14.2 Iron

The main impurities in iron and steel are the non-metallic elements $\underline{O}$, $\underline{H}$, $\underline{N}$, $\underline{S}$, and $\underline{P}$. These are 'surface active' and tend to accumulate in the interdendritic liquid during solidification. This leads to the formation of non-metallic inclusions, oxides, sulphides, oxysulphides, and so on, with iron or with the various alloy components. Porosity may be caused by the nucleation and growth of H_2 or CO bubbles. An example of a further problem due to the increased interdendritic content of surface active elements is temper embrittlement. These effects are discussed in Chapters 1 and 9. Here it is only pointed out that they are particularly complicated to analyse for iron and steel. The reason is that

TABLE 3.11 *Physical properties of the iron group metals*[2,67,68]

	Fe	Co	Ni
Melting point (°C)	1536	1495	1453
Boiling point (°C)	2859	2925	2911
Specific density at 20°C (kg/m^3)	7873	8900	8908
Youngs modulus *E* (G Pa)	211	211	200
Poisson's ratio	0.27–0.3	0.32	0.312

TABLE 3.12 *Physical properties of the iron group metals*[2,67,68]

	Ru	Rh	Pd	Os	Ir	Pt
Melting point (°C)	2250	1960	1552	3027	2443	1772
Boiling point (°C)	4146	3694	2961	5008	4424	3823
Specific density at 20°C (kg/m^3)	12 300	12 410	12 020	22 480	22 650	21 450
Youngs modulus *E* (G Pa)	432	379	121	559	528	170
Poisson's ratio	0.25	0.26	0.39	0.25	0.26	0.39

there is such a vast number of iron-based alloys in use.[49] There are about 1500 low-alloy steels to deal with and most of the fifty-six transition elements are used as additions. Furthermore the solubility of carbon, which is usually the main alloying element in iron and steel, depends on the presence of carbon as graphite or cementite, Fe_3C.[50,51] The various alloy additions used to manipulate the properties of the steel (for instance, to reduce the critical cooling rate necessary to quench austenite to martensite) change the solubility of carbon in austenite and the shape and extent of the austenite phase. Here no attempt is made to review the various Fe-based alloys, and reference must be made to the specialized literature. Some alloys will be mentioned in the section on nickel.

In Chapter 2 various thermodynamic aspects of the refining of steel were treated and thermodynamic data was given for molten iron alloys. Precipitation deoxidation was discussed together with the sulphur and phosphorus equilibrium with the slag. In this section some methods in the refining of steel not mentioned in the other chapters are briefly presented.

In Chapter 2 it was pointed out that sulphur could be transferred to a molten basic slag. Alternatively, pig iron may be desulphurized by the injection of CaC_2 using N_2 or argon as carrier gas. This process can be described by the following reaction equation:

$$CaC_2 + \underline{S} = CaS + 2\underline{C}. \qquad (3.15)$$

The yield of CaC_2 in terms of the equation is low. The reason is that the kinetics of the reaction are slow since both CaC_2 and CaS are solids (see Section 4.9).

Magnesium may also be used to remove sulphur in iron or steel. The reaction may be given as

$$Mg + \underline{S} = MgS. \qquad (3.16)$$

The yield of Mg is about 50 per cent.[52,53] Magnesium is lost partly due to the escape of magnesium in the gaseous phase and partly due to oxidation.[52] Due to the high price of magnesium relative to CaC_2, it is not obvious which method is to be preferred. Sometimes Mg and CaC_2 are injected into steel at the same time.

Sodium carbonate may be employed to remove $\underline{S}$ and $\underline{P}$ simultaneously from pig iron at low temperatures.[54,55] Soda (Na_2O) is a very efficient dephosphorizer.[64] The trick is to adjust the oxygen activity, a_O, at an intermediate value so that it is low enough to remove sulphur and still high enough to suppress phosphorus.

The principles of hydrogen and CO removal by gas purging or vacuum treatment are discussed in some detail in Chapter 4. Compared with gas purging, vacuum units have the disadvantage that mixing is poorer and vacuum-melt contact areas are smaller. In Fig. 3.10 the vacuum circulation[56,57] (also called Hereaus–Rheinstahl (RH)) process is sketched. In the RH unit two legs enter the metal in the holding ladle. A stream of purified argon is injected into one

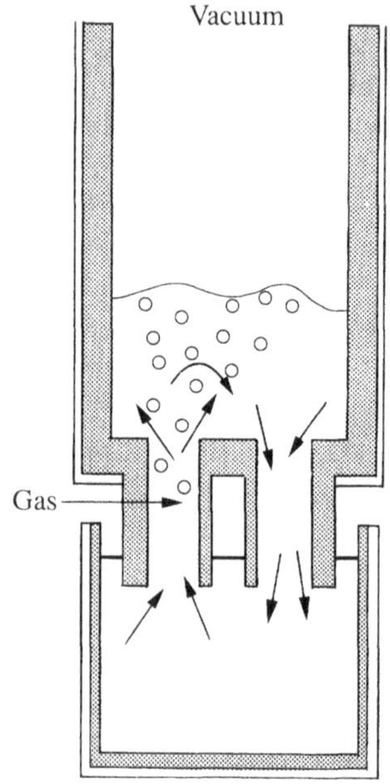

FIG. 3.10 Sketch of vacuum circulation unit.[56]

leg, causing the metal to rise, flow through the chamber, and return to the ladle via the second leg. The ferrostatic pressure in the chamber is low, and the stream of argon bubbles provides additional sites for CO formation. The RH process is discussed in some detail in Chapter 7. The Dortmund–Hörder process operates by varying the pressure in the chamber so that the metal alternately rises and falls in the leg. The metal boils vigorously as it first enters and then subsides when pressure is increased.

Refining and remelting units deserve to be discussed. These are similar to the consumable electrode melting furnaces mentioned in the section on the group 4b metals. In the remelting units the solidification structure is controlled by matching the melting to the solidification operation. In addition refining is carried out. The excellent as-cast structure and high density means that little or no hot work is required to take advantage of the properties. Frequently the material can be used in the as-cast condition. Energy for remelting the consumable electrode may be supplied in various ways: as a plasma beam or arc, using an electron beam, or employing a vacuum arc.[58,59,60] Usually the heating system is combined with a vacuum. When plasma heating is used, the pressure may vary from several atmospheres down to 10^{-4} bar. Figure 3.11 gives an example of such a unit. The molten metal droplets are exposed to a controlled atmosphere or vacuum. The difference in principle from conven-

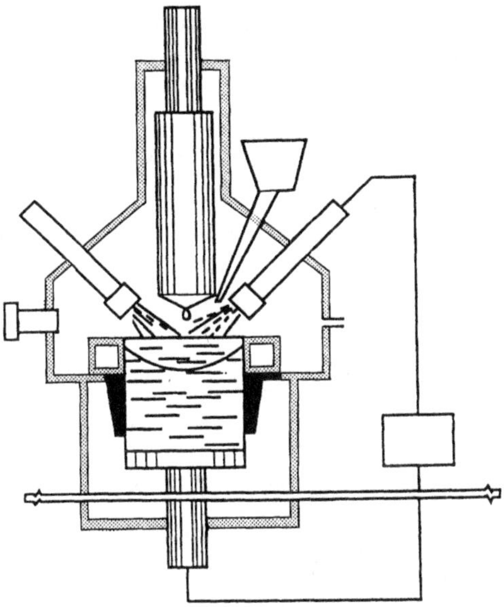

FIG. 3.11 Plasma arc remelting installation.[60]

tional refining where the extracting phase (injected slag, bubbles) traverses the bath should be noted. Here metal drips through the vacuum or gas phase. Hydrogen, CO, and impurities with a high vapour pressure such as Bi and Pb may be removed.[58]

In the electroslag remelting (ESR) process (see Fig. 3.12) the vacuum or gas phase is replaced by a refining slag. Energy is supplied by electrical resistance heating in the slag. Resistivity in the slag must be within narrow limits and in practice a compromise must be struck between chemical (refining) and electrical requirements. To attain a sufficiently high electrical conductivity, ESR slags contain calcium fluoride in addition to standard slag components such as CaO, Al_2O_3, SiO_2, and MgO. The presence of CaF_2 in ESR slags has encouraged a number of workers to study the ternary system CaF_2–CaO–Al_2O_3.[61,62] Since it seems to be unavoidable that the slags contain hydrogen (in the form of hydroxyl ions), the ESR process cannot usually remove hydrogen. If hydrogen is a problem, the consumable electrode may be vacuum degassed in a previous step and special care must be taken in the preparation of the ESR slag.

3.14.3 Cobalt

Polished cobalt is silver white with a faint bluish tinge. A close-packed hexagonal structure is stable below 427°C while a face-centred cubic is stable at high temperatures. It is ferromagnetic up to 1121°C, the highest known Curie point of any metal or alloy. About 25 per cent of the world's cobalt production

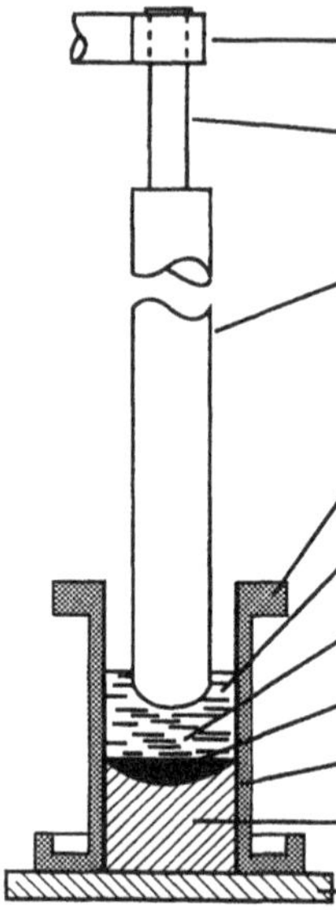

FIG. 3.12 Schematic drawing of electroslag refining process.

goes into magnets. Together with iron and nickel, cobalt exhibits magnetic properties at room temperature. These three elements are found in a large number of alloys for permanent magnets.

Another 25 per cent of the world's cobalt output is used in the field of high-temperature steel alloys. Required in gas turbines, jet engines, and similar applications, alloys with cobalt in the range 5–65 per cent retain their strength to about 650°C (see also the next section, on Ni).

Cobalt improves the wear resistance of many alloys. The addition of up to 12% Co to tool steel enables it to be used more effectively on hard materials for which deep cuts and high speeds are required.

In the production of cemented carbides, such as tungsten carbide, a briquetted mixture of the carbide and up to 25 per cent cobalt powder is heated at a temperature above the melting-point of cobalt. The latter melts and binds the carbides and gives the necessary toughness and resistance to shock for machine tools, drill bits, dies, and saws.

Even the best cobalt ores contain only low concentrations of the element. Most cobalt is recovered as a by-product from the ores of other metals, particularly copper and nickel.[63]

3.14.4 Nickel

Nickel is a silver-white metal with high electrical and thermal conductivity, and its alloys with copper are substantially white. Nickel and Ni–Cu alloys are widely used as coinage. The ability of nickel to form strong ductile alloys with many metals, including iron, chromium, cobalt, copper, and gold, is widely utilized in industry. Nickel is resistant to corrosion by fluorine, alkalis and a variety of organic materials. It remains bright indoors, but corrodes slowly outdoors. It is essential as an undercoat on chromium-plated automobile trim.

About half the nickel produced is used in alloys with iron. The first major market for nickel was in the production of nickel and nickel–chromium steels for armour plate. Later a whole new group of high-strength steels containing 0.5–5% Ni along with Cr, Mo, and so on, were developed.

Recently, steels for ultralow temperature use with liquified gases, containing at least 9% nickel, have come into demand. These steels rely on carbon for hardening by heat treatment. Nickel strengthens the ferritic body-centred structure. The transformation temperatures to bainite or martensite are depressed and the transformation to hardened bainitic or martensitic structures is extended to much larger sections.

The largest single use of nickel is in stainless steel of which the 18/8 (18% Cr and 8% Ni) variety is the best known. The steels have a stable austenitic structure and hence cannot be hardened by heat treatment, but they can be hardened by cold working. While the major hardening is then due to the deformation, some is caused by the partial decomposition of austenite to martensite.[65] The standard carbon content of austenitic stainless 18/8 steels is now usually below 0.05 per cent but may be as high as 0.1 per cent, and this leads to precipitation of chromium carbides at the grain boundaries on heating in the range 450–550°C. The result is susceptibility to intergranular corrosion and possible 'weld decay'. This problem is minimized by reducing the carbon content down to 0.01 per cent, or by the addition of small amounts of strong carbide formers such as niobium or titanium. In Chapter 6 on decarburization, the problem of carbon removal from stainless steel is treated in detail.

Ultrahigh-strength steels (maraging steels) develop their exceptional strength by adding the hardening effect of the precipitation of intermetallic compounds to the normal martensitic hardening of a transformed low-carbon steel. An austenitic matrix consisting of about 18% Ni, 8–12% Co, and 3–5% Mo, transforms directly to martensite on cooling. On reheating, intermetallic compounds formed from small additions of titanium and aluminium are precipitated. Extremely high tensile strengths up to over 2400 MN/m^2 can be achieved combined with good ductility and fracture toughness.

'Superalloys' is a term applied to Ni–Cr alloys developed for service under highly stressed conditions at high temperature. The primary requirement is high creep resistance. Creep is due to the thermally-activated movement of dislocations through the crystal lattice. Therefore, one desires a high elastic modulus and a low diffusion rate at elevated temperatures. Although nickel is not exceptional in these respects, it has a wide tolerance for other elements in solid solution. Nickel-based superalloys have contents of one or more of the following elements: Cr, Co, Fe, Mo, W, and Nb.

The second major strengthening mechanism is the presence of a stable finely dispersed second phase such as the intermetallic compound $Ni_3(Al, Ti)$. Other dispersed phases, particularly carbides, can contribute to lattice strengthening.

The metallurgy of nickel is complicated in its details, many of which vary widely according to the particular ore being processed. In general the ore is transformed into Ni_2S_3 which is roasted in air to give NiO which is then reduced with carbon to give Ni.[12,64] Some high-purity Ni (99.9 per cent) is made by carbonyl processes. In the vapo-refining process[65] CO reacts with Ni at about 100°C to produce the volatile carbonyl $Ni(CO)_4$, which on further heating to 200–300°C decomposes to give the purified metal and CO (see problem 3.5). Although Fe and Co also form volatile carbonyls, the formation of $Ni(CO)_4$ is highly selective. Ni pellets of about 8 mm diameter are grown by letting $Ni(CO)_4$ decompose on preheated pellets. In the pressure carbonyl process used by Inco in Canada, a Ni powder with an average particle size in the range 2–10 μm is produced.

There are indications that inhalation of nickel or dust fumes over long periods of time may lead to some forms of cancer. Direct skin contact with alloys or jewellery that contains Ni can give allergic reactions.

3.14.5 The platinum group metals

Platinum, (Pt) and palladium, (Pd) (see Table 3.12 and Table 2.13) are far more important than the other members of the platinum group of metals. These metals are notable for their wide range of applications, many of which are based on their high melting-points, resistance to corrosion, and catalytic properties.[66] Platinum and Pd are soft and ductile; these can be worked hot or cold. Pure platinum and Pd are too soft for many applications and are hardened by alloying with other metals, usually from the platinum group (for example, Ir or Ru).

Pt–10% Rh thermocouples are routinely used for the measurement of temperatures up to about 1500°C. For higher temperatures, Pt–7% Rh and Pt–30% Rh thermocouples are available.

Platinum alloyed with 3.5% Rh is the standard material for crucibles in the laboratory where they are exposed to a variety of corrosive melts, including molten slags and salts. A large-scale use is for Pt–Rh alloys for lining tanks that contain high-quality glass. Even more important is the use of Pt–10% Rh alloy for bushings through which molten glass flows in producing glass fibres. Platinum is employed in the exhaust system in cars to catalyse the reaction of NO_x and CO to N_2 and CO_2.

Palladium and Pt are capable of absorbing large volumes of hydrogen. At 80°C and one atmosphere Pd will absorb up to 900 times its own volume. Hydrogen diffuses rapidly through Pd. It is used industrially to separate high-purity hydrogen from mixed gases. Platinum is widely used as a catalyst in the petroleum industry.

The platinum group of metals often occur in nature in the free state but not pure. In working secondary (alluvial) deposits, the high density of the platinum metals is relied upon for separation, as is the case with gold. Platinum and palladium are also found as sulphides in Cu–Ni sulphide deposits. The platinum group of metals are then finally recovered in the anode slimes from the production of electrolytic nickel or copper.[63]

REFERENCES

1. Ohse, R. W. (ed.) (1985). *Handbook of thermodynamic and transport properties of alkali metals*. International Union of Pure and Applied Chemistry, Chem. Data Series 30. Blackwell, Oxford.
2. Goetz, P. W. (ed.) (1986). *The New Encyclopedia Britannica*. The University of Chicago, USA.
3. Gatenby, K., Loretto, M. M., Smallman, R. E., and Miller, W. S. (1985). Texture and properties of Al-Li based alloys. In *Proceedings of the 7th International Conference on the Strength of Metals and Alloys*, pp. 311–822. Pergamon Press, Oxford.
4. Field, O. J., Butler, E. P., and Scamans, G. M. (1980). High temperature oxidation studies. In *Proceeding of the First International Aluminium Lithium Conference*, (ed. T. H. Sanders and E. A. Starke), pp. 325–46. AIME, Warrendale, Pa, USA.
5. Peterson, D. T. (1987). Purification of alkaline earth metals. *Journal of metals*, **39**, No. 5, 20–3.
6. Hausner, H. H. (ed.) (1965). *Beryllium in metallurgy and its properties*. University of California Press, Berkeley, USA.
7. Simensen, C. J. (1981). Impurities in light metals. *Støperitidende*, **5** pp. 176–84. (in Norwegian)
8. Emley, E. F. (1966). *Principles of magnesium technology*, Pergamon Press, Oxford.
9. Jarrett, N. (1981). Advances in the smelting of aluminium. In *Metallurgical treatises*, (ed. J. K. Tien and J. F. Elliott) pp. 137–57. USA-China Bilateral Conf., AIME, Warrendale, Pa, USA.
10. Grandfield, J. F. (1989). Sources of inclusions and the defects they generate. In *Aluminium melt refining and alloying, theory and practice*, (ed. M. Nilmani) pp. K1-6, G.K. Williams Laboratory, University of Melbourne, Australia.
11. Sigworth, G. K. and Engh, T. A. (1982). Refining of liquid aluminium—a review of important chemical factors. *Scand. J. of Metallurgy*, **11** 143–9.
12. Rosenqvist, T. (1983). *Principles of extractive metallurgy* (2nd edn). McGraw-Hill International London.
13. Wiese, U. (1983). Der Reinstoff Germaniumtetrachlorid in Lichtwellenleitern. In *Sondermetalle, Gewinnung—Verarbeitung—Anwendung*, pp. 269–88. Verlag Chemie Weinheim, Basel.
14. Wardel, M. P. and Davidson, C. F. (1987). Acid leaching extraction of Ga and Ge. *J. of Metals*, **39**, No. 6, 39–45.
15. Hoffman, J. E. (1987). Extracting and refining germanium. *J. of Metals*, **39**, no. 6, 42–5.
16. Davey, T. R. A. (1980). Advances in lead, zinc and tin technology—projections for the 1980s. In *Lead-zinc-tin'80*, (ed. J. M. Cigan, T. S. Mackey, and T. J. O'Keefe) pp. 48–65. Warrendale, Pa, USA.

17. Strachan, A.M. and Linton, I.A. (1990). Electric smelting of tin concentrates at Capper Pass. In *Lead-zinc'90*. (ed. T.S. Mackey and R.D Prengaman) pp. 841-56. Warrendale, Pa, USA.
18. Pearce, S.C. (1980). Developments in the smelting and refining of tin. In *Lead-zinc-tin'80*. (ed. J.M. Cigan, T.S. Mackey, and T.J. O'Keefe) pp. 754-70. Warrendale, Pa, USA.
19. Hansen, M. (1958). *Constitution of binary alloys*. McGraw-Hill, New York.
20. Pawlek, F. (1983). *Metallhüttenkunde*. Walter de Gruyter, New York.
21. Davey, T.R.A. (1986). Lead production. In *Encyclopedia of materials science and engineering*, (ed. M.B. Bever), pp. 2523-6. Pergamon Press, Oxford.
22. Lee, A.Y., Cole, E.R. Jr. and Paulson, D.L. (1984). Electrolytic method for recovery of lead from scrap batteries. Bureau of Mines, report of investigations, RI 8857, USA.
23. Schenker, G. (1990). Lead recycling from battery scrap and other raw materials in Metaleurop's lead smelting plant in Oker. In *Lead-zinc'90*, (ed. T.S. Mackey and R.D. Prengaman) pp. 979-1000. Warrendale, Pa, USA.
24. Hayes, P. (1985). *Process selection in extractive metallurgy* Hayes Publishing Co., Brisbane, Australia.
25. Alcock, C.B. (1976). *Principles of pyrometallurgy*. Academic Press, London.
26. Themelis, N.J., Mäkinen, J.K., and Munroe, N.D.H. (1985). Rate phenomena in the Outokumpu flash smelting reaction shaft. In *Physical chemistry of extractive metallurgy*, (ed. V. Kudryk and Y.K. Rao) pp. 289-310. The Metallurgical Soc. of AIME, Warrendale, Pa, USA.
27. Goyal P., Themelis, N.J., and Zanchuk, W.A. (1982). Gaseous refining of anode copper. *Journal of metals*, December, 22-8.
28. Lehner, T. and Lindkvist, G. (1984). Injection metallurgy for the refining of matte and blister. Paper presented at CIM, Quebec, Canada.
29. Biswas, A.K. and Davenport, W.G. (1980). *Extractive metallurgy of copper*. International Series on Materials Science and Technology, Vol. 32. Pergamon Press, Oxford.
30. Hultgren, R., Desai, P., Hawkins, D.T., Gleiser, M., and Kelley, K.K. (1973). *Selected values of the thermodynamic properties of binary alloys*. American Soc. for Metals, Metals Park, Ohio.
31. Fugleberg, S., Järvinen, A., and Sipilä V. (1980). Solution purification at the Kokkola zinc plant, In *Lead-zinc-tin'80*, (ed. J.M. Cigan, T.S. Mackey, and T.J. O'Keefe) pp. 157-171. Warrendale, Pa, USA.
32. Chizhikov, D.M. (1966). *Cadmium*. Pergamon Press, Oxford.
33. Mills, K.C. (1974). *Thermodynamic data for inorganic sulphides, selenides and tellurides*. Butterworths, London.
34. Kilbourn, B.T. (1988). Metallurgical applications of yttrium and the lanthanides. *J. of Metals*, May, 22-5.
35. Jingsheng, Y.U., Zongsen, Y., and Chengjian, W. (1988). Rare earths and the mechanical properties of steel. *J. of Metals*, May, 26-31.
36. Gongyi, G., Yuli, G., and Yu, L. (1988). Solvent extraction of scandium from wolframite residue. *J. of Metals*, July, 28-31.
37. Bartholomè, E., Bickert, E., Hellmann, H., Ley, H., Wiegert, W., Weise, E. (Eds) (1982). Seltene Erden. In *Ullmanns Encyklopedie der Technischen Chemie*, (4th edn.). Verlag Chemie, Weinheim, 247-61.
38. Helcorsky, J. and Leveque, A. (1978). Separation et purification des terres rares. *Sciences et tecnhiques*, **52**, June, E1-E4.

39. Kubaschewski, O. and Alcock, C.B. (1979). *Metallurgical thermochemistry*, (5th edn). Pergamon Press, Oxford.
40. Sundaram, C.V., Gupta, C.K., and Garg, S.P. (1981). Extractive metallurgy of refractory metals. A thermodynamic analysis of oxide reduction processes. In *Thermodynamics and kinetics of metallurgical processes*, (ed. M.M. Rao, K.P. Abraham, G.N.K. Iyengar, and R.M. Mallya) pp. 129–44. Bangalore Press, Bangalore.
41. Reddy, S.N.S. and Rapp, R.A. (1988). The electrometallurgy of titanium. In *Terkel Rosenqvist Symposium*, (ed. S.E. Olsen and J.Kr. Tuset). pp. 47–58. The Norwegian Inst. of Technology, Trondheim.
42. Sears, J.W. (1990). Current processes for the cold-wall melting of titanium. *J. of Metals*, March, 17–21.
43. Schulze, K., Bach, O., Lupton, D., and Schreiber, F. (1981). Purification of niobium. In *Niobium, proceedings of the international symposium*. (ed. H. Stuart) pp. 163–224 AIME, Warrendale, Pa, USA.
44. Köck, W. and Paschen, P. (1989). Tantalum—processing, properties and applications. *J. of Metals*, October, 33–9.
45. Feng, C. and Kumov, P. (1989). Correlating microstructure and texture in cold-rolled Ta ingot. *J. of Metals*, October, 40–5.
46. Lauprecht, W.F. and Fairhurst, W. (1983). Verbrauch und Verfügbarkeit von Legierungsmetallen. In *Sondernmetalle, Gewinnung—Verarbeitung—Anwendung*, pp. 103–33. Verlag Chemie, Weinheim.
47. Volkert, G. and Frank, K.-D. (1972). *Metallurgie der Ferrolegierungen*, (2nd edn). Springer Verlag, Berlin.
48. Lenel, F.V. (1981). Powder metallurgy. In *Metallurgical treatises*, (ed. J.K. Tien and J.F. Elliot). pp. 345–60. The Metallurgical Soc. of AIME, Warrendale, Pa, USA.
49. Unterweiser, P.M. and Cobb, H.M. (1987). *Worldwide guide to equivalent irons and steels* (2nd edn). ASM International, Metals Park, Ohio.
50. Lupis, C.H.P. (1983). *Chemical thermodynamics of materials*. North Holland, New York.
51. Darken, L.S., Gurry, R.W., and Bever, M.B. (1953). *Physical chemistry of metals*. McGraw-Hill, New York.
52. Irons, G.A. and Guthrie, R.I. (1981). The kinetics of molten iron desulfurization using magneisum vapor. *Met. Trans. B.*, **12B**, 755–67.
53. Engh, T.A., Kristensen, B., Venaas, K. and Lövold, K. (1980). Desulphurization of hot metal by injection of magnesium granules. In *Light metals proceedings*, pp. 1011–29. AIME, Warrendale, Pa, USA.
54. Iwai, H. and Kunisada, K. (1989). Desulfurization and simultaneous desulfurization and dephosphorization of molten iron by Na_2O-SiO_2 and Na_2O-CaO-SiO_2 fluxes. *ISIJ International*, **29**, No. 2, 135–9.
55. Murayama, T. and Wada, H. (1984). Desulfurization and dephosphorization reactions of molten iron by soda ash treatment. In *Second International Symposium on metallurgical slags and fluxes* (ed. H. A. Fine and D. R. Gaskell) pp. 135–52. The Metallurgical Soc. of AIME, Warrendale, Pa, USA.
56. Kuwabara, T., Umezawa, K., Mori, K., and Watanabe, H. (1988). Investigation of decarburization behavior in RH-reactor and its operation improvement. *Trans. ISIJ*, **28**, 305–14.
57. Tsujino, R., Nakashima, J., Hirai, M., and Sawada, I. (1989). Numerical analysis of molten steel flow in ladle of RH process. *ISIJ International*, **29**, No. 7, 589–95.

58. Pridgeon, J.W., Darmara, F.N., Huntington, S.S., and Sutton, W.H. (1981). Principles and practices of vacuum induction melting and vacuum arc remelting. In *Metallurgical treatises*, (ed. J.K. Tien and J.F. Elliott) pp. 261–76. The Metallurgical Soc. of AIME, Warrendale, Pa, USA.
59. Falk, L., Jardy, A., Ablitzer, D., and Paillière, P. (1987). Thermal modelling of vacuum arc remelting of titanium or zirconium alloys. In *Mathematical modelling of material processing operations*, (ed. J. Szekely, L.B. Hales, H. Henein, N. Jarrett, K. Rajamani, and I. Samarsekera) pp. 1041–54. The Metallurgical Soc. of AIME, Warrendale, Pa, USA.
60. Feinman, J. (ed.) (1987). *Plasma technology in metallurgical processing*. Iron and Steel Soc. Inc., Warrendale, Pa, USA.
61. Hoyle, G. (1983). *Electroslag processes, principles and practice*. Applied Science Publ., London.
62. Susaki, K., Maeda, M., and Sano, N. (1990). Sulfide capacity of CaO–CaF_2–Al_2O_3 slags. *Metall. Trans. B.*, **21B**, 121–30.
63. Woodcock, J.T. (1980). *Mining and metallurgical practices in Australasia*. Monograph Inst. of Mining and Metallurgy, Parkville, Victoria, Australia.
64. Reynolds, J.O. and Flavelle, I.V. (1985). Copper and nickel extraction. In *Scientific and technological developments in extractive metallurgy*. Symposium Series, No. 43, pp. 33–66. The Australasian Inst. of Mining and Metallurgy, Parkville, Victoria, Australia.
65. Betteridge, W. (1984). *Nickel and its alloys*. Ellis Horwood, New York.
66. Edelmetall-Taschenbuch (1967). Degussa, Frankfurt am Main.
67. Brandes, E.A. (1983). *Smithells metals reference book*, (6th edn). Butterworths, London.
68. Barin, I. (1989). *Thermochemical data of pure substances* Parts I and II. VCH Verlagsgesellschaft, Weinheim, Germany.

4

REMOVAL OF DISSOLVED IMPURITY ELEMENTS FROM MOLTEN METALS

4.1 INTRODUCTION

Impurities are already present in the ores and raw materials used in the extractive metallurgical processes. Impurities and inclusions are also introduced from refractories and atmosphere during the production of the metal and even in the refining processes. Thus oxygen, hydrogen, nitrogen, and carbon often enter the molten metal and affect the mechanical properties of the cast metal.

For many extractive processes it is not practical and economic to try to protect the melt from the environment. Instead, the impurities are removed in the final refining operations. Basically in the refining stages we reverse the transfer of impurities. For instance, hydrogen and nitrogen are eliminated by gas purging and vacuum treatment. Then the pressure of H_2 and N_2 is so low in the gas that $\underline{H}$ and $\underline{N}$ leave the melt and enter the bubble or vacuum phase. For the less noble metals, such as iron, oxygen cannot be removed as gas. Instead an element with a high affinity to oxygen, such as aluminium, is introduced into the melt. The resulting reaction product (Al_2O_3 inclusions) must then be transferred out of the melt. Similarly carbon dissolved in molten aluminium from the refractories may react with the melt as it cools down to give solid aluminium carbide inclusions. In many cases these inclusions must be removed before casting the metal. This removal of dissolved elements and inclusions is discussed in this chapter and in Chapter 5. We have discussed in Chapter 1 the harmful effects of impurity elements and inclusions.

This chapter starts out by presenting the various methods that are available for removal. The concept of mixing or stirring is introduced. Then the important question of what determines the rate of removal is brought in. Are the components at an interface in the refining reactor in chemical equilibrium or is removal limited by slow mass transfer or slow chemical reaction rates? There may even be an inert layer that covers the interface, blocks mass transfer, and prevents the final attainment of equilibrium.

The principles of the various reactors that are available for refining purposes are given. Some important applications concerning impurity removal in steel and aluminium are included as examples of the various methods and principles.

The methods available for removal of dissolved elements may be listed as follows

1. The impurity x is transferred to another phase where x has a high solubility. This second phase is not soluble in the molten metal. An example is the transfer of sulphur from molten iron to slag.

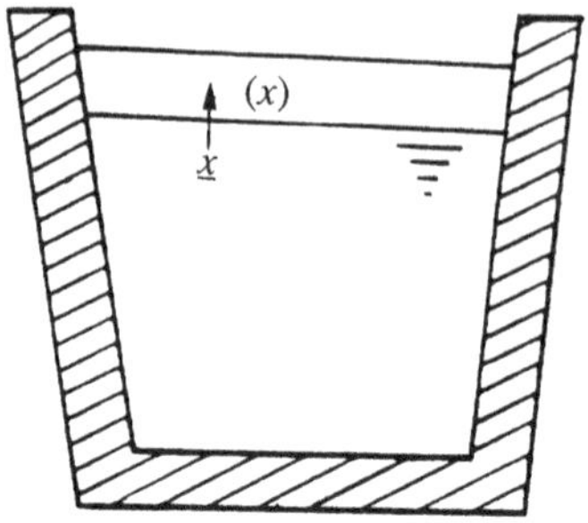

FIG. 4.1 Transfer of dissolved element x from melt to top slag.

2. x reacts with a second phase that may be solid, molten, or gaseous. Figure 4.1 illustrates methods 1 or 2. For instance, sodium is removed from molten aluminium by burning with oxygen in the air at the bath surface.
3. x reacts with another element (added to the melt). The reaction product whether solid, liquid, or gaseous, has to be removed from the melt. This removal may be produced by the effect of gravity or buoyancy forces, inertial forces, or forces generated by stirring the melt. A well-known example is precipitation deoxidation of steel with aluminium. The oxidation product Al_2O_3 must be removed.
4. An electric potential is applied to the melt to remove x by electrolysis.
5. The molten metal is allowed to solidify partially or a solid is partially melted. x is enriched in either the solid or liquid phase; this phase is removed.

All the methods outlined in points 1–5 above are applied industrially. 1 and 2 are the most commonly used. An example of 3 is precipitation deoxidation in steelmaking discussed in Chapter 2. In the reviews in Chapter 3 reference was made to electrochemical methods of purification. Method 5 is not widely used in the purification of metals. It is included because it throws considerable light on various problems in the related field of solidification and casting. Method 5 is discussed in Chapter 9.

The essential point is that usually two or more phases are to react. These phases can be metal, slag, sulphide, gas, and so on. A dissolved component (impurity) or inclusion is removed from the metal and transferred to the slag or gas (atmosphere) phase.

An impurity element can never be completely eliminated from the metal. For instance, when sulphur in molten steel is removed by extracting it to the slag phase, some sulphur will always remain in the steel. The sulphur content may be as low as ten parts per million. Though low, this sulphur level may still play a role in determining the physical and mechanical properties of the final product as we saw in Chapter 1.

Now what determines the final levels of an impurity in a melt? Is it the equilibrium between the impurity content in the metal and in the phase that is extracting the undesirable component, or is it mass transfer? This is one of the main problems that we are addressing here. Whatever the case, some impurity will remain. If the equilibrium is controlling, the chemical activity of the impurity element x is the same in the two phases. To take the example of sulphur; even if one succeeds in obtaining a slag with a very low activity coefficient for sulphur (high sulphur capacity), some $\underline{S}$ must remain in the steel due to the equilibrium. If removal is determined by mass transfer, the removal rates are governed by the driving force. In the most favourable case it is proportional to concentration [% S]. We realize that as [% S] is reduced in a batch melt, removal becomes slower and slower and the time required for removal may be so long that the cost of attaining low impurity levels becomes prohibitive.

4.2 MIXING

Chemical reaction rates increase very strongly with temperature.[1] We remember the well-known Arrhenius relation that tells us that rates increase roughly exponentially with temperature. Therefore in purification operations involving molten metals it is usually assumed that chemical reaction rates are not rate limiting.

Intuitive reasoning suggests that purification of molten metals is speeded up by increased stirring or 'mixing'. Mixing may be performed by the use of:

(1) mechanical stirrers;

(2) oscillating ladles;

(3) gas stirring (bubbling);

(4) electromagnetic stirring;

(5) pouring from one ladle to another (Perrin process);

(6) pumps (mechanical or vacuum).

Mechanical stirrers (impellers) in steel melts are usually of a ceramic material. Often they will not stand up to the combined chemical and erosion attack by the slag and metal phases. However, in an aluminium melt at temperatures below 800°C we find that graphite impellers may operate satisfactorily.

Oscillating ladles are usually no longer employed. The mechanical problems involved in moving the whole ladle with its large mass of ceramic material and the wear on bearings, axles, and so on are too troublesome.

Gas stirring (gas injection) in a ladle using a porous plug or a lance is an effective and well-established operation, but has various limitations. For instance we usually cannot employ this method to mix top-slag down into the melt. In converters, however, where the gas flows are very great, molten metal and slag are mixed.

Electromagnetic stirring often requires a high capital expenditure. Nevertheless this method may be competitive due to the fact that the stirrer is not in direct contact with the melt (see also p. 396).

The Perrin process is effective, but it often does not give reproducible results. The melt is exposed to the atmosphere.

Mixing by pumps seems to be a very interesting field in high-temperature metallurgy, though it does not seem to have been properly explored as yet.

It should be mentioned that as a side-effect of all methods unwanted 'stirring' is generated at the melt–refractory interfaces. This increases refractory wear.

In the following discussions of purification it will be assumed that mixing or stirring is sufficient. This means that the concentration of dissolved elements throughout the bulk metal bath is constant, that is, concentration does not depend on spatial coordinates. For a batch reactor, however, concentrations will change with time. Often to characterize 'mixing', the concept of mixing time $\tau_{\rm m}$ is introduced:[2]

$$\tau_m \sim 100\left(\frac{L^2}{\dot{\varepsilon}}\right)^{\frac{1}{3}}. \tag{4.1}$$

Equation (4.1) assumes that mixing takes place by 'eddy diffusion' and that the largest eddies play the dominant role in this process. Thus $\tau_{\rm m}$ depends on a characteristic vessel dimension, L, for instance the diameter. Also it is assumed that turbulence is homogeneous and isotropic. Then turbulence may be characterized by the energy dissipation $\dot{\varepsilon}$. Electromagnetic stirring of a steel ladle typically has $\dot{\varepsilon} \sim 0.05\ \mathrm{W/kg} = 0.05\ \mathrm{m^2/s^3}$. For a ladle with $L = 1$ m this gives $\tau_{\rm m} = 270\ \mathrm{s} = 4\frac{1}{2}$ min. If the treatment time for the melt is considerably greater than $\tau_{\rm m}$, we often employ the assumption that the melt is completely mixed. This means that we disregard resistance to mass transfer in the bulk melt (see also Section 7.2).

At the interface between two phases, such as metal and slag, movement is

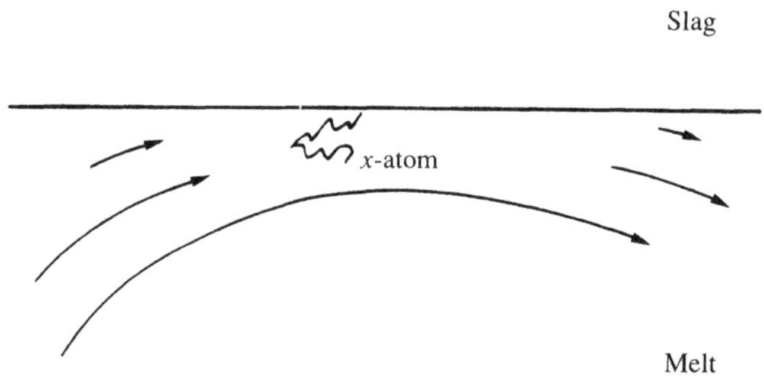

FIG. 4.2 Velocities are low along the interface so that the transfer of x depends on the slow mechanism of diffusion.

along the interface and not normal to it. This is illustrated in Fig. 4.2. Now the main mechanism for transport of a dissolved component or an inclusion is no longer convection or turbulent edies. Instead transport to a large degree must depend on the slow mechanism of diffusion. Then the mass transfer coefficient for the melt, k, may be estimated by relations of the type found in penetration theory[3,4] (see Chapter 7).

$$k \sim Dt^{-\frac{1}{2}} \tag{4.2}$$

where D is the (binary) diffusion coefficient and t is the *residence time* of the melt at the interface.

It will be seen in the following that often the product of k and the interfacial contact area A determines the kinetics of the purification process. This is the case when kA is too small to allow us to reach equilibrium.

For transfer between two phases we must in general take into account the resistance to transfer on both sides of the interface. This is discussed in the next section.

4.3 THE TOTAL MASS TRANSFER COEFFICIENT, k_t

There is resistance to flow in the 'films' on both sides of the interface between the metal and the extractive phase (here taken as slag). Here it will be seen that the two resistances can be combined into a total resistance, k_t. By analogy with electrical circuit theory one might believe that the two resistances given as the inverse of the mass transfer coefficients $1/k$ and $1/k_{slag}$, could simply be added together. However we will see that the procedure is not quite that simple. The reason is that the driving forces (voltages) in the rate equations for the two phases are not given in the same units.

For the metal phase, the molar flux of component x is in kmol/(m^2 s):

$$\dot{n} = \frac{k\rho}{100M_x}([\%\,x] - [\%\,x]_i). \tag{4.3}$$

Here M_x is the molar weight of component x in kg/kmol and ρ is the density of the metal. The metal and slag concentrations at the interface, $[\%\,x]_i$ and $(\%\,x)_i$, are assumed to be in equilibrium (due to the high chemical reaction rates at the high temperature):

$$K = \frac{(\%\,x)_i\gamma_x}{[\%\,x]_i f_x}. \tag{4.4}$$

Equation (4.3) can be rewritten in terms of the driving force (potential, voltage) thus:

$$[\%\,x] - [\%\,x]_i = \frac{\dot{n}100M_x}{k\rho}. \tag{4.5}$$

Similarly for the slag phase one has

$$\dot{n} = \frac{k_{\text{slag}}\,\rho_{\text{slag}}}{100M_x}\left((\%\,x)_i - (\%\,x)\right).$$

Again in terms of the driving force one obtains

$$(\%\,x)_i - (\%\,x) = \frac{\dot{n}100M_x}{k_{\text{slag}}\rho_{\text{slag}}}. \tag{4.6}$$

As already pointed out the units for the driving force are mass per cent metal in eqn (4.5) and mass per cent slag in eqn (4.6). However, we notice that if we divide by the partition coefficient

$$\frac{f_x K}{\gamma_x} = \frac{(\%\,x)_i}{[\%\,x]_i}$$

eqn (4.6) becomes

$$[\%\,x]_i - \frac{(\%\,x)\gamma_x}{Kf_x} = \frac{\dot{n}100M_x\gamma_x}{k_{\text{slag}}\rho_{\text{slag}}Kf_x}. \tag{4.7}$$

Now the molar flux $\dot{n}$ can be obtained in terms of the sum of the driving forces in the two phases. We sum equations (4.5) and (4.7) to give

$$[\%\,x] - \frac{(\%\,x)\gamma_x}{Kf_x} = \dot{n}100M_x\left(\frac{1}{\rho k} + \frac{\gamma_x}{\rho_{\text{slag}}k_{\text{slag}}Kf_x}\right). \tag{4.8}$$

We can finally define a total resistance $1/(\rho k_t)$ equal to the sum of the resistance for the metal phase $1/(\rho k)$ and for the slag phase $\gamma_x/(\rho_{\text{slag}}k_{\text{slag}}Kf_x)$. Equation (4.8) becomes

$$\dot{n} = \frac{\rho\,k_t}{100M_x}\left([\%\,x] - \frac{(\%\,x)\gamma_x}{Kf_x}\right). \tag{4.9}$$

$(\%\,x)\gamma_x/(Kf_x)$ is the hypothetical concentration in the metal in equilibrium with the concentration in the slag phase. Figure 4.3 sketches the concentration profiles in the two phases. Equation (4.9) plays an important role in modelling purification processes. It is derived for metal and slag, but applies to any two-phase system. For instance, it is used in the following section for the melt-bubble interface.

4.4 EQUILIBRIUM OR MASS TRANSFER CONTROL IN GAS PURGING; REACTIVE GAS

In Chapter 2, thermodynamic factors were considered. We now consider the rate at which a dissolved component x moves to the extracting phase. As an

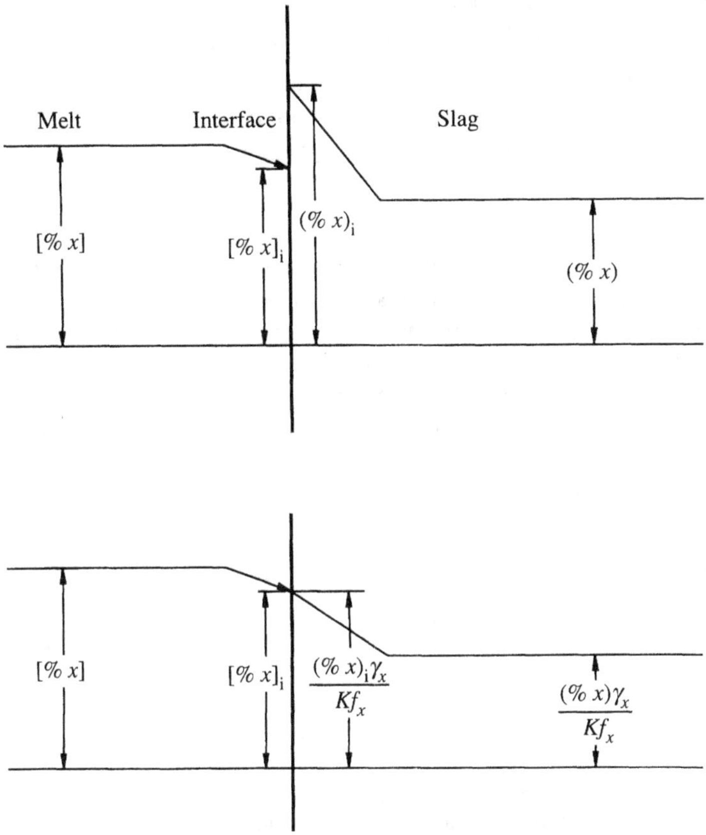

FIG. 4.3 The concentration in the slag is converted to concentration in the melt by dividing with the partition coefficient Kf_x/γ_x.

example, we study removal to ascending inert gas bubbles (gas purging). In general x is transferred by the following series of steps:

(1) transport in the melt to the vicinity of a gas bubble by a combination of convection and diffusion;

(2) diffusive transport through a thin layer of fluid, called a boundary layer, surrounding the bubble;

(3) chemical adsorption on to the outside and subsequent desorption from the inside of the bubble surface;

(4) diffusion of x or x_2 as a gaseous species inside the bubble of purge gas.

We now present a mathematical analysis of removal based on use of the total mass transfer coefficient. Four assumptions are made.

1. The metal bath is well stirred so that the concentration is essentially uniform.
2. The concentration is uniform inside the gas bubble.
3. The gas bubble surface area is not a function of height in the bath. This is tantamount to assuming that the total pressure in the gas bubble is constant.
4. The partial pressure of x in the bubbles is very small compared to the total pressure.

Assumption 1 should be met in any well-designed commercial process and is discussed in Chapter 7. Assumption 3 applies to systems where the bath is shallow. This is often the case, for instance, when aluminium melts are treated.

The third and fourth conditions do not apply for vacuum systems. These two assumptions allow us to discuss equilibrium and mass transfer in a simple manner. Once the main points are understood, it is not difficult using a computer to remove assumptions 3 and 4. We consider the change in gas composition as a bubble ascends through the melt. Figure 4.4 shows a horizontal section of a reactor. If we construct a mass balance equation for component x leaving this control volume of the melt and entering the gas phase, we find:

Moles of x transferred per second to bubbles in control volume
= Increase of flow of x carried by bubbles through control volume

or

$$\frac{k_t \rho([\%\, x] - [\%\, x]_e)\Delta A}{100 M_x} = \dot{G}\Delta\left(\frac{p_x}{p_{\text{inert}}}\right). \tag{4.10}$$

In accordance with assumption 4, p_{inert} = constant. We represent mass transfer with the melt-side total mass transfer coefficient k_t, having units of metres

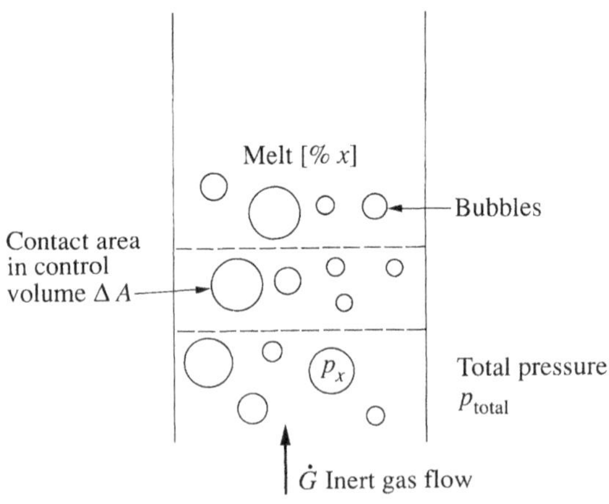

FIG. 4.4 Gas purging control volume.

per second, the surface area of the bubbles ΔA, and the driving force for transfer $[\%\,x] - [\%\,x]_e$. Here $[\%\,x]_e$ is the hypothetical x concentration in the melt in equilibrium with the partial pressure p_x in the bubble. p_x is uniform throughout the bulk gas in the bubble

$$[\%\,x]_e f_x = \frac{p_x}{K}. \tag{4.11}$$

ΔA is the bubble surface area present in the volume element under consideration (see Fig. 4.4). According to the gas law the number of moles of x is proportional to p_x. Thus $\dot{G}\Delta(p_x/p_{\text{inert}})$ is the incremental change in the gas composition of x between bubbles entering and leaving the volume element. $\dot{G}$ is the constant molar flow of inert gas.

We notice that the driving force for the bubble changes as it rises up through the melt since $[\%\,x]_e$ in eqn (4.11) changes from zero to p_x^0/Kf_x traversing the melt. We move the terms that depend on p_x over to the right-hand side of eqn (4.10) and integrate from bottom to top surface:

$$\frac{k_t \rho A p_{\text{inert}}}{100 M_x \dot{G}} = \int_0^{p_x^0} \frac{\mathrm{d}p_x}{[\%\,x] - p_x/(f_x K)}.$$

This gives:

$$\frac{k_t \rho A p_{\text{inert}}}{100 M_x \dot{G} f_x K} = \ln\left(\frac{[\%\,x]}{[\%\,x] - p_x^0/(f_x K)}\right).$$

The left-hand side of the equation is a dimensionless group ϕ_A, for the contact area A. We relate the exit pressure p_x^0 to the value at equilibrium, $f_x\,[\%\,x]K$, '*the efficiency of gas purging*':

$$Z = \frac{p_x^0}{f_x[\%\,x]K}. \tag{4.12}$$

p_x^0 cannot exceed the value $f_x\,[\%\,x]\;K$ for equilibrium with the melt. Thus $z < 1$. The antilogarithm of the previous equation gives

$$Z = 1 - \exp(-\phi_A)$$

where

$$\phi_A = \frac{k_t \rho A p_{\text{inert}}}{100 M_x \dot{G} f_x K}.$$

Figure 4.5 shows the efficiency Z as a function of ϕ_A (dimensionless contact area).

To approach equilibrium ϕ_A must be greater than 3 say. For $\phi_A = 3$, $Z = 0.95$. If $\phi_A = 2$, then $Z = 0.86$. ϕ_A increases with the ratio between contact area and gas flow rate $k_t A/\dot{G}$. Also it is apparent that ϕ_A increases

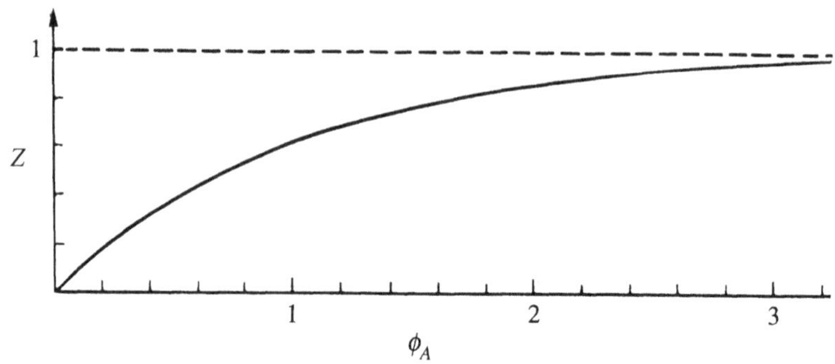

FIG. 4.5 Efficiency Z versus dimensionless contact area ϕ_A.

when the equilibrium constant K is small. When K is small, the bubbles at equilibrium will have a low partial pressure of x and the 'x-capacity' of the bubbles is low. Similarly ϕ_A increases with p_{inert} which is approximately equal to the total pressure. When the total pressure is high, the gas volume available to be 'filled' with gaseous x is low.

When the dimensionless group for contact area ϕ_A is less than, say, 1/20, $\exp(-\phi_A) \approx 1 - \phi_A$
and

$$Z \approx \phi_A .$$

Then from the definitions of Z and ϕ_A

$$\dot{G}_X = \frac{\dot{G}p_x^0}{p_{\text{inert}}} = \frac{k_t \rho A\,[\%\,x]}{100M_x} . \tag{4.13}$$

This is the case when the counter-driving force term $[\%\,x]_e$ in eqn (4.10) can be neglected; the driving force is equal to the concentration $[\%\,x]$. In summary, for $\phi_A > 3$ equilibrium between bubble and melt controls removal by gas purging. For $\phi_A < 1/20$, mass transfer as given by eqn (4.13) controls removal.

By way of illustration, we give a numerical example: sodium is purged from aluminium with an inert gas.

$$\dot{G} = 2\,\text{Nm}^3 \text{ per hour} = \frac{2}{22.4 \times 3600}\,\text{kmol/s}$$

$$\frac{1}{Kf_{\text{Na}}} = 0.48 \text{ at } 1100\,\text{K (see Chapter 2)}$$

$$M_{\text{Na}} = 23\,\text{kg/kmol}$$

$$k_t = 5 \times 10^{-4}\,\text{m/s (see Chapter 7)}$$

$$A = 0.3\,\mathrm{m}^2$$

$$p_{\mathrm{inert}} = 1$$

$$\rho = 2350\,\mathrm{kg/m^3}.$$

This gives

$$\phi_A = \frac{5 \times 10^{-4} \times 2350 \times 0.3 \times 1 \times 0.48 \times 22.4 \times 3600}{100 \times 23 \times 2} = 2.97$$

which gives an efficiency for the purge gas of

$$Z = 1 - \exp(-2.97) = 0.95.$$

The contact area A of $0.3\,\mathrm{m}^2$ is sufficient to give an efficiency of 95 per cent. We will see in the following that the bubble contact area A decreases if the bubble size increases.

Sodium is also removed at the top surface; sometimes we can even see sodium flames. This mechanism of removal is included later. Eqn (4.13) is valid also when a gas is added to the purge gas that reacts with x. For instance, Cl_2 will react with $\underline{\mathrm{Na}}$ to give NaCl:

$$\underline{\mathrm{Na}} + \tfrac{1}{2}\mathrm{Cl_2} \rightarrow \mathrm{NaCl}.$$

Then the counter-force term $[\%\,x]_e$ in eqn (4.10) can be neglected since the partial pressure of Na in equilibrium with NaCl and Cl_2 (actually $AlCl_3$) is very small. Of course this is the case only if enough Cl_2 is added to the purge gas to react with all the sodium that can be transferred to the bubbles.

If we inject Cl_2 into an aluminium alloy that contains Mg, we must expect that $AlCl_3$ first reacts with Mg since the content of Mg is much greater than for Na:

$$2\mathrm{AlCl_3} + 3\underline{\mathrm{Mg}} = 3\mathrm{MgCl_2} + 2\mathrm{Al}.$$

Then $MgCl_2$ takes part in an exchange reaction with sodium:

$$\mathrm{MgCl_2} + 2\underline{\mathrm{Na}} = 2\mathrm{NaCl} + \underline{\mathrm{Mg}}$$

since sodium has much greater affinity to chlorine than Mg (see Chapter 2 on thermodynamics and Chapter 3). The rate of production of $MgCl_2$ depends on the bubble surface area, while removal of Na depends on the $MgCl_2$/melt contact area.

When stainless steel is decarburized down to very low levels by blowing O_2 and argon, the behaviour is analogous. Oxygen first reacts with iron and chromium and then, if the contact area between the oxides and carbon is sufficient, carbon is removed.

4.5 BUBBLE CONTACT AREA

To simplify matters we assume initially that we have spherical bubbles of radius R. Also we do not take into account that, according to the gas law, the bubble volume increases with decreasing pressure as the bubble rises up through the melt.

If the flow of gas is $\dot{Q}_g$ m^3 per second, the number of bubbles created per second is

$$\dot{N} = \frac{\dot{Q}_g}{4\pi R^3/3}.$$

With a rise time (residence time) of τ_b this gives for the number of bubbles present in the melt:

$$N = \dot{N}\tau_b.$$

The contact area now becomes

$$A = N4\pi R^2 = \frac{3\dot{Q}_g\tau_b}{R}. \tag{4.14}$$

This equation is important since it demonstrates that the contact area, A, is inversely proportional to the bubble size, R. Also A is proportional to the residence time τ_b. Often τ_b in the literature is obtained by assuming that bubble velocities are much greater than convective velocities and that the bubbles move independently of each other. Then

$$\tau_b = \frac{H}{v_b} \tag{4.15}$$

where H is the vertical distance from the gas inlet to the bath surface and v_b is the velocity of a single bubble. Typically $H = 0.5$ m and $v_b = 0.5$ m/s so that $\tau_b = 1$ second.

If we assume that the gas instantly reaches the melt temperature in our previous example with the aluminium melt, then

$$\dot{Q}_g = \frac{2 \times 1100}{273 \times 3600} = 0.0022\ \text{m}^3/\text{s}.$$

If bubbles with $R = 0.010$ m are produced, then

$$A = \frac{3 \times 0.0022 \times 1}{0.01} = 0.66\ \text{m}^2.$$

As seen in the previous example this contact area is sufficient to allow the sodium gas in the bubbles to approach equilibrium. To increase the contact area, one should, according to eqn (4.14), either decrease the bubble size R or, increase the residence time, τ_b, of the bubbles. Injection of gas through a

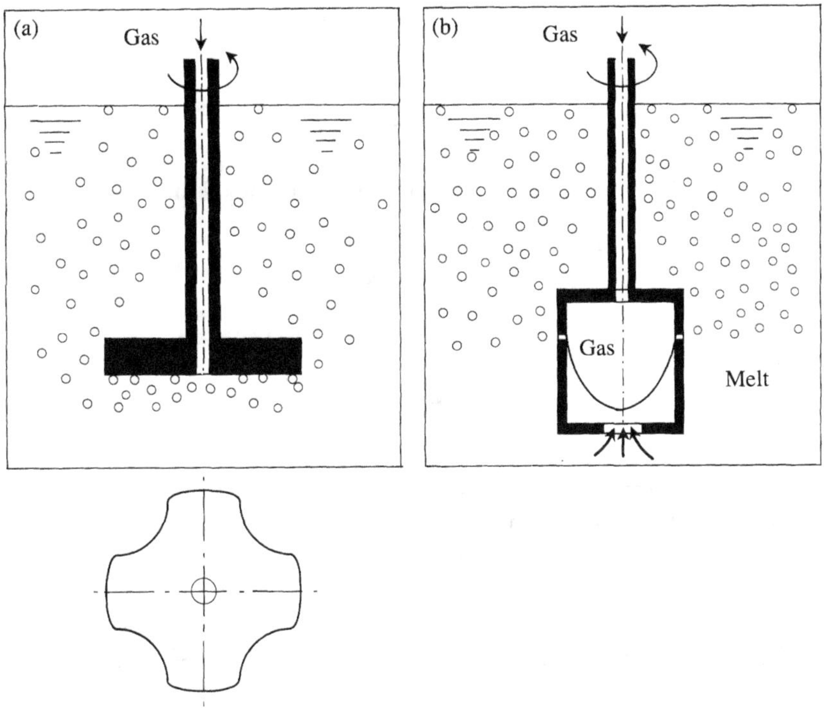

FIG. 4.6 (a) Rotor for creating and dispersing bubbles in aluminium; (b) a rotor-pump system for creating and dispersing bubbles in the melt.[10]

lance is not a solution since the bubbles roughly have a diameter equal to the outer lance diameter.[5,6] A porous plug may give bubbles of size around 1 mm at low gas volume flow rates[7] in water. However, in molten metals, the material in the porous plug is often not wetted by the melt. Then the bubbles may become much larger.[8] Various impeller and jetting systems have been developed that stir the melt and produce small bubbles.[9-12] Figure 4.6 gives two examples of such systems.

In eqn (4.14) it was assumed that all the bubbles were of the same size. In the real case there is a distribution of bubble sizes; a fraction f_R of the bubbles produced have size R. Then the number of bubbles of size R produced per second is

$$\dot{N}_R = \dot{N} f_R. \tag{4.16}$$

If these bubbles have a residence time τ_R, the number of bubbles of size R in the melt becomes

$$N_R = \dot{N} f_R \tau_R$$

and the corresponding contact area is

$$A_R = \dot{N} f_R \tau_R 4\pi R^2.$$

The total contact area A is obtained by summation over all bubble sizes

$$A = \dot{N} \sum f_R \tau_R 4\pi R^2.$$

$\dot{N}$ is equal to the volume flow divided by the volume of the bubbles

$$\dot{N} = \frac{\dot{Q}_g}{\sum f_R 4\pi R^3/3}.$$

Inserted in the previous equation this gives for the contact area

$$A = \frac{3\dot{Q}_g \sum f_R \tau_R R^2}{\sum f_R R^3}. \tag{4.17}$$

f_R can be measured by employing electrical probes,[13] using a pressure pulse technique,[14] or f_R may be obtained by photographic methods from water model studies. Equation (4.17) does not take into account the effect of coalescence and splitting of bubbles or the effect of pressure changes up through the melt.

Present theory and experimental methods do not seem sufficiently developed to allow us to consider coalescence and splitting in practice.

4.6 CONTINUOUS BACK-MIX REACTORS

Continuous reactors are conceptually simpler than batch reactors. Therefore we discuss them first even though batch reactors are employed in most metal purification processes. Continuous reactors are perhaps mainly used in aluminium cast houses, for instance, for gas purging.

Usually it is a good approximation to assume that the melt is completely mixed as mentioned initially in this chapter. (This point is discussed further in Chapter 7). To proceed we now refer to Fig. 4.7. Notice that x changes discontinuously from $[\%\,x]_{\text{in}}$ in entering the reactor to $[\%\,x]$ constant throughout the melt. The mass balance for x is then:

x (in kmols/second) following melt into reactor
  = x (in kmols/second) following melt out of reactor
    + x removed from melt per unit time

or

$$\frac{\dot{M}[\%\,x]_{\text{in}}}{100M_x} = \frac{\dot{M}[\%\,x]}{100M_x} + \dot{G}_x. \tag{4.18}$$

Here $\dot{M}$ is the flow of metal into the reactor in kg/s.

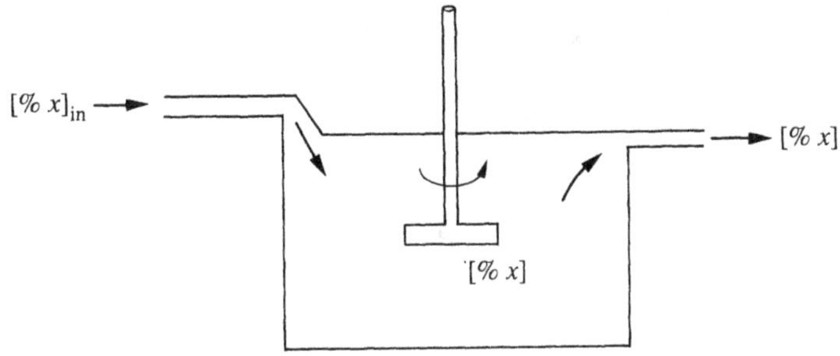

FIG. 4.7 Reactor where melt is completely mixed.

Removal, $\dot{G}_x$, is often by several mechanisms. For instance when gas-purging aluminium, Na is removed partly as gas in the bubbles and partly by evaporation from the top surface and subsequent burning with oxygen in the atmosphere. Na has a great affinity to oxygen so that the reaction

$$2\,\underline{\mathrm{Na}} + \tfrac{1}{2}\,\mathrm{O_2} \rightarrow \mathrm{Na_2O(s)} \tag{4.19}$$

is displaced completely over to the right. Then $[\%\,x]_e \approx 0$ and the driving force for the transfer of $\underline{\mathrm{Na}}$ through the surface is taken to be equal to the concentration in the bath $[\%\,x]$.

According to the gas law, removal to the bubbles is equal to $\dot{G}p_x^0/p_{\mathrm{inert}}$. This removal added to transfer at the top surface gives (see the definition of Z)

$$\dot{G}_x = \frac{\dot{G}Kf_x[\%\,x]Z}{p_{\mathrm{inert}}} + \frac{k_{\mathrm{ts}}A_s\rho[\%\,x]}{100M_x}. \tag{4.20}$$

Here k_{ts} is the total mass transfer coefficient for the bath top surface. A_s is the bath top surface area.

We notice that both terms in the equation for $\dot{G}_x$ are proportional to $[\%\,x]$. This allows us to calculate the reduction ratio $[\%\,x]/[\%\,x]_{\mathrm{in}}$ for the reactor. Equations (4.20) and (4.18) give

$$\frac{[\%\,x]}{[\%\,x]_{\mathrm{in}}} = \left(1 + \frac{\dot{G}f_xKZ100M_x}{p_{\mathrm{inert}}\dot{M}} + \frac{k_{\mathrm{ts}}A_s\rho}{\dot{M}}\right)^{-1}. \tag{4.21}$$

Due to the moderate vapour pressure of Na, we will see that the term for the effect of the bubbles in the parenthesis above is smaller than the last term for removal from the top surface. In this case the main effect of gas purging is to stir the top surface and there give a high value of the mass transfer coefficient k_{ts}. It should be noted that with increasing throughput $\dot{M}$ the reduction ratio becomes less favourable (increases).

It may be useful to expand the numerical example used previously:

$$\dot{G} = 2\,\mathrm{Nm^3/hr} = 2.48 \times 10^{-5}\,\mathrm{kmol/s}$$

$$M_{\mathrm{Na}} = 23\,\mathrm{kg/kmol}$$

$$\frac{1}{Kf_{\text{Na}}} = 0.48$$

$$p_{\text{inert}} = 1 \text{ bar}$$

$$Z = 0.95$$

$$\rho = 2350 \text{ kg/m}^3$$

$$A_s = 1 \text{ m}^2$$

$$\dot{M} = 0.5 \text{ kg/s}$$

$$k_{ts} = 1.5 \times 10^{-4} \text{ m/s (see Chapter 7).}$$

This gives, on insertion in eqn (4.21),

$$\frac{[\%\,\text{Na}]}{[\%\,\text{Na}]_{\text{in}}} = (1 + 0.23 + 0.71)^{-1} = 0.52.$$

The term 0.23 gives the removal by gas purging, while 0.71 is due to the top surface. We notice that inert gas purging does in fact play a role. However, in a practical system (see Chapter 7) the gas purging efficiency Z will often be lower than indicated in this example.

Removal by gas purging can be enhanced if a reactive gas is added and the contact area is increased. In this case the driving force for transfer to bubbles is equal to the concentration $[\%\,x]$ (eqn (4.13) applies). $k_t = 5 \times 10^{-4}$ m/s and, say, $A = 3\text{ m}^2$ then gives

$$\frac{[\%\,\text{Na}]}{[\%\,\text{Na}]_{\text{in}}} = (1 + 7.1 + 0.71)^{-1} = 0.11.$$

The term for the bubbles is 7.1 as compared with 0.23 when only inert gas is employed. This value 7.1 is impressive compared with 0.23 even though sodium has a high vapour pressure compared to most other dissolved impurities. For impurities with a low vapour pressure such as Li or Ca, a reactive gas must be added if the impurities are to be removed by bubbles.

In some cases, one wants to increase production (the throughput $\dot{M}$) and retain the low levels of sodium in the output. Equation (4.21) then tells us that $k_{ts}A_s\rho/\dot{M}$ should remain constant when $\dot{M}$ is increased. However, k_{ts} only increases slightly with an increase in the purge gas flow $\dot{G}$. Therefore the top surface contact area A_s must be increased roughly as $\dot{M}$.

Often it is not practical to rebuild the purification unit in order to increase A_s, perhaps it is possible to install a similar unit in series with the first. Let us look at the case that $\dot{M}$ is to be doubled from $\dot{M} = 0.5$ kg/s to $\dot{M} = 1$ kg/s and that two identical units are to be installed. Then

$$\frac{[\%\,\text{Na}]_1}{[\%\,\text{Na}]_{\text{in}}} = \left(1 + \frac{0.23}{2} + \frac{0.71}{2}\right)^{-1} = 0.68$$

where [% Na]$_1$ is the output from the first unit. The output from the second unit is similarly

$$\frac{[\%\ \mathrm{Na}]_2}{[\%\ \mathrm{Na}]_1} = 0.68.$$

When multiplied, these two equations give

$$\frac{[\%\ \mathrm{Na}]_2}{[\%\ \mathrm{Na}]_{\mathrm{in}}} = 0.68 \times 0.68 = 0.46.$$

Thus if we double the throughput and install an extra unit, we not only retain the quality but in fact improve it: from [% Na]/[% Na]$_{\mathrm{in}}$ = 0.52 down to the ratio of 0.46.

It may be fruitful to speculate a little regarding the above result. Why is it better to treat the melt in two units in series instead of one? Removal is proportional to the concentration [% x]. When we mix the input [% x]$_{\mathrm{in}}$ into the back-mix tank, our concentration is reduced from [% x]$_{\mathrm{in}}$ to [% x]. Thus when we carry out the mixing operation, we lose the driving force for the removal of x. With two back-mix tanks in series, the concentration is reduced in two steps: from [% x]$_{\mathrm{in}}$ to [% x]$_1$ and then to [% x]$_2$ instead of directly to [% x]. Therefore, for the same total contact areas and purge gas flows, two back-unit tanks in series are better than one.

What happens in the idealized case when we have N instead of two identical back-mix tanks in series? In this case the flow of inert gas $\dot{G}$ is distributed to N tanks and the contact area is A_s/N in each tank. Then eqn (4.21) is replaced by

$$\frac{[\%\ x]}{[\%\ x]_{\mathrm{in}}} = \left(1 + \frac{a\dot{G}}{N\dot{M}} + \frac{bA_s}{N\dot{M}}\right)^{-N} \tag{4.22}$$

where [% x] is the output from Nth unit and

$$a = \frac{f_x Z 100 M_x K}{p_{\mathrm{inert}}} \qquad b = k_{\mathrm{ts}}\rho.$$

If we allow N to increase to infinity, a well-known mathematical result is that

$$\left(1 + \frac{a\dot{G}}{N\dot{M}} + \frac{bA_s}{N\dot{M}}\right)^{-N} \rightarrow \exp\left(-\frac{a\dot{G}}{\dot{M}} - \frac{bA_s}{\dot{M}}\right). \tag{4.23}$$

In the numerical example, [% x]/[% x]$_{\mathrm{in}}$ = exp(−0.94) = 0.39. Thus introduction of an infinite number of tanks instead of two has given a further improvement in purification from 0.46 down to 0.39. Now, this result does not seem to be very impressive. However, Fig. 4.8 shows that if reduction ratios [% x]/[% x]$_{\mathrm{in}}$ less than $\frac{1}{4}$ are desired, the system with $N \rightarrow \infty$ gives a far better performance than a single back-mix tank.

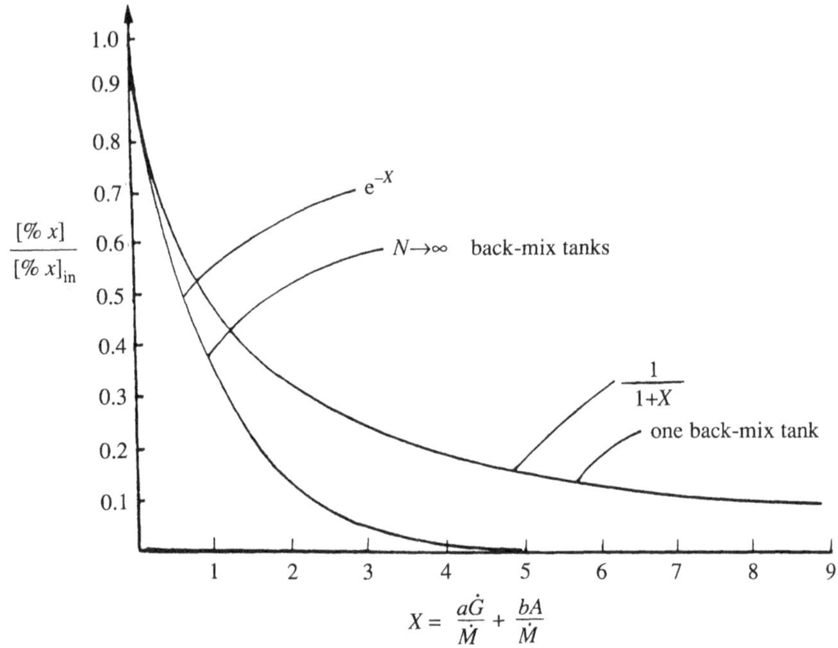

FIG. 4.8 Ratio between concentration [% x] and original concentration [% x]$_i$ as a function of X.

The effect of a large number of back-mix tanks is to give the same residence time for all elements of fluid:

$$\tau = \frac{M}{\dot{M}}$$

M is the total mass of the melt in all the N tanks. The behaviour is the same as for a reactor where the metal flows plug-like with residence time τ (or for a batch reactor operated for a time τ).

In practical high-temperature refining metallurgy, one hardly ever finds systems that behave like the plug-flow reactor or like N tanks in series. However, for systems where a low ratio [% x]/[% x]$_{in}$ is required, there is a strong economic incentive to attempt an approach to the *constant residence time continuous reactor*. Thus in refining aluminium three units in series are in use.

4.7 BATCH REACTORS

As mentioned previously, batch reactors are still the most commonly used reactors in metal refining operations. For instance in steelmaking all industrial units handle a batch of metal.

Our treatment of batch reactors is very similar to that given for continuous reactors. However, the mass balance given in eqn (4.18), now becomes:

Reduction of x (in kmols/second) contained in the melt per unit time
= x transferred (in kmols/second) from melt per unit time

or

$$-\frac{M}{100\,M_x}\frac{\mathrm{d}[\%\,x]}{\mathrm{d}t} = \dot{G}_x. \tag{4.24}$$

If we return to the example with gas purging

$$\dot{G}_x = (a\dot{G} + bA_s)\,\frac{[\%\,x]}{(100\,M_x)}. \tag{4.25}$$

Equation (4.24) then becomes

$$\frac{\mathrm{d}[\%\,x]}{[\%\,x]} = -(a\dot{G} + bA_s)\frac{dt}{M}$$

or integrating from $[\%\,x] = [\%\,x]_i$ at time $t = 0$ to $[\%\,x]$ at time t

$$\int_{[\%\,x]_{in}}^{[\%\,x]} \frac{\mathrm{d}[\%\,x]}{[\%\,x]} = -\int_0^t \frac{(a\dot{G} + bA_s)\mathrm{d}t}{M}.$$

For constant operating conditions so that $a\dot{G} + bA_s$ does not change with time, the above equation gives

$$\ln\frac{[\%\,x]}{[\%\,x]_{in}} = -(a\dot{G} + bA_s)\frac{\mathrm{t}}{\mathrm{M}}$$

or

$$\frac{[\%\,x]}{[\%\,x]_{in}} = \exp\left\{\frac{(-a\dot{G} - bA_s)t}{M}\right\}. \tag{4.26}$$

The reduction ratio drops exponentially with time.

This result may be compared with eqn (4.23) valid for a plug-flow (constant residence time) continuous reactor. If we set $t = \tau = M/\dot{M}$ in eqn (4.26), then eqn (4.23) is obtained. The reduction ratio is the same whether the fluid element is in a batch reactor or moving through a continuous plug-flow reactor.

We have seen that removal in a batch reactor is equal to the best attainable in a continuous system. Why then is there a strong trend in industry towards the development and use of continuous purification systems? The answer is probably that batch processes are messy and labour consuming due to the discontinuous operations of filling and tapping the refining ladle. Also, batch systems are difficult to instrument and control due to the inherent time dependence involved.

4.8 TRADITIONAL SLAG-METAL REFINING IN A BATCH REACTOR (LADLE)

One problem with refining metal by extraction to a second (slag) phase is the problem of mixing in the slag phase. Often the slag phase is relatively viscous so that it is difficult to obtain an even distribution of the impurity element in the slag. However, here we assume that also the slag phase is 'completely mixed'.

Equation (4.24) then becomes

$$-M\frac{\mathrm{d}[\%\,x]}{\mathrm{d}t} = k_{\mathrm{t}}\rho A_{\mathrm{s}}([\%\,x] - [\%\,x]_{\mathrm{e}}). \tag{4.27}$$

$[\%\,x]_{\mathrm{e}}$ is the hypothetical concentration in the metal in equilibrium with the actual concentration in the slag $(\%\,x)$:

$$[\%\,x]_{\mathrm{e}}Kf_x = \gamma_x(\%\,x). \tag{4.28}$$

To integrate eqn (4.27) over time $[\%\,x]_{\mathrm{e}}$ must be replaced by a function of $[\%\,x]$. $[\%\,x]_{\mathrm{e}}$ may be obtained in terms of $[\%\,x]$ if we note that whatever leaves the melt enters the slag:

$$M([\%\,x]_{\mathrm{in}} - [\%\,x]) = M_{\mathrm{s}}(\%\,x)$$

where M_{s} is the amount of slag. Here we assume that the slag originally did not contain any component x. Thus the above equation together with eqn (4.28) gives

$$[\%\,x]_{\mathrm{e}} = \frac{\gamma_x}{f_xK}\frac{M}{M_{\mathrm{s}}}([\%\,x]_{\mathrm{in}} - [\%\,x]). \tag{4.29}$$

The driving force in eqn (4.27) becomes

$$[\%\,x] - [\%\,x]_{\mathrm{e}} = [\%\,x]\left(1 + \frac{\gamma_xM}{Kf_xM_{\mathrm{s}}}\right) - \frac{\gamma_xM}{Kf_xM_{\mathrm{s}}}[\%\,x]_{\mathrm{in}}. \tag{4.30}$$

The lowest value of $[\%\,x]$ attainable is when the driving force given by eqn (4.30) becomes zero.
Then

$$[\%\,x] = [\%\,x]_{\infty} = \frac{M[\%\,x]_{\mathrm{in}}}{M + Kf_xM_{\mathrm{s}}/\gamma_x}. \tag{4.31}$$

$[\% x]_\infty$ is the value of $[\% x]$ when equilibrium between slag and melt is finally reached (at time $t \to \infty$). In terms of $[\% x]_\infty$ we may write for the driving force:

$$[\% x] - [\% x]_e = \left(1 + \frac{\gamma_x M}{K f_x M_s}\right) ([\% x] - [\% x]_\infty). \tag{4.32}$$

This equation, introduced into eqn (4.27), gives

$$\int_{[\% x]_{in}}^{[\% x]} \frac{d[\% x]}{[\% x] - [\% x]_\infty} = -\int_0^t \frac{k_t \rho A_s}{M} \left(1 + \frac{\gamma_x M}{K f_x M_s}\right) dt. \tag{4.33}$$

This gives, on integration and assuming that k_t, γ_x, f_x and so on do not change with time:

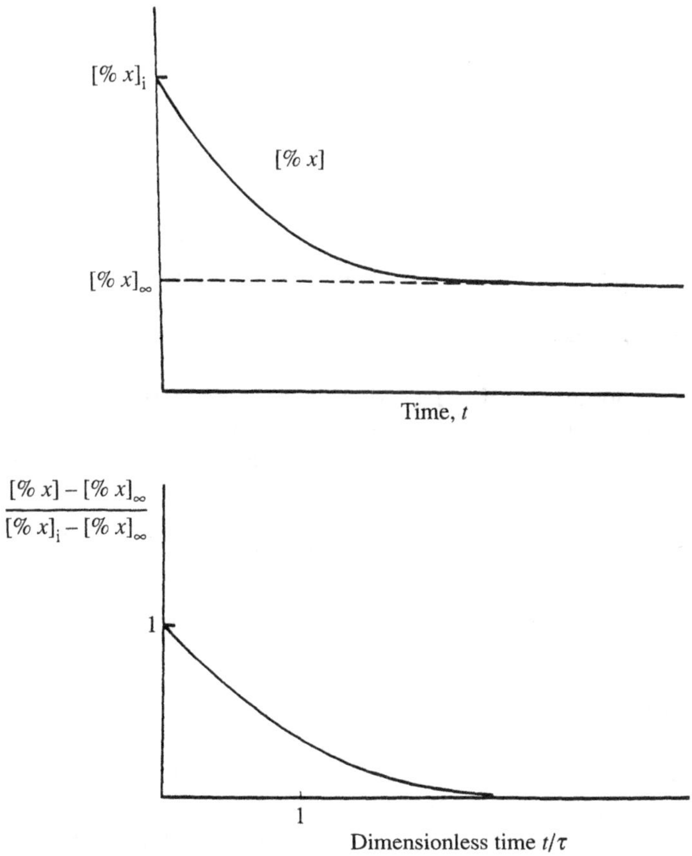

FIG. 4.9 Concentration in the melt $[\% x]$ drops exponentially down to $[\% x]_\infty$.

$$\frac{[\% x] - [\% x]_\infty}{[\% x]_{\text{in}} - [\% x]_\infty} = \exp\left\{-\frac{k_t \rho A_s t}{M}\left(1 + \frac{\gamma_x M}{K f_x M_s}\right)\right\}. \tag{4.34}$$

In Fig. 4.9 we see that $[\% x]$ drops exponentially down to the value $[\% x]_\infty$ given by the equilibrium between x in the melt and in the slag. When

$$t = \tau = \frac{M}{k_t \rho A_s \left(1 + \frac{\gamma_x M}{K f_x M_s}\right)} \tag{4.35}$$

the difference between $[\% x]$ and the final value $[\% x]_\infty$ is reduced to $1/e = 0.37$ of the difference at time t = 0. As one would expect, τ is inversely proportional to the total mass transfer coefficient k_t. In problem 4.3, eqns (4.34) and (4.35) are applied to sulphur removal in steel.

If complete mixing in the slag phase is not attained, the analysis above does not apply. $[\% x]$ will then remain higher than given by eqn (4.34).

4.9 INJECTION

We have seen that mass transfer from the melt to the gas phase can be speeded up by bubbling gas down into the melt. Transfer to the bubbles may be much greater than through the top surface of the melt. Similar advantages may be obtained for slag–melt processes by injection of powder into the melt.[15-20] Usually in refining, this powder becomes a molten phase separate (immiscible) from the melt. This phase initially has its full thermodynamic driving force for removal.

There is an important difference in principle between melt–slag and melt–gas processes, however. Impurities in the gas phase are usually totally removed from the system, while for powder injection the impurities remain in the slag phase. From this phase they can revert to the melt.

Powder injection is usually pneumatic (often using inert gas such as argon). If high-pressure systems are employed (1–3 bar) it is possible to operate at very low mass flow ratios between powder and gas (down to 1–3 Nl per kg of powder). Usually the powder is conveyed down into the melt through a lance (see Fig. 4.10).

Injection of powder has been employed in commercial operations in industries ranging from steel and ferroalloys to aluminium. However, it seems that often the flow rates of powder and gas are not controlled and measured properly. Furthermore, in some cases the position of the lance tip is not well defined. Sometimes there are problems with clogging of the lance, vibrations of the lance, splashing, variations in yield, refining efficiencies, and so on.

The injected particles can either penetrate through the bubble–melt interface, be trapped at the interface, or remain suspended in the gas phase. This is illus-

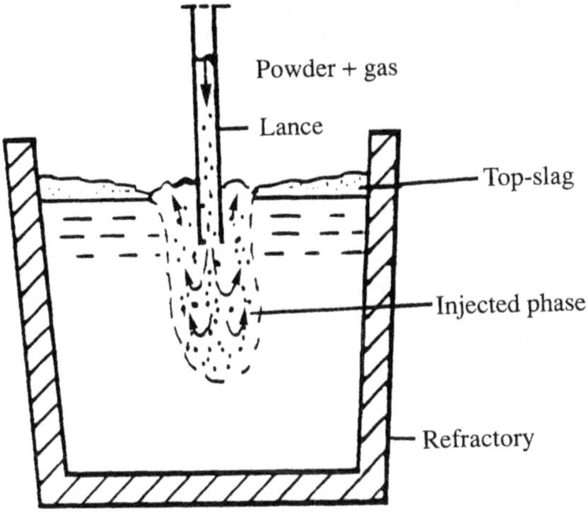

FIG. 4.10 Injection with a lance.

trated in Fig. 4.11. If particles remain in the gas phase, they are lost. In fact they will present a problem in the off-gas system from the reactor. Usually only particles in the size range less than 10 μm will remain in the gas. If particles are 'wetted'[21,22] by the melt, they will enter it. Non-wetting particles will penetrate the interface if their momentum on impact is sufficiently high. This is discussed in Chapter 7. For the case when particles remain in the interface and wet it, the melt–particle contact area may be taken to be equal to the melt–bubble contact area.[15-23]

In the following we calculate the removal of a dissolved element to the

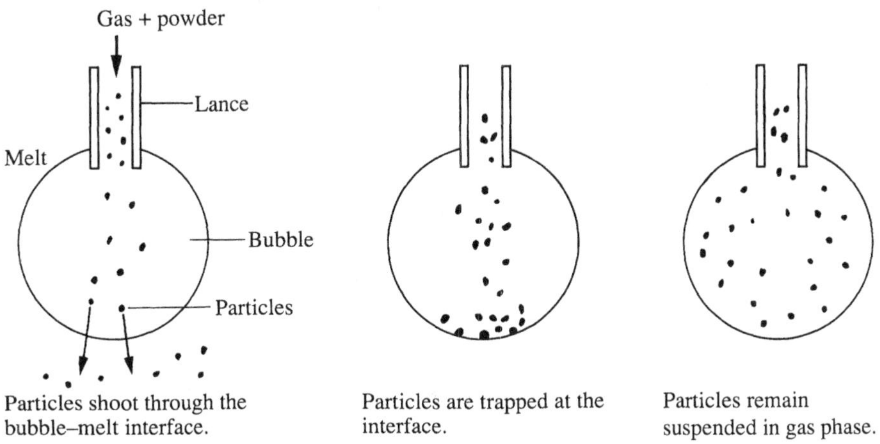

FIG. 4.11 The three possible modes of behaviour for injected particles.

injected phase. The treatment is analogous to removal by bubbling. We assume that there are no concentration gradients in the melt or in the injected phase. Also there is initially no x in the powder. Then the term on the right-hand side of eqn (4.10) is replaced by

$$\frac{\dot{M}_p}{100}\frac{\Delta(\%\,x)_p}{M_x}$$

where $\dot{M}_p$ is the rate of injection of powder in kg/s and $[\%\,x]_p$ is the weight fraction impurity in the injected phase. Thus the mass balance for the control volume (see Fig. 4.4) becomes

$$k_{tp}\rho([\%\,x] - [\%\,x]_e)\Delta A = \dot{M}_p\Delta(\%\,x)_p \tag{4.36}$$

where

$$[\%\,x]_e f_x = \gamma_{xp}\frac{(\%\,x)_p}{K_p}. \tag{4.37}$$

As for bubbling, we obtain an efficiency of injection

$$Z = \frac{(\%\,x)_p^0\gamma_{xp}}{f_x K_p[\%\,x]}. \tag{4.38}$$

Here $(\%\,x)_p^0$ is the x-concentration in the powder when it leaves the metal phase:

$$Z = 1 - \exp(-\phi_p) \tag{4.39}$$

where

$$\phi_p = \frac{k_{tp}\rho A_p\gamma_{xp}}{\dot{M}_p f_x K_p}. \tag{4.40}$$

Removal (in kmol/s) is

$$\dot{G}_x = \frac{\dot{M}_p(\%\,x)_p^0}{100\mathrm{M}_x}. \tag{4.41}$$

ϕ_p is a dimensionless contact area. If we write for eqn (4.40)

$$\phi_p = \frac{k_{tp}\rho A_p[\%\,x]}{\dot{M}_p f_x K_p[\%\,x]/\gamma_{xp}} \tag{4.42}$$

it is seen that

$$\phi_p = \frac{\text{Maximum possible removal by mass transfer}}{\text{Maximum capacity of powder for } x}.$$

As for gas purging we have that for $\phi_p > 3$, $Z \approx 1$. Then removal, $\dot{G}_x$, by the injected phase is given by the equilibrium so that

$$\dot{G}_x = \frac{f_x K_p \dot{M}_p [\%\, x]}{100 M_x \gamma_{xp}} \tag{4.43}$$

For $\frac{1}{20} < \phi_p < 3$,

$$\dot{G}_x = \frac{f_x K_p \dot{M}_p [\%\, x] Z}{100\, M_x \gamma_{xp}} \tag{4.44}$$

and for $\phi_p < \frac{1}{20}$,

$$\dot{G}_x = \frac{k_{tp} A \rho [\%\, x]}{100 M_x}. \tag{4.45}$$

For the batch reactor one obtains from eqn (4.24) (compare also eqn (4.27)):

$$-M \frac{d[\%\, x]}{dt} = \frac{f_x K_p \dot{M}_p [\%\, x] Z}{\gamma_{xp}} + k_{ts} A_s ([\%\, x] - [\%\, x]_e) \tag{4.46}$$

where $[\%\, x]_e$ depends on the slag concentration ($\%\, x$) as given by eqn (4.28). Now $[\%\, x]_e$ is eliminated by employing the mass balance for x from eqn (4.29). It is pointed out that the amount of slag now is not constant but increases due to the injection $\dot{M}_p$ of powder (at a constant rate):

$$M_s(t) = M_s(t = 0) + \dot{M}_p t. \tag{4.47}$$

Equation (4.29) introduced to eqn (4.46) gives

$$-\frac{d[\%\, x]}{dt} = \left(\frac{1}{\tau} + \frac{\phi_s}{\tau_p + t}\right) [\%\, x] - \frac{\phi_s}{\tau_p + t} [\%\, x]_{in} \tag{4.48}$$

where

$$\frac{1}{\tau} = \frac{f_x K_p \dot{M}_p Z}{M \gamma_{xp}} + \frac{k_{ts} \rho A_s}{M} \tag{4.49}$$

$$\frac{1}{\tau_p} = \frac{\dot{M}_p}{M_s(t = 0)}$$

$$\phi_s = \frac{k_{ts} \rho A_s \gamma_x}{\dot{M}_p K f_x}. \tag{4.50}$$

The equation is similar to that for top-slag removal only, except for the injection term $f_x K M_p Z / (M \gamma_{xp})$.

τ is a time constant for the combined removal to powder and top-slag. τ_p is a time that compares the injection rate $\dot{M}_p$ with the initial mass of slag. ϕ_s is a dimensionless number similar to ϕ_p. However, ϕ_s compares mass transfer at the slag–metal interface to the capacity for removal by the injection of powder.

The time constant for the transient τ_0 may be given by

$$\frac{1}{\tau_0} = \frac{1}{\tau} + \frac{\phi_s}{\tau_p}. \tag{4.51}$$

The transient term vanishes when $t > 3\tau_0$. Then

$$\frac{[\%\,x]}{[\%\,x]_{in}} \approx \frac{\phi_s}{\left(\dfrac{(\tau_p + t)}{\tau} + \phi_s\right)}. \tag{4.52}$$

Equation (4.52) is obtained by setting the right-hand side of eqn (4.48) equal to zero. This gives a rough approximation for the balance between removal and reversion from the top-slag valid for $3\tau_0 < t$. In this case sufficient time has elapsed for the transient to die out. The first term in the denominator takes into account that the amount of slag increases with time due to the injection. Reversion from the top slag and removal by the injected powder is illustrated in Fig. 4.12.

We have assumed that there is complete mixing in the injected phase. This may be reasonable for gases or even for a powder that melts during injection, but the assumption is not applicable when the injected phase remains in the solid state. For instance, when CaC_2 is injected into steel, sulphur is removed only as a thin layer of CaS that covers the CaC_2 particles. In this case k_{tp} depends on diffusion in the solid state so that k_{tp} is small and $\phi_p \ll 1$. Then eqn (4.45) must be employed. In Chapter 7 we discuss how to determine k_{tp} and the contact area A_p.

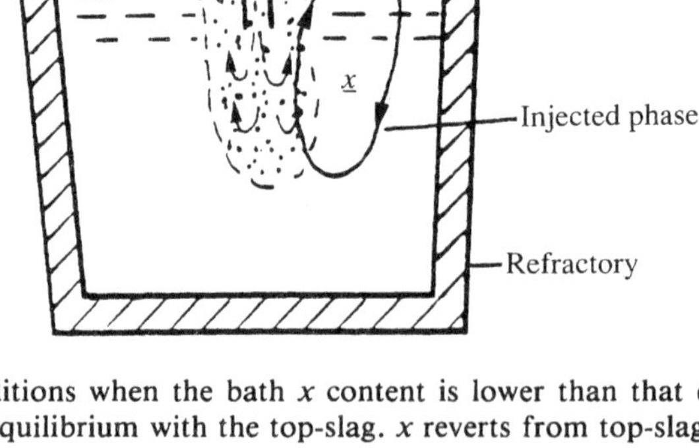

FIG. 4.12 Conditions when the bath x content is lower than that corresponding to equilibrium with the top-slag. x reverts from top-slag.

FIG. 4.13 Example of H_2 pores in an aluminium alloy in a 'gassy' sample before filtration compared with after filtration.

4.10 HYDROGEN REMOVAL; DIATOMIC GASES

Hydrogen is probably the main source of the gas problem in the casting of aluminium. The gas present in metal prior to casting has a strong influence on how the casting feeds near the end of solidification. Gas content also influences the amount and location of shrinkage and porosity. The effect of gas on porosity is exaggerated by solidification under vacuum as shown in Fig. 4.13.

The reason for our gas problem lies in the chemical and physico-chemical properties of aluminium discussed in Chapter 2. Aluminium is highly reactive and easily decomposes the oxide of hydrogen by this reaction:

$$H_2O(g) + \tfrac{2}{3}Al(1) \rightarrow \tfrac{1}{3}Al_2O_3 + 2\underline{H}(\text{in Al}). \tag{4.53}$$

This means that gas may enter the metal if steam or water vapour (in the air as humidity) is in contact with the melt. Secondly, the gas solubility in solid aluminium is much smaller than in the liquid, so most of the hydrogen present in the liquid prior to casting is evolved as gas porosity during solidification.

The problem for most foundrymen is to remove the gas to acceptably low levels. As will be seen below, several properties of aluminium determine the rate of degassing, and understanding these principles allows degassing to become much more effective.

Although what follows is addressed primarily to aluminium the principles of gas removal apply to other metals as well. Hydrogen is removed by the reaction

$$\underline{H} = \tfrac{1}{2}H_2(\text{gas}) \tag{4.54}$$

with (inverse) equilibrium constant

$$K_H = \frac{p_{H_2}^{\frac{1}{2}}}{f_H[\%\,H]} \tag{4.55}$$

given by[24], (see also Table 2.1).

$$\ln K_{\mathrm{H}} = \frac{5869}{T} + 3.282. \tag{4.56}$$

The activity coefficient, f_{H}, is introduced to account for the effect of other elements (such as Si, Cu, Mg, and Fe) on the solubility of hydrogen. General equations that can be used to calculate f_{H} are presented in Chapter 2. In Table 4.1 f_{H} is given for common alloys. Silicon and copper both reduce hydrogen solubility in liquid aluminium, as do manganese and nickel. Chromium and iron have no effect, while magnesium increases the hydrogen solubility very slightly.

The hydrogen solubility in pure aluminium is shown as a function of temperature in Fig. 4.14. Note that only about 5 per cent of the gas present in the liquid remains in the solidified metal; the other 95 per cent is rejected into the liquid until the concentration increases to the point where gas bubbles form. Also note that the scale used is semi-logarithmic. This means that the hydrogen content, at saturation, doubles for each 110°C (200°F) increase in the molten metal temperature. This dramatic increase is responsible for the increase in degassing problems at high melt temperatures. Low temperatures are an advantage not only in order to prevent hydrogen pick-up, but also with hydrogen removal in mind. With a decrease in temperature at a given [% H] the equilibrium value of p_{H_2} increases.

Important consequences can be drawn from Fig. 4.14 and eqn (4.56). Our ability to degas to low hydrogen content can be no better than thermodynamic solubility will allow. In other words, we admit purge gas (nitrogen or argon) to the bottom of the melt, whereupon hydrogen starts to diffuse to ascending purge gas bubbles, and the hydrogen pressure in the bubbles (p_{H_2}) increases. To remove all the hydrogen possible (the case of 100 per cent efficiency) p_{H_2} in the bubble as it exits the melt must be in equilibrium with the metal. Then the volume of purge gas required to remove a certain volume of hydrogen can

TABLE 4.1 *Hydrogen activity coefficients in aluminium and in common casting alloys*

f_{H}	Composition of alloy
1	pure aluminium
1.61	Al–7% Si–0.3% Mg
1.60	Al–7% Si–0.6% Mg
1.36	Al–4.5% Cu
1.93	Al–6% Si–3.5% Cu
1.05	Al–4% Mg–2% Si

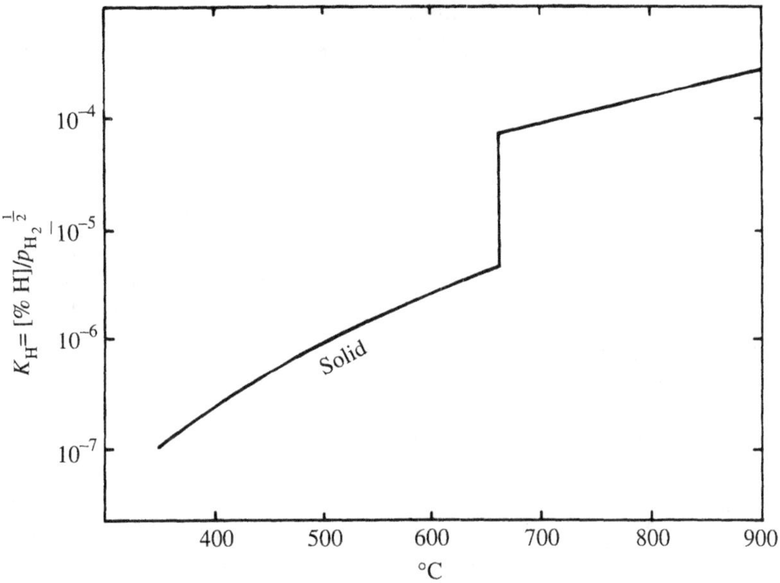

FIG. 4.14 Solubility of hydrogen (at one bar pressure) in pure aluminium as a function of temperature.

be easily calculated according to a simple mass balance. Hydrogen removed from the melt equals hydrogen in the escaping bubbles:

$$-\frac{M\mathrm{d}[\%\,\mathrm{H}]}{100\mathrm{d}t\,M_\mathrm{H}} = \frac{2\dot{G}p^0_{\mathrm{H}_2}}{p_\mathrm{inert}}. \tag{4.57}$$

Notice that there are two hydrogen atoms for each H_2 molecule. For reasonable contents of [% H] the exit pressure of hydrogen $p^0_{\mathrm{H}_2} \ll p_\mathrm{atmosphere}$. At atmospheric pressure then $p_\mathrm{inert} \approx p_\mathrm{atm}$. Equation (4.55) inserted into eqn (4.57) then gives

$$-\frac{\mathrm{d}[\%\,\mathrm{H}]}{[\%\,\mathrm{H}]^2} \approx \frac{2\times 100 M_\mathrm{H}\dot{G}K^2_\mathrm{H}f^2_\mathrm{H}\mathrm{d}t}{Mp_\mathrm{atm}}. \tag{4.58}$$

If $[\%\,\mathrm{H}] = [\%\,\mathrm{H}]_\mathrm{in}$ at time $t = 0$, integration results in:

$$\frac{1}{[\%\,\mathrm{H}]} = \frac{1}{[\%\,\mathrm{H}]_\mathrm{in}} + \frac{2\times 100 M_\mathrm{H}K^2_\mathrm{H}f^2_\mathrm{H}\dot{G}t}{Mp_\mathrm{atm}}. \tag{4.59}$$

The rate of removal increases with the flow of the purge gas $\dot{G}$ and with K^2_H (decreasing temperature).

It is instructive to consider the effect of decreasing hydrogen content in the metal on the purge gas requirements for effective gas removal. As $\underline{\mathrm{H}}$ drops, the pressure of hydrogen in the exiting purge gas bubbles decreases. This means that towards the end of degassing a lot of purge gas is required to remove a little bit of hydrogen. This ratio of the two gases, called the gas removal ratio, is

$$r = \frac{\text{volume of purge gas}}{\text{volume of } H_2 \text{ removed}} = \frac{1 - p_{H_2}}{p_{H_2}}. \tag{4.60}$$

In Fig. 4.15 the gas removal ratio is plotted versus hydrogen concentration. r is numerically equal to the volume of purge gas required to remove one volume of hydrogen. It becomes obvious from this figure that it is difficult to obtain low gas levels, since a large volume of nitrogen or argon is required to remove a small amount of hydrogen.

Actually, thermodynamic equilibrium was assumed in the calculation of the curve shown in Fig. 4.15, but in many cases this case of 100 per cent efficiency is never obtained. We realize that, when the bubbles of purge gas are large (the contact area between gas and melt is small), then kinetic factors determine the hydrogen removal rate, and we operate far below 100 per cent efficiency.

To study this case, we return to the section concerning gas purging. Equation (4.10) for the kmole balance in the control volume becomes

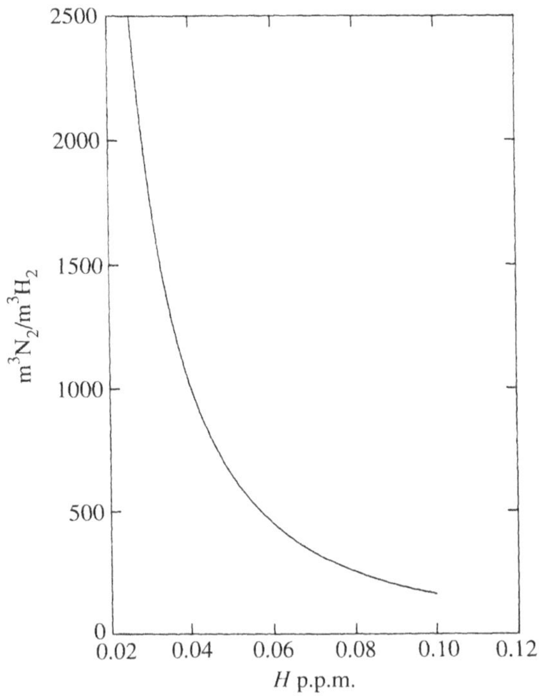

FIG. 4.15 Gas removal Ratio, r, at thermodynamic equilibrium in aluminium at 760°C. Initial concentration of H is 0.1 p.p.m.

$$\frac{k_t \rho([\% \mathrm{H}] - [\% \mathrm{H}]_e)\Delta A}{100 M_\mathrm{H}} = 2\dot{G}\Delta\left(\frac{p_{\mathrm{H}_2}}{p_{\mathrm{inert}}}\right). \tag{4.61}$$

Again the number 2 on the right-hand side reflects the fact that hydrogen is removed as a diatomic gas. The counter-driving force term in eqn (4.11) is replaced by

$$[\% \mathrm{H}]_e f_\mathrm{H} = \frac{p_{\mathrm{H}_2}^{\frac{1}{2}}}{K_\mathrm{H}}. \tag{4.62}$$

$[\% \mathrm{H}]_e$ from eqn (4.62) is seen to depend on the pressure p_{H_2} in the bubble. p_{H_2} changes from zero to $p_{\mathrm{H}_2}^0$ in traversing the melt.

We move the terms that depend on p_{H_2} over onto the right-hand side of eqn (4.61) and integrate from bottom to top surface:

$$\frac{k_t \rho [\% \mathrm{H}] A p_{\mathrm{inert}}}{100 M_\mathrm{H} 2\dot{G}} = \int_0^{p_{\mathrm{H}_2}^0} \frac{\mathrm{d}p_{\mathrm{H}_2}}{1 - \dfrac{p_{\mathrm{H}_2}^{\frac{1}{2}}}{f_\mathrm{H}[\% \mathrm{H}]K_\mathrm{H}}}. \tag{4.63}$$

This equation may be made dimensionless by relating the pressure $\mathrm{p}_{\mathrm{H}_2}$ in the bubble to the value, $f_\mathrm{H}^2[\% \mathrm{H}]^2 K_\mathrm{H}^2$, at equilibrium:

$$Z^2 = \frac{p_{\mathrm{H}_2}}{f_\mathrm{H}^2[\% \mathrm{H}]^2 K_\mathrm{H}^2}. \tag{4.64}$$

$Z = 0$ when the inert gas bubble enters the melt. If the bubble attains saturation $Z = 1$.

When Z is introduced instead of p_{H_2} in the integral, eqn (4.63) becomes

$$\frac{\psi}{[\% \mathrm{H}]} = \int_0^{Z^0} \frac{Z\mathrm{d}Z}{(1 - Z)}. \tag{4.65}$$

The group on the left-hand side is called the dimensionless contact area:

$$\frac{\psi}{[\% \mathrm{H}]} = \frac{k_t \rho A p_{\mathrm{inert}}}{400 M_\mathrm{H} \dot{G} f_\mathrm{H}^2 [\% \mathrm{H}] K_\mathrm{H}^2}. \tag{4.66}$$

It is a simple matter to show that the integral is equal to $-Z^0 - \ln -(1 - Z^0)$. Z^0 is obtained by setting $p_{\mathrm{H}_2} = p_{\mathrm{H}_2}^0$ in eqn (4.64):

$$(Z^0)^2 = \frac{p_{\mathrm{H}_2}^0}{f_\mathrm{H}^2[\% \mathrm{H}]^2 K_\mathrm{H}^2}. \tag{4.67}$$

Equation (4.65) gives the degassing efficiency $(Z^0)^2$ as a function of $\psi/[\% \mathrm{H}]$. Figure 4.16 shows this relationship graphically. Notice from eqn (4.66) that if [% H] is high, a large value of the contact area is needed in order to give $\psi/[\% \mathrm{H}]$ a value greater than 2. This reflects the fact that a large contact area

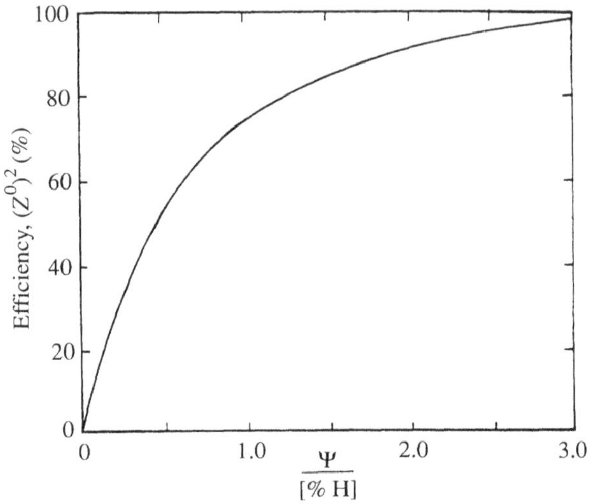

FIG. 4.16 Degassing efficiency $(Z^0)^2$ versus the dimensionless bubble contact area $\psi/[\%\text{H}]$.

is required in order to saturate the bubbles in this case. The reason for this is that p_{H_2} at equilibrium is proportional to the second power of [% H].

For low values of $\psi/[\%\text{ H}]$, less than about 0.3, we have diffusion of hydrogen limiting the removal rate, and eqn (4.13) applies. At larger values of $\psi/[\%\text{ H}]$, greater than about 2, we reach the case of thermodynamic equilibrium ($Z^0 = 1$). Then $p^0_{H_2}$ is in equilibrium with the melt, so that eqn (4.62) may be used for $p^0_{H_2}$. In between there is a transition regime.

What actually determines the performance of the purge gas is the diameter of the purge gas bubbles, which influences the total surface area, A, in eqn (4.66). Not only do smaller bubbles give a larger surface area for a given volume of purge gas, but they rise more slowly (see Fig. 4.17). As bubble size decreases, $\psi/[\%\text{ H}]$ increases sharply and we move closer to the case of 100 per cent efficiency.

To determine the dimensionless contact area $\psi/[\%\text{ H}]$, we must determine the surface area of the bubbles in the melt as a function of bubble radius. If we assume that we have a single bubble size, and that they rise in the melt at the speed shown in Fig. 4.17 without interfering with one another and neglect convection, then as shown previously from eqns (4.14) and (4.15)

$$A = \frac{3\dot{Q}_g H}{v_b R}. \tag{4.68}$$

Taking v_b from Fig. 4.17, we can calculate $\psi/[\%\text{ H}]$ as a function of bubble radius, which in turn allows us to determine the degree to which we approach chemical equilibrium. The results of the calculation are shown in Fig. 4.18 for

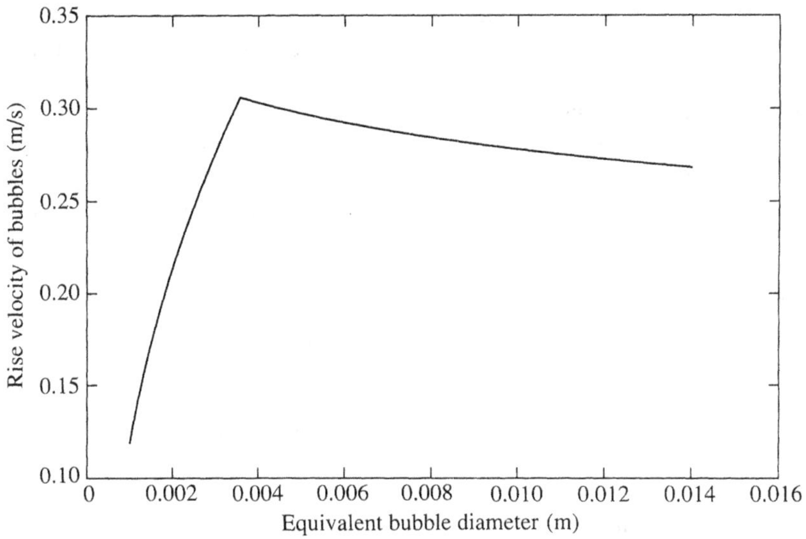

FIG. 4.17 Steady state free rise velocity of gas bubbles in aluminium at 760°C. The bubble size is plotted as equivalent spherical diameter, that is, the diameter of a sphere having the same volume as the bubble.

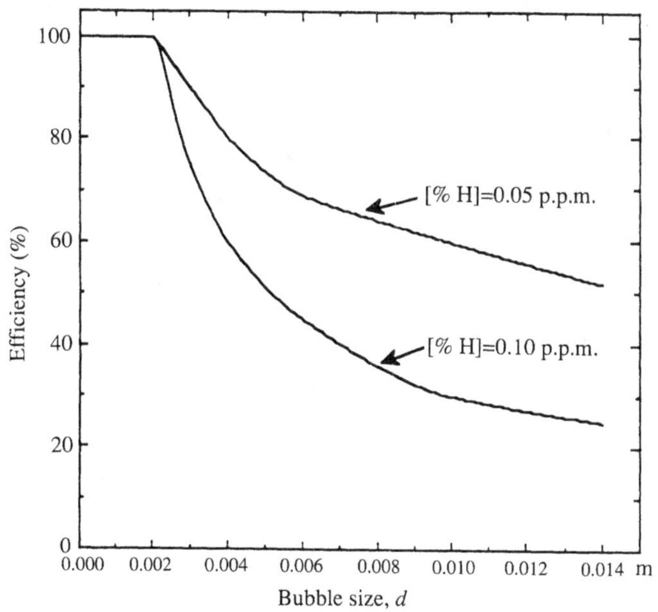

FIG. 4.18 Calculated degassing efficiency $(Z^0)^2$ as a function of gas bubble size. The result is for two different gas contents: 0.05 p.p.m. and 0.1 p.p.m. For 0.1 p.p.m. H and $d = 0.5 \times 10^{-2}$ m, $\psi/[\%H] = 0.5$.

$(Z^0)^2$. The importance of having small bubble sizes becomes very clear with this plot. Also note that we move to higher efficiencies as the hydrogen content drops—a fortunate event.

It should be noted that the calculations shown in Fig. 4.18 are approximate, since each bubble was assumed to rise at its steady state velocity in stagnant aluminium, as shown in Fig. 4.17. Usually, however, the upward velocity is greater owing to convection. Thus when a flux tube or a porous plug is used, the bubbles rise in a plume above their point of entry into the melt. As a result, the melt develops a recirculating flow pattern, which makes the bubbles rise faster, thereby lowering the degassing efficiency.

In Fig. 4.18 it is somewhat confusing that the efficiency $(Z^0)^2$ increases with decreasing hydrogen content. This fact ties in with Fig. 4.16 where we again see that efficiency increases with decreasing [% H]. The explanation is in the definition of the efficiency, eqn (4.67), that arises naturally in the mathematical solution, eqns (4.63) and (4.64). In the definition of efficiency, the H_2 pressure is compared with the equilibrium pressure $f_H^2 K_H^2 [\% H]^2$. For instance, for the 1.4 cm diameter bubbles in Fig. 4.18 the efficiency is 25 per cent at hydrogen levels of 0.1 p.p.m. and about twice as much at 0.05 p.p.m. hydrogen. However, the ratio between the equilibrium pressures is 4. Thus at 0.1 p.p.m. dissolved hydrogen, the bubbles have a hydrogen pressure twice that of bubbles in a melt that contains 0.05 p.p.m. This rather complex behaviour is due to the fact that hydrogen is a two-atomic gas and follows Sievert's law. At low hydrogen levels the bubbles do not need large contact areas to be saturated with H_2.

Undoubtedly some of the results of the above calculations would come as no surprise to foundrymen, who instinctively know that lower temperatures and

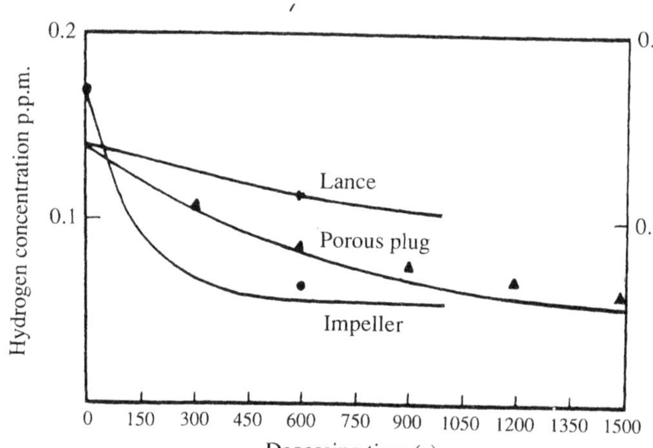

FIG. 4.19 Results obtained with three different degassing techniques in a 200 kg (500 lb) furnace.[25]

smaller bubbles give better results. Of primary importance at the higher hydrogen levels is the gas bubble size. If one uses a lance, large spherical cap bubbles (often several inches in diameter) are obtained, and the degassing process is inefficient, as seen from the results given in Fig. 4.18. Using a porous plug improves the efficiency by reducing the bubble size somewhat, but the bubbles rise in a plume above the plug, where they tend to coalesce to larger sizes even at moderate flow rates. The best results are perhaps obtained with a rotary impeller degasser, since such a unit may produce fine bubbles. The bubbles are also spread out more evenly along the bottom of the furnace. With this even dispersion efficiencies closer to 100 per cent are obtained. The results obtained with the three methods, a flux tube, a porous plug, and a rotary impeller,[25] are compared in Fig. 4.19. Three 200 kg batches of Al–7% Si–0.6% Mg alloy were degassed using the same flow rate of purge gas, and all conditions were maintained constant except for the method of gas introduction. The relevant data and physical constants are given in Table 4.2. The porous plug works fairly well, but it takes nearly five times as long to produce the results obtained with a rotary impeller.

Many foundrymen use chlorine, or freon, in their degassing cycle. The benefits of chlorine are poorly understood. Thermodynamics tell us that $\underline{H}$ and Cl_2 (actually $AlCl_3$) cannot react to HCl at these temperatures.

If the degassing process is mass transfer controlled, that is, if bubbles are large, the use of chlorine or freon may reduce the surface tension, bubble size, and rise velocity and give better results. If very small purge gas bubbles are present, however, chlorine has no effect. In this case, only the amount of gas is important (the mass transfer coefficient is not important because the efficiency is already nearly 100 per cent).

Many foundrymen are also aware of the effect oxides have on casting. One experiment in degassing is to take vacuum gas samples before and after a filter box. The result is similar to that shown in Fig. 4.13. The sample before the filter may be 'gassy'. Melters in the plant often see this and believe that the gas is 'filtered out' of the metal. However, pin samples taken at the same time and analysed for hydrogen by vacuum extraction will show that the filtered and non-filtered metal have the same gas content. Oxides therefore

TABLE 4.2 *Data used in example calculations*

Metal temperature	= 760°C = 1033 *K*
K_H	$= 0.78 \times 10^4$
f_H	= 1.75
k_t	$= 2.3 \times 10^{-4}$ m/s
ρ	= 2350 kg/m³
$\dot{G}$	$= 4.5 \times 10^{-6}$ kmol/s
Height of bath (*H*)	= 0.5 m

appear to help nucleate gas bubbles during solidification.

This effect may perhaps be illustrated in a simple manner by carefully pouring a glass of carbonated beverage (beer is best) and then adding a small amount of table salt to the liquid. The supersaturated solution forms numerous bubbles on the salt crystals.

When porosity in a casting is found, this is probably due to both oxides and hydrogen. Thus, filtration of inclusions may help. If it is hot and humid it may be difficult to reduce the hydrogen concentration. This is due to the effect of 'pick-up' from the atmosphere through the top surface. The top surface transfer may be described by

$$\frac{k_{ts} A_s \rho([\%\,H] - [\%\,H]_v)}{100 M_H}$$

where $[\%\,H] - [\%\,H]_v$ is the driving force and $[\%\,H]_v$ depends on p_{H_2O} in the atmosphere. H_2O is reduced to H_2 at the molten aluminium surface. A relation for $[\%\,H]_v$ is given in Chapter 7.

Often the atmosphere above the melt will be protected with a lid so that there should be little or no hydrogen pick-up from the air. In this case the gas composition is equal to that of the escaping bubbles so that $[\%\,H]_v = [\%\,H]Z^0$.

By blowing inert gas on the melt surface (combined with a lid), $[\%\,H]_v \approx 0$. $[\%\,H]_v$ can be determined experimentally for a given melt by letting it stand for a long time with no gas purging. Then $[\%\,H]_v = [\%\,H]$.

The mass balance equation for a batch reactor becomes:

Reduction of content of hydrogen in batch
= Hydrogen transferred to bubbles
+ Hydrogen transferred to atmosphere at bath surface.

Thus:

$$-\frac{M}{100M_H}\frac{d[\%\,H]}{dt} = \frac{2[\%\,H]^2 f_H^2 K_H^2 \dot{G} Z^2}{p_{inert}} + \frac{k_{ts} A_s \rho([\%\,H] - [\%\,H]_v)}{100M_H}. \tag{4.69}$$

Finally it should be pointed out that the mass transfer coefficients k_t and k_{ts} are not known very accurately. There are several reasons for this. The mass transfer coefficient is proportional to the square root of the diffusivity of hydrogen, D_H, which is known only roughly. Also the mass transfer coefficients are proportional to the square root of the surface velocity which depends on geometry, the position of the impeller, and so on (see Chapter 7).

Figure 4.20 shows theoretical curves for hydrogen removal[25] with (a) the surface completely exposed to air with moisture; (b) with a lid; and (c) blowing inert gas on the surface. We see from case (b) in Fig. 4.20 that a lid greatly

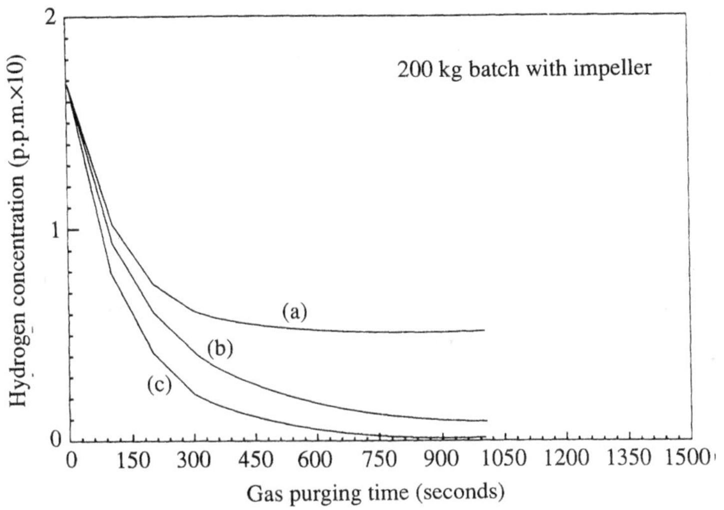

FIG. 4.20 Calculated curves for hydrogen removal with (a) surface completely exposed to air; (b) a lid; and (c) blowing inert gas on the surface (combined with a lid).

improves efficiency. Blowing inert gas on the surface seems to be interesting only if extremely low hydrogen levels are required.

4.11 METAL TO GAS (VACUUM) TRANSFER

Dissolved elements with a high vapour pressure may be removed by vacuum or purge gas treatment. Previously we have studied an example concerning Na removal by gas purging. Examples of vacuum refining are inductively stirred vacuum melts[26] and stream degassing.[27]

Vacuum treatment is carried out in order to give a reduction of the concentration, c_j, of the impurity element (here given index j). Often the solvent metal also evaporates. For vacuum treatment to be interesting, treatment must result in lower levels of the impurity. Then

$$\frac{\dot{n}_j}{\dot{n}_1} > \frac{c_j}{c_1} \tag{4.70}$$

where $\dot{n}_j$ and $\dot{n}_1$ are kmoles of impurity and solvent metal distilled off per second and per square metre, and c_j and c_1 are their concentrations (in kmol per m^3) in the melt.

Often the literature tabulates the vapour pressures p_j^0 and p_1^0 over the pure components j and 1. Then

$$p_j = a_j p_j^0$$
$$p_1 = a_1 p_1^0 \tag{4.71}$$

where a_j and a_1 are the activities

$$a_j = \gamma_j x_j$$
$$a_1 = \gamma_1 x_1.$$

Here γ_j and γ_1 are the Raoultian activity coefficients and x_j and x are the mole fractions, respectively. Usually the metals contain only small amounts of impurities and alloying elements so that $\gamma_1 \approx 1$ and $x_1 \approx 1$ and $x_j = c_j M_1/\rho$. Then the partial pressure in the gas in equilibrium with the melt is

$$p_j = \gamma_j x_j p_j^0 = \gamma_j \frac{M_1 p_j^0 c_j}{\rho} \tag{4.72}$$

or

$$p_j = \gamma_j K_j c_j \tag{4.73}$$

where $K_j = M_1 p_j^0/\rho$. Kinetic gas theory tells us that there is a flux:

$$\frac{p_j}{RT}\left(\frac{RT}{2\pi M_j}\right)^{\frac{1}{2}} \text{ moles/s}$$

of component j in one direction across any plane in a gas. M_j is the molecular weight and p_j the partial pressure of component j in the gas. Under equilibrium conditions when p_j is the same throughout, the same number of moles cross in the opposite direction. Then there is no net transport of the j component through the plane. We place a melt below an imagined horizontal plane; this does not change the flux downwards of component j. If the pressure in the gas is p_{ji} near the melt surface and the partial pressure of j in equilibrium with the melt is p_j, then the net flux from the surface is given by

$$\dot{n}_j = \frac{(p_j - p_{ji})}{RT}\left(\frac{RT}{2\pi M_j}\right)^{\frac{1}{2}} \tag{4.74}$$

or, employing eqn (4.73),

$$\dot{n}_j = (\gamma_j K_j c_{ji} - p_{ji}) \frac{1}{(2\pi M_j RT)^{\frac{1}{2}}} \tag{4.75}$$

where p_{ji} is the actual partial pressure of j in the gas at the interface, and c_{ji} is the actual concentration in the liquid at the interface. Notice that if p_{ji} and c_{ji} are in equilibrium so that $p_{ji}/\gamma_j c_{ji} = K_j$, then there is no net flux, $\dot{n}_j = 0$. Conversely, if there is a net flux of component j, the pressure in the gas, p_{ji}, and the concentration in the melt at the interface, c_{ji}, are not in equilibrium.

The flux $\dot{n}_j$ has to pass through resistances both in metal and gas films on the two sides of the interface.

$$\dot{n}_j = k(c_j - c_{ji}) \tag{4.76}$$

$$\dot{n}_j = \frac{k_g(p_{ji} - p_{bj})}{RT}. \tag{4.77}$$

p_{bj} is the pressure in the bulk gas away from the interface. k_g depends on the total pressure and increases with decreasing total pressure (see Chapter 7). As in our calculation of mass transfer from metal to slag we wish to obtain $\dot{n}_j$ in terms of the total driving force for transfer from the metal with concentration c_j to the gas with concentration p_{bj}/RT. Then we rewrite eqns (4.75)–(4.77) in terms of the driving forces:

$$\frac{\dot{n}_j}{k} = c_j - c_{ji}$$

$$\frac{\dot{n}_j(2\pi M_j RT)^{\frac{1}{2}}}{K_j\gamma_j} = c_{ji} - \frac{p_{ji}}{K_j\gamma_j}$$

$$\frac{\dot{n}_j RT}{k_g K_j\gamma_j} = \frac{p_{ji}}{K_j\gamma_j} - \frac{p_{bj}}{K_j\gamma_j}.$$

Summation of the two sides of the three equations above gives

$$\dot{n}_j\left(\frac{1}{k} + \frac{(2\pi M_j RT)^{\frac{1}{2}}}{K_j\gamma_j} + \frac{RT}{k_g K_j\gamma_j}\right) = c_j - \frac{p_{bj}}{K_j\gamma_j}. \tag{4.78}$$

We see that a total mass transfer resistance, k_t, may be defined thus:

$$\frac{1}{k_t} = \frac{1}{k} + \frac{(2\pi M_j RT)^{\frac{1}{2}}}{K_j\gamma_j} + \frac{RT}{k_g K_j\gamma_j}. \tag{4.79}$$

$1/k_t$ is equal to the sum of the three resistances for the melt boundary layer, evaporation, and gas boundary layer. The resistance in the gas boundary layer can be reduced by lowering the total pressure, that is, by vacuum treatment. In Section 7.8 it is shown that k_g is inversely proportional to the total pressure. The ultimate total pressure in the vacuum chamber determines the final value of the gas film resistance, $1/k_g$. The relation for $1/k_t$ is similar to that developed for metal to slag mass transfer except that here an additional term appears for resistance to evaporation. If the vapour pressure of j is low (K_j is low), then this resistance is high and we have a low total mass transfer coefficient, k_t. The resistance in metal and gas in fact also depend on the physical properties of component j (on the diffusivity D_j). However, we have found that it is not helpful to include the index j for k_t, and k_g.

Equations (4.78) and (4.79) give for the flux of component j

$$\dot{n}_j = k_t\left(c_j - \frac{p_{bj}}{K_j\gamma_j}\right). \tag{4.80}$$

$p_{bj}/(K_j\gamma_j)$ is the hypothetical concentration in the melt in equilibrium with the partial pressure p_{bj} in the bulk gas phase.

If the vapour pressure of component j is low, that is, if $K_j\gamma_j$ is low, then the evaporation and gas-side resistances in eqn (4.79) become important. This result may be compared with Section 4.3 which deals with mass transfer from melt to slag. There, if the equilibrium constant K_j has a low value, resistance in the slag dominates. In Section 7.8 we shall look at the role of the gas film. Sometimes when gas purging processes are used resistance in the gas film is important. This is discussed in Chapter 7. If the gas side is important, it may be practical to apply a vacuum or to use a reactive gas instead of vacuum treatment.

Vacuum treatment differs from melt–slag mass transfer in that evacuation gives a net convective flux, $\dot{n}$. To take this properly into account we set up the transport equations for component j in the gas film

$$\dot{n}_j = -D_j c\frac{\partial y_j}{\partial z} + \dot{n}y_j. \tag{4.81}$$

The first term on the right is the Fick's law diffusion term and the second is the convection term. D_j is the diffusivity of component j and $\dot{n}$ is the total molar flux. The total concentration is

$$c = \frac{p_{tot}}{RT}$$

where p_{tot} is the total pressure in the gas phase. y_j is the mole fraction of component j. The first term on the right-hand side accounts for diffusion and the second for convection. z is the spatial coordinate normal to the interface. We assume that no chemical reaction takes place in the film; then $\dot{n}_j$ and $\dot{n}$ are constant. Furthermore the total pressure is constant so that c is constant. Also D_j is taken as constant. Equation (4.81) then becomes a separable differential equation:

$$-\frac{\mathrm{d}y_j}{\dot{n}_j - \dot{n}y_j} = \frac{\mathrm{d}z}{D_j c}. \tag{4.82}$$

As boundary conditions we have

$$\text{at } z = 0,\ y_j = y_{ji}$$

$$\text{at } z = \delta,\ y_j = y_{j\delta}$$

where δ is the *diffusion boundary layer thickness*. y_{ji} is the concentration (in mole fraction) at the interface, and $y_{j\delta}$ is the mole fraction in the bulk gas phase (see Fig. 4.21).

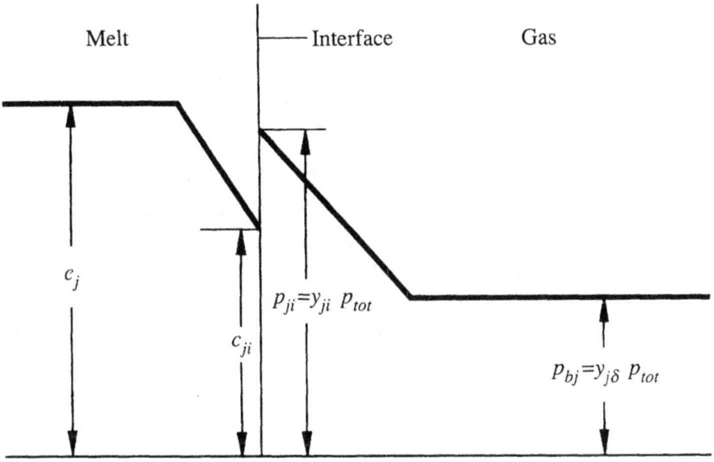

FIG. 4.21 Concentrations c_j and c_{ji} in the melt and pressures p_{ji} and p_{bj} in the gas phase.

Integration of eqn (4.82) gives

$$\ln\left(\frac{\dot{n}_j - \dot{n}y_j}{\dot{n}_j - \dot{n}y_{ji}}\right) = \frac{\dot{n}z}{D_j c}. \tag{4.83}$$

At $z = \delta$

$$\frac{\dot{n}_j - \dot{n}y_{j\delta}}{\dot{n}_j - y_{ji}} = \exp\left(\frac{\dot{n}\delta}{D_j c}\right).$$

This gives

$$\frac{\dot{n}_j}{\dot{n}} = \frac{\exp\left(\frac{\dot{n}\delta}{D_j c}\right) y_{ji} - y_{j\delta}}{\exp\left(\frac{\dot{n}\delta}{D_j c}\right) - 1}. \tag{4.84}$$

The total flux removed through the gas film, $\dot{n}$, is

$$\dot{n} = \sum_j \dot{n}_j. \tag{4.85}$$

Here the summation is over all the components removed, including removal of the solvent metal by evaporation. $\dot{n}$ and the total pressure p_{tot} are determined by the geometry of the evacuation chamber and by the vacuum pump system.

If $\dot{n}$ is low, so that $\dot{n}\delta/(D_j c) < 1$,

$$\exp\left(\frac{\dot{n}\delta}{D_j c}\right) \sim 1 + \frac{\dot{n}\delta}{D_j c'}$$

eqn (4.84) reduces to

$$\dot{n}_j \sim \frac{D_j c}{\delta}(y_{ji} - y_{bj}) = \frac{k_g}{RT}(p_{ji} - p_{bj}). \tag{4.77}$$

This is our previous eqn (4.77) for mass transfer in the gas film since $k_g = D_j/\delta$. Equation (4.77) applies, very roughly, when the vacuum system has attained a low pressure so that $\dot{n}$ is also low. Strictly, eqn (4.77) only applies if $\dot{n}\delta/(D_j c) \ll 1$. However, usually in a vacuum system the $\dot{n}_j$ in eqn (4.85) will all have the same sign so that it is not possible to fulfil this requirement. Nevertheless, eqn (4.77) is often employed in the literature.

At the high pumping rates at the beginning of evacuation $\dot{n}\delta/(D_j c)$ may be greater than 1 so that eqn (4.84) simplifies to

$$\frac{\dot{n}_j}{\dot{n}} \approx y_{ji} = \frac{p_{ji}}{p_{tot}}. \tag{4.86}$$

In this case only the convection term in eqn (4.81) is taken into account. Then summing eqns (4.75), (4.76), and (4.86) we have

$$\frac{1}{k_t} = \frac{1}{k} + \frac{(2\pi M_j RT)^{\frac{1}{2}}}{K_j \gamma_j} + \frac{p_{tot}}{\dot{n} K_j \gamma_j} \tag{4.87}$$

and

$$\dot{n}_j = k_t c_j. \tag{4.88}$$

Applications of eqns (4.79), (4.80), (4.87), and (4.88) are given in the literature.[28-31]

To sum up, when evacuation is initiated it is usually the case that $\dot{n}\delta/D_j c \gg 1$. Then eqns (4.87) and (4.88) apply. However, $\dot{n}$ finally becomes small, the resistance on the gas side increases, and eqns (4.78) and (4.79) may be employed.

When neither diffusion nor convection can be neglected, eqns (4.84) and (4.85) replace eqn (4.77). By an iterative procedure eqns (4.84) and (4.85) together with eqn (4.76) are used to calculate corrected values of y_{ji}, c_{ji}, and $\dot{n}_j$. The starting values in the calculation may be obtained using eqns (4.87), and (4.88). Strictly, δ in eqn (4.84) should not be the same for the various components j.

Calculation of vacuum refining is illustrated in the following simple example.

Example Can Cu be extracted from a melt of pure iron by evacuation? Assume that there is complete mixing in the melt and that eqns (4.79) and (4.80) apply. The Cu pressure in the bulk gas is set equal to zero. Determine k_t, and the Cu concentration after ten hours of operation. The initial Cu concentration is $C^0_{Cu} = 0.1$ wt%.

The Raoultian activity coefficient is $\gamma_{Cu} = 8$. The mass transfer coefficient in the melt boundary layer is $k = 10^{-4}\,\mathrm{m/s}$ and $k_g = 1\,\mathrm{m/s}$. The vapour pressure of pure Cu at 2000 K is $p^0_{Cu} = 430\,\mathrm{Pa}$. The atomic weights of Cu and Fe are 63.5 and 55.85 kg/kmol, respectively. The density of the molten steel is $7030\,\mathrm{kg/m^3}$. The gas constant R is equal to $8.314 \times 10^3\,\mathrm{J/kmol\,K}$. The melt surface area is $A = 0.1\,\mathrm{m^2}$ and the volume $V = 0.01\,\mathrm{m^3}$.

Solution Equations (4.79) and (4.80) give for the flux of Cu, in $\mathrm{kmol/m^2 s}$, $\dot{n}_{Cu} = k_t c_{Cu}$, and for the total resistance:

$$\frac{1}{k_t} = \frac{1}{k} + \frac{(2\pi M_{Cu} RT)^{\frac{1}{2}}}{K_{Cu}\gamma_{Cu}} + \frac{RT}{k_g K_{Cu}\gamma_j}.$$

K_{Cu} is the equilibrium constant given in eqn (4.73):

$$\gamma_{Cu} K_{Cu} = \frac{\gamma_{Cu} M_{Fe} p^0_{Cu}}{\rho_{Fe}} = \frac{8 \times 55.85 \times 430}{7030}$$

$$= 27.3 \frac{\mathrm{Nm}}{\mathrm{kmol}}$$

$$\frac{1}{k_t} = 10^4 + \frac{(2\pi \times 63.5 \times 8.314 \times 10^3 \times 2000)^{\frac{1}{2}}}{27.4} + \frac{8314 \times 2000}{1 \times 27.3}$$

$$= 10^4 + 2.97 \times 10^3 + 609 \times 10^3$$

or

$$k_t = 0.16 \times 10^{-5}\,\mathrm{m/s}.$$

Now the mass balance for Cu gives that the rate of decreases of Cu in the melt is equal to the mass flow of Cu from the surface, or

$$-V\frac{\mathrm{d}c_{Cu}}{\mathrm{d}t} = A_s k_t c_{Cu}.$$

Integrating we have

$$\int_{c_{Cu_{in}}}^{c_{Cu}} \frac{dc_{Cu}}{c_{Cu}} = -\frac{A_s k_t t}{V}$$

or

$$\ln\frac{c_{Cu_{in}}}{c_{Cu}} = -\frac{A_s k_t t}{V}.$$

In ten hours

$$\frac{A_s k_t t}{V} = \frac{0.1 \times 0.16 \times 10^{-5} \times 36000}{0.01}$$

$$= 0.58$$

Then $c_{Cu} = 0.1 \times e^{-0.58} = 0.056$ mass%.

This meager reduction indicates that it is not practical to remove Cu from Fe by vacuum treatment (see also Section 7.8). However, impurities such as Zn and Cd can be removed by vacuum treatment of iron melts. The vapour pressures of these impurities are higher than for Cu so that $K_j \gamma_j$ is higher than 27.4.

4.12 DISTILLATION

For many metals, both on a laboratory and industrial scale, purification by distillation is important. An example is separation of Zn, Pb, and Cd by distillation. Some indication is given here of how an absorption or distillation tower is designed. The goal in design is to describe the number of trays, their dimensions, the distance between each tray, and so on.[5, 31–34] Here the discussion will be confined to the determination of the number of trays required to give the specified composition of the exit streams. The most important tool that we need is thermodynamic data that relates the gas composition of component j, y_j, to the composition in the melt, x_j. Ideally the gas above a tray has been in intimate contact with the melt and is in equilibrium with it. Usually the assumption is made that the melt in a tray is completely mixed and that the gas above the tray is also completely mixed. The composition, x_n, in tray number n is different from tray $n+1$ (the index j for component j has been dropped to simplify the notation). Similarly, the gas compositions above trays n and $n+1$ are different. From Fig. 4.22 it is seen that y_n can be given in terms of x_{n+1} by setting up a mass (or molar) balance for the lower part of the tower:

$$\dot{G}_{in} y_{in} + \dot{M}_{n+1} x_{n+1} = \dot{G}_n y_n + \dot{M}_1 x_1 \tag{4.89}$$

or

$$y_n = \frac{\dot{M}_{n+1}}{\dot{G}_n} x_{n+1} + \frac{\dot{G}_{in} y_{in} - \dot{M}_1 x_1}{\dot{G}_n}. \tag{4.90}$$

$\dot{G}_n$ is obtained from the overall material balance for the lower part of the tower

$$\dot{G}_{in} + \dot{M}_{n+1} = \dot{G}_n + \dot{M}_1. \tag{4.91}$$

Equation (4.90) relates gas composition above plate n to composition in the melt in the plate above, $n+1$. Equations (4.90) and (4.91) together with the equilibrium data and mass balance equations for the melt and gas at the various plates

$$\sum_{j=1}^{k} x_j = 1 \tag{4.92}$$

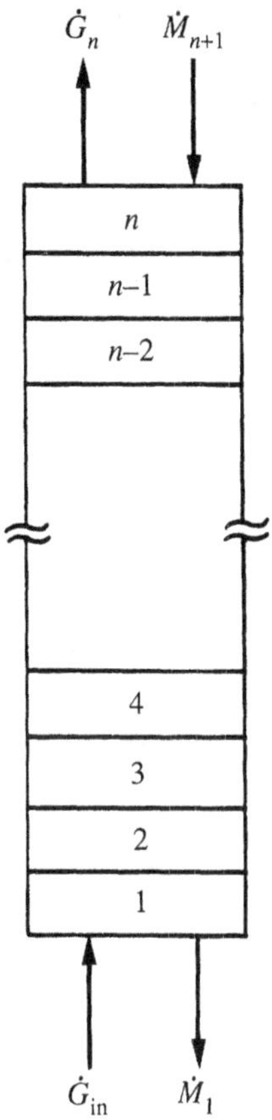

FIG. 4.22 Schematic diagram of absorption tower.

$$\sum_{j=1}^{k} y_j = 1 \tag{4.93}$$

where k is the number of components, describe the distillation tower. Obviously, extensive numerical calculations employing a powerful computer are needed to handle eqns (4.90)–(4.93) for a tower with a large number of trays. In fact, no mention has been made that the heat balance for each tray must also be taken into account to calculate the temperatures. In turn, the temperature determines y_j in equilibrium with x_j.

If it is assumed that the temperature is constant and that the $\dot{G}_n$ and $\dot{M}_n$ are constant and equal to $\dot{G}$ and $\dot{M}$, the number of trays can be determined very simply from a diagram such as that shown in Fig. 4.23. In this case eqn (4.90) reduces to

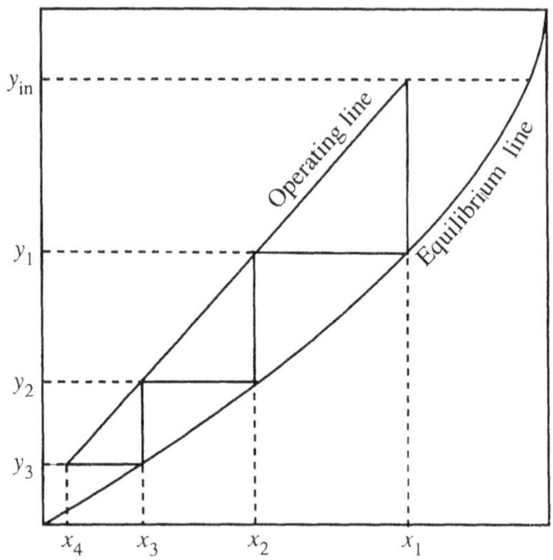

FIG. 4.23 Graphical calculation of the required number of stages in gas absorption.

$$y_n = \frac{\dot{M}}{\dot{G}} x_{n+1} + y_{in} - \frac{\dot{M}}{\dot{G}} x_1 . \tag{4.94}$$

In Fig. 4.23 the gas composition is given on the y-axis and the composition in the melt on the x-axis.

Equation (4.94) is called the operating line. The slope of the operating line $\dot{M}/\dot{G}$ can often (within limits) be chosen. Once $\dot{M}/\dot{G}$ has been chosen, the operating line is determined by the process conditions and specifications. For instance in absorption of a component from a gas to the melt, the inlet composition of the gas, y_{in}, is known. Also the maximum allowed exit composition of the gas, y_{exit}, is specified. Furthermore, the inlet composition of this component in the melt, x_{in}, is known. As seen from Fig. 4.23, this data together with the slope determines the operating line. y_{in} determines the composition on the first plate, x_1. Now the melt composition on the second plate, x_2, is fixed by the gas composition below, y_1, according to the mass balance eqn (4.94) (the operating line). y_1 is also assumed to be in equilibrium with x_1. Therefore, since (x_2, y_1) is on the operating line, we may proceed to draw a horizontal line from x_1 to x_2 as shown on the figure. We draw a line from x_2 on the equilibrium line to x_3 on the operating line and so forth. If $y_3 = y_{exit}$, only three trays are required. In practice, more than three trays would be needed. The reason is that the assumptions made here concerning attainment of equilibrium and complete mixing are only rough approximations.

REFERENCES

1. Glasstone, S., Laidler, K. J., and Eyring, H. (1941). *Theory of rate processes.* McGraw-Hill, New York.
2. Nakanishi, K., Fujii, T., and Szekely, J. (1975). Possible relationship between energy dissipation and agitation in steel processing operations. *Ironmaking and Steelmaking*, **3**, 193–7.
3. Higbie, R. (1935). The rate of absorption of a pure gas into a still liquid during short periods of exposure. *Trans AIChEJ*, **31**, 365389.
4. Danckwerts, P. V. (1951). Significance of liquid-film coefficients in gas absorption. *Ind. Engr. Chem.*, **43**, 1460–7.
5. Szekely, J. and Themelis, N. J. (1971). *Rate phenomena in process metallurgy,* Wiley-Interscience, New York.
6. Clift, R., Grace, J. R., and Weber, M. E. (1978). *Bubbles, drops and particles.* Academic Press, New York.
7. Qin, G., Weian, L., Xingyan, C., and Keren, C. (1984). 2nd Chin. Metal. Science Meeting, Shenyang, China, **5**, 483.
8. Mori, K., Ozawa, Y., and Sano, M. (1982). Characterization of gas jet behaviour at a submerged orifice in liquid metal. *Trans ISIJ*, **22**, 377–84.
9. Szekely, A. G. (1976). The removal of solid particles from molten aluminium in the spinning nozzle inert flotation process. *Trans AIME*, **7B**, 259–70.
10. Myrbostad, E., Pedersen, T., Venås, K., and Johansen, S.T. (1986). The ÅSV inline system for refining of aluminium. In *Light Metals*, pp. 861–72. The Metallurgical Society of AIME, Warrendale, Pa.
11. Nagata, S. (1975). *Mixing, principles and applications.* Wiley, New York.
12. Bopp, J. T., Neff, D. V., and Stankiewicz, E. P. (1987). Degassing multicast filtration system—New technology for producing high quality metal. In *Light Metals*, pp. 729–36. The Metallurgical Society of AIME, Warrendale, Pa.
13. Kawakami, M., Kitazawa, Y., Nakamura, T., Miyake, T., and Ito, K. (1985). Dispersion of bubbles in molten iron and nitrogen transfer in the bubble dispersion zone at 1250°C. *Trans ISIJ*, **25**, 394–402.
14. Mori, K., Sano, M., Sato, T. (1979). Size of bubbles formed at single nozzle in molten iron. *Trans ISIJ*, **9**, 553–8.
15. Engh, T. A., Sandberg, H., Hultkvist, A., and Nordberg, L. G. (1972). Si deoxidation of steel by injection of slags with low SiO_2 activity. *Scand. J. of Metallurgy*, **1**, 103–14.
16. Monden, S., Tanaka, J., and Hisaoka, I. (1983). Impurity eliminations from molten copper by alkaline flux injection. In *Advances In Sulphide Smelting* vol. II, pp. 901–15. AIME, Warrendale, Pa.
17. Lehner, T. and Carlson, G. (ed.) (1977), (1980), (1983), (1986). Scaninject I, II, III, IV. Proceedings organized by Jernkontoret, Mefos Luleå, Sweden.
18. Proceedings of Shanghai symposium on injection metallurgy, Nov. 1982, Shanghai, China.
19. Engh, T. A., Larsen, K., and Venås, K. (1979). Penetration of particle-gas jets into liquids. *Ironmaking and Steelmaking*, **6**, 268–73.
20. Cordier, J., Chaussy, J. P., Lhussier, G., Bienvenu, Y., and Dernier, G. (1981). Désulfuration et traitement de la fonte par injection de magnésium: le procédé USIRMAG 2. *Revue de Métallurgie CIT*, March, 200–12.
21. Mehrotra, S. P. and Chaklader, A. C. D. (1985). Interfacial phenomena between molten metals and sapphire substrate. *Trans AIME*, **16B**, 567–75.

22. Rohatgi P.K. (1986). Interfacial phenomena in cast metal–ceramic particle composites, In '*Interfaces in Metal-Matrix Composites*' (Ed. A.K. Dingra and S.G. Fishman) 185–202. The Metallurgical Soc. Inc., Warrendale, Pa.
23. Ohguchi, S. and Robertson, D.G.C. (1984). Kinetic model for refining by submerged powder injection. Part 1 Transitory and permanent contact reactions. *Ironmaking and Steelmaking*, **11**, No. 5, 262–82.
24. Sigworth, G.K. and Engh, T.A. (1982). Chemical and kinetic factors related to hydrogen removal from aluminium. *Trans AIME*, **13B**, 447–60.
25. Engh, T.A. and Pedersen, T. (1984). Removal of hydrogen from molten aluminium by gas purging. In *Light Metals*, pp. 1329–44. The Metallurgical Society of AIME, Warrendale, Pa.
26. Inouye, M. (ed.) (1982). Proceedings of the Seventh International Conference on Vacuum Metallurgy, Nov., ISIJ, Keidanren Kaikan, Tokyo.
27. Turkdogan, E.T. (1980). *Physical chemistry of high temperature technology*. Academic Press, New York.
28. Harris, R. and Davenport, W.G. (1982). Vacuum distillation of liquid metals. Part 1 Theory and experimental study. *Trans AIME*, **13B**, 581–8.
29. Ozberk, E. and Guthrie, R.I.L. (1986). A kinetic for the vacuum refining of inductively stirred copper melts. *Trans AIME*, **17B**, 87–103.
30. Geiger, G.H. and Poirer, D.R. (1973). *Transport phenomena in metallurgy*. Addison-Wesley, London.
31. Li, Ruiqing and Harris, R. (1991). Vacuum refining of metallurgical melts. Extraction and Processing Division, (ed. D.R. Gaskell) pp. 387–404. TMS, Warrendale, Pa.
32. King, C.J. (1980). *Separation processes*. McGraw-Hill, New York.
33. Nisenfeld, A.E. and Seeman, R.C. (1981). *Distillation columns*. Monograph, Series 2, Instrument Soc. of America, NC 27709.
34. Third International Symposium on Distillation (Organising Commitee, P.R. Perkins and H.A. Day) (1979), Vol. 1 and 2, The Institution of Chem. Engineers Symposium Series No. 56. Rugby, UK.

5

REMOVAL OF INCLUSIONS

5.1 INTRODUCTION

The solubility of dissolved impurities in molten metals was seen in Chapter 2 to decrease with decreasing temperature. During solidification as the temperature drops and solubility decreases, impurities are precipitated as a second phase (non-metallic inclusions or intermetallic phases) in the metal. Oxides and sulphides in steel are examples of inclusions. Such particles are usually very harmful to the mechanical and other properties of the metal (Chapter 1). Therefore one attempts to keep the amount of impurities at low levels so that inclusions are not produced. Partly the approach is to employ stable refractories and a suitable protective slag. For instance, in steelmaking a basic reducing slag prevents the pick-up of oxygen from the atmosphere. Furthermore, alloy additions should be chosen that contain low levels of impurities and inclusions. Combined with these preventative measures dissolved impurity levels may be suppressed by the addition of elements with a high affinity for the impurity. An example of this is precipitation deoxidation of steel with aluminium. However, the reaction product (alumina inclusions) must be removed from the melt before casting. Thus in addition to the particles present in the melt previously, we produce inclusions during the refining operation! Various methods for removal of inclusions are discussed in this chapter. The methods are settling, flotation, filtration, and removal to walls and top-slag by stirring (Section 5.8–5.14). A serious problem can be re-entrainment of inclusions.

Sections 5.4 and 5.5 concerning nucleation may present some difficulty for the reader and can be skipped initially. Some previous knowledge concerning fluid flow and turbulence is an advantage in the study of Section 5.14.

5.2 ORIGIN OF INCLUSIONS

Inclusions can arise from the mechanical entrapment of furnace and ladle dross or slag particles and from the chemical and mechanical erosion of ladle and casting refractories. For steel, entrained air in the tundish and in the mould pool may be a source of inclusions. Air can be entrained even when submerged entry tundish nozzles are employed. Top-slag from the tundish and mould powder may also contribute to inclusions. The occurrence of such exogeneous inclusions is sporadic, but due to their large size they may have a deleterious effect on steel properties. Good handling practices and experience today make the avoidance of large exogeneous inclusions more of an art than a science. Exogeneous inclusions are therefore only briefly referred to in the following, except that Chapter 3 included some material concerning the content of inclusions in alloying additions.

In this chapter we discuss initially how inclusions can nucleate and grow in the melt (endogeneous inclusions).

5.3 MEASUREMENT OF INCLUSIONS

A number of methods exist to determine the content of inclusions: metallographic examination, centrifuge methods,[1] ultrasonic examination, and finally

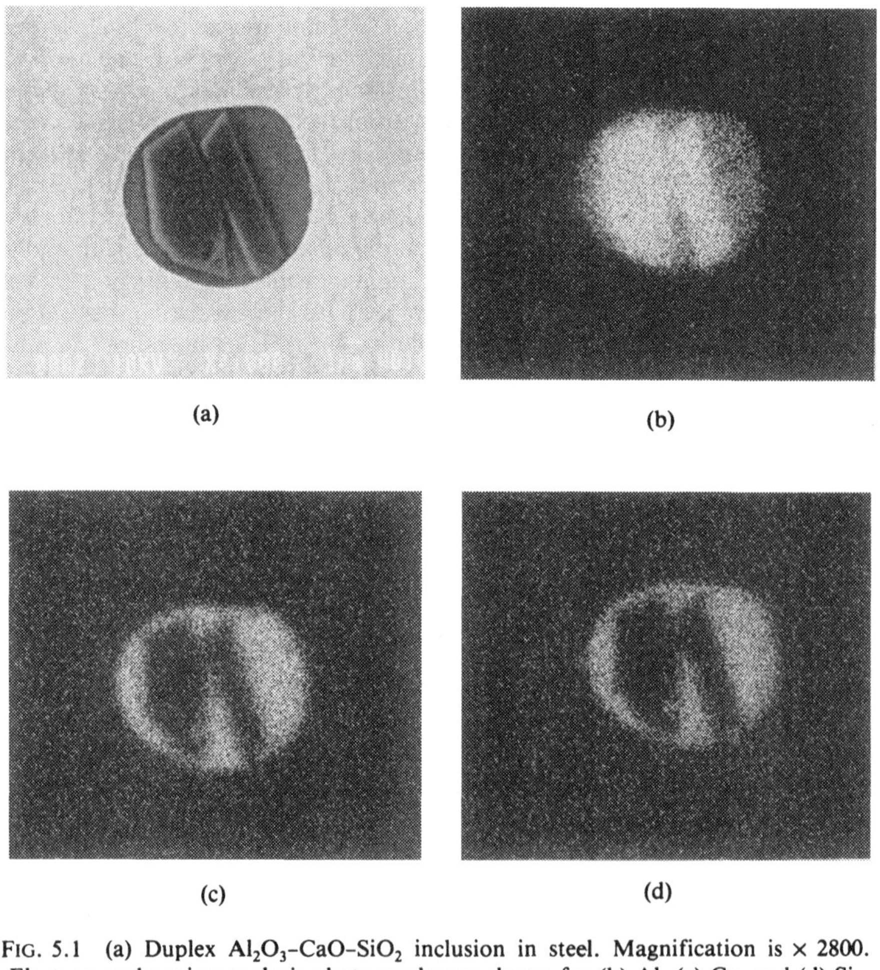

FIG. 5.1 (a) Duplex Al_2O_3–CaO–SiO_2 inclusion in steel. Magnification is × 2800. Electron probe microanalysis photographs are shown for (b) Al, (c) Ca, and (d) Si.

the Coulter counter method. In the metallographic techniques samples are taken of the melt prior to casting or from the cast product. Alternatively, processed material (rolled or extruded product) may be examined. The volume of metal that may easily be analysed by metallographic examination is very small ($\ll 1$ g) so that a significant analysis of metal quality requires many hours of laboratory work.[2,3] The sensitivity or detection limit of this method is good, since it is limited only by the quality of metallurgical preparation and microscopy. The photographs in Fig. 5.1 show an electron microprobe analysis of an inclusion in steel. The two jagged particles are Al_2O_3 surrounded by a molten inclusion of SiO_2–CaO. Figure 5.2 shows an example of oxide films in Al–Mg alloy.

Often in the theoretical description of inclusion removal one resorts to the assumption that the inclusions are spherical. In fact, however, the inclusions may be elongated and irregular in shape. In aluminium one may find Al_2O_3 films torn off from the surface of the bath (see Fig. 5.2). A fundamental problem is that the metallographic picture is only two-dimensional. The three-dimensional shape must be derived from what is seen in two dimensions.

The volume of metal examined may be increased by using sampling filters.[4–6] Such filters operate by concentrating the inclusions either as a 'cake' on the inlet side of the filter or as a deposit inside the filter. More information concerning the distribution of inclusions is obtained when the particles are deposited inside the filter (in-depth filtration). On the other hand when samples are prepared from inside the filter, considerable care must be taken to conserve both the filter and metal material. Also when counting the inclusions, the filter material must be excluded. To speed up the counting of inclusions automatic image analysis may be employed.[7]

The ultrasonic method[8] detects 'discontinuities' in the molten aluminium. In a commercial device[9] a hollow titanium electrode is immersed into the melt, and an ultrasonic signal is transmitted and reflected between parallel faces within the probe. In practice only particles larger than 100 μm are detected. Poor melt quality or detrimental handling practices are immediately apparent. The method can also be used in the control of melts of a relatively good quality, especially if the inclusion size distributions have been determined by other methods. For very high-quality products the probe may be too insensitive even though a large amount of metal can be analysed, of the order of 5 per cent of the casting.

The Coulter method instantaneously counts particles as they intercept the probe. The method has recently been applied to molten metals.[10,11] The Coulter probe detects particles by measurement of the changes in electrical resistance of metal when a suspended particle is carried with the melt through an orifice. The main difficulty in the technique lies in selecting materials capable of withstanding the melt environment. The probe measures particles of size down to 10 μm with a probe orifice of approximately 250 μm diameter. Sample size is small, representing less than 0.01 per cent of the melt during casting. The orifice through which the melt is sampled can be plugged by a

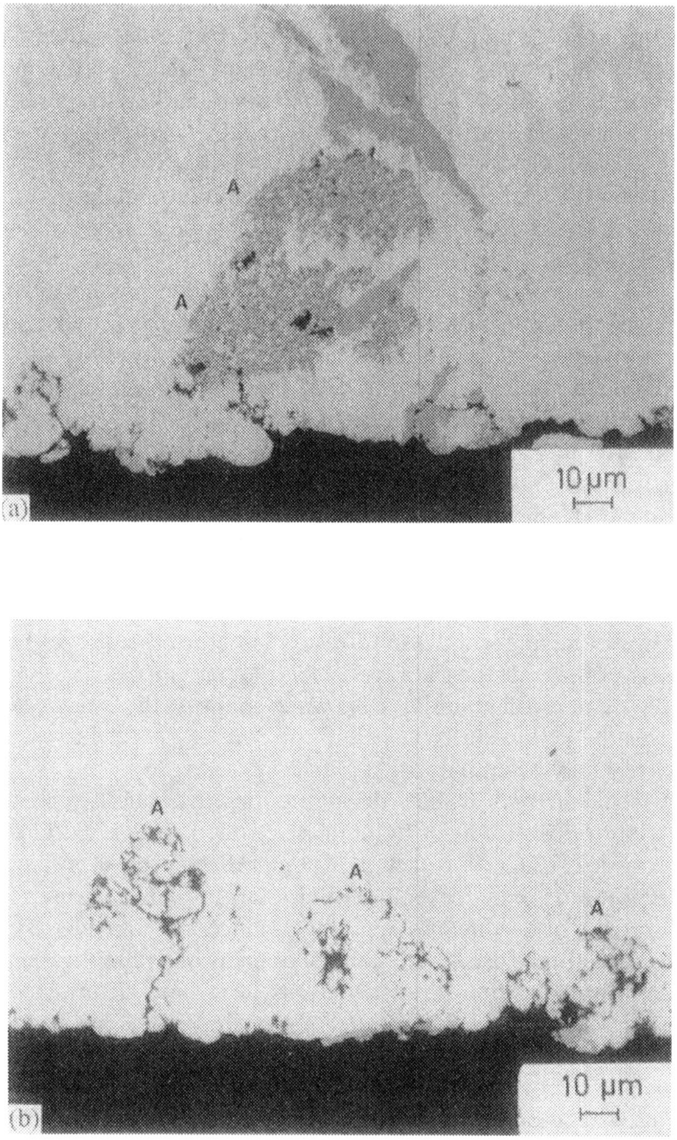

FIG. 5.2 Oxide films in an aluminium–magnesium alloy (a) sedimented Al_2MgO_4 clusters (A) in an aluminium–magnesium alloy (b) sedimented oxide films (A) in an aluminium–5 wt% magnesium alloy.

single particle. The technique at this stage is useful as a development and research tool.

5.4 HOMOGENEOUS NUCLEATION

When a solid forms within its own melt without the aid of foreign materials, it is said to nucleate homogeneously. Nucleation in this manner requires a large driving force due to the relatively large contributions of the surface energy to the total Gibbs energy of very small particles. Precipitation of solid particles may take place due to undercooling of the melt,[12] or precipitation may be caused by the addition or presence of melt components that react to a second phase. For instance, when aluminium is added in order to deoxidize steel, alumina inclusions are precipitated. This case is pursued in the following. To simplify the treatment it is assumed here that the temperature is constant. The effect of a reduction in temperature is briefly discussed in Chapter 9.

If aluminium is added to a molten steel bath, we know from thermodynamics (Section 2.1.8) that the thermodynamic driving force ΔG in J/mol for producing the reaction product is $RT\ln\,(a_{\mathrm{O}}^{3}a_{\mathrm{Al}}^{2}) + \Delta G^{0}$ where ΔG^{0} is the molar Gibbs energy (for 1 mass% reference state) for the reaction

$$Al_2O_3 = 2\,\underline{Al} + 3\,\underline{O}. \tag{5.1}$$

As usual, in accordance with our chosen reference state of mass per cent, activities in the melt are given relative to 1 mass%. The greater the value of ΔG^{0}, the more reaction (5.1) is displaced over to the left. Also higher concentrations of $\underline{Al}$ and $\underline{O}$ move (5.1) over to the left. However, in these considerations two facts have been disregarded. When Al_2O_3 is precipitated homogeneously from a melt, only small particles (clusters) form initially. For these clusters surface energy plays an important role. Then the total Gibbs energy for the cluster is

$$\Delta G_{\mathrm{n}} = -\frac{\Delta G}{V_m}\frac{4\,\pi r^3}{3} + \sigma 4\pi r^2 \tag{5.2}$$

Here ΔG in the first term on the right-hand side is the bulk Gibbs energy per mole. Bulk Al_2O_3 may form, so this term is negative. The last term is the surface energy, which is positive. r is the 'radius', V_{m} is the molar volume, and σ is the melt–particle interfacial tension. Equation (5.2) is sketched on Fig. 5.3. It is noticed that for $r > r^*$ the negative slope increases with r. Then $-\Delta G_{\mathrm{n}}$ attains a higher value for every molecule added. Thus a particle (called a nucleus) with radius r^* or greater can grow, while a particle smaller than r^* (called a cluster) tends to shrink since $-\Delta G_{\mathrm{n}}$ decreases with decreasing r.

To find r^* the maximum for ΔG_{n} is determined:

$$\frac{\partial \Delta G_{\mathrm{n}}}{\partial r} = 0.$$

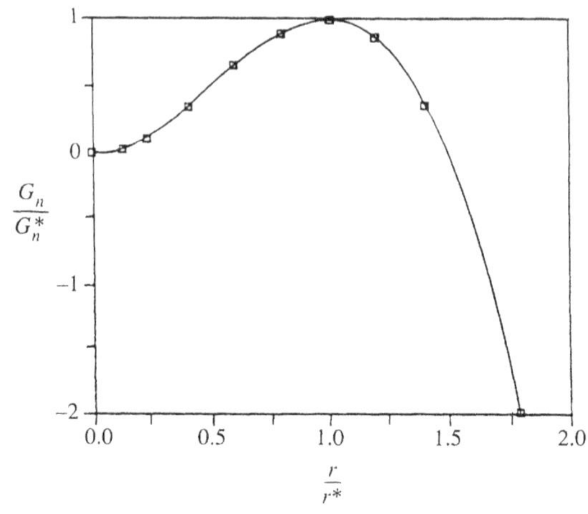

FIG. 5.3 Gibbs energy, ΔG_n, of cluster/nuclei as a function of radius, r.

This gives

$$r = r^* = \frac{2\sigma V_m}{\Delta G} = \frac{2\sigma V_m}{RT\ln(a_O^3 a_{Al}^2) + \Delta G^0}. \tag{5.3}$$

For Al_2O_3 in steel the interfacial tension[13] $\sigma \sim 1.5\ J/m^2$ for $[\%\ O] = 0.004$. ΔG^0 at 1873 K is found from Appendix 1 and Table 2.2: $\Delta G^0 = 476\,300$ J/mol. (See also precipitation deoxidation in Section 2.1.10). $V_m = 32 \times 10^{-6}\ m^3/mol$. Then for $[\%\ Al] = 0.08$ and $a_0 \approx 0.002$ one obtains from eqn (5.3)

$$r^* = \frac{2 \times 1.5 \times 32 \times 10^{-6}}{8.31451 \times 1873 \ln(0.002^3 \times 0.08^2) + 476\,300}$$

$$= 8.95 \times 10^{-10}\ m = 8.95\ Å.$$

For other nuclei, for instance silicates created from $\underline{Si}$ and $\underline{O}$, $V_m = 28 \times 10^{-6}\ m^3/mol$ and[14] $\sigma \approx 0.9\ J/m^2$ and $\Delta G = RT\ln a_O^2 a_{Si} + \Delta G^0$ where ΔG^0 now pertains to the reaction $SiO_2 = \underline{Si} + 2\underline{O}$. ΔG^0 at 1873 K is 161 800 J/mol. $[\%\ O] = 0.017$ and $[\%\ Si] = 0.3$ give

$$r^* = \frac{2 \times 0.9 \times 28 \times 10^{-6}}{8.3145 \times 1873 \ln(0.017^2 \times 0.3) + 161\,800}$$

$$= 31.2 \times 10^{-10} = 31.2\ Å.$$

Thus r^* is much greater than in the example for deoxidation with Al even though both oxygen and silicon concentrations are roughly four times higher.

The values for r^* may be compared with the metallic radii of aluminium, and silicon.[15] $r_{Al} = 1.4$ Å, $r_{Si} = 1.3$ Å. Also $r_O = 0.6$ Å. The ionic radii are $r_{Al^{3+}} = 0.5$ Å, $r_{Si^{4+}} = 0.4$ Å, and $r_{O^{2-}} = 1.3$ Å. It is realized that in the examples the critical nucleus is formed from a large number of atoms. For alumina the volume of the critical cluster is

$$V^* = \frac{4\pi}{3} r^{*3} = \frac{4\pi}{3} \times 8.95^3 = 3000 \text{ Å}^3.$$

This may be compared with the volume of an 'Al_2O_3 molecule' in the nucleus:

$$\frac{V_m}{N_A} = \frac{32 \times 10^{-6}}{6.022 \times 10^{23}} = 53 \times 10^{-30} \text{ m}^3$$

$$= 53 \text{ Å}^3.$$

The critical nucleus then contains about 56 'Al_2O_3 molecules' or 281 atoms of Al and O. Only when a cluster happens to have attained this size can it grow. Clusters form continuously in the melt caused by collisions between Al and O atoms. Energy is provided partly by the reduction in free energy once the solid phase is formed and partly by thermal motion of the atoms. As Darken and Gurry[16] state for the analogous case of freezing of a liquid: 'a liquid anticipates its freezing in that it is continually producing potential nuclei of the solid phase—trying them out, as it were—to see if they will grow'. If the system attains a reduction in Gibbs energy, the nuclei will grow. The higher the concentrations of oxygen and aluminium, the smaller r^* becomes. Conversely, if the concentrations of $\underline{O}$ and $\underline{Al}$ are so low that they are at thermodynamic equilibrium with bulk Al_2O_3, r^* is very large (infinite). Then we certainly cannot expect spontaneous (homogeneous) formation of Al_2O_3 in the melt. The critical Gibbs energy of nucleation, ΔG_n^* is obtained from eqns (5.2) and (5.3):

$$\Delta G_n^* = \frac{4\pi\sigma r^{*2}}{3}. \tag{5.4}$$

It should be noted that ΔG_n^* is positive; ΔG_n^* is the potential barrier to the creation of a nucleus. The critical free energy of nucleation is equal to the surface energy of the critical nucleus divided by 3. Strictly, eqn (5.4) is valid only for liquid nuclei. If we disregard the fact that the surface forces in a solid depend on the various crystal directions, eqn (5.4) may also be applied to solid nuclei. Then σ is an 'average' interfacial tension between melt and solid nuclei.

It is useful in some cases to write eqn (5.4) in the form

$$\Delta G_n^* = \frac{2\sigma}{r^*}\frac{V^*}{2}. \tag{5.5}$$

For a fluid nucleus this is the product of the excess pressure $2\sigma/r^*$ due to the interfacial tension multiplied by half the critical volume V^*.

In spite of the potential barrier to the creation of nuclei, nuclei may in fact be formed in a melt. We determine the concentration (number per m^3) of nuclei of critical size, c^*, where c^* is taken to be the concentration of nuclei that gives a minimum of the total Gibbs energy. The Gibbs energy of the reaction from $\underline{Al}$ and $\underline{O}$ to Al_2O_3 nuclei, including the surface Gibbs free energy is initially given by $c_n \Delta G_n^*$. The effect of the increase of entropy ΔS_n due to the presence of the nuclei must be taken into account in order to determine the Gibbs energy for the total system. To perform this calculation of entropy, we assume that the c_n Al_2O_3 nuclei are created in three steps. First two $c_n Al$ particles and three $c_n O$ particles are formed. Then these react to form $5c_n$ particles. Single species nucleation[12] takes into account the increase of entropy due to the presence of inclusions in addition to the dissolved species. This gives, using eqn (2.4), if it is assumed that aluminium atoms and particles are randomly distributed:

$$\Delta S_n = -k\left\{2c_n \ln\left(\frac{2c_n}{2c_n + c_{Al}}\right) + c_{Al}\ln\left(\frac{c_{Al}}{2c_n + c_{Al}}\right)\right\}.$$

Similarly for oxygen:

$$\Delta S_n = -k\left\{3c_n \ln\left(\frac{3c_n}{3c_n + c_O}\right) + c_O\ln\left(\frac{c_O}{3c_n + c_O}\right)\right\}.$$

When the two $c_n Al$ particles and three $c_n O$ particles form c_n Al_2O_3 nuclei, assuming that Al and O atoms are randomly distributed within the nuclei, the configurational increase of entropy is:

$$\Delta S_n = k \ln \frac{(5c_n)!}{(2c_n)!\,(3c_n)!}.$$

On employing Stirling's approximation (see problem 2.1)

$$\Delta S_n = kc_n(5\ln 5 - 2\ln 2 - 3\ln 3).$$

Since Al and O tend to group together in a stoichiometric ratio this expression overestimates the entropy. However, this entropy contribution is small compared with the two other entropy terms if $c_n \ll c_{Al}$ and $c_n \ll c_O$.

Summation of all three entropy terms gives for the entropy of mixing and for the transfer of aluminium and oxygen to the nuclei containing Al_2O_3:

$$\Delta S_n = -k\left\{2c_n \ln\left(\frac{4c_n}{2c_n + c_{Al}}\right) + c_{Al}\ln\left(\frac{c_{Al}}{2c_n + c_{Al}}\right)\right.$$
$$\left. 3c_n \ln\left(\frac{9c_n}{3c_n + c_O}\right) + c_O\ln\left(\frac{c_O}{3c_n + c_O}\right) - 5c_n \ln 5\right\}. \qquad (5.6)$$

The Gibbs energy change per unit volume, ΔG_v, on forming c_n nuclei is

$$\Delta G_\mathrm{v} = c_\mathrm{n}\Delta G_\mathrm{n}^* - T\Delta S_\mathrm{n}.$$

c^* is now obtained from the condition that at equilibrium

$$\frac{\partial \Delta G_\mathrm{v}}{\partial c_\mathrm{n}} = 0.$$

This leads to the following equation for $c_\mathrm{n} = c^*$:

$$\Delta G_\mathrm{n}^* + kT\left\{2\ln\left(\frac{4c_\mathrm{n}}{2c_\mathrm{n} + c_\mathrm{Al}}\right) + 3\ln\left(\frac{9c_\mathrm{n}}{3c_\mathrm{n} + c_\mathrm{O}}\right) - 5\ln 5\right\} = 0. \quad (5.7)$$

This equation shows that $\Delta G_\mathrm{v} < 0$. It is also found that $\partial^2\Delta G_\mathrm{v}/\partial c_n^2 > 0$ so that $c = c^*$ gives a negative minimum for ΔG_v. Thus while the Gibbs energy, ΔG_n^*, for the formation of a nucleus is positive, the Gibbs energy for the combined system nuclei and solute components, ΔG_v, is negative. If $c_\mathrm{n} \ll c_\mathrm{Al}$ and $c_\mathrm{n} \ll c_\mathrm{O}$, eqn (5.7) reduces to

$$c^* = \left(\frac{c_\mathrm{Al}}{4}\right)^{\frac{2}{5}}\left(\frac{c_\mathrm{O}}{9}\right)^{\frac{3}{5}} 5\exp\left(-\frac{\Delta G_\mathrm{n}^*}{5kT}\right). \quad (5.8)$$

For the silica inclusions the same approach gives

$$c^* = c_\mathrm{Si}^{\frac{1}{3}}\left(\frac{c_\mathrm{O}}{4}\right)^{\frac{2}{3}} 3\exp\left(-\frac{\Delta G_\mathrm{n}^*}{3kT}\right).$$

In the following a numerical example is presented of the use of eqn (5.8). Since c_Al and c_O are concentrations given in atoms per m^3 we have

$$c_\mathrm{O} = \frac{\rho[\%\,\mathrm{O}]N_\mathrm{A}}{100M_\mathrm{O}}$$

and

$$c_\mathrm{Al} = \frac{\rho[\%\,\mathrm{Al}]N_\mathrm{A}}{100M_\mathrm{Al}}$$

where $N_\mathrm{A} = 6.022 \times 10^{23}$ atoms/mol. Thus, for $[\%\,\mathrm{O}] = 0.0040$ and $[\%\,\mathrm{Al}] = 0.08$

$$c_\mathrm{O} = \frac{7030 \times 0.004N_\mathrm{A}1000}{100 \times 16}\ \text{atoms/m}^3$$

and

$$c_\mathrm{Al} = \frac{7030 \times 0.08N_\mathrm{A}1000}{100 \times 27}\ \text{atoms/m}^3.$$

Then from eqns (5.4) and (5.8)

$$c^* = 70\,300\left(\frac{0.08}{27 \times 4}\right)^{\frac{2}{5}}\left(\frac{0.004}{16 \times 9}\right)^{\frac{3}{5}} 5N_A \exp\left(\frac{-4\pi \times 1.5 \times 8.95^2 \times 10^{-20}}{3 \times 5 \times 1.38 \times 1873 \times 10^{-23}}\right)$$

$$= 2.7 \times 10^8 \text{ nuclei/m}^3.$$

It is seen that $c^* \ll c_{Al}$ and $c^* \ll c_O$. According to the calculation the average distance between nuclei is $(2.7 \times 10^8)^{-\frac{1}{3}} = 1.5 \times 10^{-3}$ m or 1.5 mm.

This example is pursued in the following. The nuclei grow into inclusions due to the oxygen supersaturation. As the inclusions grow, the probability increases that they float up or are removed to walls. Figure 2.8 sketches how inclusions grow and are removed. Nucleation can also take place heterogeneously, that is, on the walls or slag or on particles already in the melt. The problem of heterogeneous versus homogeneous nucleation seems to elude any simple treatment. Nevertheless, due to its importance heterogeneous nucleation is briefly discussed in the following.

5.5 HETEROGENEOUS NUCLEATION; WETTING

Wetting is important for the contact between phases, mass transfer, and filtration and to describe nucleation and solidification phenomena. Usually the contact angle as defined here is employed to describe the wetting properties of a three-phase system.

In homogeneous nucleation a second phase is created surrounded by the original phase. In heterogeneous nucleation three phases are involved, and a new phase is formed at the interface between the two original phases. When

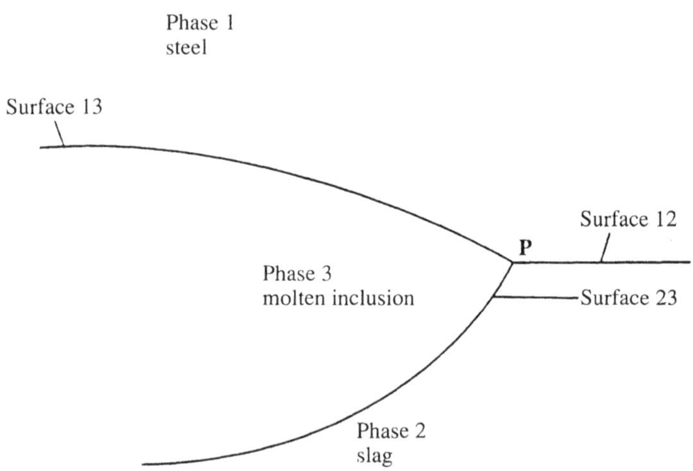

FIG. 5.4 Plane normal to the line of contact **p** of a three-phase system.

three liquid phases meet along three surfaces, they do so in general along a line of contact. Figure 5.4 represents a section of the surfaces 13, 12, and 23 in a plane normal to **P**. Surface 13 separates fluid phases 1 and 3, surface 12 separates phases 1 and 2, and so on. For instance 1 is the metal phase, 2 is the slag phase and 3 is a molten inclusion. If the line **P** is to be in mechanical equilibrium, the vectorial equation

$$\vec{\sigma}_{13} + \vec{\sigma}_{12} + \vec{\sigma}_{23} = 0 \tag{5.9}$$

must be satisfied. This equation is often called Neumann's triangle since it can describe a triangle where the three vectors give the three sides. Thus it is seen that the interfacial tension is necessarily smaller than the sum of the two others. If this condition is not fulfilled the three liquid phases cannot coexist.

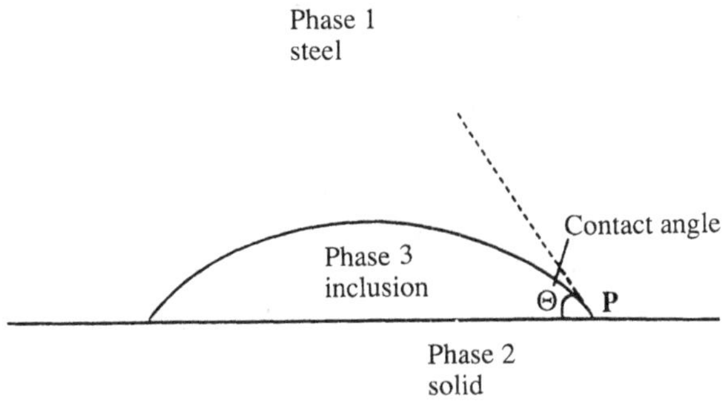

FIG. 5.5 Section through three phases; one phase is a solid.

If one of the phases is a solid, for instance phase 2 (see Fig. 5.5), surfaces 12 and 23 are forced to follow the surface of the solid. In this case Neumann's vector equation is only obeyed for the force component parallel to the solid surface. This then gives the so-called Young–Dupré equation.[17]

$$\sigma_{12} - \sigma_{23} - \sigma_{13}\cos\theta = 0 \tag{5.10}$$

where θ is called the contact angle. Again, if the three phases are to coexist $0 < \theta < 180°$. In general in this situation, we do not have chemical equilibrium between the three phases. If the experiment to determine the contact angle is prolonged over a long time, the shape of the solid close to **P** may change with time.[18] Values of θ measured as a function of temperature may depend on whether the temperature is rising or falling.[19]

Let us consider nucleation of a lens of liquid or gas (phase 3) at the surface separating fluid phases 1 and 2 (Fig. 5.6). The lens will minimize its surface area to volume ratio. The lens, whose volume we denote V_3, therefore consists of two spherical caps. It may be shown that the radius of curvature r_{13} of the lens surface that contacts phase 1 is the same as the critical radius r_{13}^* for a spherical nucleus of phase 3 surrounded by phase 1.[20] Similarly, the lens

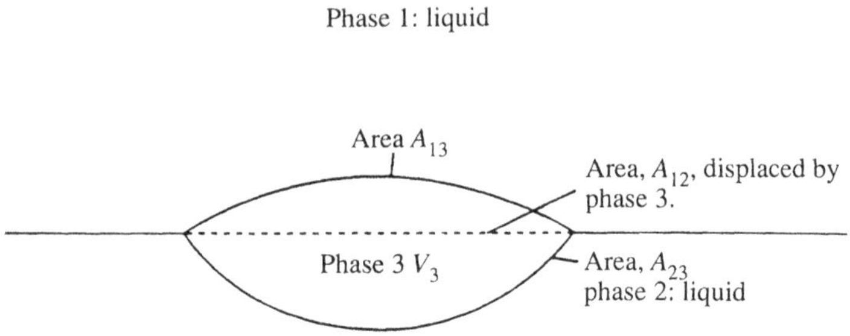

FIG. 5.6 Lens of liquid or gas separating two liquids.

surface in contact with surface 2 has a radius of curvature r^*_{23}. The pressure inside the lens exceeds the external pressure by an amount Δp:

$$\Delta p = \frac{2\sigma_{13}}{r^*_{13}} = \frac{2\sigma_{23}}{r^*_{23}}. \tag{5.11}$$

It is useful to compare this equation with eqn (5.3). Then $2\sigma/r^* = \Delta G/V_m$. Δp is equal to the free energy of formation of phase 3 per unit volume. It should be noted that ΔG in eqn (5.3) is referred to 1 mass% dissolved components in the molten metal. For instance, if $CaO–Al_2O_3–SiO_2$ nucleates at the interface between steel and slag, ΔG refers to $\underline{Ca}$, $\underline{Al}$, $\underline{Si}$, and $\underline{O}$ given in mass per cent. Then $\sigma_{23}/r^*_{23} = \sigma_{13}/r^*_{13}$ is correct only if the chemical potential of $\underline{O}$, $\underline{Ca}$, $\underline{Al}$, and $\underline{Si}$ in the slag, are the same as in the steel bath. In practice this will not be the case, for instance just after the addition of the deoxidizers. However, after a time, equilibrium may be attained between metal and slag. Then eqn (5.11) should apply. If we have equilibrium between phases 1 and 2, the critical Gibbs energy (Gibbs energy barrier) is only

$$\Delta G^*_{\text{lens}} = \frac{\sigma_{13}}{r^*_{13}} V_3 = \frac{\sigma_{23}}{r^*_{23}} V_3 \tag{5.12}$$

where V_3 is the volume of the lens. This equation may be compared with eqn (5.5). The lenticular nucleus must have a volume smaller than spheres with the critical radius r^*_{13} or r^*_{23}. Therefore the formation of a critical nucleus is favoured by the presence of an interface between two fluids so that the nucleus can have a common interface with both phases.

The fact that the nucleation Gibbs energy barrier, ΔG^*_n, is smaller at an interface than in the bulk should not be taken as proof that this heterogeneous nucleation always dominates. The reason is kinetic, in that heterogeneous nucleation only takes place along a wall surface. By comparison homogeneous nucleation can occur throughout the whole melt volume. However, due to the dominance of the exponential term containing $-\Delta G^*_n$ in relations for the number of nuclei and nucleation rates,[21] heterogeneous nucleation takes place in preference to homogeneous nucleation if the Gibbs energy for heterogeneous nucleation is much less than that for homogeneous nucleation:

$$\Delta G^*_{\text{lens}} \ll \Delta G^*_{\text{n}} . \tag{5.13}$$

To study the problem of heterogeneous versus homogeneous nucleation we use a relation for ΔG^*_{lens}[20] analogous to eqn (5.4):

$$\Delta G^*_{\text{lens}} = (\sigma_{13} A_{13} + \sigma_{23} A_{23} - \sigma_{12} A_{12})/3. \tag{5.14}$$

Here A_{13} is the spherical surface (with radius r^*_{13}) between phases 1 and 3, A_{23} is the area of the spherical cap between phases 2 and 3, and A_{12} is the reduction of interfacial area between phases 1 and 2 caused by the presence of the lens (see Fig. 5.6).

It is instructive to study the special but important case that the sum of the interfacial tensions $\sigma_{13} + \sigma_{23}$ is only barely greater than σ_{12}, so that the lens tends to become flat. Then A_{13} and A_{23} are only slightly greater than A_{12} and the expression on the right-hand side of eqn (5.14) tends to become very small. Thus in this case heterogeneous nucleation takes place if the reaction product, 3, fulfils the condition

$$\sigma_{13} + \sigma_{23} - \sigma_{12} < \sigma_{13} . \tag{5.15}$$

The nucleus is formed on the melt–slag interface instead of homogeneously in the bulk melt. The left-hand side of eqn (5.15) may be regarded as the work required to separate melt and slag in order to 'make room' for the inclusion. This work must be much less than the surface energy of the inclusion in the bulk melt, σ_{13}. It is stressed that the left-hand side of eqn (5.15) must be positive or zero. If $\sigma_{13} + \sigma_{23} = \sigma_{12}$ the inclusions (phase 3) stretch out as a film with surfaces 13 and 23 between phases 1 and 2. If $\sigma_{13} + \sigma_{23} < \sigma_{12}$, Neumann's equation cannot be satisfied and the three phases cannot coexist. In this case heterogeneous nucleation should not occur at the interface.

It may be shown[20] that eqn (5.14) is also valid for nucleation at a solid

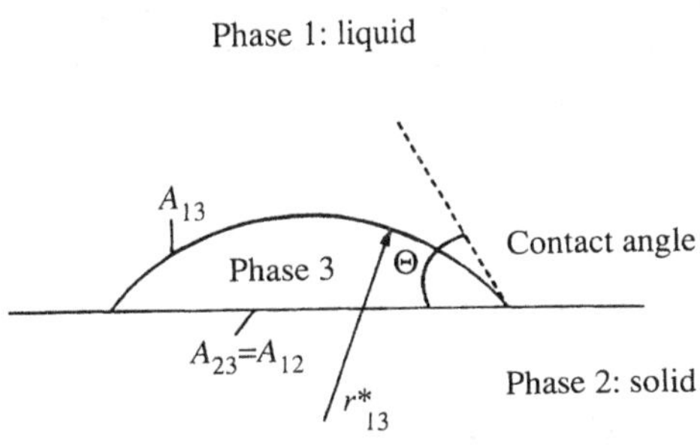

FIG. 5.7 Inclusion at the interface between solid and liquid.

surface. This is illustrated in Fig. 5.7, where 1 is the melt phase, 3 is the nucleated inclusion (liquid, gaseous, or even a solid), and 2 is the solid surface (for instance, a refractory wall). As is seen from the figure, A_{13} is a curved surface (part of a sphere), while A_{23} and A_{12} are parts of a plane. Also

$$A_{12} = A_{23}. \tag{5.16}$$

This equation and Young's equation inserted into eqn (5.14) gives

$$\Delta G^*_{\text{lens}} = \frac{\sigma_{13}}{3}(A_{13} - A_{23}\cos\theta). \tag{5.17}$$

The spherical cap surface A_{13} and the plane surface A_{23} may be obtained in terms of the radius of curvature r^*_{13}:

$$A_{13} = 2\pi r^*_{13}h = 2\pi r^{*2}_{13}(1 - \cos\theta)$$

and

$$A_{23} = \pi r^{*2}_{13}\sin^2\theta.$$

These areas inserted into eqn (5.17) give

$$\Delta G^*_{\text{lens}} = \frac{2\pi\sigma_{13}r^{*2}_{13}}{3}\left(1 - \cos\theta - \cos\theta\frac{\sin^2\theta}{2}\right). \tag{5.18}$$

This equation gives a measure of the supersaturation required to create phase 3 between phases 1 and 2 under equilibrium conditions. In particular, when phase 3 covers the interface, that is:

when $\theta = 0$ then $\Delta G^*_{\text{lens}} = 0$;

when $\theta = 90°$ then $\Delta G^*_{\text{lens}} = \Delta G^*_{\text{n}}/2$

Finally if phase 3 barely touches the interface,

when $(\theta = 180°)$ then $\Delta G^*_{\text{lens}} = \Delta G^*_{\text{n}}$.

In the last case phase 3 cannot exist at the interface and what is in fact homogeneous nucleation of phase 3 within phase 1 must take place.

We now prove that if $0 < \theta < 180°$ the Gibbs energy of formation of a critical nucleus is less than that of formation of a spherical nucleus in the bulk phase (homogeneous nucleation). Then we have that

$$\frac{2\pi r^{*2}_{13}}{3}\left(1 - \cos\theta - \cos\theta\frac{\sin^2\theta}{2}\right) \leq \frac{4\pi}{3}r^*_{13}$$

or $1 + \cos\theta + \frac{1}{2}\cos\theta\sin^2\theta \geq 0$ or

$$1 + \tfrac{3}{2}\cos\theta - \tfrac{1}{2}\cos^3\theta \geq 0$$

since $\sin^2\theta = 1 - \cos^2\theta$. By differentiation it is found that $\cos\theta = -1$ or $\theta = 180°$ gives a minimum for the left-hand side of the inequality. The minimum value is zero. We saw that this corresponds to homogeneous nucleation. Therefore for nucleation at the interface (heterogeneous nucleation $0 < \theta < 180°$):

$$\Delta G^*_{\text{lens}} < \Delta G^*_{\text{n}}. \tag{5.19}$$

The smaller the wetting angle θ, the smaller the left-hand side becomes compared to the right-hand side. When θ is small the walls of a container can catalyse freezing or condensation of a liquid, and boiling or nucleation of inclusions at the walls. The walls become more effective the better the nucleating phase wets the solid. The more readily the solid is wetted, the flatter the lens and the smaller is ΔG^*_{lens}. If the nucleating phase wets the solid perfectly ($\theta = 0°$), then $\Delta G^*_{\text{lens}} = 0$ and, under equilibrium conditions, it is impossible to maintain supersaturation of melt components that nucleate the reaction product at the walls. Thus θ can describe the supersaturation with regard to phase 3 under equilibrium conditions.

As for heterogeneous nucleation at the slag–melt interface, nucleation at the walls may be a slow process.

It could be argued that heterogeneous nucleation at the walls is desirable since no nuclei are created for new inclusions in the bulk. On the other hand if the reaction product inclusions are rapidly removed to the walls, nucleation in the bulk melt may not be a problem.

It is pointed out that nucleation in the bulk melt does not have to be homogeneous. Inclusions may nucleate on solid particles already present in the melt, for instance, other types of inclusions already there or present in alloying additions.

The problem of heterogeneous nucleation is discussed in Chapter 9 in connection with grain refining in which nuclei obey eqn (5.19), that is they wet the grain refiner (nucleant). With great numbers of nucleants added, a large number of nuclei are created.

5.6 THE NUMBER-SIZE DISTRIBUTION OF INCLUSIONS

Inclusions normally grow due to diffusion of a supersaturated component to inclusions already present in the melt. Inclusions may also grow due to collisions between the inclusions.[22,23] For instance, for precipitation deoxidation, collisions may occur just after the addition of the deoxidizer before dissolution and mixing have properly taken place. In Appendix 2 it is shown that collisions should not normally be an important mechanism because the cleaner the melt the less important is the role of collisions. Collisions may occur due to the presence of large exogeneous inclusions. However, this problem is not discussed here and the mechanism of collision is not included in the following.

A typical example of growth is the diffusion of oxygen in a steel melt to oxide inclusions.[24] In this case if we assume spherical inclusions with radius a, that an inclusion contains ρ_O oxygen moles per unit volume, and that the mass transfer coefficient for oxygen is k_O, then the growth rate of the inclusion, $\mathrm{d}a/\mathrm{d}t$, is given by the kmole balance

$$\rho_O 4\pi a^2 \frac{\mathrm{d}a}{\mathrm{d}t} = \frac{4\pi a^2 k_O \rho \Delta[\%\,\mathrm{O}]}{16 \times 100}. \tag{5.20}$$

Here $\Delta[\%\,\mathrm{O}]$ is the driving force for diffusion of oxygen (also called supersaturation). $\Delta[\%\,\mathrm{O}] = [\%\,\mathrm{O}] - [\%\,\mathrm{O}]_{\mathrm{ie}}$ where $[\%\,\mathrm{O}]_{\mathrm{ie}}$ is oxygen in the melt in equilibrium with oxygen in the inclusions (compare Section 4.3). 16 in the denominator is the atomic weight of oxygen. The right-hand side gives the number of kmoles of oxygen per unit time that is transferred to the inclusion. The left-hand side is the rate of accumulation of kmoles of oxygen in the inclusion since $4\pi a^2\,\mathrm{d}a/\mathrm{d}t$ is the volume increase per unit time. Equation (5.20) reduces to

$$\frac{\mathrm{d}a}{\mathrm{d}t} = \frac{k_O \rho \Delta[\%\,\mathrm{O}]}{\rho_O 100 \times 16}. \tag{5.21}$$

In Chapter 7 relations for the mass transfer coefficient k_O will be given. ρ_O depends on the density and composition of the inclusions. It will be shown later that the supersaturation $\Delta[\%\,\mathrm{O}]$ is obtained from an oxygen balance for the melt.

Starting with eqn (5.21), one finds that the number of inclusions per unit volume of melt of size a, to $a + \Delta a, f_N(a)\Delta a$ can be determined. The approach is to employ a number balance ('population balance') in the size interval from a to $a + \Delta a$: the rate of reduction of the number of inclusions of size a to $a + \Delta a$ is equal to the number of inclusions of size a to $a + \Delta a$ that are removed from the melt per second, or

$$-\frac{\mathrm{d}(Vf_N(a)\Delta a)}{\mathrm{d}t} = f_N(a)\Delta a A k_t. \tag{5.22}$$

$Vf_N(a)\Delta a$ is the number of inclusions in the size interval a to $a + \Delta a$. k_t is the mass transfer coefficient for removal of an inclusion of size a; it has dimensions of velocity. A is the contact area between the melt and refractories and top slag. In the following sections relations for k_t will be derived.

The left-hand side of eqn (5.22) may be split into two parts. One part takes into account the change of the distribution function with time. The other part is due to the growth of the inclusions with time. Thus the left-hand side of eqn (5.22) becomes

$$V\Delta a\left(\frac{\partial f_N(t,a)}{\partial t} + \frac{\partial f_N(t,a)}{\partial a}\frac{\mathrm{d}a}{\mathrm{d}t}\right).$$

If there is little or no supersaturation of oxygen, that is, $\Delta[\%\,\mathrm{O}] \approx 0$ so that

$da/dt \approx 0$, only the time-dependent part remains. Solutions when only the time-dependent part is important are given in problems 5.3, 5.4, 5.5, and 5.7. $\partial f_N(t, a)/\partial t = 0$ under stationary conditions, so that $f_N(a)$ does not change with time. An example of a near-stationary situation may be when an Al-deoxidized melt has been standing for a time without any additions or changes being made. Then dissolved oxygen and oxygen in the inclusions has reached steady state (see next section). The excess of aluminium keeps the oxygen supersaturation, $\Delta[\%\,O]$, at its constant low level. Then eqn (5.22) simplifies to

$$V\frac{df_N(a)}{da}\frac{da}{dt} = -f_N(a)Ak_t. \tag{5.23}$$

For a continuous steady state backmixed reactor V in eqn (5.23) is replaced by $\dot{q}\,\tau$ where τ is the mean residence time in the reactor, $\tau = V/\dot{q}$. $\dot{q}$ is the volume flow. When eqn (5.21) is used to eliminate da/dt, one obtains

$$\frac{df_N(a)}{f_N(a)} = -\frac{da A k_t \rho_O 16 \times 100}{\rho k_O \Delta[\%\,O]\,V}. \tag{5.24}$$

If k_t/k_O does not depend on a (see eqns (7.64) and (7.52) and the final sections of this chapter), integration gives

$$\frac{f_N(a)}{f_N(a_c)} = \exp\left(-(a - a_c)\frac{k_t A \rho_O 16 \times 100}{\rho k_O \Delta[\%\,O]\,V}\right) \tag{5.25}$$

where $f_N(a_c)$ is the number of particles of size a_c. It is seen that the number–size distribution of the inclusions is a decreasing exponential. Figures 5.8, 5.9, and 5.10 give examples from the literature[25-28] that, within the size-range measured, show the negative exponential dependence of the number–size distribution. The exponential dependence does not necessarily prove that k_t/k_O is independent of particle size since often the particle size range studied is narrow.

The negative slope increases with the surface-to-volume ratio A/V and with increasing ratio of the coefficients k_t/k_O. If the mass transfer coefficient, k_t, to the walls is large, the inclusions are essentially removed while they are still small. Methods of attaining this goal will be discussed in the following. If some of the inclusions are re-entrained, k_t is reduced accordingly. In the last section of this chapter methods of increasing k_t are considered. It should be noted that the number of large inclusions can also be reduced by decreasing the supersaturation $\Delta[\%\,O]$.

Eqn (5.25), valid for component i, becomes:

$$\frac{f_N(a)}{f_N(a_c)} = \exp\left(-\frac{(a - a_c)k_t A \rho_i M_i 100}{\rho k_i \Delta[\%\,i]\,V}\right) \tag{5.26}$$

when i is the diffusing component that limits growth of the inclusions. k_i is the mass transfer coefficient for dissolved component i to the inclusion, M_i is the

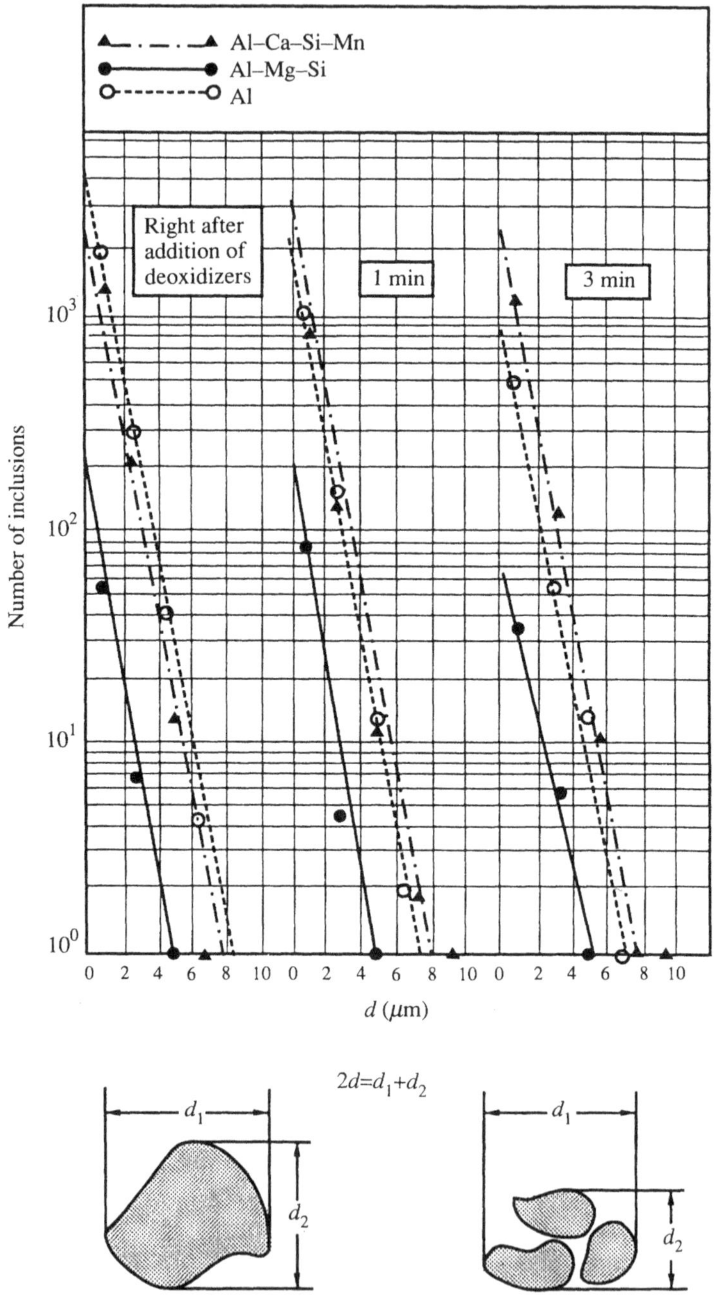

FIG. 5.8 Number of inclusions in steel of size d. d is measured as indicated.[26]

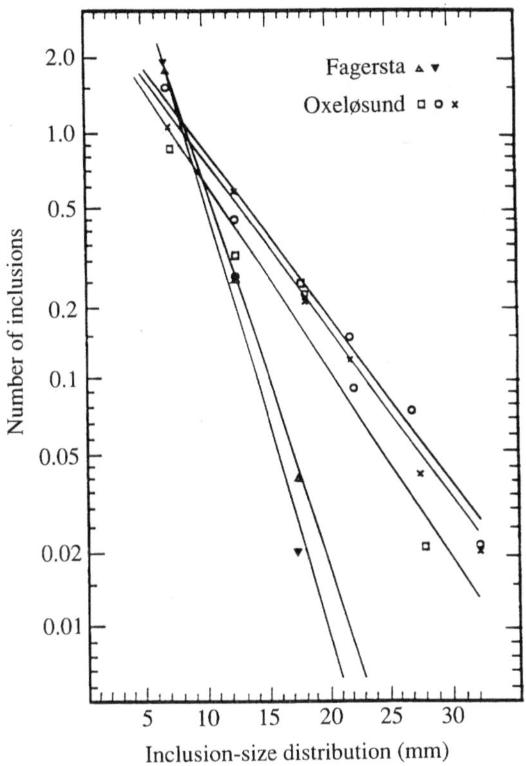

FIG. 5.9 Number–size distributions of inclusions in stainless steel (Fagersta) and in carbon steel (Oxelösund).[28]

mole weight, and ρ_i is the number of moles of i in the inclusion per unit volume.

In general $f_N(a_c)$ and $\Delta[\%\, i]$ depend on time. The interdependence of $f_N(a_c)$ and $\Delta[\%\, i]$ is briefly addressed in the following section. In principle $\Delta[\%\, i]$ can be determined from a balance for component i for the melt as is shown in the next section. Often k_t/k_i depends weakly on inclusion size. Strictly, then, the size distribution is not decreasing exponentially. However, even when k_t/k_i depends on a, eqn (5.26) may be useful, over a limited size range, to discuss qualitatively the interaction between inclusions and dissolved elements.

5.7 THE OXYGEN BALANCE

Since the deoxidation of steel is so well known, an oxygen balance is employed to illustrate how a balance for a melt component i is derived. It will be seen that

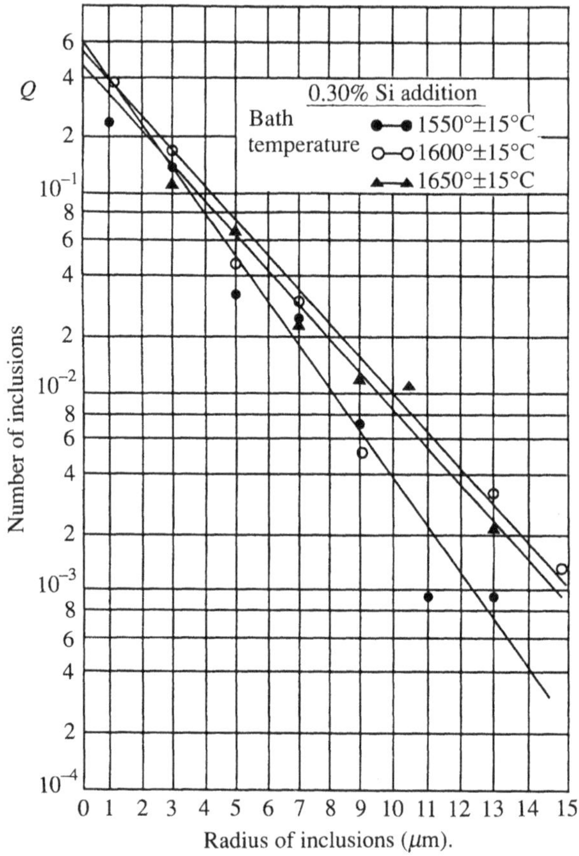

FIG. 5.10 Number distribution of SiO_2 inclusions of size d in steel.[25]

the presence of component i also as a solute in the melt complicates matters somewhat. As already mentioned oxygen is an example of a component present both as a dissolved element and as part of the inclusions. Oxygen is picked up by the melt from oxygen and moisture in the atmosphere and from the refractories (see Fig. 2.1). Sulphur is another example of a component present both in solution and in inclusions in steel. In aluminium melts carbon is dissolved in the melt from carbon refractories and is found also in inclusions such as Al_4C_3.

Oxygen pick-up from slag or dross, refractories, and atmosphere by the melt may be described by

$$\dot{n}_j = \frac{\rho k_j}{100 \times 16}([\%\,O]_{je} - [\%\,O]) \tag{5.27}$$

where $j = 1, 2, 3$ refers to top slag, refractory, and air interfaces, respectively, and k_j is the total two-film mass transfer coefficients introduced in Chapter 4. $[\%\,O]_{je}$ is the hypothetical dissolved oxygen concentration in the melt in equilibrium with the bulk oxygen concentration in phase j. As usual $[\%\,O]$ is the dissolved bulk oxygen concentration in the steel. In terms of the number-size distribution function, $f_N(a)$, the moles of oxygen present as inclusions are

$$\frac{\rho[\%\,O]_i}{100 \times 16} = \rho_O \int_0^\infty f_N(a)\,\frac{4\pi a^3}{3}\,da. \tag{5.28}$$

If aluminium has been added in order to deoxidize, oxygen is removed in two steps: by transfer of oxygen to the inclusions and by removal of the inclusions to wall and top-slag (see Fig. 2.8). A balance for moles of oxygen in a batch melt now gives:

Accumulation of oxygen in the melt (as dissolved oxygen and as inclusions)
= Moles of oxygen picked up
− Moles of oxygen transferred with the inclusions out of the melt

Per unit time this becomes:

$$\frac{d}{dt}\left(\frac{V\rho([\%\,O] + [\%\,O]_i)}{100 \times 16}\right) = \dot{n}_1 A_1 + \dot{n}_2 A_2 + \dot{n}_3 A_3 - \frac{\rho[\%\,O]_i}{100 \times 16}\,(k_{t1}A_1 + k_{t2}A_2). \tag{5.29}$$

The inclusion mass transfer coefficients are k_{t1} and k_{t2} for the top-slag and the wall respectively. If eqn (5.27) is introduced in eqn (5.29), one obtains

$$V\frac{d}{dt}([\%\,O] + [\%\,O]_i) = \sum_{j=1}^{3} k_j A_j([\%\,O]_{je} - [\%\,O]) - (k_{t1}A_1 + k_{t2}A_2)\,[\%\,O]_i. \tag{5.30}$$

This equation contains two unknown variables, dissolved oxygen $[\%\,O]$ and oxygen in the form of inclusions $[\%\,O]_i$. However, $[\%\,O]_i$ depends on $[\%\,O]$ as may be seen from eqns (5.25) and (5.28). Thus in principle eqn (5.30) can be integrated to give $[\%\,O]$ as a function of time. Then also $[\%\,O]_i$ can be calculated from eqns (5.25) and (5.28).

It may perhaps be instructive to start with the special case that there are no inclusions, that is $[\%\,O]_i = 0$. Then eqn (5.30) reduces to

$$V\frac{d[\%\,O]}{dt} = \sum_{j=1}^{3} k_j A_j[\%\,O]_{je} - \sum_{j=1}^{3} k_j A_j[\%\,O]. \tag{5.31}$$

This equation may be transformed to

$$V\frac{d[\%\,O]}{dt} = \sum_{j=1}^{3} k_j A_j([\%\,O]_\infty - [\%\,O]) \tag{5.32}$$

where

$$[\%\,O]_\infty = \frac{\sum_{j=1}^{3} k_j A_j [\%\,O]_{je}}{\sum_{j=1}^{3} k_j A_j}. \tag{5.33}$$

The $[\%\,O]_\infty$ is a mean value of the hypothetical oxygen concentration in steel in equilibrium with the oxygen activity of the three phases (corresponding to $j = 1, 2$, and 3) in contact with the melt. It is seen that the hypothetical concentrations $[\%\,O]_{je}$ are weighted with the kinetic factor $k_j A_j / \Sigma_{j=1}^{3} k_j A_j$.

Since $[\%\,O]_\infty$ is constant, eqn (5.32) integrates to give

$$\frac{[\%\,O] - [\%\,O]_\infty}{[\%\,O]_{in} - [\%\,O]_\infty} = \exp\left(-\sum_{j=1}^{3} k_j A_j t/V\right). \tag{5.34}$$

Here $[\%\,O]_{in}$ is the initial value at time $t = 0$, of the dissolved oxygen. For $t \to \infty$, the exponential approaches zero and $[\%\,O] \to [\%\,O]_\infty$ (compare Fig. 4.9). $[\%\,O]_\infty$ is the final steady state value. $[\%\,O]_\infty$ could have been obtained directly from eqn (5.31), by noting that at steady state the left-hand side of the equation is zero. From the right-hand side one then obtains $[\%\,O] = \Sigma_{j=1}^{3} k_j A_j [\%\,O]_{je} / \Sigma_{j=1}^{3} k_j A_j$ which is identical to eqn (5.33).

In the general situation that we do in fact have inclusions, no attempt is made to derive a general relation for the time dependence of $[\%\,O]$ and $[\%\,O]_i$. If $[\%\,O]_i \gg [\%\,O]$, it is often a simple matter to calculate the time-dependent behaviour of $[\%\,O]_i$. Examples are given in the problems at the end of the book. Dissolved oxygen often drops to the steady state value in a very short time compared to the time it takes $[\%\,O]_i$ to reach the final value.

Here we look at the case when a steady state has been attained so that $[\%\,O]$ and $[\%\,O]_i$ no longer change with time. Then setting the right-hand side of eqn (5.30) to zero gives

$$\sum_{j=1}^{3} k_j A_j [\%\,O]_{je} - [\%\,O] \sum_{j=1}^{3} k_j A_j = (k_{t1} A_1 + k_{t2} A_2)\,[\%\,O]_i. \tag{5.35}$$

This equation can be solved for $[\%\,O]$ and $[\%\,O]_i$ employing eqns (5.25) and (5.28). Eqn (5.35) says that oxygen pick-up from top-slag, walls, and atmosphere is equal to oxygen removed in the form of inclusions to wall and top-slag. In terms of the supersaturation $\Delta[\%\,O] = [\%\,O] - [\%\,O]_{ie}$, eqn (5.35) becomes

$$\sum_{j=1}^{3} k_j A_j([\%\,O]_{je} - [\%\,O]_{ie}) - \Delta[\%\,O] \sum_{j=1}^{3} k_j A_j = [\%\,O]_i (k_{t1} A_1 + k_{t2} A_2). \tag{5.36}$$

Here $[\% \text{O}]_{je} - [\% \text{O}]_{ie}$ is the driving force for the transfer of oxygen from phase j to the inclusions. The first terms depend on the refractories, top-slag conditions, and atmosphere above the melt. For a given ladle and deoxidation practice these terms may be taken to be constant. $\Delta[\% \text{O}]\, \Sigma_{j=1}^{3} k_j A_j$ gives the transfer of oxygen from the melt to the inclusions. The last term in eqn (5.36) takes into account the transport of oxygen as inclusions out of the melt.

It is seen that the supersaturation, $\Delta[\% \text{O}]$, increases with decreasing content of inclusions, $[\% \text{O}]_i$. On the other hand the number of nuclei increases with increasing supersaturation, $\Delta[\% \text{O}]$. If there is a disturbance so that $\Delta[\% \text{O}]$ is increased compared to the original steady state value, then the number of nuclei and the resulting number of inclusions starts to increase with time. However, in this situation more oxygen is picked up by the inclusions, and then removed by the inclusions as they are transferred to top-slag and wall. This in turn tends to reduce the supersaturation $\Delta[\% \text{O}]$ and finally the original constant supersaturation level is recovered. If we start with a supersaturation that is lower than the original value, similar arguments return us to the steady state situation.

To obtain a relation for the number of nuclei c^* in terms of the supersaturation, we return to eqn (5.25). There we use the number of nuclei of size r^* to determine the proportionality constant. Then eqn (5.25) is written in the form

$$f_N(a) = b_1 \exp(-a/b_2). \tag{5.37}$$

By comparison it is seen that

$$b_1 = f_N(a_c) \exp(a_c/b_2)$$

and

$$\frac{1}{b_2} = \frac{A k_t \rho_O 16 \times 100}{\rho k_O \Delta[\% \text{O}] V}. \tag{5.38}$$

In eqn (5.37) the particle number–size distribution has been extrapolated down to include clusters in spite of the fact that the clûster size distribution is not known. However the size distribution should be continuous at $a = r^*$. It will be seen that for $a \sim r^*, f_N(a) \approx b_1$. It is assumed that only the size distribution near $a \approx r^*$ is important. To determine b_1 the condition is employed that:

Volume of nuclei = Volume of hypothetical particles less than or equal to r^*

or

$$c^* \frac{4\pi r^{*3}}{3} = \int_0^{r^*} f_N(a) \frac{4\pi a^3}{3} \, da. \tag{5.39}$$

It will be seen in the following that the main contribution to the integral is at $a = r^*$. The total mass of inclusions is given by eqn (5.28).

The right-hand side of eqn (5.28) is equal to $\rho_O 6 b_1 b_2^4 4\pi/3$ according to eqn (5.37). Similarly the right-hand side of eqn (5.39) is equal to $b_1 b_2^4 \int_O^{r^*/b_2} e^{-v} v^3 dv 4\pi/3$. If eqn (5.28) is divided by eqn (5.39) one obtains

$$\frac{\rho[\%\,O]_i}{100 \times 16\rho_O} = \frac{\dfrac{c^* 4\pi r^{*3} 6}{3}}{\displaystyle\int_O^{r^*/b_2} \exp(-v) v^3 dv}. \tag{5.40}$$

The factor $6/\int_O^{r^*/b_2} \exp(-v) v^3 dv$ gives the ratio between the total mass of inclusions and the mass of the nuclei. If $r^*/b_2 \ll 1$ then $\exp(-v) \approx 1$ and $f_N(a) \approx b_1$ for $a \lesssim r^*$. The factor simplifies to $6/\int_O^{r^*/b_2} v^3 dv = 24 \times b_2^4/r^{*4}$. The main contribution to the integral is at the upper limit $v = r^*/b_2$ corresponding to $a = r^*$.

Equation (5.40) becomes

$$\frac{\rho[\%\,O]_i}{100 \times 16\rho_O} \approx 24c^* \frac{4\pi}{3} \frac{b_2^4}{r^*}. \tag{5.41}$$

At steady state eqns (5.41), (5.39), (5.36), (5.25), (5.8), (5.4), and (5.3) determine $[\%\,O]_i$ and the supersaturation $\Delta[\%\,O]$ for an Al deoxidized steel bath.

To illustrate eqn (5.41) numerically, we return to the previous example with aluminium deoxidation. For Al_2O_3 inclusions of density 3200 kg/m^3, $\rho_O = 3200 \times 3/102 = 94$ kmol/m^3. The supersaturation $\Delta[\%\,O] = 0.004 - 0.0004 = 0.0036$ where the equilibrium oxygen concentration has been taken as 0.0004% (see Section 2.1.10). For the mass transfer coefficient of oxygen we take[29] $k_O = 5 \times 10^{-4}$ m/s (see Section 7.6) and for the inclusions $k_t = 4 \times 10^{-3}$ m/s. A melt of depth 1 m in a ladle with diameter 1 m has $A/V = 6\,m^{-1}$. Now according to eqn (5.38)

$$b_2 = \frac{5 \times 10^{-4} \times 7030 \times 0.0036}{4 \times 10^{-3} \times 94 \times 16 \times 100 \times 6} = 3.506 \times 10^{-6}\,m.$$

Since $r^* = 8.95 \times 10^{-10}$ m it is found that $r^*/b_2 \ll 1$ and $f_N(a) \approx b_1$ for $a \ll r^*$ as assumed previously. Then $r^* = 8.95 \times 10^{-10}$ m and $c^* = 2.7 \times 10^8\,m^{-3}$ in eqn (5.41) give

$$[\%\,O]_i = \frac{100 \times 16 \times 94 \times 24 \times 2.7 \times 10^8 \times 4\pi \times 3.51^4 \times 10^{-24}}{7030 \times 3 \times 8.95 \times 10^{-10}}$$

$$= 0.1 \quad \text{or} \quad 1000\,\text{p.p.m.}$$

The oxygen supersaturation in the example produces a large number of Al_2O_3 inclusions. However, $[\%\,O]_i$ is very sensitive to the oxygen supersaturation. A slight decrease in the supersaturation gives a very strong reduction in the number of nuclei c^*. In the example, if it is possible to suppress oxygen down to $[\%\,O] = 0.003$, then $r^* = 10.2$ Å and $c^* = 2000$ nuclei/m^3.

In this case very few new Al_2O_3 inclusions are formed. Such a reduction may be attained by the addition of a deoxidizer stronger than aluminium, such as calcium. Addition of a stronger deoxidizer prevents the formation of nuclei of the weaker one. After calcium addition the melt still contains the inclusions present before the addition (plus the new inclusions formed by the stronger deoxidizer). These inclusions must be removed by stirring; that is, by increasing the transfer coefficient for the inclusions, k_t. Simultaneously, dissolved oxygen declines down to $[\%\,O]_\infty$ even faster than the rate given by eqn (5.34) since some of the dissolved oxygen in the melt is also transferred to the inclusions. From eqn (5.33) it is seen that oxygen in equilibrium with the j interface is weighted by the kinetic factor $k_jA_j/\sum_{j=1}^{4} k_jA_j$ where 4 refers to the inclusion phase. Only when the number of large inclusions is reduced to very low levels should casting be carried out. The times involved in inclusion removal for various types of stirring are discussed in the final section of this chapter.

To obtain a melt with a low inclusion content before casting it is important that stable refractories and a reducing slag are employed so that the $[\%\,O]_{je}$ values in eqn (5.36) are low. This gives a low value of the supersaturation $\Delta[\%\,O]$ and a correspondingly low number of large inclusions according to eqn (5.25).

During casting care must be taken not to expose the melt to the atmosphere. Also entrapment of top-slag in the tundish must be prevented. If large exogeneous inclusions are created during teeming, all our previous efforts may have been in vain. Equation (5.25) only gives the inclusion distribution if metal cleanliness can be retained during the casting operation.

5.8 REMOVAL BY SETTLING

Inclusions may be lighter than the molten metal (this situation is usually found in steelmaking) or they may be heavier than the melt as is often the case for aluminium. The inclusions that float up are collected by a slag or dross phase while heavy inclusions accumulate as a slurry in the lower parts of the furnace (a holding or settling furnace). In either case the removal mechanism is gravity coupled with the density difference (negative or positive) between the particle and the surrounding melt. The term 'settling' is used both when particles sink or float up.

The gravity force acting on a particle is given by

$$F_g = \Delta\rho g V \tag{5.42}$$

where V is the volume of the particle, $\Delta\rho$ is the density difference, and g is the acceleration of gravity. The drag F_d on the particle is usually calculated in terms of an empirical friction factor ϕ:

$$F_d = \rho u_r^2 A_p \phi. \tag{5.43}$$

(When studying the literature, one should check the definition employed for the friction factor.) Here u_r is the relative velocity between the surrounding melt and the particle. A_p is the area of the cross-section transverse to the direction of flow. The terminal velocity is determined by the balance between the opposing forces F_g and F_d:

$$\Delta\rho g V = \rho u_r^2 A_p \phi$$

or

$$u_r^2 = \frac{\Delta\rho g V}{\phi\rho A_p}. \tag{5.44}$$

For spherical particles

$$V = 4\pi a^3/3, \qquad A_p = \pi a^2 \qquad \text{and} \qquad \phi = 12/\text{Re} \tag{5.45}$$

if the Reynolds number Re is $\text{Re} = u_r 2a/\nu < 2$ and

$$\phi \approx 12(1 + 0.15\ \text{Re}^{0.687})/\text{Re} \tag{5.46}$$

for $2 < \text{Re} < 750$.[48] For $750 < \text{Re} < 2 \times 10^5$, $\phi \approx 0.22$.

Inclusions are usually so small that Re < 2. Then for spherical particles

$$u_r = \frac{2\Delta\rho g a^2}{\rho\nu 9}. \tag{5.47}$$

This is the well-known Stokes' equation valid for Reynolds number Re < 2. For instance in aluminium $\nu = 0.4 \times 10^{-6}\ \text{m}^2/\text{s}$, $2a = 10^{-4}\ \text{m}$ (100 μm), $\Delta\rho/\rho = 0.1$, and $g = 9.81\ \text{m/s}^2$. Therefore eqn (5.47) gives: $u_r = 0.136 \times 10^{-2}\ \text{m/s}$ and $\text{Re} = 0.136 \times 10^{-2} \times 10^{-4}/(0.4 \times 10^{-6}) = 0.34$. Thus it is seen that if the relative density difference $\Delta\rho/\rho$ is not too great, the Stokes' equation is still valid for particles of diameter 100 μm.

5.9 STIRRING AND COMPLETE MIXING

Particles are removed not only by gravity effects but also by various other mechanisms: Brownian diffusion, inertial and turbulent removal to walls, and so-called *hydrodynamic effects*. These mechanisms are discussed in the following sections. If only gravity effects are important, then the particle mass transfer coefficient k_t is simply equal to the velocity u_r. In agitated steel melts experimental results[31-33] indicate that k_t is much greater than u_r. It will be seen that the turbulent removal part of k_t increases with the stirring energy input $\dot{\varepsilon}$. Equation (5.26) shows that a large k_t tends to give an inclusion size distribution that drops strongly with increasing particle size. However, it is also found (in Section 7.5) that increased stirring gives a higher value of the coefficient for mass transfer to the inclusions k_O. This partly offsets the effect of an increase in k_t.

The molten metal bath may be stirred mechanically,[34,35] by bubbling,[29,36,37] or electromagnetically.[33,38,39] Also temperature gradients or concentration gradients can induce convection. Laboratory experiments[40] and plant studies often show that the melt is 'completely mixed' especially if attempts are made to stir the bath. This means in practice that one usually cannot detect gradients in the distribution of inclusions throughout the bulk melt.[25,41] This may also be the case in holding or settling furnaces for the melt above the inclusion slurry or sludge zone and below the top layer of dross[27] or slag. As found in the previous chapter, the assumption of complete mixing greatly simplifies the mathematical treatment of refining operations.

5.10 RE-ENTRAINMENT OF PARTICLES

In water model experiments with polystyrene particles slightly heavier than the water[42] it is observed that particles deposited on the bottom of the ladle may be re-entrained from areas where velocities and turbulence levels are high. Entrainement is caused by drag and lift from the flow of metal along the bottom. To prevent re-entrainment it is important that velocities and turbulence levels are sufficiently low. The critical velocities are determined as follows. The shear, τ_0, or lift forces acting on a particle on the bottom of the ladle (or holding furnace) are of the order of magnitude.

$$\tau_0 \pi a^2 = \rho v_b^2 \phi \pi a^2. \tag{5.48}$$

v_b is the velocity outside the boundary layer, where for a plane surface[43]

$$\phi = 0.03(\mathrm{Re}_x)^{-\frac{1}{5}} \tag{5.49}$$

$$\mathrm{Re}_x = \frac{v_b x}{\nu} \tag{5.50}$$

where x is the distance from the leading edge of the plane. For a pipe ϕ is given later in eqn (5.108).

The gravity force acting on the spherical particle is $\Delta\rho g\, 4\pi a^3/3$. As a rough approximation we assume that τ_0 and the gravity force are equal and opposite. Then to avoid suspending a particle

$$v_b^2 \phi < \frac{\Delta\rho g 4a}{3\rho} \tag{5.51}$$

or using eqn (5.49)

$$v_b < \left(\frac{\Delta\rho g 4a x^{\frac{1}{5}}}{3\rho \times 0.03 \nu^{\frac{1}{5}}}\right)^{\frac{5}{9}}. \tag{5.52}$$

For a pipe the corresponding relation is

$$v_b < \left[\frac{4\Delta\rho g a (2R)^{\frac{1}{4}}}{3\rho \times 0.04\nu^{\frac{1}{4}}}\right]^{\frac{4}{7}}.$$

To avoid suspending the particle with diameter $2a = 100\,\mu$m discussed in the section on settling, we find in aluminium for $\Delta\rho/\rho = 0.1$ and velocity $v_b = 1$ m/s, taking $x = 1$ m:

$$v_b < \left(\frac{0.1 \times 9.81 \times 4 \times 50 \times 10^{-6} \times 1}{3 \times 0.03 \times (0.4 \times 10^{-6})^{\frac{1}{5}}}\right)^{\frac{5}{9}} = 0.17\,\text{m/s}.$$

Under the same conditions for the pipe, taking $2R = 0.3$ m, we obtain $v_b < 0.18$ m/s.

As seen from these calculations, quite a large particle may be re-entrained at what must be considered to be a moderate or even low velocity, v_b, in the bulk melt outside the boundary layer. An alternative interpretation is to state that for a given velocity, v_b, particles smaller than a certain critical size may be entrained. Usually particles of this size are simultaneously deposited and re-entrained. These combined effects depend not only on velocities and turbulence in the bath but also on the adhesion, sintering, and wetting properties of the melt, inclusions, and refractory. Some remarks are made concerning these effects in the section on filtration.

In steel melts, problems with entrainment of inclusions are not usually encountered. There seem to be several reasons for this. One is that the density ratio $\Delta\rho/\rho$ is much higher, about $\frac{1}{2}$, so that the critical velocity is correspondingly higher. Furthermore, the inclusions that float up react or are absorbed in the top-slag layer. Usually $\Delta\rho/\rho$ for the top-slag in steel is $\frac{1}{2}$ or more. The main reason that re-entrainment does not seem to occur in steel, however, is probably that bath temperatures are high and the inclusions react chemically with, or sinter to, the refractory. Also, in an aluminium bath, inclusions can react with the refractory. However, such phenomena are expected to occur only to a limited extent at the relatively low temperatures around 700°C.

5.11 FLOTATION

In the section on stirring and complete mixing it was mentioned that inclusions could be transferred to the bottom or walls by various effects such as gravity and turbulence. If gas is bubbled through the melt, the particles may come into contact with the bubbles by the same mechanisms. Furthermore if these particles or inclusions are wetted by the bubbles but not by the melt, there is a high probability that particles remain trapped at the bubble–melt interface. They are then carried by the bubbles up to the dross or top-slag layer. This operation called flotation is described at some length in the following.

The theory of flotation is also useful as a background for understanding filtration.

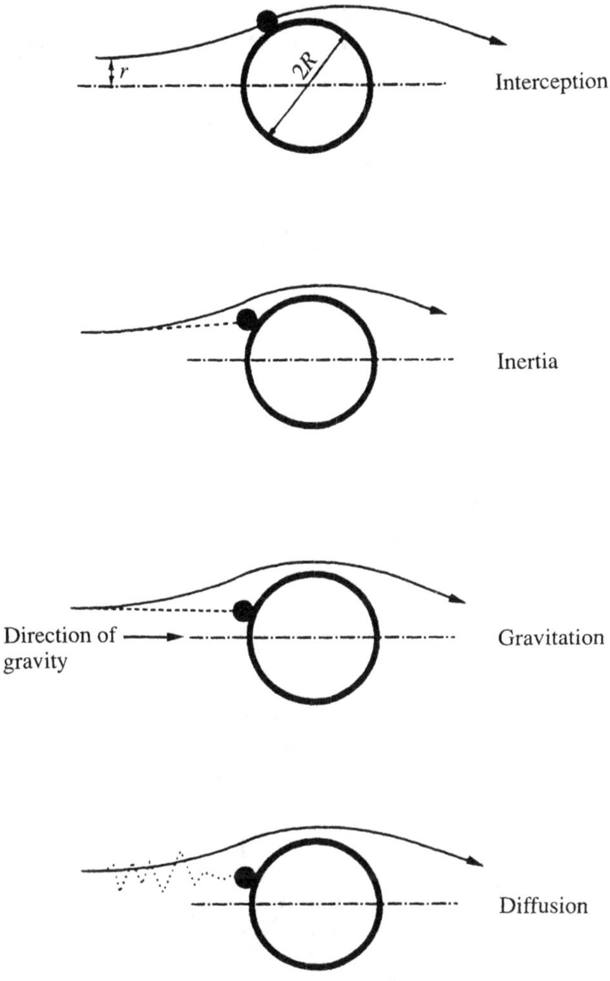

FIG. 5.11 Various mechanisms for the transfer of (spherical) particles to a spherical collector (bubble).

A mechanism more important for flotation than for removal to the collector walls or filters is *interception* illustrated in Fig. 5.11. According to this mechanism, the centre of the particle follows the streamlines along the bubble, but the outer edge of the particle close to the bubble intercepts the bubble–melt interface. Since velocities are high along a bubble as opposed to along a solid wall, a large number of particles pass close to the bubble so that removal in this manner is more important for bubbles than for a solid collector.

To develop a theory for flotation, we first assume that any particle is removed that 'stands in the path' of a rising bubble. Afterwards, it will be taken into account that a particle in fact is partly swept sideways out of the way of the rising bubble. The volume $\dot{V}_s$ that the bubbles sweep through per second is given by

$\dot{V}_s$ = cross-section of bubbles transverse to the direction of bubble flow × relative velocity between bubble and melt

or (for spherical bubbles of radius R):

$$\dot{V}_s = \pi R^2 N v_b$$

where N is the number of bubbles present in the melt and v_b is the velocity of the bubble relative to the melt. If the inclusions did not move sideways and stuck to the bubbles once contact with the interface was established, the expression that would give the volume purified per second is

$$\dot{V}_s = \pi R^2 N u_{br}$$

where u_{br} is the relative velocity between bubbles and particles in the melt. However, Fig. 5.12 illustrates that only inclusions approaching the bubble within a cross-sectional area πr^2 attain contact. Thus the effective volume purified is only

$$\dot{V}_e = \pi r^2 N u_{br}. \tag{5.53}$$

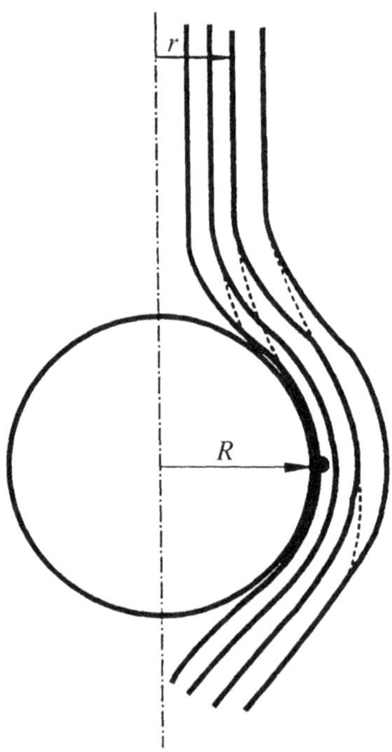

FIG. 5.12 Collision cross-section πr^2 for particles removed by a spherical collector.

As shown in eqn (4.14), N is equal to $3\dot{Q}\tau_b/4\pi R^3 = A/4\pi R^2$. In terms of the bubble–melt contact area A one obtains

$$\dot{V}_e = \frac{r^2}{R^2}\frac{Au_{br}}{4}. \tag{5.54}$$

Here $r^2/R^2 = \pi r^2/\pi R^2 = \eta$ is called the collision efficiency. η is the fraction of the cross-sectional area of the bubble where particles collide with the bubble. A more correct definition based on the particles instead of on the fluid is

$$\eta = \frac{\text{Number of particles that collide with the bubble per unit time}}{\begin{array}{l}\text{Number of particles per unit volume} \times \\ \text{Cross-sectional area of the bubble normal to the particle velocity} \times \\ \text{Velocity of particles relative to the bubble.}\end{array}} \tag{5.55}$$

If the particles have the same velocity as the melt, the definition reduces to

$$\eta = \frac{\begin{array}{l}\text{Volume flow of melt containing particles} \\ \text{that will collide with the bubble}\end{array}}{\text{Volume flow approaching the bubble}} \tag{5.55a}$$

To illustrate the use of this equation η is calculated for the interception mechanism. Since the flow around a bubble is only slightly perturbed from irrotational, we assume that we have potential flow. Furthermore, for small bubbles we may presume that the form is spherical. Then the stream function is given by [44]

$$\psi = \frac{v_b r^2 \sin^2\theta}{2}\left(1 - \frac{R^3}{r^3}\right) \tag{5.56}$$

where r is the distance from the centre of the sphere and θ is the polar angle. The curves ψ = constant are streamlines. The solution is chosen such that $\psi = 0$ for $r = R$. The flow of liquid in an annulus between r and R is equal to $2\pi\psi$.[44] Particles of radius a will colllide with the bubble if they lie within a distance $r = R + a$ (see Fig. 5.11). The volume flow approaching the spherical bubble is $v_b \pi R^2$. Then according to the definition of η

$$\eta = \frac{\sin^2\theta}{R^2}\left\{(R + a)^2 - \frac{R^3}{(R + a)}\right\} \approx \frac{3a\sin^2\theta}{R}$$

for $a \ll R$. Inclusions collide with the bubbles from the 'pole' at $\theta = 0$ to the 'equator' $\theta = \pi/2$. This gives

$$\eta = \frac{3a}{R} \tag{5.57}$$

for the collision efficiency of interception. η is usually much smaller than 1. For instance, inclusions of diameter $2a = 100\,\mu\text{m} = 10^{-4}\,\text{m}$ and bubbles of radius $R = 10^{-2}\,\text{m}$ give $\eta = 3 \times 10^{-4}/(2 \times 10^{-2}) = 0.015$.

Alternatively eqn (5.57) can be derived directly from the definition of η in eqn (5.55*a*):

$$\eta = \frac{2\pi Rav_\theta(\theta = \pi/2, r = R + a)}{\pi R^2 v_b}$$

where the tangential velocity is given by

$$v_\theta(\theta = \pi/2, r = R + a) = \frac{1}{r\sin\theta}\frac{\partial\psi}{\partial r} \approx \tfrac{3}{2}v_b$$

for $a \ll R$.

It is seen that the tangential velocity is higher than the approach velocity; the velocity must increase for the melt to get around the bubble.

$$\eta = \frac{2\pi Ra\frac{3}{2}v_b}{\pi R^2 v_b} = \frac{3a}{R}.$$

Thus for the interception mechanism, the definition (5.55a) may be replaced by

$$\eta = \frac{\text{Circumference of the bubble normal to the direction of flow} \times \text{Particle radius} \times \text{Velocity in the direction of flow along the circumference}}{\text{Volume flow of melt approaching bubble}} \qquad (5.55b)$$

The interception mechanism must often be augmented by other mechanisms such as removal by gravity or buoyancy. This will be discussed later. For a filter, settling may even be the dominating effect.

The number of particles that collide with spherical bubbles per unit time is

$$\dot{V}_e c = \eta \frac{A}{4} u_{br} c$$

where c is the number of particles in unit volume of the melt. $A/4$ is the projected area of the bubble. From this one can relate η to the coefficient for transfer of particles, k_t, introduced previously. k_t is defined so that the number of particles transferred to the bubble surface A is $k_t Ac$. This is equal to $\dot{V}_e c$. Then by comparison with the previous expression one finds that

$$k_t = \frac{\eta u_{br}}{4}. \qquad (5.58)$$

Here u_{br} is the particle velocity relative to the bubbles. Particles are only removed by the relative movement, u_{br}, between particles and spherical bubbles in the vertical direction. This is the reason for the factor $\frac{1}{4}$. If the bubble is not spherical, the factor $\frac{1}{4}$ in eqn (5.58) must be replaced by the factor b equal to the ratio between the projected area normal to the particle velocity and the surface area.

By analogy relations derived for mass transfer of a dissolved component also apply to flotation if the mass transfer coefficient k_t is replaced by $\eta u_{br}/4$ and the concentration in moles/m^3 is replaced by the number of inclusions per m^3. Examples are given in the next section. These results are utilized to describe what happens to the inclusions in completely mixed melts.

The concept of collision efficiency is also very useful in order to describe filtration. There is considerable interest in comparing the relative merits of filtration and flotation. A detailed discussion of the various collision mechanisms is presented in the section on filtration.

5.12 MODELS OF INCLUSION REMOVAL FROM A BATH WITH COMPLETE MIXING

As mentioned in Chapter 4 both continuous and batch refining units are in use. Due to its simplicity a steady state, back-mixed, continuous refining unit is treated first. Melt flows in and out at a volume flow rate $\dot{q}$. The content of inclusions in the inlet is c_{in} (number/m^3) in the reactor, and in the outlet it is c. A number balance for inclusions is (compare Fig. 4.7):

Number of inclusions in

= Number of inclusions out

+ Inclusions removed

or

$$\dot{q}c_{in} = \dot{q}c + k_t Ac \tag{5.59}$$

where k_t is the coefficient for transfer of particles to the contact surface, A. A is the surface area of the bubbles.

The output/input ratio is found from eqn (5.59):

$$\frac{c}{c_{in}} = \frac{1}{1 + k_t A/\dot{q}} = \frac{1}{1 + k_t AV/(V\dot{q})} \tag{5.60}$$

where V is the volume of the melt. $V/\dot{q} = \tau$ is the residence time of the melt in the refining reactor. A/V is the contact area per unit volume of melt. Equation (5.60) is the same as that obtained for dissolved elements (see Chapter 4) except that now k_t replaces the mass transfer coefficient k. To attain a significant removal of inclusions

$$\frac{k_t A}{\dot{q}} > 1. \tag{5.61}$$

A depends on the flow rate $\dot{Q}$, of gas, (see eqn (4.14). Equation (5.60) then becomes

$$\frac{c}{c_{\text{in}}} = \frac{1}{1 + 3k_t \dot{Q}\tau_b/(R\dot{q})}. \tag{5.62}$$

$\tau_b = H/v_{bw}$ where H is the depth down to the gas inlet and v_{bw} is the velocity of the bubble relative to the wall. By way of illustration we assume that $H = 1$ m, $\dot{Q} = 5 \times 10^{-3}$ m^3/s, $R = 0.005$ m, $\dot{q} = 0.01$ m^3/s (85 tonne/h Al), and $\eta = 0.015$. Then from eqn (5.58)

$$\frac{c}{c_{\text{in}}} = \frac{1}{1 + 3\eta \dot{Q} H u_{br}/(4\dot{q}Rv_{bw})} \tag{5.63}$$

or if $u_{br} \approx v_{bw}$

$$\frac{c}{c_{\text{in}}} \approx \frac{1}{1 + 3 \times 0.015 \times 5 \times 10^{-3}/(4 \times 0.01 \times 0.005)} = 0.47.$$

This removal is not very impressive. If other mechanisms in addition to interception are taken into account, removal is slightly better. Removal by flotation may be only marginal, especially for the smaller inclusions.

In general we have a distribution of inclusion sizes and k_t and η depend on a. Then, the volume of inclusions per unit volume of the exit stream is, on introducing the residence time $\tau = V/\dot{q}$,

$$c = \int_0^\infty \frac{f_{N_{\text{in}}}(a)\, 4\pi a^3/3}{1 + 3\eta(a)\dot{Q}H\tau u_{br}/(4RVv_{bw})}\, \mathrm{d}a \tag{5.64}$$

Notice that $f_{N_{\text{in}}}(a)\,\Delta a$ is the number of inclusions per unit volume of size a to $a + \Delta a$ in the feed. As previously we assume that the inclusions are spherical.

It is found as in Chapter 4 that the relation for the batch reactor is given by

$$\frac{f_N(a)}{f_{N_{\text{in}}}(a)} = \exp\left(-\frac{k_t(a)\,At}{V}\right) \tag{5.65}$$

where an exponential function now describes $f_N(a)/f_{N_{\text{in}}}(a)$ and the refining time t replaces the residence time $\tau = V/\dot{q}$ for the melt in the continuous reactor. $f_{N_{\text{in}}}(a)$ is the number of inclusions of size a per unit volume initially (at $t = 0$). Similarly, eqn (5.64) is replaced by

$$c = \int_0^\infty f_{N_{\text{in}}}(a)\,\frac{4\pi a^3}{3}\exp\left(-\frac{3\eta(a)\dot{Q}Htu_{br}}{4RVv_{bw}}\right)\mathrm{d}a \tag{5.66}$$

where c is the volume fraction of inclusions that remains after a time t.

5.13 FILTRATION

5.13.1 Introduction

Here filtration is defined as the process of removing inclusions by forcing molten metal through a porous material. The most important characteristic of a filter is the filtration efficiency E defined by

$$E = 1 - \frac{c_o}{c_{in}} \tag{5.67}$$

where c_{in} is the number of inclusions per unit volume that enters the filter, while c_o is the number per unit volume that leaves.

There are at least three participants in the filtration process: the inclusions, the molten metal, and the filter material. The solid or liquid inclusions are described by their size distribution, composition, density, and shape. The properties of the fluid depend on alloy composition, kinematic viscosity, and density. The filter material is characterized by its geometrical dimensions: the size of the structural units, their distribution and arrangement in the filter, and by their porosity.

The forces at the three interfaces (interfacial tensions) between the melt and filter material, melt and inclusions, and inclusions and filter play an important role. Also a gas or vacuum phase formed between the filter material and inclusion may influence the filtration properties.

The pressure drop, Δp, through a filter is important. In many applications in the aluminium industry today the metallostatic head at our disposal is limited. Often 0.1 or 0.2 m of metal head must suffice to force the metal through the filter. Initially the pressure drop depends only on the properties of the fluid and the filter. As filtration proceeds the pressure drop becomes dependent also on the properties of particles deposited on or in the filter. If the particles are deposited on the filter it is said that we have '*cake mode*' filtration. In this case particles deposited previously act as a filter for the last particles approaching the filter. When the inclusions are deposited inside the filter we speak of deep bed filtration. If the size of the inclusions are comparable to the filter pore size, cake filtration is inevitable. Even if the inclusions are one or two orders of magnitude smaller than the pore size, the section of the filter closest to the inlet gradually fills up so that the mechanism here more and more resembles cake mode filtration.

Filter capacity is defined as the quantity of deposited particles (usually expressed in grams or kilograms) which the filter is capable of accumulating before reaching a certain pressure drop (loss of metallostatic head). The filter capacity clearly depends on the type and size of the particles.

5.13.2 Cake mode filtration

Cake mode filtration as a 'quality assurance', is useful, for instance, in aluminium melt refining should there be a surge of inclusions. This may be the case when metal is recycled or remelted and when 'slurry' from the bottom section of a settling or holding furnace suddenly appears downstream. Certain alloys may contain a large number of inclusions. Cake mode filtration results in the development of a superficial layer of inclusions separated on the substrate filter. Such a cake may lead to a high pressure drop. The cake is responsible for and is also a consequence of particle capture. The behaviour of the cake depends on the size and size distribution of the inclusions and the pressure gradient experienced by the inclusions in the cake. Filter cakes can be broadly classified into two types: incompressible and compressible. An incompressible cake is one in which the void fraction, ε, of the cake does not change within the pressure limits. Then an increase in pressure drop with time is essentially caused by increases in cake thickness. Compressible cakes typically develop when the separated solids are deformable and represent a broad size distribution.

It is important to be able to predict the pressure drop in cake mode filtration for a given alloy. Measurements of flow rates can be carried out in the laboratory at a given (constant) pressure.[45] This allows resistances for the cake, r_c, and for the filter medium, r_f, to be determined. From the Blake–Kozeny equation valid for viscous flow[30] one obtains

$$A\frac{dp}{dz} = \dot{q}\rho\nu r_c \tag{5.68}$$

for the pressure gradient in the cake and similarly

$$A\frac{dp}{dz} = \dot{q}\rho\nu r_f \tag{5.69}$$

for the pressure gradient in the filter. Initially the pressure drop at our disposal, Δp, is due to the filter medium only. Then the volume flow, $\dot{q}$, is given by

$$\frac{A\Delta p_f}{H} = \dot{q}\rho\nu r_f \tag{5.70}$$

where H is the filter thickness.

As the cake builds up, an increasing part of the pressure drop is due to the cake. If the cake is incompressible so that r_c does not depend on dp/dz, eqn (5.68) can be integrated to give a pressure drop over a thickness of cake z:

$$A\Delta p_c = \dot{q}\rho\nu r_c z. \tag{5.71}$$

Addition of the two resistances gives

$$A\Delta p = \dot{q}\rho\nu(r_f H + r_c z). \tag{5.72}$$

The growth rate of the cake dz/dt depends on the metal flow rate $\dot{q}$, on the content of inclusions in the melt, c, (given in volume fraction) and on the volume fraction of inclusions constituting the cake, c_c. Then if it is assumed that all inclusions are captured, a volume balance gives

$$c_c A \frac{dz}{dt} = \dot{q}c. \tag{5.73}$$

From this equation it is seen that the cake thickness z for constant c and c_c is proportional to the volume of melt V that has passed through the filter:

$$z = \frac{c}{c_c A}\int_0^t \dot{q}\,dt = \frac{cV}{c_c A}. \tag{5.74}$$

If z is eliminated from eqn (5.72), $\dot{q} = dV/dt$ is inserted, and the integration over time t is carried out, one obtains for V:

$$A\Delta p t = V\rho\nu\left(r_f H + \frac{r_c cV}{2c_c A}\right). \tag{5.75}$$

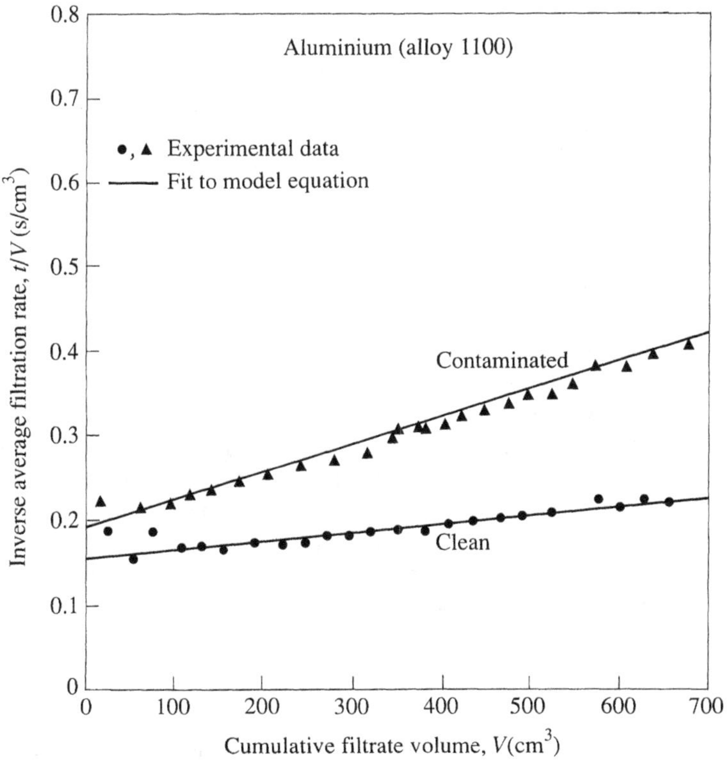

FIG. 5.13 Inverse average filtration rate t/V as a function of the cumulative filtrate volume, V, for both clean and contaminated aluminium (1100 alloy) melts.[45]

An example of the use of this equation is shown in Fig. 5.13. There the inverse average filtration rate t/V is given as a function of V. $\rho\nu r_f H/(A\Delta p)$ is obtained from the intercept of the straight lines with the y-axis, while the slope gives $\rho\nu r_c c/2A^2\Delta p c_c$. The slope is seen to be much greater for the contaminated melt.

It may be illuminating to apply the above analysis to an industrial situation where the required head of aluminium is calculated.[45] r_f and r_c have been determined in the laboratory; for instance,

$$r_f H = 3 \times 10^7\,\mathrm{m}^{-1}$$

$$r_c = 2 \times 10^{11}\,\mathrm{m}^{-2}.$$

The casting conditions are as follows:

$$\text{casting rate, } \dot{q} = 4 \times 10^{-3}\,\mathrm{m^3}/s$$
$$\text{drop size, } V = 33\,\mathrm{m^3}$$
$$\text{filter area } (20'' \times 20'') = 0.2581\,\mathrm{m^2}$$
$$\nu = 0.4 \times 10^{-6}\,\mathrm{m^2/s}$$
$$\rho = 2350\,\mathrm{kg/m^3}$$
$$\frac{c}{c_c} = 10^{-5}.$$

Then eqns (5.71) and (5.73) give

$$\Delta p = \frac{\dot{q}\rho\nu}{A}\left(r_f H + \frac{r_c Vc}{Ac_c}\right)$$
$$= 4.16 \times 10^3\,\mathrm{N/m^2}$$

or an aluminium head of $4.16 \times 10^3/(9.81 \times 2350) = 0.18\,\mathrm{m}$.

This head represents a problem with the present design of cast houses. It may therefore be worthwhile taking a closer look at the resistances. Resistance in the cake, $r_c Vc/(Ac_c)$, dominates that in the filter, $r_f H$, even though the thickness of the cake $Vc/(Ac_c)$ is only about 1 mm compared to $H = 5$ cm (for a ceramic foam filter). The explanation is that resistance in the cake is governed by inclusion size while resistance in the filter depends on the size of the pores. Then resistance in the cake must be high as may be seen from the Blake–Kozeny[30] equation for a packed bed:

$$r_c \approx \frac{150(1-\varepsilon)^2}{d_p^2 \varepsilon^3} \tag{5.76}$$

where d_p^2 is the average squared particle diameter and ε is the porosity. For instance, for a cake with inclusions of diameter $d_p = 30\mu$m and a porosity of $\varepsilon = \frac{1}{2}$

$$r_c = \frac{150 \times (0.5)^2}{(30 \times 10^{-6})^2 (0.5)^3} = \frac{30}{9} \times 10^{11}\,\mathrm{m}^{-2}.$$

This happens to be close to the value used in the previous calculation of pressure drop. In a ceramic foam filter pore size is usually in the range 500–1000 μm. Thus the pressure gradient is several orders of magnitude less than for the cake.

This section is concluded by stating that cake filtration very effectively removes practically all inclusions since the inclusions are their own filter. The voids between the inclusions are small enough to trap all but the smallest particles in the size distribution. Some disadvantages are that particles may escape in the early stages since it takes some time to build up a cake, inclusions may be re-entrained in the melt from the 'back-side' of the cake, and the filter capacity is limited due to a rapid increase of pressure drop with time.

5.13.3 Deep bed filtration

In the previous section the problems with cake mode filtration were outlined: a high pressure drop and a limited filter capacity. The reasons were that the pressure drop is caused essentially by the small inclusions, deposited only on the inlet surface of the filter.

If particles can be deposited on the walls inside a filter with collector dimensions much larger than the particles, the pressure drop is drastically reduced. Also, if the total surface area within the filter is much greater than the inlet surface, filter capacity is improved.

To attain satisfactory deep bed filtration two or three problems must be considered. First, cake mode filtration should not occur and second, the inclusions must in fact be deposited within the filter. Third, inclusions may protrude into the melt from the collector surface. Such 'dendrites' will give a significant increase in flow resistance.

To prevent cake mode filtration, the dimensions of the inclusions must be much smaller than the filter pore size. This means that ideally the melt should not contain large collections of inclusions or oxide films which may tend to choke up the filter inlet. Large inclusions and films should therefore ideally be removed in a previous separation stage.

The previous separation unit may be a settling furnace, a flotation unit, or a filter with large pores and a high filter capacity. When flotation is employed, it is possible to skim off the inclusions contained in a dross or slag layer. Both settling and flotation tend to remove mainly the large inclusions. Gravity and inertia are the important mechanisms in this removal. When the density differences are small it is important not to stir the melt too much so as to disperse the inclusions again once they have been removed to the bottom or top melt layers. In steel where density differences are large, inclusions can be eliminated by the action of stirring and turbulence to walls, and top-slag. This is discussed in the final section of this chapter.

It may be said that the large inclusions (greater than 50 mm) are easy to remove if reasonable measures are taken. The small inclusions (less than 1 μm) today present no problem for the metal quality. It is the intermediate size inclusions, say from 5 to 20 μm, that give difficulty. They are small enough to elude capture and sufficiently large to damage the end metal product. In the rest of this chapter priority is given to the study of the removal of this intermediate size.

It will be seen that to assure removal of the intermediate size inclusions within the filter a large inner surface area is required. To give a large surface area the unit collector dimensions must be small. A balance must be struck which depends upon the inclusion size: filter pore dimensions should be large enough to avoid cake build up and small enough to capture inclusions down through the filter. Thus in *sandwich* type filters the inlet section has larger pores than the outlet part.

To describe the removal of inclusions in a filter, it is assumed initially that once the inclusions touch the collector, they adhere. A number balance over a differential slice of a filter (control volume) gives (see Fig. 5.14):

Flow of inclusions through the surface at position z
= Flow of inclusions through the surface at position $(z + \Delta z))$
+ Number of inclusions transferred to the collectors in unit time

or

$$c(z)Au\varepsilon = c(z + \Delta z)Au\varepsilon + k_t a_s (A\Delta z)\varepsilon c \tag{5.77}$$

where a_s is the surface area per unit volume of melt and u is the mean particle velocity. u is in the direction of z. $A\Delta z$ is the control volume and $A\Delta z\varepsilon$ is the melt volume. k_t is the coefficient for transfer of particles derived previously in eqn (5.58) for the analogous case of transfer to bubbles:

$$k_t = \eta ub. \tag{5.78}$$

Since u is an average velocity of the particles in the voids between the collectors, the approach velocities and tangential velocities to the collectors are the same. b, as defined previously, is the ratio of the collector surface projected in the flow direction to the total collector surface area. The collision efficiency for a

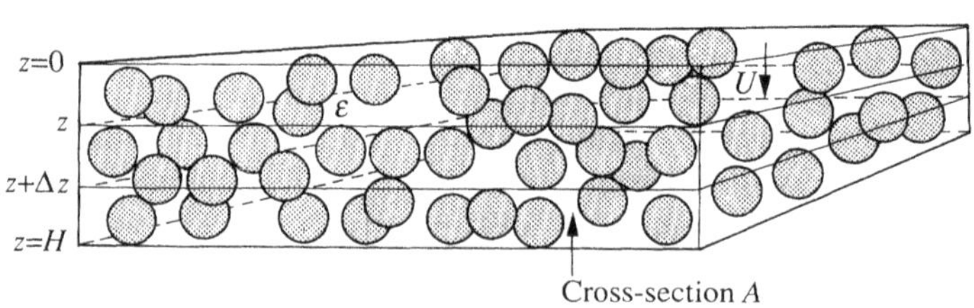

FIG. 5.14 A differential slice (control volume) in a filter consisting of spheres (for example Al_2O_3-balls).

collector in a filter may be defined as previously in eqn (5.55) for a bubble, except that the word collector replaces bubble in the definition. For instance, eqn (5.55b) for the interception mechanism gives

$$\eta = \frac{2\pi Rau}{\pi R^2 u} = \frac{2a}{R}. \tag{5.79}$$

In fact there will be a velocity boundary layer around the solid collector. Velocities are much less than u in this layer. Therefore eqn (5.79) overestimates the value of η for inclusions smaller than the boundary layer.

If eqn (5.78) is inserted into eqn (5.77), and $\Delta z \to 0$, one obtains the differential equation:

$$0 = \frac{\mathrm{d}c}{\mathrm{d}z} + \eta a_s bc \tag{5.80}$$

or on integration:

$$\frac{c(z)}{c_{\text{in}}} = \exp(-\eta a_s bz) \tag{5.81}$$

where c_{in} is the number of inclusions per unit volume of melt that enters the filter.

The calculation has been carried out for the one-dimensional case. However, the derivation applies also in two or three dimensions if z is replaced by the arc length, s, of a stream-tube (a tube where the volumetric flow is the same throughout). Strictly, in this case η depends on u and thereby on s so that

$$\frac{c(s)}{c_{\text{in}}} = \exp\left(-a_s b \int_O^s \eta(s)\,\mathrm{d}s\right). \tag{5.81a}$$

In the one-dimensional case if the filter depth is H[58]

$$\frac{c}{c_{\text{in}}} = \exp(-\eta a_s bH). \tag{5.82}$$

This gives a filtration efficiency:

$$E = 1 - \frac{c}{c_{\text{in}}} = 1 - \exp(-\eta a_s bH). \tag{5.83}$$

For a void fraction ε and a collector diameter of $2R$, the specific surface area becomes

$$a_s = \frac{\text{Surface area of collector}}{\text{Volume of melt}}$$

$$= \frac{\text{Surface area of collector}}{\text{Volume of collector}} \times \frac{\text{Volume of collector}}{\text{Volume of melt}}$$

$$= \frac{4\pi R^2}{4\pi R^3/3} \times \frac{(1-\varepsilon)}{\varepsilon} = \frac{3(1-\varepsilon)}{R\varepsilon}. \tag{5.84}$$

Then for a packed bed of spheres (where $b = \frac{1}{4}$)

$$E = 1 - \exp\left(-\frac{3\eta(1-\varepsilon)H}{4R\varepsilon}\right). \tag{5.85}$$

For the interception mechanism (with $\eta = 2a/R$)

$$E = 1 - \exp\left(-\frac{3a(1-\varepsilon)H}{2R^2\varepsilon}\right). \tag{5.86}$$

For instance

$$\begin{aligned} a &= 5\,\mu\text{m} \quad = 5 \times 10^{-6}\,\text{m} \\ R &= 0.5\,\text{mm} = 0.5 \times 10^{-3}\,\text{m} \\ H &= 0.05\,\text{m} \\ \varepsilon &= 0.5 \end{aligned}$$

gives $E = 1 - 0.22 = 0.78$. The concentration at the outlet is reduced to 22 per cent of the inlet concentration.

It is seen that the exponent in eqn (5.86) is proportional to the inclusion size a and inversely proportional to the square of the collector dimensions. Thus the efficiency improves strongly with decreasing collector dimensions. On the other hand pressure drop also increases significantly.

5.13.4 Collision mechanisms

The interception mechanism has been discussed in some detail previously. For completeness flow normal to the axis of a cylindrical collector in a filter is included. Then according to eqn (5.55b) the collision efficiency becomes

$$\eta = \frac{a}{R} \tag{5.87}$$

where R is the radius of the cylinder and a is the radius of a spherical inclusion.

Another collision mechanism that is or may become important is gravity (buoyancy forces). Also hydrodynamic forces may play a role in practice. For removal of inclusions larger than 1 μm from metals it seems that Brownian diffusion plays a minor part.[46,47] Various collector mechanisms are illustrated in Fig. 5.11. Inertial forces are important for the removal of particles (aerosols) from gases but usually not in the filtration of molten metals. The reason is essentially that the ratio of particle density to fluid density is a factor 1/1000 smaller for metal systems. The inertial force of the particles relative to the viscous force is small. Thus they do not penetrate the boundary layer surround-

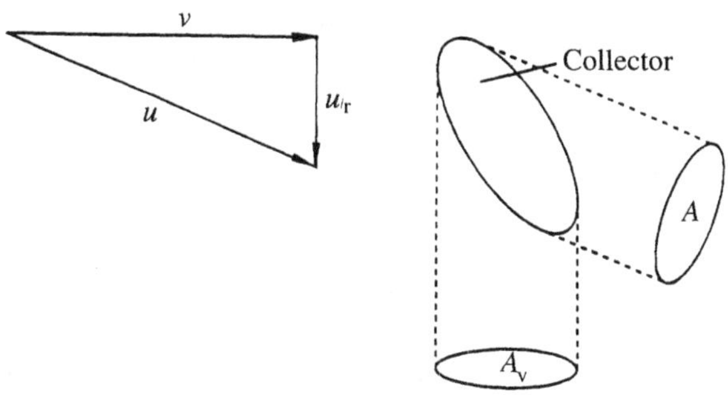

FIG. 5.15 The vertical projection A_v and the projection normal to the path of particle A, needed to calculate the collision efficiency due to gravity or buoyancy.

ing the collector. Only for velocities above 0.1 m/s and inclusion sizes of 100 μm do inertial forces become important.[47]

To determine the collision efficiency due to gravity or buoyancy, the velocity difference between fluid and particle (inclusion), u_r, is introduced. In Fig. 5.15 u_r is taken in the downwards direction. The fluid velocity is in the horizontal direction. To discuss this case, eqn (5.55) is employed except that the word 'bubble' is replaced by the word 'collector'. Then if the vertical projection of the collector is A_v and the projection normal to the mean particle flow direction is A

$$\eta = \frac{A_v u_r}{A u}. \tag{5.88}$$

For a sphere $A_v = A$ and

$$\eta = \frac{u_r}{(v^2 + u^2)^{\frac{1}{2}}} \approx u_r v. \tag{5.89}$$

This collection efficiency for the effect of gravity is often presented in the literature. Notice that v is the relative velocity between the melt and collector, while u_r is the relative velocity between particles and the melt. Equation (5.89) applies only when v and u_r are in the horizontal and vertical directions, respectively. Other configurations are left as an exercise for the reader.

Settling in a filter may be important and may even be the dominant removal mechanism. For instance, in the section on settling, a particle of diameter 100 μm gave $u_r = 0.136 \times 10^{-2}$ m/s. This may be compared with a velocity $v = 0.6 \times 10^{-2}$ m/s. Then for the settling mechanism $\eta = 0.136 \times 10^{-2}/(0.6 \times 10^{-2} + 0.136 \times 10^{-2}) = 0.185$ while the interception mechanism[59] gave $\eta = 2a/R = 2 \times 50 \times 10^{-6}/(0.5 \times 10^{-3}) = 0.2$. Even though eqn (5.79) overestimates the effect of interception, settling in the example competes with interception.

If particles are not spherical but elongated, irregular, or formed as plates, they will not follow the streamlines. They may wobble or tumble,[44,48] and this should give a high collision efficiency. These phenomena are called hydrodynamic effects. They are difficult to describe and no method for the calculation of η is proposed for this case.

It is often necessary to include more than one mechanism in order to determine the total collision efficiency, η_t. For instance, often both interception and gravity should be taken into account. Then for $\eta \ll 1$ we assume that the various mechanisms are independent of each other and regard the η as probabilities. The probability that a given particle approaching a collector does not collide with it is given by

$$1 - \eta = (1 - \eta_i)(1 - \eta_r) \tag{5.90}$$

where $1 - \eta_i$ is the probability that it does not collide due to interception. $1 - \eta_r$ is the probability that collision is not caused by gravity. The equation overestimates the total collision efficiency since sometimes the same particle may be removed by both mechanisms. For η_i and $\eta_r < 0.1$ eqn (5.90) simplifies to[46]

$$\eta \approx \eta_i + \eta_r. \tag{5.91}$$

5.13.5 Re-entrainment of inclusions

In the previous section it was assumed that once an inclusion collided with the collector, it remained trapped at the point of collision. In this case the collection efficiency η in eqn (5.83) is equal to the collision efficiency. Various experimental results indicate, however, that inclusions in aluminium can in fact be released from a filter.[10,49] This may seem rather surprising since the material used in the filters has a high interfacial tension relative to the melt. There is a net reduction of Gibbs energy when an inclusion resides at the interface between filter and melt. Since the interfacial tension is of the order of 1 N/m, the forces holding back an inclusion are of the order of magnitude a, where a is the inclusion radius. The viscous drag forces on the particle are roughly $a\nu\rho v$. $\nu\rho v$ may be about 10^{-5} N/m. Thus seemingly the forces restraining the particles should be a factor 10^5 greater than the forces that re-entrain. An explanation is that both the particles and the filter have rough surfaces on a microscopic scale and that the true contact area between the particle and filter is only roughly 0.01 per cent of the apparent area.[50] One might believe that a remedy against re-entrainement would be to encourage the creation of liquid inclusions that could attain close contact with the collector or to cover the filter surface with a molten salt. However, a second liquid phase tends after a time to run through the filter.[49] One recourse is to prevent fluctuations in the flow. Certainly, surges of metal through the filter are to be prevented. Also, if the filter depth is ample, re-entrained inclusions can be recaptured downstream.

It may be mentioned that there does not seem to be any evidence of re-entrainment of solid inclusions from alumina filters for steel. On the contrary such filters have tended to choke up so that the pore sizes used must be larger than for aluminium.[51] Probably, the inclusions that hit the filter medium sinter or react chemically with the filter at temperatures around 1600°C.

The use of a greater pressure drop is obviously of interest in filtration, especially if 'dendrites' are formed. One problem is that the pressure gradients or velocity may become too high so that particles are re-entrained as described above. Another difficulty-for flow upwards-with bed filters is that if the pressure gradient exceeds $(\rho_b - \rho)g(1 - \varepsilon)$ where ρ_b is the density of the collector medium, then fluidization or channelling may take place. This is obviously not a problem for rigid media filters. However, these are more expensive.

5.14 TURBULENT REMOVAL OF INCLUSIONS TO WALLS

5.14.1 Introduction

The fluid mechanics of transfer of inclusions to the walls of a ladle is treated. In this section it is assumed that once an inclusion touches a wall it adheres to it. As mentioned previously this assumption may be fully valid for steel melts.

Due to turbulence the inclusions are completely mixed in the bulk melt. According to classical hydrodynamic theory a boundary layer is present at the melt–wall interface (also at the melt–slag interface). As shown schematically in Fig. 5.16 for a smooth wall the velocity drops from v_b to zero through the boundary layer of thickness δ_b. Following a theory originally proposed by Levich[43] the boundary layer is split up into three parts: the turbulent wall layer, the restricted turbulence layer at distance δ_1 from the wall, where turbulence is strongly damped but where turbulent diffusion is greater than diffusion by Brownian motion, and finally the inner viscous (laminar flow) layer, where the inclusions are transferred only by Brownian motion. It is shown in an example in Section 7.3 that the thickness δ_2 of the inner layer for the cases of interest may be about 1 μm. At $y = \delta_2$ the turbulent diffusion coefficient is so small that it is equal to the Brownian diffusion coefficient. If inclusions are larger than δ_2, the viscous boundary layer can be neglected. Inclusions greater than δ_b penetrate simply due to their size so that there is no problem in removing them. In the following the behaviour of particles is studied only for the inclusion size range between δ_2 and δ_b. Removal of inclusions smaller than δ_2 is discussed in Section 7.3 concerning dissolved elements.

5.14.2 Pipe flow

Removal of inclusions in fully developed pipe flow is treated first and in detail. There are several reasons: flow in a pipe is well known from the literature and

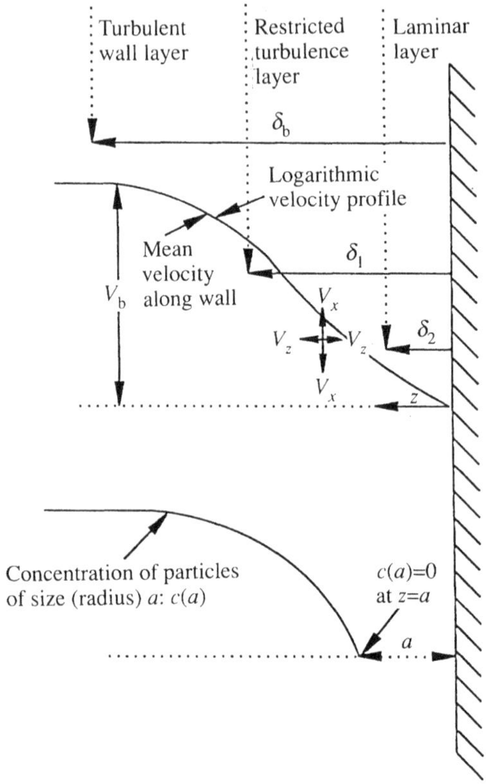

FIG. 5.16 Velocity and particle concentration along a smooth ladle wall.

this case can be used to understand the mechanism of inclusion removal for more complex geometries. Possibly in filtration the pipe flow case can be of direct use. Also light is thrown on the phenomenon of nozzle blockage even though removal in this case relates to entrance effects in a pipe and not to developed flow further down the pipe.

For the boundary layer region between δ_1 and δ_2 the distribution of the average velocity $\bar{v}_x$ along the wall has the same form as in laminar flow.[43]

$$\bar{v}_x \propto z \tag{5.92}$$

where z is the distance from the wall.

If the usual assumption[43] is made that there is proportionality between eddy velocities and average velocities, we obtain for the eddy velocity (see also Chapter 7)

$$v_x \propto z. \tag{5.93}$$

In view of the continuity equation we have for the boundary layer:

$$\frac{\partial v_x}{\partial x} + \frac{\partial v_z}{\partial z} = 0. \tag{5.94}$$

The normal component of the eddy velocity is

$$v_z \sim -\int_0^z \frac{\partial v_x}{\partial x} \mathrm{d}z \propto z^2. \tag{5.95}$$

Shear stress at the wall τ_0, the density ρ, and the kinematic viscosity ν determine v_z. From dimensional analysis it is found that v_z depends on τ_0, ρ, and ν as follows:

$$v_z \propto \left(\frac{\tau_0}{\rho}\right)^{\frac{3}{2}} \frac{z^2}{\nu^2}.$$

The proportionality constant must be determined experimentally. It has been estimated to be[43] $0.36 \times 10^{-2} v_0^3/\nu^2$ for the case of flow in a smooth tube where the mean velocity fluctuations in the axial direction are about $\frac{1}{4}$ of the mean velocity. v_0 is the so-called shear velocity, $v_0 = (\tau_0/\rho)^{\frac{1}{2}}$. The shear velocity plays a very important role when conditions near a solid wall are described. Now the mass diffusivity by eddies D_E may be obtained using the Prandtl mixing length model[43] (see section 7.11.2):

$$D_\mathrm{E} = v_z l \tag{5.96}$$

where l is the Prandtl mixing length. $l = 0.4\,z$, which gives

$$D_\mathrm{E} = \frac{0.14 \times 10^{-2} v_0^3 z^3}{\nu^2}. \tag{5.97}$$

Probably, this relation gives an upper limit to the turbulent diffusivity in a smooth pipe.

For eddy diffusion close to a solid surface Fick's first law of diffusion may be written in the form

$$\dot{n}_a = -D_\mathrm{E} \frac{\mathrm{d}c(a)}{\mathrm{d}z} \tag{5.98}$$

where $\dot{n}_a$ is the flux of inclusions of diameter $2a$ and $c(a)$ is the concentration of inclusions of diameter $2a$ at position z. One boundary condition is:

(1) At $z = \delta_1$, $c(a) = c_{\delta_1}(a)$: (5.99)

Furthermore an interception boundary condition is introduced:

(2) When an inclusion touches the wall or a protrusion on the wall it adheres. Then at

$$z = a \text{ we get } c(a) = 0 \tag{5.100}$$

for a perfectly smooth wall

or at

$$z = e \text{ we have } c(a) = 0 \tag{5.101}$$

where $e - a$ is the height of a protrusion (see Fig. 5.17).

The assumption is that the protrusions do not significantly change D_E. This means that the height $e - a$ is small and the number of protrusions is limited so that flow is not noticeably modified. On the other hand the number of protrusions should be sufficient to collect all inclusions present in the melt at a distance less than e from the wall. In Appendix 3 the boundary condition eqn (5.101) is discussed further. This boundary condition is used in the following.

If eqn (5.97) is inserted into eqn (5.98) one obtains for the flux

$$\dot{n}_a = \frac{-0.14 \times 10^{-2} v_0^3 z^3}{\nu^2} \frac{\mathrm{d}c(a)}{\mathrm{d}z}. \tag{5.102}$$

This equation may be integrated by writing

$$-\frac{\dot{n}_a \nu^2 \mathrm{d}z}{0.14 \times 10^{-2} v_0^3 z^3} = \mathrm{d}c(a). \tag{5.103}$$

Use of the boundary conditions (5.99) and (5.101) gives on integration from $z = e$ to $z = \delta_1$:

$$-c_{\delta_1}(a) = \frac{\dot{n}_a \left(\dfrac{1}{e^2} - \dfrac{1}{\delta_1^2}\right) \nu^2}{v_0^3 \times 0.28 \times 10^{-2}}. \tag{5.104}$$

Of course this equation only applies when the protrusion height e is less than the boundary layer thickness, δ_1. For the region $\delta_1 < z < \delta_b$, one may use, as a first approximation, the equations of the Reynolds analogy

$$-\frac{\dot{n}_a}{\tau_0} = \frac{c_b(a) - c_{\delta_1}(a)}{\rho(v_m - v_{\delta_1})} \tag{5.105}$$

where $c_b(a)$ is the concentration in the bulk melt and v_m is the mean velocity in the pipe and v_{δ_1} is the velocity at $z = \delta_1$. The basis for equation (5.105) is the assumption that the mean eddy lengths for the transfer of mass and momentum are the same. Thus concentration and velocity vary proportionally to each other. $\dot{n}_a$ is proportional to $c_b(a) - c_{\delta_1}(a)$ and $-\tau_0$ is proportional to $\rho(v_m - v_{\delta_1})$. Then the ratio between $\dot{n}_a$ and $-\tau_0$ is given by eqn (5.105).

Now the velocity, v_{δ_1}, between the restricted turbulence layer and the turbulent wall layer is equal to $5v_0$. Furthermore the thickness of the restricted turbulence layer δ_1 decreases with v_0, $\delta_1 = 5\nu/v_0$.[43] If $c_{\delta_1}(a)$ and v_{δ_1} are eliminated from eqns (5.104) and (5.105), one obtains

$$\dot{n}_a = \frac{v_\text{m} c_\text{b}(a)\phi}{1 + \phi^{\frac{1}{2}} \left(\dfrac{353\nu^2}{e^2 v_0^2} - 19\right)} \tag{5.106}$$

where

$$\phi = \frac{v_0^2}{v_\text{m}^2} = \frac{\tau_0}{\rho v_\text{m}^2} \tag{5.107}$$

is the friction factor for the pipe.

$$\left\{1 + \phi^{\frac{1}{2}} \left(\frac{353\nu^2}{e^2 v_0^2} - 19\right)\right\} \Big/ v_\text{m}\phi$$

gives the total resistance in the two boundary layers. $1/(v_\text{m}\phi)$ is the resistance from the turbulent wall layer.

Blasius found empirically for $\text{Re} = v_\text{m} 2R/\nu$ from 3000 to 10^5 (for smooth pipes) that[43]

$$\phi \approx 0.04\text{Re}^{-\frac{1}{4}}. \tag{5.108}$$

Resistance in the restricted turbulence layer dominates when the protrusions are small, for small e. Then eqn (5.106) reduces to

$$\dot{n}_a \approx \frac{v_\text{m} c_\text{b}(a)\phi^{\frac{1}{2}} v_0^2 e^2}{353\nu^2}$$

or, using eqn (5.107),

$$\dot{n}_a = \frac{v_\text{m}^3 \phi^{\frac{3}{2}} e^2 c_\text{b}(a)}{353\nu^2} \tag{5.109}$$

Resistance to particle transfer in the turbulent wall layer prevails for large e and we obtain from the analogy between momentum and mass transfer in the turbulent wall layer

$$\dot{n}_a \approx v_\text{m}\phi c_\text{b}(a). \tag{5.110}$$

5.14.3 Nozzle blockage and removal of inclusions by stirring

To illustrate the use of the equations derived in the previous section the case of nozzle blockage[51,52] is studied. Then removal due to stirring in general is discussed.

The experimental conditions are given by $R = 0.003$ m, $v_\text{m} \approx 2$ m/s, $T = 1873$ K, $\rho = 7030$ kg/m^3, and $\nu = 0.8 \times 10^{-6}$ m^2/s so that Re = 15 000 and $\phi = 0.0036$. It is assumed that the nozzle walls were originally smooth and that the protrusions are single inclusions already deposited on the walls (see Fig. 5.17). Then choosing $e = 3a = 10^{-5}$ m, eqns (5.109) and (5.106) give,

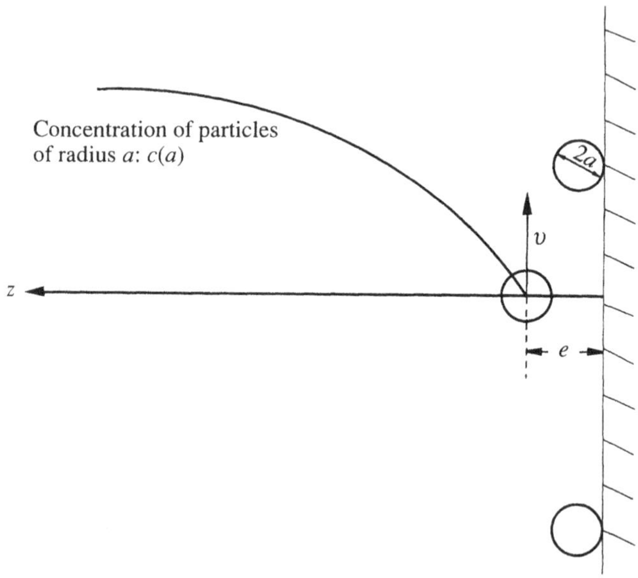

FIG. 5.17 Protrusions created by particles deposited on the wall.

respectively: $\dot{n}_a/c_b(a) = 0.76 \times 10^{-3}$ m/s and 0.77×10^{-3} m/s. For a choice of $e = 3a = 3 \times 10^{-5}$ m eqns (5.110) and (5.106) give $\dot{n}_a/c_b(a) = 7.2 \times 10^{-3}$ m/s and 7.9×10^{-3} m/s, respectively. In the example, $e = 3a = 10\,\mu$m is only a small protrusion so that resistance in the restricted turbulence layer dominates, while $e = 3a = 30\,\mu$m turns out to be large enough to penetrate the restricted turbulence layer. This may also be seen by comparing e and $\delta_l = 5\nu/v_0 = 5\nu/(v_m)\phi^{\frac{1}{2}} = 5 \times 0.8 \times 10^{-6}/(2 \times 0.0036^{\frac{1}{2}}) = 3.3 \times 10^{-5}\,\text{m} = 33\,\mu\text{m}$. With large protuberances only resistance in the turbulent wall layer remains.

The deposits often have a dendritic form mainly composed of particles of size 1–10 μm. Deposition is *autocatalytic*. Particles pile up where deposition has been initiated. A small amount of deposit can then clog the nozzle. For instance, for the experimental $2R = 0.006$ m nozzle[52] the volume of deposit that causes blockage was very small, about $2.8 \times 10^{-8}\,\text{m}^3$. With about 150 p.p.m. (by volume) inclusions, blockage occurred in about three minutes. If the surface area of the nozzle is taken as $0.6 \times 10^{-3}\,\text{m}^2$, the volume of the deposit (for particles with $a = 3.3\,\mu$m) in three minutes becomes $0.6 \times 10^{-3} \times 180 \times 0.77 \times 10^{-3} \times 1.5 \times 10^{-4} = 1.2 \times 10^{-8}\,\text{m}^3$. Thus the theory given here is in rough agreement with the experiments. The discrepancy is probably partly due to the fact that the inclusions pile up so that the protrusions soon become larger than $2a$. Wilson *et al.*[52] present some results concerning the calculation of inclusion transfer to walls by numerical methods. For instance, with the FLUENT program[53] it is possible to calculate inclusion trajectories to the nozzle walls.

For a melt stirred with an impeller turbulent fluctuations are roughly twice as great as for pipe flow.[43] Then the factor 353 in eqn (5.106) is replaced by 353/2 = 176. Equations (5.109) and (5.106) are replaced by

$$\dot{n}_a = \frac{v_m^3 \phi^{\frac{3}{2}} e^2 c(a)}{176\nu^2} \tag{5.111}$$

for transfer in the zone of restricted turbulence, while eqn (5.110) for the transfer resistance in the turbulent wall layer remains unchanged. The friction factor ϕ may be calculated using the circulation velocity v_b and eqn (5.49). For instance, $x = 1$ m, $v_b = 2$ m/s, and $\nu = 0.8 \times 10^{-6}$ m^2/s gives Re $= 2.5 \times 10^6$ and $\phi = 1.6 \times 10^{-3}$. Also for impeller stirring, inclusions as small as $a = 10\,\mu$m ($e = 3a = 30\,\mu$m) are removed in accordance with eqn (5.110) valid for resistance only in the turbulent wall layer. In this case eqn (5.110) gives $n_a/v_b c(a) = \phi = 1.6 \times 10^{-3}$, the same as that obtained from eqn (5.106) using the factor 176 instead of 353.

At points where the flow leaves the wall, for instance at the wall–slag line, the wall shear stress[54] will be very large, the turbulence level increases, and transport is enhanced. In general to determine the relative resistances of the turbulent wall layer and the zone of restricted turbulence, extensive numerical calculations must be carried out. It will be seen that an indication of their relative importance may be obtained from measurements of inclusion removal versus 'stirring power'. Mass transfer in the restricted turbulence boundary layer increases as the third power of the velocity, v_b, and the thickness δ_l of the layer decreases with increasing velocity. Thus at sufficiently high velocities (sufficient stirring) resistance in the restricted turbulence layer becomes negligible. Ideally, velocities and stirring should be sufficient to remove transfer resistance in this layer. This leaves only the resistance in the turbulent wall layer (eqn (5.110)).

Stirring may be obtained not only with an impeller but also electromagnetically.[28, 33, 38, 39] Either a travelling magnetic field or a pulsating stationary field may be employed. In Chapter 4 we have given examples of the use of gas to stir a melt. The flow and turbulence induced by bubbling is mentioned briefly in Chapter 7 as an illustration of the use of numerical methods. At lower temperatures there is a choice of stirring method so that for magnesium and aluminium all three can be used. For steel, impellers, cannot usually be employed due to the mechanical and chemical attack on the impeller material at the high temperatures. We have seen that stirring can serve not only to mix the bulk bath but also to transfer inclusions to the walls of the crucible and to the top-slag. On the other hand, stirring must not be so intense that slag, dross, slurry, or inclusions already deposited are re-entrained.

5.14.4 Stirring energy

In many industrial applications numerical calculations such as those mentioned in Chapter 7 are not available. It may not even be a simple matter to apply eqn (5.111) to an impeller-stirred system since we usually do not know the roughness of the crucible wall, that is, the protrusion height, e. Whatever the type of stirring, we expect turbulent diffusion in the zone of restricted turbulence to limit inclusion transfer at lower stirring rates, while the turbulent wall layer (eqn 5.110) limits transfer at high stirring rates. It is a problem that for bubble and electromagnetic stirring the proportionality constant in eqn (5.111) cannot be specified without extensive numerical calculations. Furthermore, the friction factor in eqn (5.110) is not known.

Often the specific stirring power input, $\dot{\varepsilon}$, to the melt can be determined whatever the method of stirring.[57] In this case, in order to obtain a rough estimate for the inclusion transfer one may, according to Kolmogorov,[43] assume that $\dot{\varepsilon}$ is proportional to the flow of kinetic energy per unit mass, that is, to the third power of the mean velocity. In terms of the shear velocity this gives the relation

$$\dot{\varepsilon} = \frac{v_0^3}{l_e} \tag{5.112}$$

where l_e is a characteristic length of the energy containing eddies. Equation (5.112) says that $\dot{\varepsilon}$ is proportional to v_0^3 with proportionality constant $1/l_e$. For pipe flow it is found experimentally that l_e is proportional to the diameter d and has a weak dependence on the Reynolds number.[43]

$$l_e = 0.05\, d\, \mathrm{Re}^{-\frac{1}{8}} \tag{5.113}$$

where $\mathrm{Re} = v_m d/\nu$. Then eqn (5.109) for resistance dominated by the restricted turbulence layer gives for the mass transfer coefficient $k_t = \dot{n}_a/c_b(a)$ on employing eqn (5.107):

$$\frac{\dot{n}_a}{c_b(a)} = \frac{\dot{\varepsilon} e^2 \mathrm{Re}^{-\frac{1}{8}}\, d\, 142 \times 10^{-6}}{\nu^2} \tag{5.114}$$

while eqn (5.110), valid when resistance is determined only by the turbulent wall layer, becomes

$$\frac{\dot{n}_a}{c_b(a)} = 0.37\, \dot{\varepsilon}^{\frac{1}{3}}\, \phi^{\frac{1}{2}}\, \mathrm{Re}^{-\frac{1}{24}}\, d^{\frac{1}{3}}. \tag{5.115}$$

If the dependence of the Re and ϕ terms on the stirring power $\dot{\varepsilon}$ is disregarded, resistance in the restricted turbulence layer is proportional to $\dot{\varepsilon}$ while the contribution from the turbulent wall layer is proportional to $\dot{\varepsilon}^{\frac{1}{3}}$. The same dependence exists for impellers, bubbling, and electromagnetic stirring. If both resistances must be taken into account, it is found from eqns (5.114) and (5.115) in analogy with eqn (5.106) that

$$\frac{\dot{n}_a}{c(a)} = \frac{0.37\,\dot{\varepsilon}^{\frac{1}{3}}\phi^{\frac{1}{2}}\,\mathrm{Re}^{-\frac{1}{24}}\,d^{\frac{1}{3}}}{1 + \phi^{\frac{1}{2}}\left(\dfrac{353\nu^2\mathrm{Re}^{\frac{1}{12}}}{e^2\dot{\varepsilon}^{\frac{2}{3}}\,0.136d^{\frac{2}{3}}} - 19\right)}. \tag{5.116}$$

It is seen that the total mass transfer coefficient $k_t = \dot{n}_a/c_a$ varies as $\dot{\varepsilon}^{\frac{1}{3}}$ for high values of $\dot{\varepsilon}$ and as $\dot{\varepsilon}$ for lower values of $\dot{\varepsilon}$. Thus, as shown previously at high stirring rates (high velocities), particle transfer is determined by resistance in the turbulent wall layer (eqn (5.115)), while at lower stirring rates resistance in the restricted turbulence layer, eqn (5.114), dominates. For very low values of $\dot{\varepsilon}$, k_t varies as $\dot{\varepsilon}^{\frac{7}{24}}$ (see Section 7.3 and Fig. 5.18).

5.14.4.1 *Mechanical stirring*

In impeller systems the power input may be measured with a torque meter. The power input is simply the torque multiplied by 2π times the frequency of rotation. If this quantity is divided by the mass M, of the bath, $\dot{\varepsilon}$ is obtained. Experiments in steel[55] show that transfer of inclusions to a refining slag is greatly enhanced by mechanical stirring (tapping and stirring by an impeller). The reasons are that the coefficients for transfer of inclusions to a refining slag are greatly increased. Furthermore, if stirring is sufficient to disperse part of the buoyant refining slag down into the melt, contact areas are extended. Figure 5.19 shows that the total oxygen content (inclusion content) decreases

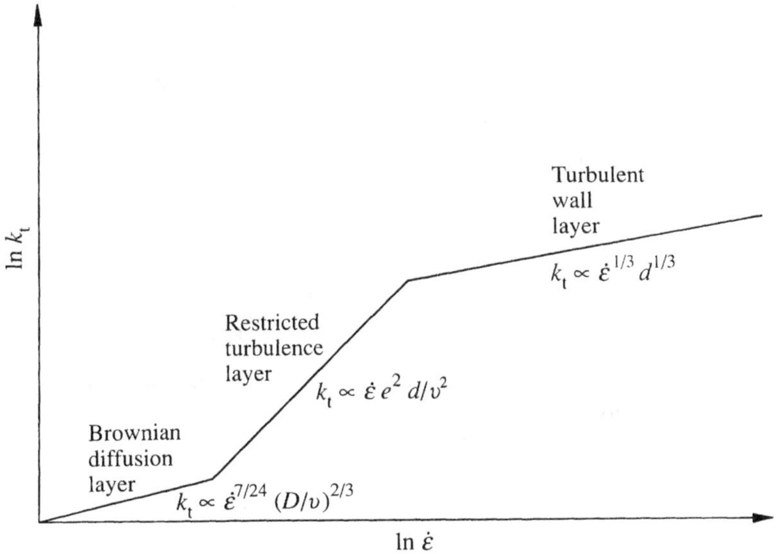

FIG. 5.18 The mass transfer coefficient for inclusions, k_t, as a function of the specific stirring power $\dot{\varepsilon}$ in the three regimes. d is the pipe diameter, e the protrusion height, and D is the Brownian diffusion coefficient. ν is kinematic viscosity.

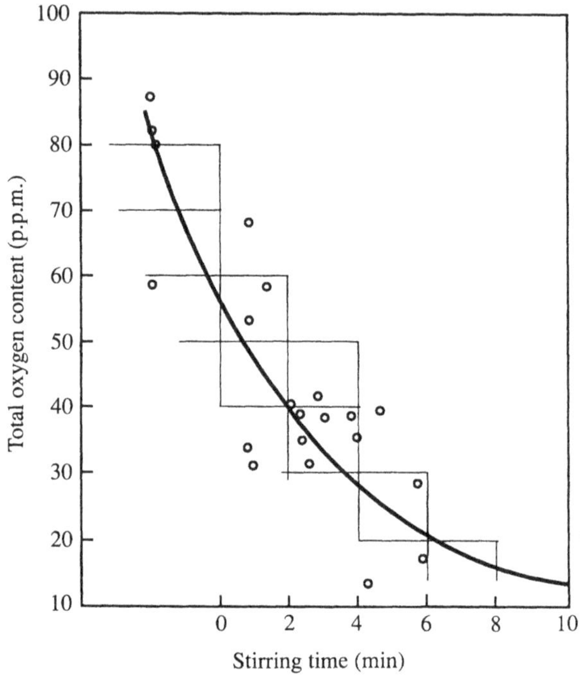

FIG. 5.19 Total oxygen content of steel covered by refining slag and stirred by an impeller as a function of stirring time.[55]

exponentially with time. This result is in accordance with theory since a molar balance for oxygen in inclusions gives:

Rate of decrease of moles of oxygen in inclusions in the bath
= Moles of oxygen in inclusions removed to the slag per second

or

$$-\frac{M}{100 \times 16}\frac{\mathrm{d}[\%\,\mathrm{O}]_{\mathrm{i}}}{\mathrm{d}t} = \frac{k_{\mathrm{t}}A\rho[\%\,\mathrm{O}]_{\mathrm{i}}}{100 \times 16} \qquad (5.117)$$

where k_t, A, and ρ are the mass transfer coefficient, contact area, and melt density, respectively. M is the mass of the bath.

Integrated, eqn (5.117) becomes

$$\frac{[\%\,\mathrm{O}]_{\mathrm{i}}}{[\%\,\mathrm{O}]_{\mathrm{i}}\,(\text{at time } t = 0)} = \exp\left(-\frac{t}{\tau}\right) \qquad (5.118)$$

where the inverse time constant, $1/\tau = k_t A\rho/M$ increases with $\dot{\varepsilon}$ as discussed in the previous section $1/\tau$ also increases with the contact area A, between slag and melt. τ is proportional to M. The amount of dissolved oxygen is disregarded in the calculation on the premise that [% O]$_i$ is very much greater than the dissolved oxygen, [% 0].

5.14.4.2 *Bubble stirring*

Bubble-stirred systems are discussed in Chapter 7 as well as in Chapter 4. In addition to flotation, inclusions may be removed by transfer to the top-slag and to the walls. All three mass transfer coefficients k_{tb}, k_{ts}, and k_{tw} increase with stirring. A molar balance for a batch of metal gives

$$-\frac{M}{100 \times 16}\frac{d[\%\,O]_i}{dt} = \frac{\rho}{100 \times 16}[\%\,O]_i\,(k_{tb}A_b + k_{ts}A_s + k_{tw}A_w)$$

where the indices b, s, and w refer to bubbles, top-slag, and walls, respectively. Integration gives the same exponentially decreasing function of time as eqn (5.118). The inverse time constant is

$$\frac{1}{\tau} = \frac{\rho}{M}(k_{tb}A_b + k_{ts}A_s + k_{tw}A_w). \qquad (5.119)$$

Figure 5.20 depicts experiments concerning the deoxidation of steel with aluminium for different inert gas stirring rates.[56] The results may indicate that the k_t's are proportional to $\dot{\varepsilon}^{\frac{1}{3}}$. Transfer resistance in the turbulent wall layer dominates in this case.

$\dot{\varepsilon}$ is equal to the rate of expansion work of the gas divided by the mass of the melt. The expansion work is composed of two parts. One part arises from the decrease of static pressure when a bubble rises from the inlet to the point of exit. The other part is due to the thermal expansion of the gas from the inlet temperature to the melt temperature. Thus, on integration of $p\,dV$ using the ideal gas law, one obtains

$$\dot{\varepsilon} = \frac{\dot{G}}{M}\left[RT\ln\left(\frac{p_{in}}{p_o}\right) + R(T - T_{in})\right] \qquad (5.120)$$

where p_{in} and p_o are the inlet and outlet pressures, respectively. The last term has often been disregarded in the literature. However, in steelmaking where T is around 1900 K the last term may be greater than the first. To give an impression of what the stirring power input may be, an industrial example is given.[28] 30 Nl/min of gas at a temperature of 25°C is injected at a depth of 1.7 m into a 60 tonne stainless steel melt at 1600°C. The density of the steel is taken to be 7030 kg/m^3. Then:

$$\dot{\varepsilon} = \frac{30 \times 8.314}{60\,000 \times 22.4 \times 60}\left\{1873\ln\left(1 + \frac{7030 \times 9.81 \times 1.7}{1.013 \times 10^5}\right) + 1873 - 298\right\}$$

$$= 0.0093\ \mathrm{m^2/s^3}.$$

This corresponds to a stirring power input $0.0093 \times 7030 = 65\ \mathrm{W/m^3}$. This is the same order of magnitude as found in systems with electromagnetic stirring.

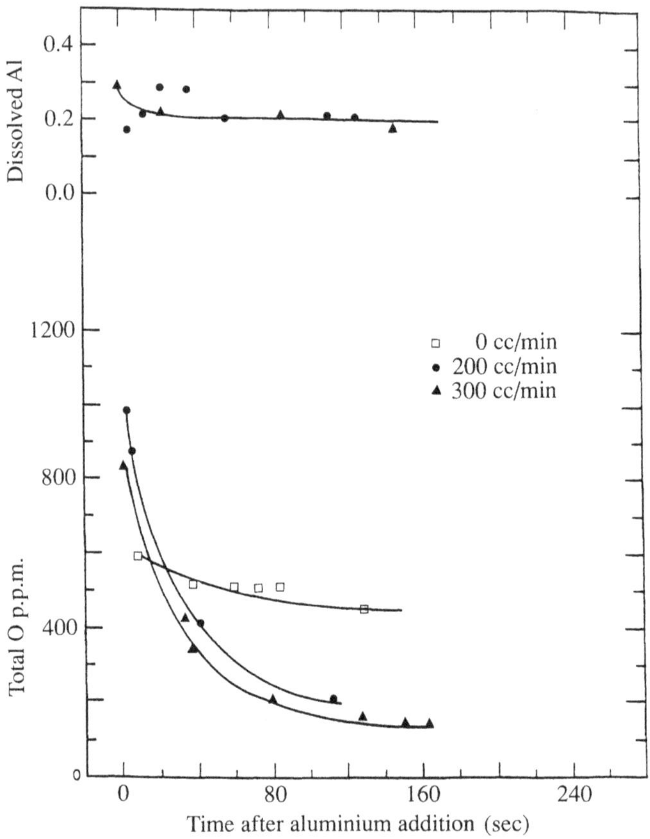

FIG. 5.20 Deoxidation in a gas-stirred melt at different gas flow rates. Changes in total oxygen and dissolved Al levels with time are shown.[56]

The choice between gas stirring and electromagnetic stirring in steel refining does not seem to be a simple matter. In some plants porous plugs for stirring and gas purging are standard equipment. Bubbling will also stir the melt–top-slag interface. Dependent on the process, this may be desirable or undesirable. Some production people have found that gas from a porous plug is not dispersed properly and that maintenance is troublesome. Then electromagnetic stirring may be preferable even though capital costs are higher than for porous plugs, and a non-magnetic steel structure is required.

5.14.4.3 *Electromagnetic stirring*

The net power input to the melt from electromagnetic stirrers can be calculated both for a pulsating stationary field or a travelling field.[38,39] Calculations of the transfer of inclusions through the wall boundary layers do not seem to have been attempted. In this situation we take recourse to eqn (5.114). Inclusions are removed to the walls, top-slag, and bottom surface of the ladle. It will prove practical to replace the diameter with the volume-to-surface ratio V/A where A includes all three surfaces. If the diameter-to-height ratio is taken to be 1,

$$\frac{V}{A} = \frac{\pi d^3/4}{\pi dd + 2\pi d^2/4} = \frac{d}{6}.$$

Then eqn (5.114) becomes

$$\frac{\dot{n}_a}{c(a)} = \frac{\dot{\varepsilon} e^2 6V 142 \times 10^{-6}}{\mathrm{Re}^{\frac{1}{8}} A \nu^2}. \tag{5.121}$$

Here $\mathrm{Re}^{\frac{1}{8}}$ is determined by the convective velocities induced by the electromagnetic stirring. These velocities are roughly 1 m/s. Also the dimensions of the ladle are about 1 m. Then $\mathrm{Re}^{\frac{1}{8}} \sim (1 \times 1/10^{-6})^{\frac{1}{8}} = 5.6$. The height of the protrusions, e, is determined by the roughness of the wall. Thus e depends on the surface properties of the refractory and on the inclusions deposited.

Now, if it is assumed that the addition of a deoxidizer has suppressed dissolved oxygen so that few or no new inclusions are created by nucleation and that growth of inclusions is negligible, a number balance for the batch of melt becomes:[28]

Rate of decrease of the number of inclusions

= Number of inclusions removed per unit time.

$$-V\frac{\mathrm{d}f_N(a)}{\mathrm{d}t} = \frac{\dot{\varepsilon} e^2 6V 142 \times 10^{-6} A f_N(a)}{\mathrm{Re}^{\frac{1}{8}} A \nu^2} \tag{5.122}$$

or, integrated,

$$\frac{f_N(a)}{f_N(a)\,(t=0)} = \exp(-t/\tau) \tag{5.123}$$

where the inverse time constant for the removal of inclusions is

$$\frac{1}{\tau} = \frac{\dot{\varepsilon} e^2 0.85 \times 10^{-3}}{\mathrm{Re}^{\frac{1}{8}} \nu^2}. \tag{5.124}$$

$f_N(a)(t = 0)$ in eqn (5.123) is the distribution of inclusions when stirring is initiated. As discussed previously in Section 5.6 the size distribution is often a decreasing exponential function of the size a.

The number balance, eqn (5.122), can be replaced by a molar balance for oxygen in the inclusions. Then eqn (5.123) is substituted by

$$\frac{c}{c(t=0)} = \exp(-t/\tau) \tag{5.125}$$

where c is the mass per cent of inclusions. A value of roughly $e = 50\,\mu\mathrm{m}$ is obtained from the measurements in the Oxelösund steel works and $e \sim 20\,\mu m$ from the experiments in the Fagersta (see Fig. 5.9). steel works.[28] These may be taken as purely experimental quantities. However, it should be noted that the e are roughly 3 times the mean inclusion sizes (cf. Fig. 5.9). The data indicates that $1/\tau$ is proportional to stirring energy and thus confirms the assumption that in this example transfer resistance is governed by the restricted turbulence boundary layer.

REFERENCES

1. Simensen, C. J. (1981). Sedimentation and analysis of inclusions in aluminium and magnesium. *Trans. AIME*, **12B**, 733–43.
2. Irwin, D. W. (1989). Sampling to detect inclusions in molten aluminium. In *Aluminium melt refining and alloying—theory and practice*, (ed. M. Nilmani). University of Melbourne, Australia.
3. Gustavson, S. and Mellberg, P. O. (1981). Inclusion picture and inclusion origin in a calcium treated steel. In *Swedish symposium on non-metallic inclusions in steel*. Hagfors, Sweden.
4. Simensen, C. J. (1982). Analysis of inclusions and hydrogen in aluminium and magnesium. Dr. techn. Thesis, NTH, Trondheim, Norway.
5. Levy, S. A. (1985). Evaluation of molten metal quality for aluminium alloys. *Trans. American Foundryman*, **93**, 889–94.
6. Bates, DA. and Hutter, L. C. (1981). An evaluation of aluminium filtering systems using a vacuum-filtration sampling device. In *Light Metals*, pp. 707–21. Warrendale, Pennsylvania.
7. Bathen, E. (1985). Investigation of inclusions in aluminium melts by image analysis. In *Proceedings of the international seminar on refining and alloying of liquid aluminium and ferro-alloys*, (ed. T. A. Engh, S. Lyng and H. A. Øye, NTH, Trondheim, Norway), pp. 173–91. Aluminium Verlag, Düsseldorf.
8. Mansfield, T. L. (1982). Ultrasonic technology for measuring molten aluminium quality. In *Light Metals*, pp. 969–82. Warrendale, Pennsylvania.
9. Mansfield, T. L. (1984). Molten aluminum quality measured with Reynolds 4M TM system. In *Light Metals*, pp. 1305–28. Warrendale, Pennsylvania.
10. Guthrie, R. I. L. and Doutre, D. A. (1985). On line measurement of inclusions in liquid melts. In *Proceedings of the international seminar on refining and alloying of liquid aluminium and ferro-alloys*, (ed. T. A. Engh, S. Lyng and H. A. Øye, NTH, Trondheim, Norway), pp. 145–64. Aluminium Verlag, Düsseldorf.
11 Simensen, C. J. (1986). Qualitätssicherung beim Stranggiessen von Aluminiumwerkstoffen. In *Stranggiessen* (ed. G. E. Lossack), pp. 305–18. D. G. M. Informationgesellschaft Verlag, Adenauer Alle 23, 6370 Oberursel 1, BRD.
12 Flemings, M. C. (1974). *Solidification processing*. McGraw-Hill, New York.
13. Popova, E. A. and Lepinskikh (1981). Nucleation and growth in non-metallic inclusions in bearing and tube steels. *Steel in the USSR*, April, 191–93.
14. Ogino, K., Hara, S., Miwa, T., and Kimoto, S. (1984). The effect of the oxygen content in molten iron on the interfacial tension between molten iron and slag. *Trans. ISIJ*, **24**, 522–31.
15. Aylward, G. H. and Findlay, T. J. V. (1974). *SI chemical data*. Wiley, Hong Kong.

16. Darken, L. S. and Gurry, R. W. (1953). *Physical chemistry of metals*, McGraw-Hill, New York.
17. Lupis, C.H.P. (1983). *Chemical thermodynamics of materials*. North Holland, New York.
18. Brennan, J.J. and Pask, J.A. (1968). Effect of nature of surfaces on wetting of sapphire by liquid aluminium. *J. Am. Ceram. Soc.*, **51**, 569.
19. Livey, D.T. and Murray, P. (1956). The wetting properties of solid oxides and carbides by liquid metals. Second Plansee Seminar, pp. 375–404.
20. Defay, R., Prigogine, I., and Bellemans, A. (1951). *Surface tension and absorption*, Longmans, Green and Co., London.
21. Turnbull, D. and Fisher, J.C. (1949). Rate of nucleation in condensed systems. *J. Chem. Phys.*, **17**, 71.
22. Lindborg, U. and Torsell, K. (1968). A collision model for the growth and separation of deoxidation products. *Trans. AIME*, **242**, 94–102.
23. Torsell, K. (1967). Removal of inclusions resulting from deoxidation of iron with silicon. *Jernkontorets Annaler*, **151**, 890–949.
24. Turkdogan, E.T. (1980). *Physical chemistry of steelmaking*, Academic Press, New York.
25. Kawa, T. and Ohkubo, M. (1968). A kinetic study on the deoxidation of steel. *Trans. ISIJ*, **8**, 203–19.
26. Asano, K. and Nakano, T. (1972). Deoxidation of molten steel with deoxidizer containing aluminium. *Trans. ISIJ*, **12**, 343–9.
27. Martin, J.P., Dubé, G., Frayce, D., and Guthrie, R. (1989). Settling phenomena in casting furnaces: A fundamental and experimental investigation. In *Aluminium melt refining and practice*, (ed. M. Nilmani). University of Melbourne, Australia.
28. Engh, T.A. and Lindskog, N. (1975). A fluid mechanical model of inclusion removal. *Scand. J. Metallurgy*, **4**, 49–58.
29. Nagata, S. (1975). *Mixing, principles and applications*. Wiley, New York.
30. Bird, R.B., Stewart, W.E., and Lightfoot, E.N. (1960). *Transport phenomena*. Wiley, New York.
31. Iyengar, R.K. and Philbrook, W.O. (1972). Mathematical model to predict the growth and elimination of inclusions in liquid steel stirred by natural convection. *Metall. Trans.*, **3B**, 1823–30.
32. Lindskog, N. and Sandberg, H. (1973). Studies on the deoxidation mechanism in a stirred steel bath, using radioactive isotopes. *Scand. J. Metallurgy*, **2**, 71–8.
33. Szekely, J. and Nakanishi, K. (1975). Stirring and its effects on aluminium deoxidation in the ASEA-SKF furnace: Part II. Mathematical representation of the turbulent flow field and of tracer dispersion. *Metall. Trans.*, **6B**, 245–56.
34. Gariepy, B., Dubé, G., Simoneau, C., and Leblanc, G. (1984). The TAC process: A proven technology. In *Light Metals*, pp. 1267–79. Warrendale, Pennsylvania.
35. Davey, T.R.A. (1968). Pyrometallurgical methods for refining metals. In *Techniques of materials preparation and handling part 2*, (ed. R.F. Bunshah), Chapter 12, Interscience, New York.
36. Buxmann, K. and Leconte, G.B. (1977). Verunreinigung und Reinigung von Aluminium in Giessereien. In *Verunreinigungen in Metallen*, pp. 109–27. Duetsche Gesellschaft für Metallkunde E.V., Adenauerallee 21, 6370 Oberursel 1, BRD.
37. Szekely, J. and Themelis, N.J. (1971) *Rate phenomena in process metallurgy*. Wiley-Interscience, New York.
38. Sundberg, Y. (1978). Mechanical stirring power in molten metal in ladles obtained by induction stirring and gas blowing. *Scand. J. Metallurgy*, **7**, 81–7.

39. Cremer, P. and Driole, J. (1982). Effects of the electromagnetic stirring on the removal of inclusions of oxide from liquid steel. *Metall. Trans.* **13B**, 45–52.
40. Miyashita, Y., Nishikawa, K., Kawawa, T., and Ohkubo, M. (1969). Removal of inclusions during deoxidation of steel. In *The second Japan-USSR joint symposium on the physical chemistry of metallurgical processes*, pp. 101–12. Spec. report No. 10, JSIJ, Tokyo.
41. Miyashita, Y. and Nishikawa, K. (1968). Measurement of the size distribution of nonmetallic inclusions in steel. *Trans. ISIJ*, **8**, 181–5.
42. Johansen, S.T. and Engh, T.A. (1985). Combined surface, injection and gas-stirring in ladles studies in water models. *Scand. J. Metallurgy*, **14**, 214–23.
43. Davies, J.T. (1972). *Turbulence phenomena.* Academic Press, New York.
44. Lamb, H. (1932). *Hydrodynamics*, (6th edn). Cambridge University Press.
45. Apelian, D., Eckert, C.E., Mutharasan, R., and Miller, R.E. (1985). Refining of molten aluminium by filtration technology. In *Proceedings of the international seminar on refining and alloying of liquid aluminium and ferro-alloys*, (ed. T.A. Engh, S. Lyng and H.A. Øye, NTH, Trondheim, Norway), pp. 123–43. Aluminium Verlag, Düsseldorf.
46. Rajagopalan, R. and Tien, C. (1976). Trajectory analysis of deep-bed filtration with the sphere-in-cell porous media model. *AIChE J.*, **22**, 523–33.
47. Conti, C. and Jacob, M. (1984). Depth filtration of liquid metals: application of the limiting trajectory method to the calculation of inclusion deposition in open cell structure filters. *Bulletin de la Société Chimique de France*, **No. 11–12**, I-297-313.
48. Cliff, R., Grace, J.R., and Weber, M.E. (1978). *Bubbles, drops and particles.* Academic Press, New York.
49. Apelian, D., Luk, S., Piccone, T., and Mutharasan, R. (1986). Removal of inclusions from steel melts. *Proc. Steelmaking Conf. 69*, Book 1, pp. 957–67.
50. Adamson, A.W. (1982). *Physical chemistry of surfaces*, (4th edn). Wiley, New York.
51. Saxena, S.K., Sandberg, H., Waldenström, T., Persson, Å., an Steensen, S. (1978). Mechanism of clogging of tundish nozzle during continuous casting of aluminium killed steel. *Scand. J. Metallurgy*, **7**, 126–33.
52. Wilson, F.G., Heesom, M.J., Nicholson, A., and Hills, A.W.D. (1987). Effect of fluid flow characteristics on nozzle blockage in aluminium-killed steels. *Ironmaking and Steelmaking*, **14**, No. 6, 296–309.
53. Boysan, F. and Johansen, S.T. (1985). Mathematical modelling of gas-stirred reactors. In *Proceedings of the international seminar on refining and alloying of liquid aluminium and ferro-alloys*, (ed. T.A. Engh, S. Lyng, and H.A. Øye, NTH, Trondheim, Norway), pp. 267–88. Aluminium Verlag, Düsseldorf.
54. Johansen, S.T., Boysan, F., and Engh, T.A. (1986). Numerical calculations of removal of inclusions and dissolution of refractory in a bubble stirred ladle. In *Proceedings of the fourth Japan- Nordic countries joint symposium on the science and technology of process metallurgy*, pp. 182–215. *ISIJ*, Tokyo, Japan.
55. Burghardt, V.H. (1970). Erhöhung der Stahlreinheit durch Behandlung mit synthetischer Schlacke unter gleichzeitigem mechanischem Ruhren. *Neue Hütte*, **15**, Heft 3, 148–9.
56. Nilmani, M. (1984). Alumina inclusion removal from a stirred steel ladle: An experimental study. In *Symposium on extractive metallurgy*, pp. 339–45. The Australian I.M.M. Branch, Melbourne.
57. Levich, V.G. (1962). Physicochemical hydrodynamics. Prentice-Hall, Englewood Cliffs, N.J.

58. Frisvold, F., Engh, T. A., Bathen, E. (1990). Measurement of filtration efficiency of an Alcoa 528 filter. In *Light metals* (ed. C. M. Bickert). Warrendale, Pa.
59. Frisvold, F. (1990). Filtration of aluminium. Theory, mechanisms, and experiments. Dr.ing. Thesis. The Norwegian Institute of Technology, Trondheim, Norway.

6

STAINLESS STEELMAKING IN CONVERTERS

Olle Wijk

6.1 INTRODUCTION

In converter steelmaking, a pig iron produced in a blast furnace or a raw steel with a high carbon content, usually melted in an electric arc furnace, is oxidized with gaseous oxygen in order to remove various dissolved elements. The oxygen is added to the molten metal through a water-cooled lance from the top or through tuyeres placed under the bath surface in the converter. Combinations of these two techniques supplemented by inert gas are also commonly used.

In this chapter we shall apply the principles presented in previous chapters to the production of stainless steels in converters. Normally a stainless steel contains 12–30% Cr and other alloying elements of which Ni and Mo are the most important. To give a deeper understanding, we start with a general description of the conventional production of stainless steels in converters and then continue with a more theoretical treatment of some process phenomena.

6.2 STAINLESS STEEL PRODUCTION

For many years stainless steels were produced in induction furnaces with expensive low-carbon raw materials and in electric arc furnaces where oxygen was added through a steel tube for decarburization. In the electric arc furnace very high temperatures, often above 2000°C, were needed to prevent excessive chromium oxidation. This resulted in a very heavy refractory wear, and the production was also characterized by low productivity and high consumption of expensive low-carbon ferrochromium.

The most common process route for stainless steelmaking today is shown in Fig. 6.1. Scrap and alloys are melted in an electric arc furnace while decarburization, reduction of the top-slag, sulphur removal, and adjustment of the steel composition is carried out in a converter. Very often the converter is complemented with some kind of ladle treatment for deoxidation, final trimming of the composition, and stirring for homogenization and removal of non-metallic inclusions.

Another process route includes melting in an electric arc furnace and vacuum treatment in a ladle. Both process routes show benefits and drawbacks which have resulted in combinations with a converter followed by vacuum treatment in a ladle. There also exist converter processes equipped with vacuum stations. In the following, however, the process route illustrated in Fig. 6.1 is treated.

The electric arc furnace is charged with scrap, alloys, and slag-forming agents. The carbon content of the molten metal after it has left the electric

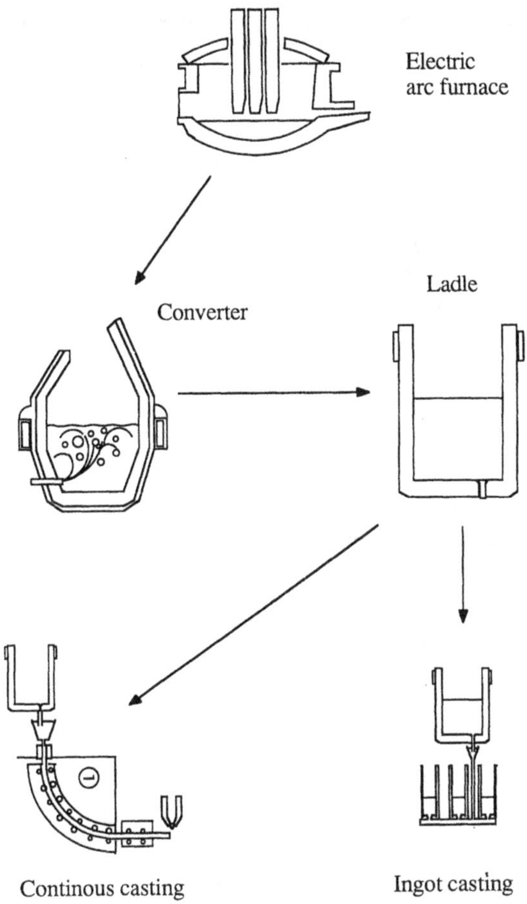

FIG. 6.1 Process route for scrap-based stainless steelmaking.

arc furnace depends on the supply of raw materials. If there is a lack of stainless steel scrap, more high-carbon ferroalloys must be used and the carbon content is increased in the raw steel. A typical composition is 1–3.5% C, 0.1–0.5% Si, and 0.6–1.2% Mn. Other elements are adjusted to meet the demands of the finished steel product. One very critical element is phosphorus, as this element cannot yet be removed from chromium alloyed steels on an industrial scale.

A comparison of the composition of the metal from the electric arc furnace with the composition of some stainless steels according to international standards (see Table 6.1) illustrates that the main metallurgical problem in stainless steelmaking is to decrease the carbon content in the steel at a high chromium content. Having a process that makes this possible is of fundamental importance for minimizing the consumption of expensive low-carbon ferroalloys.

TABLE 6.1 *Composition of some standard stainless steel grades*

Element	Standard		
	AISI 304 L	AISI 316 L	AISI 310 S
C	≤0.030	≤0.030	≤0.08
Si	≤1.0	≤1.0	≤1.5
Mn	≤2.0	≤2.0	≤2.0
P	≤0.045	≤0.045	≤0.045
S	≤0.030	≤0.030	≤0.030
Cr	17.2–20.0	16.0–18.5	24.0–26.0
Ni	9.0–12.0	11.0–14.0	19.0–22.0
Mo	-	2.0–2.5	-

If we only consider dissolved carbon and chromium, the two following reactions occur during the addition of oxygen:

$$C + \tfrac{1}{2}O_2 = CO(g) \tag{6.1}$$

$$2Cr + \tfrac{3}{2}O_2 = Cr_2O_3(\text{in slag}). \tag{6.2}$$

These two competitive reactions give the overall reaction which can be written as (see problem 2.10):

$$Cr_2O_3\,(\text{in slag}) + 3\underline{C} = 2\underline{Cr} + 3CO(g) \tag{6.3}$$

$$\ln K = \ln\left(\frac{(a_{Cr})^2(p_{CO})^3}{(a_C)^3 a_{Cr_2O_3}}\right) = -\frac{88\,704}{T} + 56.67 \tag{6.4}$$

where a_{Cr} is the chromium activity in the metal bath, p_{CO} is the partial pressure of CO in the gas phase, a_C is the carbon activity in the metal bath, and $a_{Cr_2O_3}$ is the chromium oxide activity in the top slag.

From eqn (6.4) it can be seen that the carbon oxidation according to reaction (6.3) is favoured by a low chromium activity and a high carbon activity in the metal, a high chromium oxide activity in the slag phase, and a low partial pressure of carbon monoxide in the gas phase. From the expression it is also clear that the carbon oxidation is enhanced by a high temperature.

Figure 6.2 gives the equilibrium carbon and chromium contents in liquid steel in contact with pure chromium oxide ($a_{Cr_2O_3} = 1$) at varying temperatures and partial pressures of carbon monoxide. For instance, for a steel with 15% Cr at 1550°C and $p_{CO} = 1$, it is thermodynamically possible to decrease the carbon content to 0.55% without any oxidation of chromium. If the temperature is increased to 1700°C, the corresponding carbon content is decreased to 0.18%.

If we return to Fig. 6.2, we can see that by decreasing the partial pressure

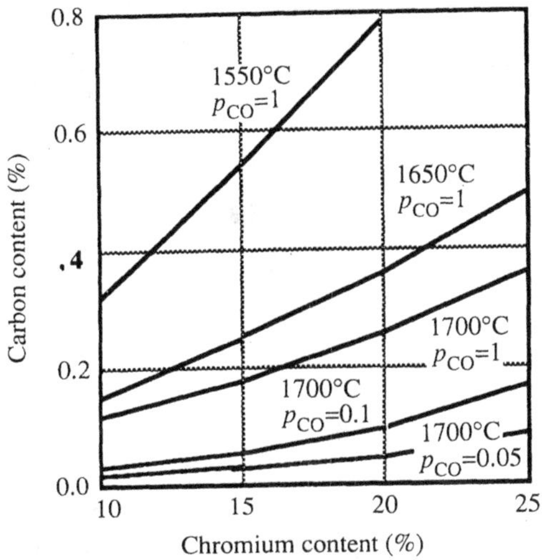

FIG. 6.2 Equilibrium contents of C and Cr in a steel bath in contact with pure chromium oxide.

of carbon monoxide in the gas phase to 0.1, it is possible to decrease the carbon content to 0.06% at 1700°C.

In the example above, it was assumed that only chromium and carbon were dissolved in the metal bath. However, when calculating the equilibrium for the carbon oxidation, it is necessary to take the interaction from all the elements dissolved in the metal bath on the carbon and chromium activities into consideration. Elements with a positive interaction on carbon, such as Ni, increase the carbon activity and enhance the decarburization reaction, while elements with a negative interaction such as Cr and Mo have the opposite effect. Table 6.2 gives the temperature-dependent second-order interaction coefficients between carbon and a third added element. The first-order interaction

TABLE 6.2 *Second-order interaction coefficients between carbon and some elements in liquid iron.*[1,2]

$$\log f_c = \sum_{j=2}^{k} e_c^j\,[\%\, j] + r_c^j\,[\%\, j]^2$$

$$r_c^c = -4.70 \times 10^{-2} + 177.2/T - 1.5 \times 10^3/T^2$$

$$r_c^{Cr} = 7.95 \times 10^{-4} - 1.462/T$$

$$r_c^{Ni} = 0$$

TABLE 6.3 *Heats of formation for various oxides from oxygen gas and elements dissolved in steel and the corresponding temperature increase in the steel bath*

Reaction	Enthalpy change (kJ/mole)	Temperature increase (theoretical per % of oxidized element (°C)
$Fe(l) + \frac{1}{2} O_2 = FeO$	−256.3	46
$\underline{C} + \frac{1}{2} O_2 = CO$	−139.8	109
$\underline{Si} + O_2 = SiO_2$	−825.9	333
$\underline{Mn} + \frac{1}{2} O_2 = MnO$	−404.3	74
$2\,\underline{Cr} + \frac{3}{2} O_2 = Cr_2O_3$	−1141.6	108

coefficients are found in Table 2.7. If we go back to the example in Fig. 6.2 and make an addition of, for instance, 8% Ni to the melt which originally contained 18% Cr, it is theoretically possible, at 1700°C and $p_{CO} = 0.1$, to further decrease the carbon content from 0.06 to 0.04%.

The chromium oxide activity in the slag also plays an important role for the decarburization. However, due to the low chromium oxide solubility in the type of slags used for steelmaking,[3] the chromium oxide activity will be close to unity during decarburization.

As shown above, the temperature is a fundamental parameter in the production of stainless steels. In converter processes, the thermal heat required is supplied by the oxidation of various elements dissolved in the metal bath. Table 6.3 gives the heat of formation for different oxides from elements dissolved in the liquid iron when oxygen gas is added. The table also gives the temperature increase per per cent of the oxidized dissolved element. It is obvious that the oxidation of carbon alone does not give sufficient heat. Oxidation of other elements such as Si and Cr, at least during some periods of the blow, is necessary for an efficient production.

The thermodynamic equilibria give the theoretically possible compositions of the reacting phases. However, in practical steelmaking, the kinetics of the different reactions play a fundamental role for the processes. This will be discussed in detail in Section 6.4.

6.3 CONVERTER PROCESSES FOR STAINLESS STEELMAKING

The first to utilize the influence of a decreased partial pressure of carbon monoxide on chromium oxidation was W.A. Krivsky who, within the Union Carbide Corporation at the Metals Research Laboratories at Niagara Falls, USA, made experiments on a 220 kg scale where injected oxygen was diluted with argon during the decarburization of stainless steel melts. This work

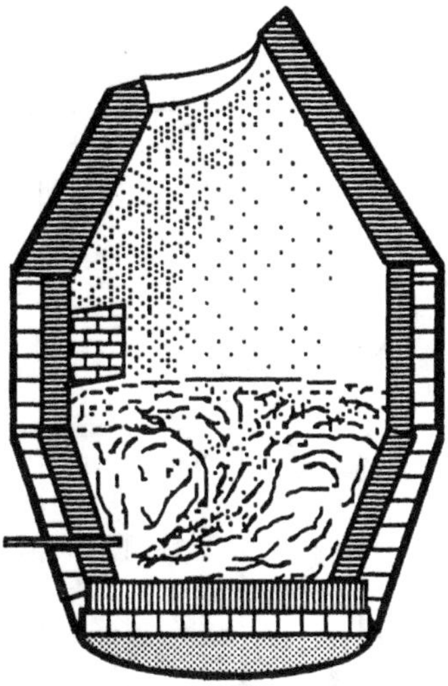

FIG. 6.3 Argon Oxygen Decarburization (AOD) converter.

resulted in the Argon Oxygen Decarburization (AOD) process which made an industrial breakthrough at the end of the 1960s. Today, this original converter process for stainless steelmaking dominates production. It has, however, been followed by other similar processes based on the same fundamental principles.

An AOD converter of original design is shown in Fig. 6.3. Oxygen gas and an inert gas for the dilution of the carbon monoxide produced is added through tuyeres placed in the wall near the bottom of the vessel. Normally around 1 Nm^3 gas per tonne of steel per minute is added through the tuyeres. In later designs, the addition of gas from the bottom was complemented by the addition of oxygen or a mixture of oxygen and an inert gas through a top lance. The impact of this extra addition is to increase the productivity of the process.

As a general background, we now continue with a description of stainless steelmaking in an AOD converter without a top lance. The processing time can be divided into three main periods: decarburization, reduction of the top-slag, and desulphurization.

6.3.1 Decarburization

The raw steel from the electric arc furnace normally has a temperature of 1550°C when it is charged into the converter. During the blow, the oxygen gas is diluted with an inert gas which is usually nitrogen in the beginning of the blow and argon towards the end. The justification for the use of expensive argon is the removal by gas-purging of nitrogen dissolved in the steel. Dilution

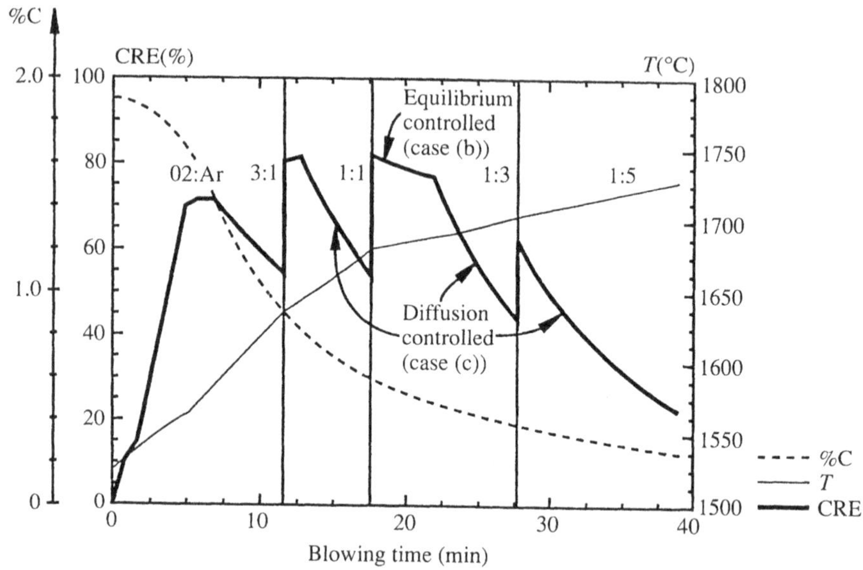

FIG. 6.4 Carbon content in the metal bath, CRE number, and temperature versus blowing time in an AOD converter. For mechanisms (b) and (c), see Section 6.4.2.

of the oxygen is usually carried out in different steps where the relation between oxygen and the inert gas is changed stepwise, for instance in the ratios 3:1, 1:1, and 1:3. Often the addition of oxygen is followed by purging with inert gas to attain a further decrease in the carbon content of the metal bath. Figure 6.4 illustrates the change of the carbon content in the metal bath, the bath temperature, and the Carbon Removal Efficiency (CRE) number during the blow. The CRE number is a measure of how much of the oxygen reacts with carbon and is defined by

$$\mathrm{CRE}(\%) = \frac{\text{Amount of oxygen reacting with carbon} \times 100}{\text{Total added oxygen}}. \qquad (6.5)$$

If we return to Fig. 6.4 we can see that the CRE number is initially low, since in the beginning of the blow the oxygen reacts with Si, Mn, and Cr. As the content of Si and Mn decreases, the slag becomes chromium-saturated and the temperature increases due to the exothermic oxidation reactions, the carbon starts to oxidize, and the CRE number increases. However, as the carbon activity in the metal bath decreases, chromium starts to oxidize and the rate of decarburization decreases. At a certain stage determined in each individual case by the temperature, which should not be too high (normally not above 1725°C due to severe problems with the refractory) or the CRE number or the carbon activity in the metal bath, the blow is turned to step 2 where the dilution of the oxygen gas is increased. At this moment, the top-slag has been overoxidized due to the low CRE number at the end of step 1. The CRE number falls during step 2 and 3 in a manner similar to step 1 until the fourth step is initiated and blowing proceeds until the specified final carbon content is attained. Very often the overoxidized slag at the end of the blow is utilized to further lower the carbon content using a final purging with inert gas.

To minimize the chromium losses during the blow, it is important to control the composition of the top slag. This is performed by the addition of lime at the beginning of the blow. The amount of lime is determined from the silicon content in the metal charged in the converter. The aim is to get a basicity measured as $CaO:SiO_2$ of around 1.7. A lower basicity gives a slag with lower viscosity, a higher solubility of chromium oxide, and a lower $a_{Cr_2O_3}$ which results in severe problems in attaining high chromium and low carbon levels in the metal bath. In practical production, the slag is more or less solid, consisting of one molten slag phase saturated with chromium oxide and one solid chromium oxide phase.

6.3.2 Reduction of the top-slag

After the decarburization period, the top-slag, usually around 70 kg per tonne of steel, has a typical average composition of 20–30% CaO, 10–15% SiO_2, 5–7% MgO (mainly due to refractory wear if MgO has not been added), 5–10% MnO, 5–10% FeO, 1–5% Al_2O_3, and 25–40% Cr_2O_3. To get an optimal yield of chromium in the process, a reducing agent such as silicon (as ferrosilicon) is added for reduction of the top-slag. The slag is reduced during argon stirring for 5–8 minutes. As an alternative to the ferrosilicon, aluminium metal is used, especially when extremely low sulphur contents are required. Also combinations of these two reducing agents are practiced. The overall reaction during reduction with silicon is

$$Cr_2O_3\,(\text{in slag}) + \tfrac{3}{2}\,\underline{Si} = 2\underline{Cr} + \tfrac{3}{2}\,SiO_2\,(\text{in slag})\,. \qquad (6.6)$$

This reaction is favoured by a low SiO_2 activity and a high Cr_2O_3 activity in the top-slag, i.e. a basic slag, low chromium activity in the metal, and a high temperature. It is difficult to influence the chromium activity since the composition of the metal bath is determined by the demands on the finished steel. The activity of silicon is increased by an extra addition normally aiming at 0.3–0.5% Si in the metal during the reduction. This, however, demands a very good process model in order to estimate the amount of chromium that has been oxidized during the decarburization so that the silicon content remains within the limits specified for the composition of the finished steel. If the reduction is carried out properly, the resulting chromium content in the slag is less than 0.5 per cent. The process model should also make it possible to estimate the amount of lime added to give the desired slag basicity ($CaO:SiO_2$) of around 1.7. It is also important to predict the chromium content after the reduction period in order to eliminate the use of expensive low-carbon ferrochromium.

6.3.3 Desulphurization

After the reduction of the top-slag, a slag-off is carried out and, if necessary, usually 10–15 kg per tonne of steel of a mixture of lime and fluorspar is added for desulphurization, followed by 5–10 minutes of stirring with argon. If the reduction of the top-slag has been carried out satisfactorily, that is, the oxygen potential in the system is low, it is possible to reach a sulphur content in the steel of less than 0.003 per cent. The use of aluminium as a reducing agent allows extremely efficient sulphur removal, and sulphur levels of 10 p.p.m. with a one-slag practice have been reported in the literature.[4]

In an AOD converter, the tap-to-tap time is normally 90–120 minutes. The decarburization takes 35–90 minutes, where the longer times occur under extreme conditions with high initial carbon contents combined with high chromium, low nickel, and low carbon content in the finished product. Typical consumption figures per tonne of steel 18% Cr, 8% Ni, initial carbon content 2% in the AOD process are: Ar = 9–12 Nm^3, N_2 = 9–11 Nm^3, O_2 = 30–38 Nm^3, lime = 35–45 kg, fluorspar = 2–3 kg, and ferrosilicon (75% Si) = 8–11 kg.

As mentioned in the text above, the different stages in the production of stainless steel are strongly coupled to each other. A good control of the decarburization steps gives a minimum of chromium oxidation which has a direct effect on the consumption of gases and reducing agent. The amount and type of reducing agent in turn influences the oxygen level in the metal bath and the sulphur level in the finished steel.

6.4 THE RATE OF DECARBURIZATION

A knowledge of the parameters that affect the rate of decarburization in a converter is important in order to optimize the process. In the following we shall derive an expression for the rate.

A reactor for stainless steelmaking is schematically shown in Fig. 6.5. The system contains three phases:

(1) molten metal with dissolved elements such as C, Si, Mn, Cr, Ni, and N;

(2) slag, which can be molten or solid, containing oxides such as CaO, Al_2O_3, SiO_2, MnO, MgO, Cr_2O_3, and FeO;

(3) inert gas and oxygen which reacts to form carbon monoxide.

The following two assumptions are made:

1. Due to the bubble stirring, no concentration gradients occur in the metal bath.

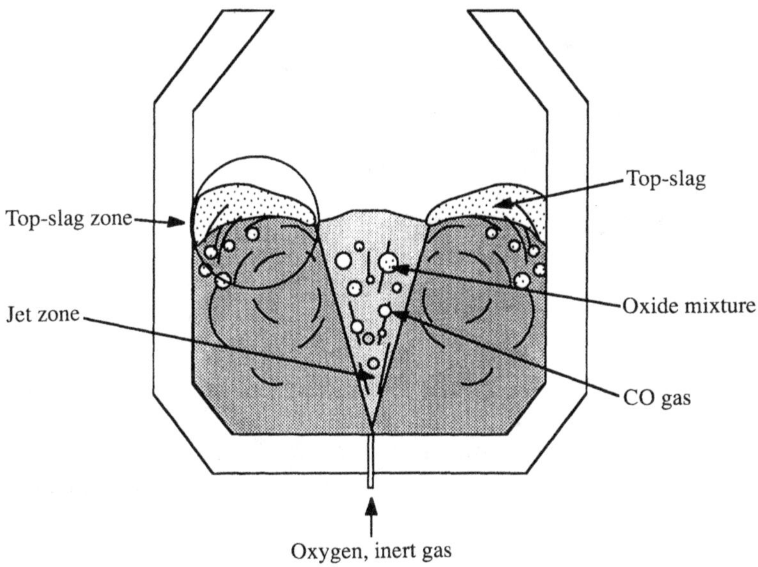

FIG. 6.5 Reaction zones in a converter for stainless steelmaking.

2. The interfacial area between gas bubbles and metal is constant during decarburization despite a successively increased addition of inert gas.

When modelling a metallurgical process it is often very convenient to divide the reactor into separate zones which are later put together to derive a model for the entire process. In Fig. 6.5 the reactor has been divided into three zones which are all active during the decarburization. These zones are the tuyere zone, the jet zone, and the top-slag zone. In the following the tuyere zone is treated separately. However, because of the very high gas flow, around 1 Nm^3 per tonne of liquid steel per minute, a defined separate top-slag zone is assumed not to exist. To eliminate uncertainties in estimating rate constants in the different zones, the jet zone and the top-slag zone are treated together and are hereafter called the oxide/metal zone.

6.4.1 Tuyere zone

In this zone, in front of the tuyeres, the injected oxygen reacts instantaneously with the elements dissolved in the metal bath. Due to a high local oxygen potential, the different elements are oxidized in amounts proportional to the atomic percentage of each element. Thus the following reactions are assumed to occur:

$$\underline{C} + \tfrac{1}{2}O_2 = CO$$

$$\underline{\text{Fe}} + \tfrac{1}{2}\text{O}_2 = \text{FeO}$$

$$\underline{\text{Mn}} + \tfrac{1}{2}\text{O}_2 = \text{MnO}$$

$$\underline{\text{Si}} + \text{O}_2 = \text{SiO}_2$$

$$\underline{\text{Cr}} + \tfrac{3}{4}\text{O}_2 = \tfrac{1}{2}\,\text{Cr}_2\text{O}_3 .$$

The amount of oxygen that reacts with the respective elements can be written:

$$\dot{Q}_{O_2}^{(j)} = \frac{c_j n_j \dot{Q}_{O_2}}{\sum c_j n_j} \tag{6.7}$$

where $\dot{Q}_{O_2}^{(j)}$ is the volume of added oxygen that reacts with the element j per unit time (Nm^3/s), c_j is the molar concentration of element j in the metal ($kmol/m^3$), $\dot{Q}_{O_2}$ is the total flow of oxygen added per unit time (Nm^3/s), and n_j is the stoichiometric factor for the various reactions. $n_j = \frac{1}{2}$ except for Si where $n = 1$ and Cr where $n = \frac{3}{4}$.

The instantaneous oxidation of the elements in the tuyere zone results in a mixture of oxides that is transported into the oxide/metal zone.

6.4.2 Oxide/metal zone

In the oxide/metal zone the oxide mixture formed in the tuyere zone is reduced by the various elements dissolved in the metal bath. This zone in which the gas and oxide mixture is transported from the tuyeres to the top surface can be regarded as an emulsion of oxides and liquid metal. To derive an expression for the rate of decarburization, the following assumptions are made.

1. Due to the high chromium content in the metal bath, formation of chromium oxide occurs at a high rate above the tuyere zone. Other dissolved elements such as Si and Mn are in equilibrium with chromium oxide, (see eqns (6.21) and (6.22)). The decarburization reaction is assumed to occur by a reaction between the oxide and carbon dissolved in the metal bath.
2. The chromium oxide activity in the oxide phase is 1 during the blow, which means that there is a residual chromium oxide content in the oxide mixture.
3. The carbon monoxide formed during the reaction between the oxide mixture and carbon is part of the gas phase and changes the partial pressure of carbon monoxide in the gas bubbles as they are transported up through the metal bath.
4. The rate-controlling step (mechanism) in decarburization is one of the following:
   (a) the flow of oxygen into the bath; or
   (b) the equilibrium given by eqn (6.4); or

(c) diffusion of carbon in the boundary layer around the bubbles in the melt (eqn (6.8) in the following).

If p_{CO} at equilibrium calculated from eqn (6.4) is greater than p_{CO} (max) and if $\dot{m}$ from eqn (6.8) should give a flow of carbon exceeding that needed to react with the flow of oxygen, then step (a) is rate limiting. p_{CO}(max) is the partial pressure of CO if all the available oxygen reacts with carbon and the CO is diluted with the inert gas added.

If p_{CO} at equilibrium is less than p_{CO}(max) and $\dot{m}$ exceeds that needed to react with the flow of oxygen, then step (b) is rate limiting (see Fig. 6.4 and problem 4.12).

If $\dot{m}$ is not sufficient to handle the flow of oxygen then (c) is rate limiting (see Fig. 6.4).

Decarburization rates are low when step (c) is rate limiting compared to (a) or (b). Step (a) is called the linear period in the following. Special attention is here given to step (c).

The mass flux of carbon to the gas phase is obtained from the expression:

$$\dot{m} = \frac{Ak_C\rho}{100}([\%\,C] - [\%\,C]_e) \tag{6.8}$$

where $\dot{m}$ is the mass flux of carbon (kg/s), A is the surface contact area (m^2), k_C is the total mass transfer coefficient for carbon (m/s), and ρ is the density of the metal kg/m^3. $[\%\,C]_e$ is the hypothetical concentration of carbon in the melt in equilibrium with chromium oxide at the partial pressure p_{CO} existing in the bubbles (see section 4.3 and eqn (6.11)) and [% C] is the carbon concentration in the bulk melt. The mass flux, $\dot{m}$, equals the rate of depletion of carbon in the bath:

$$\dot{m} = -V\frac{\rho}{100}\frac{d[\%\,C]}{dt} = \frac{Ak_C\rho}{100}([\%\,C] - [\%\,C]_e). \tag{6.9}$$

Then

$$\frac{d[\%\,C]}{dt} = -\alpha([\%\,C] - [\%\,C]_e) \tag{6.10}$$

where $\alpha = Ak_C/V$.

α is a mass transfer rate constant for the process that depends on geometry and size of the metallurgical reactor. For instance, a deeper metal bath or an increased gas flow rate increases the contact area A in the jet zone. The stirring also influences the mass transfer coefficient for carbon $k_C = D/\delta$ as the diffusion boundary layer thickness, δ, is decreased by increased stirring. If α is increased, p_{CO} in the bubbles leaving the bath will come closer to equilibrium with [% C] as given by eqn (6.4).

From the general assumption that the overall rate constant, α, is known, we shall now derive an expression for the rate of decarburization. From eqn (6.10) it is seen that if α is constant, the mass flux depends only on the difference in carbon content between the metal bulk concentration [% C] and at the oxide surface, $[\%\,C]_e$. According to the reaction given by eqn (6.3) with $a_{Cr_2O_3} = 1$, the equilibrium carbon content, $[\%\,C]_e$, can be calculated from the expression

$$[\%\,C]_e = \frac{(f_{Cr}[\%\,Cr])^{\frac{2}{3}} p_{CO}}{f_C K^{\frac{1}{3}}} \tag{6.11}$$

where f_C and f_{Cr} are the activity coefficients of C and Cr, respectively, in the metal bath.

From this expression and eqn (6.10) we see that the rate of decarburization is increased when $[\%\,C]_e$ is low, that is, if the chromium activity in the metal bath and the partial pressure of carbon monoxide in the gas phase are low and the activity coefficient for carbon is high. Also the temperature is a very important parameter, as a high temperature increases the value of the equilibrium constant, K, (see eqn (6.4)) and thereby decreases the equilibrium carbon content $[\%\,C]_e$.

Near the tuyere zone in the converter, where $[\%\,C]_e$ approaches zero, the rate of decarburization reaches its maximum value. If these conditions were valid for the whole oxide/metal zone, the rate of decarburization would be described by

$$\frac{d[\%\,C]}{dt}(\max) = -\alpha[\%\,C]. \tag{6.12}$$

However, as the gas bubbles rise through the metal bath, the partial pressure of carbon monoxide increases in the gas phase which gives a higher equilibrium carbon content at the oxide metal interface. This reduces the driving force for transfer of carbon.

To obtain an expression for the local rate of decarburization at the top surface where the partial pressure of carbon monoxide is a maximum, the treatment is the following. The gas law gives:

$$p_{CO} = \frac{\dot{Q}_{CO}}{\dot{Q}_{CO} + \dot{Q}_{Ar}} p_{tot}$$

where $\dot{Q}_{CO}$ and $\dot{Q}_{Ar}$ are the respective gas flows in Nm^3/s or

$$p_{CO} = \frac{1}{1 + \dot{Q}_{Ar}/\dot{Q}_{CO}} p_{tot}. \tag{6.13}$$

From stoichiometry the CO flow that leaves the bath is

$$\dot{Q}_{CO} = -\frac{d[\%\,C]}{dt}\frac{V\rho 22.4}{M_c 100} \tag{6.14}$$

where $M_c = 12$, the molar weight of carbon.
From eqns (6.13) and (6.14) we get

$$p_{CO} = \frac{p_{tot}}{1 - \dfrac{\dot{Q}_{Ar}\, 12 \times 100}{22.4\, V\rho \mathrm{d}[\%\,C]/\mathrm{d}t.}} \tag{6.15}$$

By combining eqns (6.15), (6.10), and (6.11) the following expression for the rate of decarburization is obtained:

$$\frac{d[\%\,C]}{dt} = -\alpha\left\{[\%\,C] - \frac{(f_{Cr}\,\%\,Cr)^{\frac{2}{3}} p_{tot}}{f_c K^{\frac{1}{3}}\left(1 - \dfrac{\dot{Q}_{Ar}\, 12 \times 100}{22.4 V\rho\, d[\%\,C]/dt}\right)}\right\}. \tag{6.16}$$

To simplify this expression we multiply the right and left side with $(1 - \dot{Q}_{Ar} 12 \times 100/(22.4V\rho\, d[\%\,C]/dt))$. After rearranging we get

$$\frac{d[\%\,C]}{dt}\left(1 - \frac{\dot{Q}_{Ar}\, 12 \times 100}{22.4V\rho\, d[\%\,C]/dt}\right) = -\alpha\left\{[\%\,C] \times \left(1 - \frac{\dot{Q}_{Ar}\, 12 \times 100}{22.4V\rho\, d[\%\,C]/dt}\right) - \frac{(f_{Cr}[\%\,Cr])^{\frac{2}{3}} p_{tot}}{f_C K^{\frac{1}{3}}}\right\}. \tag{6.17}$$

This equation can be transformed to an equation quadratic in $d[\%\,C]/dt$. Multiplication with $d[\%\,C]/dt$ gives

$$\left(\frac{d[\%\,C]}{dt}\right)^2 + \frac{d[\%\,C]}{dt}\left\{-\frac{\dot{Q}_{Ar}\, 12 \times 100}{22.4V\rho} + \alpha\left([\%\,C] - \frac{(f_{Cr}[\%\,Cr])^{\frac{2}{3}} p_{tot}}{f_C K^{\frac{1}{3}}}\right)\right\} - \frac{\alpha \dot{Q}_{Ar}\, 12 \times 100[\%\,C]}{22.4V\rho} = 0. \tag{6.18}$$

This quadratic equation has the following solution:

$$\frac{d[\%\,C]}{dt} = -\tfrac{1}{2}\left(-\frac{\dot{Q}_{Ar}\, 12 \times 100}{V\rho 22.4} + \alpha\left([\%\,C] - \frac{(f_{Cr}[\%\,Cr])^{\frac{2}{3}} p_{tot}}{f_C K^{\frac{1}{3}}}\right)\right)$$

$$- \left(\left\{ -\frac{\dot{Q}_{Ar} 12 \times 100}{V\rho 22.4} + \alpha \left([\% \mathrm{C}] - \frac{(f_{Cr}[\% \mathrm{Cr}])^{\frac{2}{3}} p_{tot}}{f_C K^{\frac{1}{3}}} \right) \right\}^2 \frac{1}{4} + \frac{\alpha \dot{Q}_{Ar} 12 \times 100 [\% \mathrm{C}]}{22.4 V\rho} \right)^{\frac{1}{2}}. \tag{6.19}$$

By studying this expression it can be seen that the sign in front of the square root should be negative as this sign gives an increased rate of decarburization when $\dot{Q}_{Ar}$ is increased. Equation (6.19) gives a minimum local decarburization rate since the driving force for transfer of carbon is smallest at the top of the reactor.

We now have two expressions describing the maximum and minimum rate of decarburization, respectively. Thus the average rate of decarburization describing the overall rate of decarburization is given by

$$\frac{d[\% \mathrm{C}]}{dt} (\text{average}) = \left\{ \frac{d[\% \mathrm{C}]}{dt} (\max) + \frac{d[\% \mathrm{C}]}{dt} (\min) \right\} b \tag{6.20}$$

where the maximum and minimum rates of decarburization are given by eqn (6.12) for the tuyere region and eqn (6.19) for the top of the converter, respectively.

b is a constant which is difficult to estimate. In the following we set $b = \frac{1}{2}$ which means that the rate of decarburization is taken to be the average of the rate at the tuyeres and at the top surface.

To calculate the rate of decarburization, the following must be known:

(1) the initial weight of liquid metal, composition, and temperature;

(2) the composition of the metal bath at every instant where it is assumed that other dissolved elements such as Si and Mn are in equilibrium with the chromium oxide so that equilibria are attained according to the reactions:

$$Cr_2O_3 + \tfrac{3}{2}\,\underline{Si} = 2\underline{Cr} + \tfrac{3}{2}\,SiO_2 \tag{6.21}$$

$$Cr_2O_3 + 3\underline{Mn} = 2\underline{Cr} + 3MnO; \tag{6.22}$$

(3) the residual minimum amount of chromium oxide in the slag (see Section 6.4.1) is assumed to be equal to the amount of chromium oxidized at the tuyeres, (see eqn (6.7));

(4) the temperature in the system which is calculated from a heat balance, that is, from data concerning the amount of elements that have been oxidized, the heat evolved during the reactions, and a model of the temperature losses in the process.

The rate of decarburization is controlled by one of the following conditions:

(1) the total amount of oxygen available for decarburization (linear period, mechanism (a));

(2) the equilibrium according to eqn (6.4) (mechanism (b));

(3) diffusion of carbon (eqn (6.20), mechanism (c)).

Starting with the initial composition and temperature, it is now possible to calculate the composition of the metal bath as a function of time.

6.4.3 Simulation of the blow

A calculation of the rate of decarburization can be carried out manually. However, the system is so complex that it is more convenient to use a computer. The results of some process simulations are given below.

The following conditions are assumed for the first calculation:

Metal weight = 60 tonnes

Initial temperature = 1550°C

Initial composition of the metal bath = 1.5% C, 18% Cr

$\alpha = 0.2\ \mathrm{min}^{-1}$ (empirical parameter)

Blowing steps ($Nm^3 O_2/Nm^3 Ar$) = 45/15; 30/30; 15/45; 10/50

Gas flow1 Nm^3/t per min.

In Fig. 6.4 we see that the rate of decarburization initially has a tendency to be at a low level after which the rate of decarburization remains at a constant and high value (mechanism (a) in Section 6.4.2). When the rate of

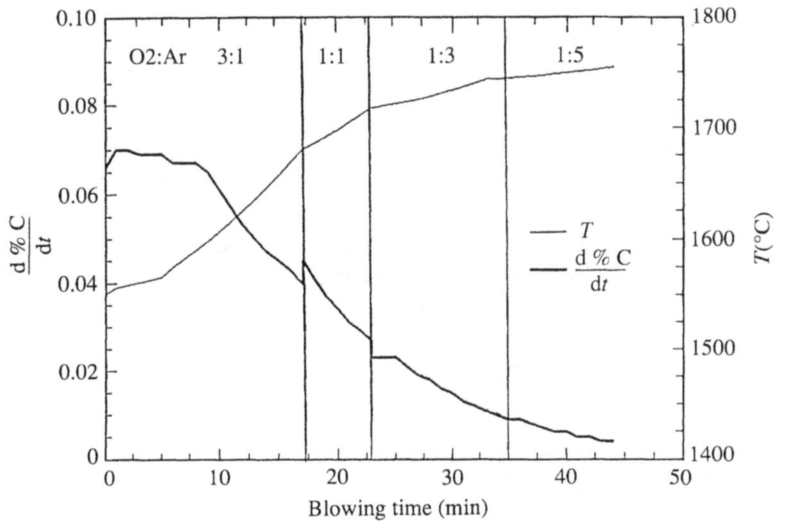

FIG. 6.6 Computer simulation of decarburization of stainless steel in a 60 tonne converter, case 1. Metal composition: 1.5% C, 18% Cr. Total gas flow rate is 60 Nm^3/min. Initial temperature is 1550°C. Formation of Cr_2O_3 = 1750 kg.

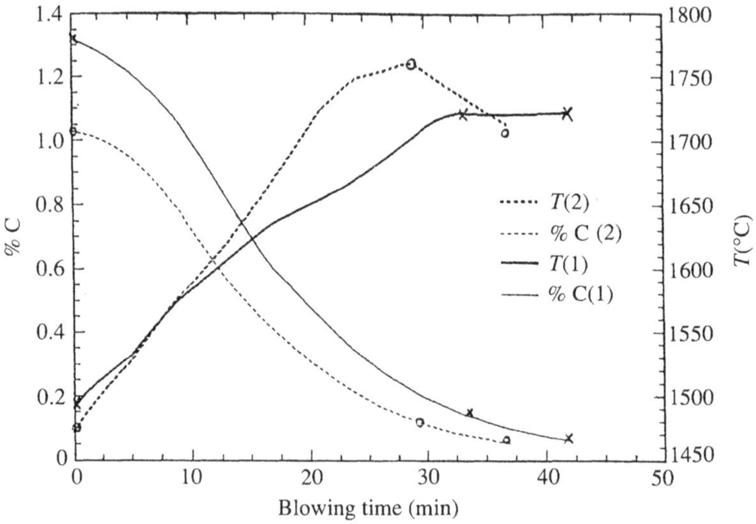

FIG. 6.7 Calculated temperature and carbon contents in the liquid steel compared with production data from a 57 tonne AOD converter. Blowing conditions:
1. Initial bath composition: 1.32% C, 0.35% Si, 1.81% Mn, 18.2% Cr, and 9.2% Ni. Total gas flow rate is 53 Nm^3/min during step I–III; Step IV is 36 Nm^3 Ar/min, 8 Nm^3 O_2/min,
2. 1.05% C, 0.20% Si, 1.85% Mn, 17.9% Cr, and 9.4% Ni. Total gas flow rate is 53 Nm^3/min during step I–III; step IV is 40 Nm^3 Ar/min.

decarburization decreases, the process is first controlled by the chemical equilibria (mechanism (b)) and then by carbon diffusion (mechanism (c)).

To test the validity of the model described above, it is convenient to calculate the carbon content in the metal bath and compare with what is obtained in the converter. This has been done in Fig. 6.7 where data from two heats in a 57 tonne AOD converter is compared with calculated curves. The converter was blown in four steps, three with an O_2/Ar ratio of 3.2:1, 1:1, and 1:2.4 and a final step with 1:4.5 or a final purging with inert gas. During the final inert gas purging all the oxygen for the decarburization is assumed to be supplied by the overoxidized slag. As seen in the figure, the correspondence between practical and calculated results is good.

6.4.4 Some operating conditions and their effect on the process

A number of variables influence the metallurgical process. It is therefore necessary to have a detailed knowledge of steelmaking practice to predetermine the effect of different actions undertaken in the production. In the following we shall utilize the model developed above to discuss what happens during the

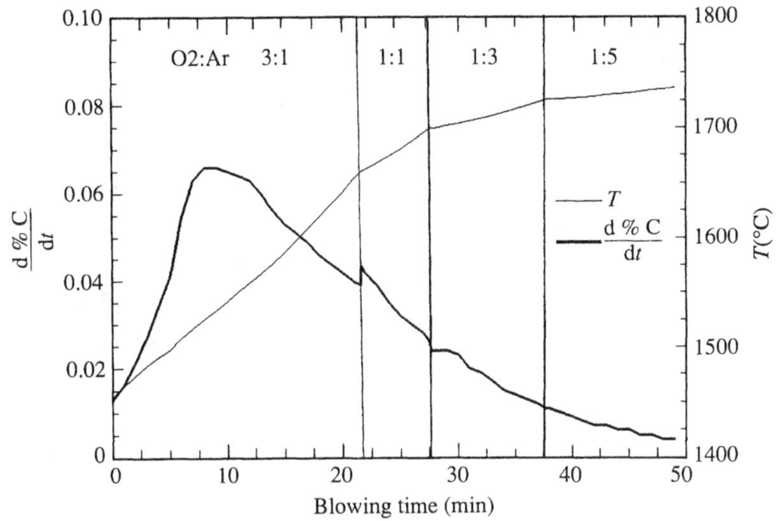

FIG. 6.8 Computer simulation of decarburization of stainless steel in a 60 tonne converter, case 2. Metal composition: 1.5% C, 18% Cr. Total gas flow rate is 60 Nm^3/min. Initial temperature is 1450°C. Formation of Cr_2O_3 = 2610 kg.

decarburization of stainless steels in a converter when some of the operating conditions are changed.

In Section 6.2, the fundamentals of stainless steel production were treated. From what has been said above, the main parameters influencing the decarburization reaction are the temperature and the carbon and chromium activity in the steel bath. All these affect the CRE number and are important in order to determine when to change the blowing steps.

As an example we return to the reference heat given in case 1, Fig. 6.6, and assume that the initial temperature in the converter is lowered to 1450°C. The low temperature can be caused by a very long waiting time in the transfer ladle between the electric arc furnace and the converter. The result of a calculation is given in case 2, Fig. 6.8. The first blowing period is prolonged and more chromium has to be oxidized to increase the temperature and to produce a more efficient decarburization. This increases the total Cr_2O_3 formation from 1750 kg to 2610 kg which gives a higher consumption of reducing agent and process gas and a longer process time.

To demonstrate the influence of the carbon activity on the rate of decarburization, case 3, Fig. 6.9, gives the calculated result when the initial metal bath composition is changed. The bath contains 8% Ni instead of no Ni as was the case in Fig. 6.6. Due to an increased carbon activity when the Ni is added, the linear decarburization period is prolonged and the total blowing

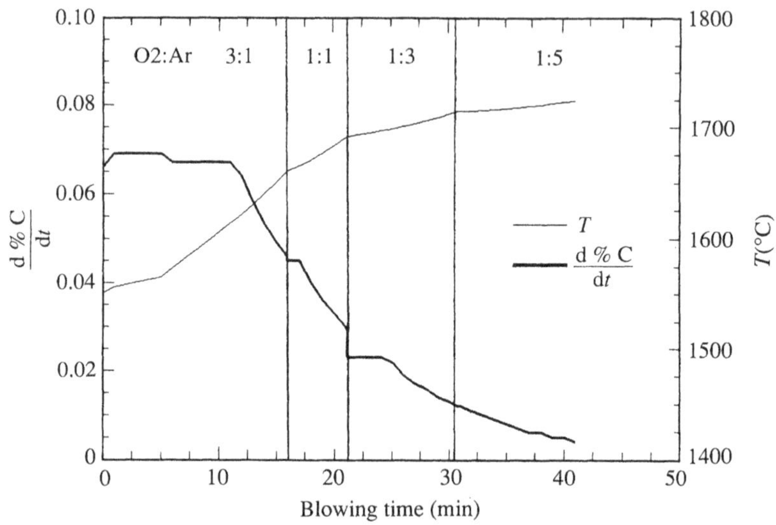

FIG. 6.9 Computer simulation of decarburization of stainless steel in a 60 tonne converter, case 3. Metal composition: 1.5% C, 18% Cr, 8% Ni. Total gas flow rate is 60 Nm^3/min. Initial temperature is 1550° C. Formation of Cr_2O_3 = 1380 kg. Blowing time = 41 min.

time is shortened. This leads to a decreased consumption of reducing agent and gases.

6.4.5 Trends in converter stainless steelmaking and their effect on the rate of decarburization

The industrial production of stainless steels has developed very much during the last decade. The productivity, especially in the electric arc furnaces, has increased considerably. In many steel shops, the converter has become the bottle-neck in the production. Therefore a great deal of work has been carried out in order to increase the productivity. In this connection, some developments will be discussed and exemplified in the following based on computer simulations.

Increased flow rate of oxygen during the first period of the blow By increasing the oxygen gas flow rate from 45 to 55 Nm^3/min, (case 4) during the first step of the blow when the oxygen flow is the rate-determining step for the process, the rate of decarburization is increased so that step 2 can be started earlier and at a higher temperature due to lower temperature losses (see Fig. 6.10). The total blowing time is decreased to 41 minutes from 45 minutes in Fig. 6.6.

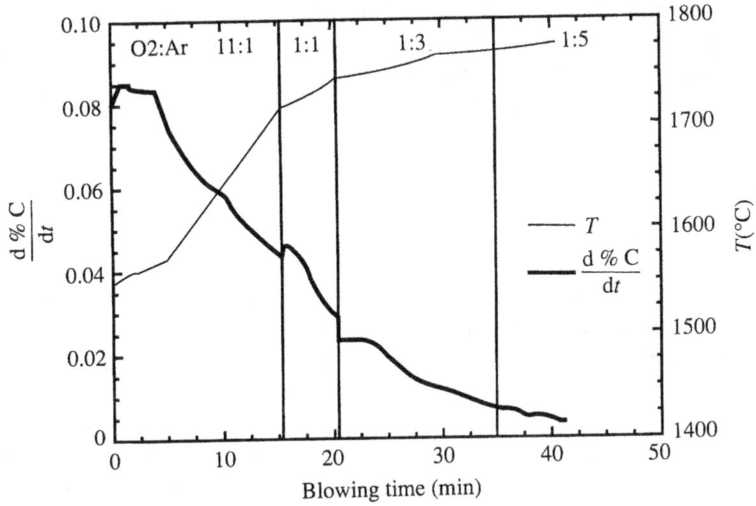

FIG. 6.10 Computer simulation of decarburization of stainless steel in a 60 tonne converter, case 4. Metal composition: 1.5% C, 18% Cr. Total gas flow rate is 60 Nm^3/min. Initial temperature is 1550° C. Formation of Cr_2O_3 = 1932 kg. Blowing time = 41 min.

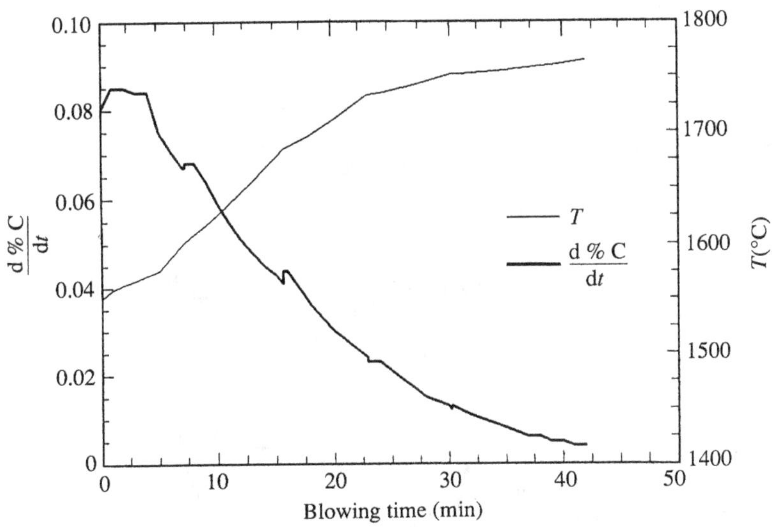

FIG. 6.11 Computer simulation of decarburization of stainless steel in a 60 tonne converter with optimized flow of inert gas, case 5. Metal composition: 1.5% C, 18% Cr. Oxygen flow rate is 45 Nm^3/min. Initial temperature is 1550° C. Formation of Cr_2O_3 = 1700 kg. Blowing time = 42 min.

Continuous dilution with inert gas One way of increasing the productivity during stainless steelmaking in converters is to optimize the blow so that the CRE number is always at a maximum level. This means that after an initial conventional blowing during the linear period, a continuous dilution aimed at keeping the CRE number on a maximum level is carried out. Case 5 (Fig. 6.11) presents a simulation of this blowing practice. A comparison with case 1 (Fig. 6.6) shows that both the process time and the amount of Cr_2O_3 that is formed during the blow are decreased.

Combined blowing A common way to increase the productivity in converters is to install a top lance for the supply of extra oxygen at higher carbon contents, usually above 0.5 per cent. The layout of the process (see Fig. 6.12) is quite similar to what has been developed for the production of low-alloyed steels in integrated steel plants. Normally, the total oxygen flow is increased by 60–100 per cent by the use of a lance, and the rate of decarburization as well as the scrap melting capacity is increased. Production figures[5] show that the blowing time can be decreased by 10 per cent at the same time as the scrap melting capacity is increased by 25 per cent.

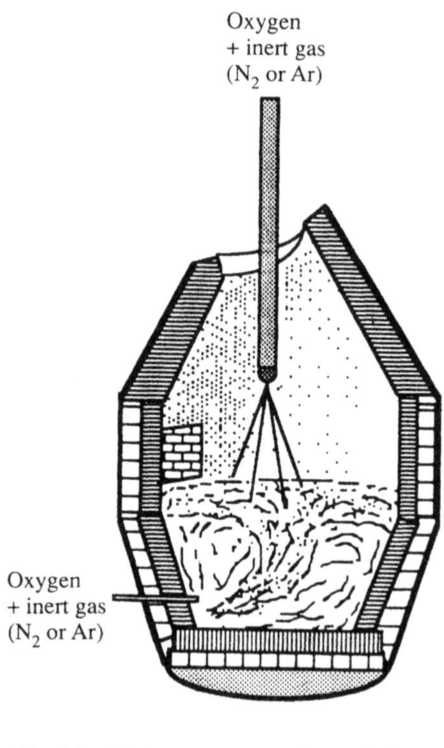

FIG. 6.12 Combined blown converter for stainless steelmaking.

If in the later blowing steps, when the rate of decarburization decreases exponentially with time, we increase the contact area between the reacting phases, a more rapid decarburization is attained. This is performed on an industrial scale in the so-called Krupp Combined Blowing Stainless (KCB-S) process[6] where, in combination with the addition of oxygen and inert gas from the bottom, inert gas and oxygen are also added through a top lance. A comparison between a conventional AOD converter without a top lance and the KCB-S process shows that the blowing time can be decreased by 50 per cent. Tap-to-tap times of around 45 minutes are reported as realistic for this process.

6.5 FINAL COMMENTS

In this chapter, the rather complex production of stainless steels has been described. Despite the use of an expression for the overall decarburization rate in Section 6.4 that is quite simple, the model obtained shows good agreement with production data from the converter. The treatment of different blowing practices in Sections 6.4.4 and 6.4.5 has indicated that substantial benefits can be attained in existing production units by optimizing the blow. In the future, computer models for stainless steelmaking in converters, combined with measurements during the blow, should play an even more important role in order to reduce the process time, consumption of gases, oxidation of chromium, and the use of reducing agents.

REFERENCES

1. Dresler, W. (1988). The effect of Ni additions on the decarburization of liquid high-carbon Ferrochromium. *Iron and Steelmaker*, January, 43–51.
2. Jörgensen, C. and Thorngren, I. (1969). *Thermodynamic Tables for Process Metallurgists*, Almkvist and Wiksell, Stockholm.
3. Turkdogan, E.T. (1983). *Physicochemical Properties of Molten Slags and Glasses*. The Metals Society, London.
4. R. Hidaka, *et al.* (1984). Development in the production of extra low sulfur grade stainless steel. Tetsu-to-Hagané, 235.
5. Tohge, T., Fujita, Y., and Watanabe, T. (1984). Some considerations on high blow rate AOD practice with top blowing system. 4th PTD Conference, April 1–4, Chicago. Ill. Conference Proceedings, ISS or AIME.
6. Röpke, W., Zörcher, H. (1989). Combined blowing process for stainless steel. *Steel Technology International*, 125–8.

7

MIXING, MASS TRANSFER, AND NUMERICAL MODELS

7.1 INTRODUCTION

Mixing and stirring are terms widely used when discussing melt treatment. These terms are useful in that they have an immediate intuitive meaning. However, a worker in refining metallurgy very early finds it necessary to give them more substance. In the following section an attempt is made to explain and clarify the meaning of these terms.

Previously mass transfer coefficients, k, at various interfaces have been introduced. In the examples, values of k were presented as purely empirical quantities. Also, it should be noted that we have assumed that at gas–melt interfaces the mass transfer coefficient in the melt is rate limiting. Here, one seeks to relate the mass transfer coefficients to quantities such as the diffusion coefficients, D, and the velocity along the interface. The effect of turbulence on mass transfer is discussed. Finally, resistances to mass transfer in the melt and gas phases are compared.

In Chapters 4 and 5 flow and mixing are described in terms of idealized extreme situations: plug flow (no mixing) and complete mixing in so-called back-mix tanks or reactors. In real life we attempt to develop reactors that approach the idealized case.

However, in the planning stage or in operation the question arises about how close we are to the idealized reactors and what steps can be taken for improvement. Then it becomes necessary to perform a mathematical analysis of the turbulent flow and to carry out extensive numerical calculations or physical model and molten metal experiments. In this chapter we are not seriously going to enter the difficult fields of fluid flow and numerical analysis. Only some basic concepts in turbulence theory are mentioned and the potential benefits of numerical calculations are sketched. Some references to the specialized literature are given.

7.2 MIXING AND CIRCULATION FLOW; FLOW MODELS

To illustrate problems concerning mixing, the addition of alloys is examined. When alloys in lump, granule, or particle form have melted and partly dissolved into a surrounding melt, they still, to a degree, retain their identity as concentrated volumes of solute surrounded by melt. They have not yet mixed properly into the melt so that the concentration c at an arbitrary point is not equal to the average concentration, c_m. It might be argued that such a state of affairs is only a problem if sufficient blending has not occurred by the time that solidification takes place, but this is not the case. For instance, most alloy additions to steel are more reactive to oxygen (and even nitrogen) in the air than the melt. Also, they may react with the slag or refractory. This causes a loss of the alloy addition and even delivers an extra penalty in the form of unwanted reaction products such as oxide inclusions. Thus ideally, flow, dissolution, and mixing in a high-temperature melt should be such that alloy concentrations near the various interfaces do not exceed c_m. Now this sounds like an impossible task, especially if one considers that convection seems to be necessary in order to attain good mixing. Furthermore, there are only three main ways of stirring: bubbling, electromagnetic, and mechanical (see Chapter 5). Moreover, at high temperatures mechanical stirring is a problem due to the attack on the impeller material.

Figure 7.1 illustrates how the difficulties as outlined here may be solved. A ceramic ring keeps the slag away from an 'eye' swept open by rising bubbles from a porous plug. The alloy additions are fed at a moderate rate into the melt through the eye. The bubbling action gives good circulation of the melt around the alloy additions and dispenses the alloy throughout the melt. Attack is concentrated on the simple refractory ring which can be replaced when

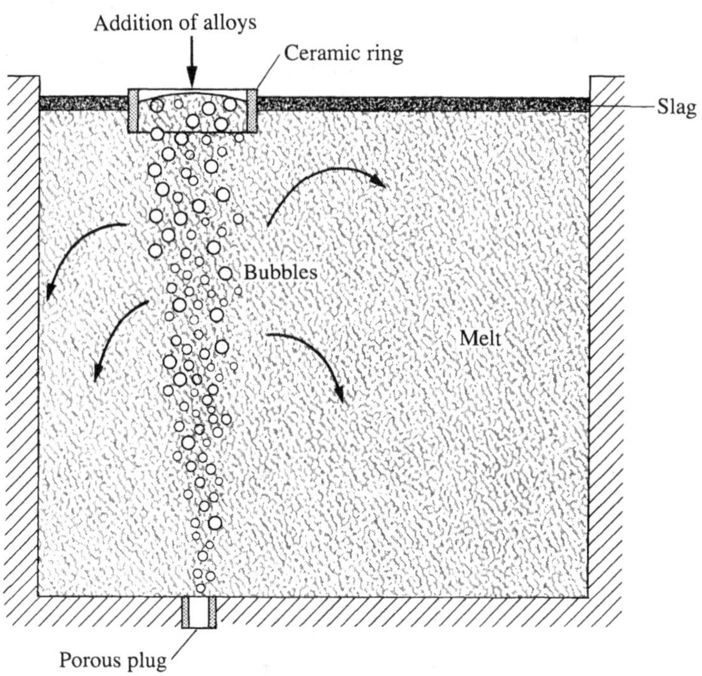

FIG. 7.1 Ceramic ring for the addition of alloys to steel.

necessary. The strongly reduced slag does not react with the alloy, and the high circulation rates of the melt and moderate addition rates of the alloy ensure that no excessively high alloy concentrations occur at any time.

The vacuum circulation RH reactor mentioned previously in Section 3.14.2 operates on a similar principle with regard to alloy additions. The increased yield is due to the controlled addition of alloys in the vacuum chamber environment. To be fair it should be mentioned that the penalty is that the RH is susceptible to attack on the refractories, especially in the legs and bottom surface. This is due either to the effect of the alloy additions or to erosion from the shear forces created by the flowing metal and rising bubbles.

When the mechanism or process that we are interested in takes place in a limited region of the melt or in a special reactor as described in the two examples given above, it is often helpful to describe mixing in terms of a circulation flow $\dot{Q}$[1,2] with units of m^3/s. We show that sometimes $\dot{Q}$ can be determined from rather simple reasoning or experiments. For the RH, $\dot{Q}$ is the flow that circulates through the vacuum unit. Additions of Si have been used to measure this $\dot{Q}$ in an RH.[3] If c_{up} and c_{down} are the Si concentrations (in kg/m^3) in the up and down leg, respectively, then

$$(c_{down} - c_{up})\dot{Q} = \dot{M}_{Si} \tag{7.1}$$

where $\dot{M}_{Si}$ is the feed rate in kg/s of Si added in the chamber in the HR. When c_{down} and c_{up} are measured, they increase linearly with time and with the same constant slope. This allows $\dot{Q}$ to be determined from eqn (7.1). Various cold model and plant experiments[2,3] indicate that

$$\dot{Q} \,\alpha \{\dot{Q}_g d^4 \ln(p_{in}/p_0)\}^{\frac{1}{3}} . \tag{7.2}$$

Here $\dot{Q}_g$ is the flow rate of gas in Nm^3/s, d is the diameter of the up-leg (snorkel), and p_{in} and p_0 are the (static) pressures where the stirring gas enters and exits, respectively.

Equation (7.2) can be derived from elementary energy considerations. The rate at which mechanical energy is irreversibly converted to thermal energy in the up and down legs is

$$\dot{E} = e\rho\dot{Q}v_m^2 \tag{7.3}$$

where the factor e depends on geometry and, slightly, on the Reynolds number. e is of the order of magnitude 1. By comparison, for a rounded 90° valve $0.4 < e < 0.9$.[4] v_m is the mean velocity over a cross-section of a leg:

$$\dot{Q} = \frac{\pi d^2}{4} v_m . \tag{7.4}$$

The stirring power $\dot{E}$ is supplied by the expansion energy of the bubbles:

$$\dot{E} = \dot{Q}_{in} p_{in} \ln(p_{in}/p_0) . \tag{7.5}$$

Here $\dot{Q}_{\text{in}}$ is the gas supply at the inlet point at pressure p_{in}. Equation (7.5) is valid for isothermal (cold) conditions. In hot conditions, the thermal expansion term must also be taken into account (see eqn (5.120)). Equations (7.3)–(7.5) give

$$\dot{Q} = \left(\frac{\pi^2}{16e}\right)^{\frac{1}{3}} \left(\frac{\dot{Q}_{\text{in}} p_{\text{in}} d^4 \ln(p_{\text{in}}/p_0)}{\rho}\right)^{\frac{1}{3}}. \tag{7.6}$$

Comparison of eqn (7.6) with plant experiments reported by Kawabara *et al.* give $e = 0.55$. However, Kawabara *et al.* not seem to have taken the thermal expansion term of the gas into account. If such a correction is performed, e would be greater.

In many cases it is difficult to obtain $\dot{Q}$ from elementary considerations. It is, for instance, a complex matter to determine $\dot{Q}$ for a bubble-stirred metal bath.[5,6] However, $\dot{Q}$ may be measured in tracer experiments.[1,7] The circulation flow $\dot{Q}$ is often used to describe mixing. Then the circulation time, t_c, is defined as

$$t_c = \frac{V}{\dot{Q}} \tag{7.7}$$

where V is the melt volume. To attain satisfactory mixing (blending), the time from the last addition, t, must be, for example, $t \geqslant 3t_c$.

The inverse value of $\dot{Q}$ may be regarded as a resistance that must be added to the mass transfer resistance. Let us illustrate this by constructing a simple example where carbon is removed by reaction with oxygen in an RH operating as part of a future continuous steelmaking processes. We assume complete mixing in the RH chamber so that the concentration in this unit is c_r and complete mixing in the ladle with concentration c (kmoles/m^3). The vacuum-melt contact area in the chamber is A with the total mass transfer coefficient for carbon equal to k_t. Then the number of moles of carbon transferred from ladle to chamber per second is

$$\dot{N} = \dot{Q}(c - c_r) \tag{7.8}$$

and the number per second removed from the melt in the chamber to the vacuum system is

$$\dot{N} = k_t A c_r. \tag{7.9}$$

If eqn (7.8) is divided by $\dot{Q}$ and (7.9) by $k_t A$, and the resulting two equations are summed, one obtains

$$\dot{N}\left(\frac{1}{\dot{Q}} + \frac{1}{k_t A}\right) = c. \tag{7.10}$$

It is seen that $1/\dot{Q}$ must be added to the mass transfer resistance $1/(k_t A)$ to give the total resistance. Thus if impurities are extracted in only part of our system, the *transportation resistance* $1/\dot{Q}$ must also be taken into account. This is one reason why vacuum units in some cases may be less efficient than gas purge units. The impurity is removed only from a surface layer. If the circulation, $\dot{Q}$ between the bulk melt and the surface is poor in the vacuum system, the resistance $1/\dot{Q}$ is large. A mass balance for carbon in our hypothetical continuous steelmaking process becomes

$$\dot{q}c_{\text{in}} = \dot{q}c + \dot{N}$$

or

$$\dot{q}c_{\text{in}} = \dot{q}c + \frac{c}{\dfrac{1}{k_t A} + \dfrac{1}{\dot{Q}}} \tag{7.11}$$

or

$$\frac{c}{c_{\text{in}}} = \frac{1}{1 + \left(\dfrac{\dot{q}}{k_t A} + \dfrac{\dot{q}}{\dot{Q}}\right)^{-1}}$$

where $\dot{q}$ is the volumetric flow of metal. This means that the 'dimensionless' resistance

$$\frac{\dfrac{1}{k_t A} + \dfrac{1}{\dot{Q}}}{\dfrac{1}{\dot{q}}} \ll 1 \tag{7.12}$$

in order to give good carbon removal. This result may be compared with the case of the simple completely mixed reactor ($\dot{Q} = \infty$) in Chapter 4. There it was only required that

$$\frac{\dot{q}}{k_t A} \ll 1. \tag{7.13}$$

In the equations in Chapter 4, $\dot{Q}$ can be taken into account by replacing $k_t A/\dot{q}$ with $(\dot{q}(1/k_t A + 1/\dot{Q}))^{-1}$.

Basically, the same results are obtained for batch reactors, for instance for a conventional RH/ladle system. However, the mathematics and models become more complex.

Both in continuous reactors and batch reactors, there will be gradients in concentration, zones which are near-stagnant where the exchange flow, $\dot{Q}$, with the other zones is small, and there will be short-circuiting effects. Short-circuiting here means that part of the metal flow circumvents the reaction zone and is not treated in the reactor (see problem[7.1]). Both the stagnant zones and short-circuiting can be determined from tracer experiments.[1,7,8] Once this is known together with the mass transfer coefficients, k_t, and contact areas, A, the conversion in an extractive reactor[6,9] or the fraction removed in a refining unit can be calculated. A fundamental problem is that the reactor must physically exist before tracer measurements can be carried out. Alternatively, tracer experiments can be carried out on a smaller-scale reactor or in a water model. This water model may be built in full scale. The measurement results must be scaled-up for the melt tracer experiments or transformed from water to molten metal in the full-scale experiments. Scaling requires a good understanding of the mechanisms and phenomena that occur and may lead to extensive calculations. Furthermore, it is often only practical to physically model the flow aspects of a process. Phenomena such as a change of phase (solidification), heat, and mass transfer can usually be included only in a mathematical model. As a result it is tempting to develop 'complete' models that include flow. Such models should apply both to small and large scales and to water and molten metal. This allows them to be checked and adjusted by comparison with model, laboratory, and industrial experimental results.

7.3 MASS TRANSFER OF SOLUTE TO WALLS; BROWNIAN DIFFUSION

In Chapter 5 when we studied the transfer of inclusions to a wall we neglected the viscous (laminar flow) boundary layer δ_2. The justification was that the inclusions penetrated this layer simply due to their size. This is of course not the case for dissolved elements. In fact for dissolved elements mass transfer is governed primarily by diffusion in δ_2 so that the mass transfer coefficient, k, for steady flow past a wall is given by:[10]

$$k = \frac{D}{1.5\,\delta_2} \tag{7.14}$$

where D is the diffusion coefficient. Equation (7.14) is only valid when δ_2 is smaller than the restricted turbulence boundary layer δ_1. Comparison of $\delta_1 = 5\nu/v_0$[10] with eqn (7.15) gives that $\delta_2 < \delta_1$ for sufficiently large Sc(hmidt) numbers

$$\mathrm{Sc} = \frac{\nu}{D} > 10.$$

In eqn (7.14) the factor 1.5 takes resistance in the other boundary layers into account.[10] $1.5\delta_2$ may be regarded as a *diffusion boundary layer*. δ_2 is found by setting D equal to the mass diffusivity by eddies for the restricted turbulence layer given in eqn (5.97)

$$D = \frac{0.14 \times 10^{-2} v_0^3 \delta_2^3}{\nu^2}$$

or

$$\delta_2 = 8.9 D^{\frac{1}{3}} \nu^{\frac{2}{3}} / v_0 \tag{7.15}$$

or, inserting in eqn (7.14),

$$k = 0.075 \left(\frac{D}{\nu}\right)^{\frac{2}{3}} v_0 . \tag{7.16}$$

Here the shear velocity $v_0 = (\tau_0/\rho)^{\frac{1}{2}}$ may be obtained directly from flow calculations, determined from the specific stirring power $\dot{\varepsilon}$ (eqn (5.112)), or we make use of equations such as (5.107) and (5.108) in Chapter 5. Then for a smooth pipe:

$$k = 0.015 \left(\frac{D}{\nu}\right)^{\frac{2}{3}} v_{\mathrm{m}} \, \mathrm{Re}^{-\frac{1}{8}} \tag{7.17}$$

where v_{m} is the mean velocity in the pipe.

Equation (7.17) applies not only to dissolved elements but also to inclusions where the diameter $d_{\mathrm{p}} < \delta_2$. To give an example of the calculation of the critical diameter where $d_{\mathrm{pc}} = \delta_2$, we return to Section 5.14.3 on nozzle blockage and use the data given there. From Section 2.2.9 we have $D = kT/3\pi\rho\nu d_{\mathrm{p}}$, so that eqn (7.15) gives on setting $d_{\mathrm{p}} = \delta_2$, and $v_0 = \phi^{\frac{1}{2}} v_{\mathrm{m}}$:

$$d_{\mathrm{pc}}^{\frac{4}{3}} = 8.9 \left(\frac{kT}{3\pi\rho\nu}\right)^{\frac{1}{3}} \frac{\nu^{\frac{2}{3}}}{\phi^{\frac{1}{2}} v_{\mathrm{m}}}$$

or

$$d_{\mathrm{pc}}^{\frac{4}{3}} = 8.9 \left(\frac{1.38 \times 10^{-23} \times 1873}{3\pi \times 7030 \times 0.8 \times 10^{-6}}\right)^{\frac{1}{3}} \frac{(0.8 \times 10^{-6})^{\frac{2}{3}}}{0.0036^{\frac{1}{2}} \times 2} .$$

This gives $d_{\mathrm{pc}} = 0.6 \times 10^{-6}\,\mathrm{m}$.

Removal of inclusions less than d_{c} in size is proportional to $v_{\mathrm{m}}^{\frac{7}{8}}$ according to eqn (7.17). Then in terms of the specific stirring power $\dot{\varepsilon}$ introduced in Section 5.14.4, k is proportional to $\dot{\varepsilon}^{\frac{7}{24}}$ when $d_{\mathrm{p}}^{\frac{4}{3}} \phi^{\frac{1}{2}} v_{\mathrm{m}} < 8.9(kT\nu/3\pi\rho)^{\frac{1}{3}}$. The exponent $\frac{7}{24}$ is roughly the same as for resistance to transfer in the turbulent wall layer. The dependence in the three layers is illustrated in Fig. 5.18.

7.4 MASS TRANSFER IN LIQUIDS TO A CLEAN FREE SURFACE

Mass transfer between a melt and a gas phase plays a very important role in extractive and refining metallurgy. For instance, vacuum treatment is based on removal of unwanted elements through a free surface. Metal in holding furnaces picks up hydrogen from moisture in the atmosphere above the melt.

In gas purging, impurity elements are transferred to the bubbles through the melt–gas interface.

In chemical engineering systems the presence of impurities at the interface can greatly influence mass transfer.[10,11] The same problems are encountered with metal melts.[12,13,14,15] In addition, metal oxide films may play a role.

In this section we attempt to take a close look at liquid-side mass transfer to a clean interface. Conditions are very different from those at a solid surface (cf. Section 5.14.2). As opposed to a solid surface, there remain (if the surface is clean) very considerable average movements along the surface. These movements persist right into the surface and may be observed clearly by sprinkling a powder on to the surface. In the z-direction normal to the horizontal surface the turbulent eddies are restrained by surface tension and gravity effects, so that at $z = 0$, we may as a first approximation set the eddy velocity $v_z = 0$. To obtain a relation for v_z, we proceed by the method of Levich as described in.[10] There is a zone of thickness δ_1 near the surface within which turbulence in the z-direction is damped.[10] At the lower limit the eddy velocity is

$$v_z = v_0 \tag{7.18}$$

where v_0 is the z-component of the eddy velocity in the bulk melt near the boundary layer. Since the velocity along the surface, v_x, does not depend on z, the continuity equation (cf. Section 5.14.2) gives that

$$v_z = \frac{z v_0}{\delta_1}. \tag{7.19}$$

Thus the eddy velocity decreases linearly with z instead of with z^2 as was the case for a solid surface. The application of eqn (7.19) depends on calculating the value of δ_1, that is, the thickness of the restricted turbulence layer and the z-component of the eddy velocity v_0 at $z = \delta_1$.

As an eddy approaches from the bulk, the surface becomes deformed. The pressure fluctuations p' are proportional to the square of the velocity fluctuations so that

$$p' = C\rho v_0^2 \tag{7.20}$$

where C is about 1.[10] For isotropic turbulence at high Reynolds numbers,[10] $C = 0.7$. In general turbulence is not isotropic close to the surface and C will depend on geometry and the method of stirring. The pressure fluctuation is restrained by surface tension and gravity effects. Thus, p' may be compared to the excess pressure in a soap bubble, for example. One obtains[10]

$$p' = \frac{2}{R}\left(\sigma + \frac{d_e^2 \rho g}{16}\right) \tag{7.21}$$

where R is the radius of curvature of the deformation, σ is the surface tension, and d_e is the diameter of the surface deformation at $1/e$ of the maximum height. Experiments show that roughly (for $C = 1$)

$$\delta_1 = \frac{R}{2}. \tag{7.22}$$

Equations (7.19)–(7.22) give

$$v_z = \frac{z\rho v_0^3}{(\sigma + d_e^2 \rho g/16)}. \tag{7.23}$$

Then from the Prandtl mixing length model with eddy length $l = 0.4\,z$ (see section 7.11.2), if it is assumed that eddy viscosity and eddy diffusivity are equal, the diffusion coefficient becomes

$$D_E = v_z l = \frac{0.4\,z^2 \rho v_0^3}{\sigma + d_e^2 \rho g/16}. \tag{7.24}$$

The thickness of the diffusion boundary layer δ_2 is found by setting D_E equal to D and solving eqn (7.24) with respect to $z = \delta_2$:

$$\delta_2 = \left(\frac{D(\sigma + d_e^2 \rho g/16)}{0.4 \rho v_0^3}\right)^{\frac{1}{2}}. \tag{7.25}$$

If there had been no convection along the surface and mass transfer had taken place only in the z-direction, the mass transfer coefficient would be given by[10]

$$k = \frac{D}{2\delta_2} \tag{7.26}$$

for Sc > 10. The factor of 2 takes resistance in the restricted turbulence layer into account. This means that the diffusion boundary layer thickness is $2\delta_2$.

However, the liquid is free to move along the surface in the x-direction. This effect must also be taken into account. Then the conservation equation for the diffusing component gives (compare eqn (2.130) and Chapter 8 concerning diffusion of heat)

$$\frac{\partial c}{\partial t} = D \frac{\partial^2 c}{\partial z^2}. \tag{7.27}$$

The left-hand side of eqn (7.27) is the rate of accumulation of component c in a unit volume of liquid moving along the surface. The right-hand side gives the net diffusion into the volume element in the z-direction. Diffusion in the other directions is neglected since concentration changes strongly only in the z-direction. Figure 7.2 illustrates the situation and gives the boundary conditions for our solution of eqn (7.27). The unit volume of fluid approaches the surface from the bulk melt at time $t = 0$. thus for

$$t = 0, \qquad c = c_b. \tag{7.28}$$

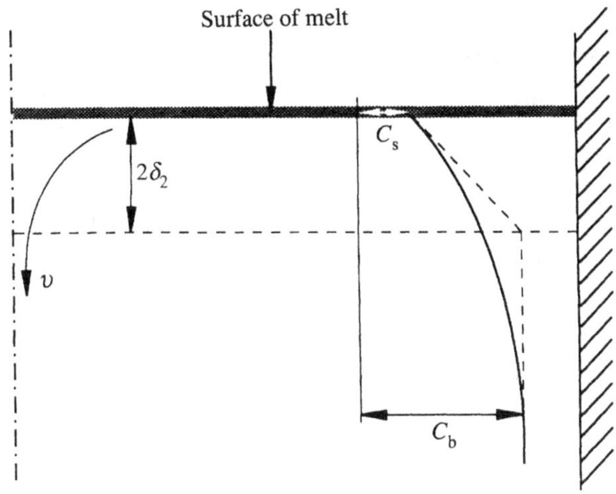

FIG. 7.2 Concentration profiles and flow of melt at the free surface. The mass transfer coefficient $k = D/2\delta_2$ where $2\delta_2$ is the diffusion boundary layer. v is the velocity of the liquid.

At the surface where

$$z = 0, \qquad c = c_S . \tag{7.29}$$

Furthermore, below $z = 2\delta_2$ the eddy diffusion coefficient is so large that the concentration is equal to that in the bulk. For

$$z \geqslant 2\delta_2, \qquad c = c_b . \tag{7.30}$$

The solution of eqn (7.27) that fulfils the boundary conditions (eqns (7.28)–(7.30) is [16, 17]

$$\frac{c - c_S}{c_b - c_S} = \operatorname{erf}\left(\frac{z}{2(Dt)^{\frac{1}{2}}}\right) + \sum_{j=1}^{\infty}\left\{\operatorname{erfc}\left(\frac{2j\,2\delta_2 - z}{2(Dt)^{\frac{1}{2}}}\right) - \operatorname{erfc}\left(\frac{2j\,2\delta_2 + z}{2(Dt)^{\frac{1}{2}}}\right)\right\} . \tag{7.31}$$

Here $\operatorname{erf}(x) = 2\pi^{-\frac{1}{2}}\int_0^x \exp(-u^2)\,du$ and $\operatorname{erfc}(x) = 1 - \operatorname{erf}(x)$. $\operatorname{erf}(\infty) = 1$. These functions and the solutions of the diffusion equation are discussed in greater detail in Chapter 8. The first term on the right-hand side of eqn (7.31) represents a solution for a semi-infinite body ($2\delta_2 = \infty$). For large δ_2 or small contact times, t, the remaining terms are negligible. Hence, at small times diffusion occurs as if the diffusion boundary layer, $2\delta_2$, were infinite. For large contact times, t, on the other hand, the remaining terms become important.

The total mass transfer rate at time t, n, from the unit volume is obtained from Fick's law (see eqn (2.129)):

$$\dot{n}(t) = -D\frac{\mathrm{d}c}{\mathrm{d}z} = \frac{D(c_b - c_S)}{(\pi Dt)^{\frac{1}{2}}}\left(1 + 2\sum_{j=1}^{\infty} \exp\left\{-\left(\frac{j\,2\delta_2}{(Dt)^{\frac{1}{2}}}\right)^2\right\}\right)$$
$$z = 0. \tag{7.32}$$

Then by integrating over time one obtains the number of moles of c per unit area which have crossed $z = 0$ between $t = 0$ and t:

$$-N = 4\,(Dt)^{\frac{1}{2}}\left\{\frac{1}{2\pi^{\frac{1}{2}}} + \sum_{j=1}^{\infty} \operatorname{ierfc}\left(\frac{j\,2\delta_2}{(Dt)^{\frac{1}{2}}}\right)\right\}(c_b - c_S) \tag{7.33}$$

where

$$\operatorname{ierfc}(x) = \frac{\exp(-x^2)}{\pi^{\frac{1}{2}}} - x\operatorname{erfc}(x).$$

For large $x \to \infty$, $\operatorname{ierfc}(x) \to 0$.

The average mass transfer coefficient is obtained by dividing N by the elapsed contact time, t, and by the driving force for diffusion, $c_b - c_S$. This gives

$$k = 4\left(\frac{D}{t}\right)^{\frac{1}{2}}\left\{\frac{1}{2\pi^{\frac{1}{2}}} + \sum_{j=1}^{\infty} \operatorname{ierfc}\left(\frac{j\,2\delta_2}{(Dt)^{\frac{1}{2}}}\right)\right\}. \tag{7.34}$$

For small contact times the infinite sum is negligible. Thus the first term

$$k = \left(\frac{4D}{\pi t}\right)^{\frac{1}{2}} \tag{7.35}$$

is commonly used to describe mass transfer at a free surface.[4, 10, 11] For instance, for mass transfer to bubbles one obtains the contact time, t, by dividing bubble diameter d by the relative velocity between bubble and melt, v:

$$t = \frac{d}{v}. \tag{7.36}$$

Then the first term in eqn (7.34) becomes

$$k = \left(\frac{4Dv}{\pi d}\right)^{\frac{1}{2}}. \tag{7.37}$$

We will now take a closer look at the remaining terms. 'Small times' $t < t_c$ is here defined by

$$\frac{\delta_2}{(Dt_c)^{\frac{1}{2}}} = 1. \tag{7.38}$$

It will be seen that the infinite sum in eqn (7.32) is, in fact, of little importance for $t < t_c$. The mean mass transfer coefficient is then obtained from eqn (7.34) retaining only the first two terms:

$$k_1(t) \approx \left(\frac{D}{t}\right)^{\frac{1}{2}} \left\{\frac{2}{\pi^{\frac{1}{2}}} + 4\,\mathrm{ierfc}\left(\frac{2\delta_2}{(Dt)^{\frac{1}{2}}}\right)\right\}$$

$$\approx \left(\frac{4D}{\pi t}\right)^{\frac{1}{2}} \left\{1 + \frac{\exp(-4\delta_2^2/(Dt))}{4\delta_2^2/(Dt)}\right\}. \qquad (7.39)$$

The correction represented by the exponential term is less than 1 per cent and can usually be neglected. Thus, eqn (7.35) may be used for $t \leqslant t_c$.

For contact times $t > t_c$ it is not convenient to sum the terms in the infinite series in eqn (7.33). Instead it is better to utilize a solution of the diffusion equation, eqn (7.27), that converges rapidly for large t. This method of solution is discussed in detail in Chapter 8. Similarly to eqn (8.19), one obtains

$$\frac{c - c_S}{c_b - c_S} = \frac{z}{2\delta_2} - \frac{2}{\pi}\sum_{j=1}^{\infty} \frac{(-1)^j}{j} \sin\left\{j\pi\left(1 - \frac{z}{2\delta_2}\right)\right\} \exp\left(-\frac{D\pi^2 j^2 t}{4\delta_2^2}\right). \qquad (7.40)$$

For $t > t_c$ one only needs to include the first term in the infinite series:

$$-\dot{n}(t) \approx \frac{D(c_b - c_S)}{\delta_2}\left\{\frac{1}{2} + \exp\left(\frac{-D\pi^2 t}{4\delta_2^2}\right)\right\}. \qquad (7.41)$$

The remaining terms are less than $\exp(-\pi^2) = 5 \times 10^{-5}$. The mean mass transfer during the period from t_c to t is found by integration of eqn (7.41) from t_c to t and then dividing by $t - t_c$. This gives

$$k_2(t) = \frac{D}{2\delta_2}\left(1 + \frac{8\exp(-\pi^2/4)}{\pi^2(t/t_c - 1)}\left(1 - \exp\left\{-\frac{\pi^2}{4}\left(\frac{t}{t_c} - 1\right)\right\}\right)\right). \qquad (7.42)$$

In eqn (7.42) $D/(2\delta_2)$ is identical to eqn (7.26), valid for the case when the effect of flow along the interface can be neglected. At large times the melt moving in the diffusion boundary layer has been depleted of component c, leaving only the linear profile due to diffusion normal to the surface. The exponential term in eqn (7.42) is a fraction 0.17 of the first term for $t = t_c$, and for $t > 2t_c$ the fraction is less than $0.07/(t/t_c - 1)$.

The mean mass transfer coefficient is the weighted time-average of the two mass transfer coefficients:

$$k = \frac{k_1(t_c)t_c + k_2(t)(t - t_c)}{t}. \qquad (7.43)$$

If $t < t_c$ then $k = k_1(t)$. Obviously it is important to determine t_c. From eqns (7.25) and (7.38) one obtains

$$t_c = \frac{\sigma + d_e^2 \rho g/16}{0.4 \rho v_0^3}. \tag{7.44}$$

The problem now is to determine the turbulent velocity in the normal direction, v_0, and the surface deformation diameter, d_e. For molten metals σ is of the order of magnitude 1 N/m. For small surface deformations, $d_e < 5$ mm,

$$t_c \approx \frac{\sigma}{0.4 \rho v_0^3}. \tag{7.45}$$

Very roughly, for impeller stirring, v_0 is given by [10]

$$v_0 \sim nL \tag{7.46}$$

where n is the number of rotations per second and L is the length of the stirrer blades. Thus

$$t_c \sim \frac{\sigma}{\rho n^3 L^3} \tag{7.47}$$

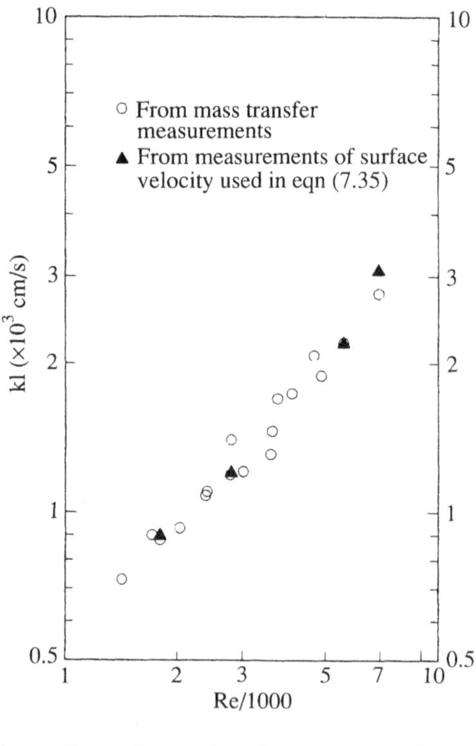

FIG. 7.3 Comparison of experimental surface mass transfer coefficient, k, for an impeller-stirred melt, with k calculated from eqn (7.35) using a time t obtained from measurements.[43] A flat paddle impeller is used.

As we increase the dimensions or rotational speed, n, the critical time, t_c, decreases strongly, and the mass transfer coefficient, k, approaches that given by eqn (7.26). For small dimensions, L, and a low rotation speed, t_c, is large and mass transfer is given by eqn (7.35). Figure 7.3 compares k measured by absorption/desorption of CO_2 in water in a small stirred laboratory unit with eqn (7.35). The residence time, t, in eqn (7.35) was determined from near-surface velocity measurements using a laser-doppler unit.[43] It is seen that the correspondence is very good.

If the melt is covered by even a thin slag layer, it may be appropriate to assume that the surface is solid and apply eqn (7.16) where v_0 is now the shear velocity, $v_0 = (\tau_0/\rho)^{\frac{1}{2}}$. τ_0 is the shear force between the slag and melt.

7.5 MASS TRANSFER IN LIQUIDS TO BUBBLES, DROPLETS, AND PARTICLES

In the previous section it was mentioned that eqn (7.37)

$$k = \left(\frac{4Dv}{\pi d}\right)^{\frac{1}{2}} \tag{7.37}$$

was very often used to describe mass transfer to bubbles. When it was derived, we did not take viscosity into account. This might indicate that it is valid only for high bubble Reynolds numbers $\mathrm{Re} = vd/\nu$, that is, large bubbles. We will see, however, that eqn (7.37) is useful as a reference even for small bubbles and small Re.

Boundary layer theory[11] only introduces a correction for the effect of viscosity:

$$k = \left(\frac{4Dv}{\pi d}\right)^{\frac{1}{2}} \left(1 - \frac{2.89}{\mathrm{Re}^{\frac{1}{2}}}\right)^{\frac{1}{2}} \tag{7.48}$$

valid for $\mathrm{Re} \geqslant 70$. At $\mathrm{Re} = 70$ the correction factor to eqn (7.37) becomes $(1 - 2.89/70^{\frac{1}{2}})^{\frac{1}{2}} = 0.81$.

For both bubbles and liquid drops with low viscosity, $\kappa = \mu_p/\mu < 2$, eqn (7.48) may be expanded to give:[11]

$$k = \left(\frac{4Dv}{\pi d}\right)^{\frac{1}{2}} \left(1 - \frac{2.89 + 2.15\kappa^{0.64}}{\mathrm{Re}^{\frac{1}{2}}}\right)^{\frac{1}{2}} \tag{7.49}$$

An empirical formula[11] applicable for $\mathrm{Re} > 1$ is

$$k = \left(\frac{4Dv}{\pi d}\right)^{\frac{1}{2}} \left\{1 - \frac{\dfrac{2 + 3\kappa}{3(1 + \kappa)}}{\left(1 + \left[\dfrac{(2 + 3\kappa)\mathrm{Re}^{\frac{1}{2}}}{(1 + \kappa)(8.67 + 6.45\kappa^{0.64})}\right]^{n}\right)^{\frac{1}{n}}}\right\}^{\frac{1}{2}} \tag{7.50}$$

where $n = 3\kappa + \frac{4}{3}$. The correction factor to eqn (7.37) for bubbles ($\kappa = 0$) at Re = 1 is 0.63.

For Re < 1 we have creeping, also called Stokes' flow. In this case, if the interfacial contamination is negligible, for bubbles or low-viscosity droplets with $\kappa \leqslant 2$,[11]

$$k = \left(\frac{4Dv}{\pi d}\right)^{\frac{1}{2}} \left(\frac{1}{1 + \kappa}\right)^{\frac{1}{2}} 0.58. \tag{7.51}$$

Bubbles less than 1 mm, bubbles in contaminated liquids, and high-viscosity droplets tend to give the same mass transfer relations as solid particles. Then the tangential velocity is zero at the interface and increases linearly with distance from the interface. The situation is very different from what we have on a free surface. Quite involved calculations are required to deal with this case. A convenient solution valid for Re < 1 is:[11]

$$\mathrm{Sh} = 1 + (1 + \mathrm{Pe})^{\frac{1}{3}} \tag{7.52}$$

where $\mathrm{Sh} = kd/D$ and $\mathrm{Pe} = \mathrm{Re}\,\mathrm{Sc} = vd/D$.

In a stagnant bath where the relative velocity $v = 0$, eqn (7.52) reduces to

$$\mathrm{Sh} = 2 \qquad \text{or} \qquad k = 2D/d. \tag{7.53}$$

This may be deduced directly from the diffusion equation for a sphere (cf. eqn (8.11)):

$$\frac{\partial\left(r^2 \dfrac{\partial c}{\partial r}\right)}{\partial r} = 0. \tag{7.54}$$

The boundary conditions are $c = c_s$ at $r = \frac{d}{2}$ and $c = c_b$ when $r \to \infty$. The solution is

$$c = c_b - (c_b - c_S)\frac{d}{2r}$$

and the molar flux is

$$\dot{n} = -D\frac{dc}{dr}\bigg|_{r=\frac{d}{2}} = -(c_b - c_S)\frac{2D}{d}.$$

Since by definition

$$\dot{n} = -k(c_b - c_S),$$

one obtains the constant terms in eqn (7.53).

For $0.25 < \mathrm{Sc} < 100$ and $1 < \mathrm{Re} \leqslant 400$ numerical solutions available for mass transfer to rigid spheres can be correlated by the expression:[11]

$$\mathrm{Sh} = 1 + \mathrm{Re}^{0.41}\left[\mathrm{Sc} + \frac{1}{\mathrm{Re}}\right]^{\frac{1}{3}}. \tag{7.55}$$

For high $\mathrm{Sc} = \nu/D$, $\mathrm{Sc} > 200$ and $100 < \mathrm{Re} < 2000$ we have

$$\mathrm{Sh} = 1 + 0.724\,\mathrm{Re}^{0.48}\,\mathrm{Sc}^{\frac{1}{3}}. \tag{7.56}$$

In extractive and refining metallurgy we usually want to obtain a high value for the product of the mass transfer coefficient, k, and the contact area, A. As shown in Section 4.5, A for bubbles is inversely proportional to the diameter d and proportional to the residence time τ_b. Thus

$$A \propto \frac{1}{d}\tau_b \propto \frac{1}{dv}. \tag{7.57}$$

Then the product kA is proportional to $d^{-\frac{3}{2}}v^{-\frac{1}{2}}$ for large bubbles (eqn (7.37)) and to $d^{-\frac{5}{3}}v^{-\frac{2}{3}}$ for small bubbles (eqn (7.52)). For small diameters, velocities v increase with diameter. In Fig. 4.17, this was illustrated for bubbles. Thus kA increases with decreasing size of the dispersed phase. Later we will discover, however, that there are practical limits to how small the diameters of the bubbles or droplets may become.

7.6 VELOCITIES OF BUBBLES, DROPLETS, OR PARTICLES AND THE CORRESPONDING MASS TRANSFER COEFFICIENTS

It is apparent from the previous section and Chapter 4 that we strive to achieve large contact areas in refining processes. This means that bubbles, droplets, or particles as the case may be should be as small as possible. Sometimes in gas-purging processes a fraction of the gas is in the form of large bubbles. The contact area associated with this ineffective fraction is relatively small, and it is not very important to describe this mass transfer accurately. For these reasons, in the following attention is mainly given to obtaining velocities relative to the melt of a dispersed phase, with relatively small diameters. The friction factor ϕ was introduced in Section 5.8 for solid particles. For bubbles and droplets

$$\phi = \frac{4}{\mathrm{Re}}\left(\frac{2 + 3\kappa}{1 + \kappa}\right) \tag{7.58}$$

is valid in the creeping flow region $\mathrm{Re} = vd/\nu < 4$. The factor that contain κ takes the drag reduction into account due to circulation of the fluid in the droplet. For bubbles where $\kappa = 0$, $\phi = 8/\mathrm{Re}$. Equation (7.58) also applies to solids for $\mathrm{Re} < 1$. For solids the viscosity ratio $\kappa = \mu_p/\mu = \infty$ so that eqn (5.45) in Section 5.8 is recovered. For $4 < \mathrm{Re} < 100$ the following correlation is suggested[11] for bubbles and droplets:

$$\phi = \frac{1.525(783\kappa^2 + 2142\kappa + 1080)\mathrm{Re}^{-0.74}}{(60 + 29\kappa)(4 + 3\kappa)}. \tag{7.59}$$

For a solid ($\kappa = \infty$) this equation would reduce to

$$\phi = 13.725\,\mathrm{Re}^{-0.74}. \tag{7.60}$$

It is interesting to compare this with eqn (5.46). At Re = 100 and Re = 4, eqn (7.59) gives $\phi = 0.455$ and $\phi = 4.92$, respectively, while eqn (5.46) in Chapter 5 gives $\phi = 0.555$ and $\phi = 4.17$. Thus, eqn (7.59) can not also be used for solids.

Above Re = 100, bubbles in water tend to oscillate and this strongly affects the drag. Oscillation depends on the interplay between gravitational, viscous, surface tension, and inertial forces. Two dimensionless groups have proved to be useful to describe the behaviour of bubbles and droplets. These are the Morton and Eötvös numbers defined respectively by

$$M = \Delta\rho\rho^2 g\frac{\nu^4}{\sigma^3} \tag{7.61}$$

and

$$\mathrm{Eo} = \frac{\Delta\rho g d^2}{\sigma}.$$

For bubbles in Al

$$M = \frac{2350^3 \times 9.81 \times (0.4 \times 10^{-6})^4}{0.9^3} = 4.5 \times 10^{-15}.$$

For bubbles of intermediate size (for instance $d = 10^{-2}$ m)

$$\mathrm{Eo} = \frac{2350 \times 9.81 \times 10^{-4}}{0.9} = 2.6.$$

Now if $M < 10^{-3}$ and Eo < 40, then for contaminated liquids[11]

$$\mathrm{Re}\,M^{0.149} + 0.857 = 3.42\,H^{0.441} \tag{7.62}$$

where

$$H = \frac{4}{3}\,\mathrm{Eo}\,M^{-0.149}\,(\mu/\mu_w)^{-0.14} \qquad \text{if } H > 59.3. \tag{7.63}$$

In practice we usually find that bubbles are so large that H > 59.3. In eqn (7.63) μ_w is the viscosity of $H_2O = 0.9 \times 10^{-3}$ Ns/m^2. In the example

$$H = \tfrac{4}{3} \times 2.6 \times (4.5 \times 10^{-15})^{-0.149}\left(\frac{0.9}{2.35 \times 0.4}\right)^{0.14} = 473.$$

Then from eqn (7.62) we obtain

$$\mathrm{Re} = 6984$$

$$v = \frac{6984 \times 0.4 \times 10^{-6}}{10^{-2}} = 0.28\,\text{m/s}.$$

Equations (7.62) and (7.63) are employed to obtain the bubble velocities sketched in Fig. 4.17.

We use Al_2O_3 inclusions in steel as an example of how to calculate the mass transfer coefficient at small Re. Using eqn (5.44), v is found to be

$$v = \left(\frac{\Delta\rho g 4d^3}{\phi\rho 6}\right)^{\frac{1}{2}} \tag{7.64}$$

where $\Delta\rho = \rho - \rho_p$. Now for small Re numbers, Re < 4, eqns (7.52) and (7.58) give

$$\text{Sh} = 1 + \left\{1 + \frac{\Delta\rho g d^3(1+\kappa)}{\rho\nu 6(2+3\kappa)D}\right\}^{\frac{1}{3}}.$$

The inclusion diameter is $d = 100 \times 10^{-6}\,\text{m}$, $\Delta\rho = 7030 - 3200 = 3830\,\text{kg/m}^3$, $\nu = 0.8 \times 10^{-6}\,\text{m}^2/\text{s}$, and $D = 0.12 \times 10^{-7}\,\text{m}^2/\text{s}$.[18] (Compare also eqn (2.132).) Thus

$$k = \frac{0.12 \times 10^{-7}}{100 \times 10^{-6}}\left(1 + \left(1 + \frac{3830 \times 9.81 \times (10^{-4})^3}{7030 \times 0.8 \times 10^{-6} \times 18 \times 0.12 \times 10^{-7}}\right)^{\frac{1}{3}}\right)$$

$$= 5 \times 10^{-4}\text{m/s}.$$

Finally we calculate the mass transfer coefficient at large Re. As an example sodium transfer to argon bubbles with diameter $d = 10^{-2}\,\text{m}$ in Al is treated. Then from eqn (2.132) at $T = 923\,\text{K}$

$$D = \frac{1.38 \times 10^{-23} \times 923}{2\pi \times 0.4 \times 10^{-6} \times 2350 \times 2 \times 1.86 \times 10^{-10}}\left(\frac{27+23}{2 \times 23}\right)^{\frac{1}{2}}$$

so $D = 6 \times 10^{-9}\,\text{m}^2/\text{s}$. The velocity was found to be $v = 0.28\,\text{m/s}$. Then eqn (7.48) gives

$$k = \left(\frac{4 \times 6 \times 10^{-9} \times 0.28}{\pi \times 0.01}\right)^{\frac{1}{2}}\left(1 - \frac{2.89}{6984^{\frac{1}{2}}}\right)^{\frac{1}{2}} = 5 \times 10^{-4}\,\text{m/s}.$$

7.7 DIMENSIONS OF PARTICLES, BUBBLES, OR DROPLETS DISPERSED IN MOLTEN METAL

7.7.1 Introduction

In the previous sections of this chapter and in Chapter 4, the importance of attaining large contact areas A has been emphasized. Large contact areas are necessary in order to speed up the kinetics and to achieve a high productivity and high yields. In Chapter 4 we saw, taking bubbles as an example, that the contact area is inversely proportional to the dimensions of the dispersed phase and proportional to the residence times. The residence time depends on the size of the particle, droplet, or bubble, the point of entry into the melt, and convection and turbulence in the bath.

For solid particles we might expect that the particle size is known from a screening analysis taken before injection into the melt. However, this is not necessarily so. In particular, the smaller size fractions may not be able to penetrate the bubble melt interface, as illustrated in Fig. 4.11. This is discussed in the following where mainly the large size fractions enter and these give smaller surface areas relative to their mass.

For bubbles and droplets the size must be determined on the basis of our knowledge of gas–liquid and liquid–liquid fluid dynamics and of the conditions in our metallurgical reactor. As we increase the amounts of gas blown into the reactor and increase the mixing and stirring of the slag and melt phases, the modelling task becomes progressively more difficult. The explanation is that research in this field has, reasonably enough, relied on the gradual accumulation of knowledge based first on single bubbles using low gas flow rates, then a train of bubbles, and finally the transition between bubbling and jetting. These are phenomena where the melt is the continuous phase and bubbles are the dispersed phase. Some process developments in the industry, however, seem to have outpaced research. For instance, decarburization reactors for stainless steel (AOD) and for low-alloy steel (LD and BOF) use such large amounts of oxygen and argon that it might be possible to describe them as units where gas is the continuous phase and metal and slag droplets are the dispersed phases. There seems to be very little research published in this area and this chapter will not deal with such systems. Fortunately, however, even without this knowledge concerning contact areas, useful process models can be developed as demonstrated in Chapter 6.

7.7.2 Penetration of solid particles into a melt

The problem of penetration through a gas–melt interface by particles has attracted a number of workers.[19] Here their work will not be described in detail; only a few major points will be outlined. Finally, some remarks are made based on the practical industrial situation.

Initially, theoretical and experimental work studied the penetration of a single particle through a planar gas–liquid interface. It was found that the particle velocity must exceed a certain critical velocity in order to penetrate the interface. This velocity is given in dimensionless form as a Weber number:

$$\mathrm{We} = \frac{\rho u^2 d_p}{2\sigma}. \tag{7.65}$$

Another dimensionless group is the particle–melt density ratio:

$$\rho^* = \frac{\rho_p}{\rho}. \tag{7.66}$$

ρ^* is the ratio between the density of the particle and the melt. We ρ^* gives a ratio between the kinetic energy of the particle and the surface tension energy. Furthermore, there is a group that characterizes the 'wetting' between a particle and the melt:

$$-\cos\theta = \frac{\sigma_{ls} - \sigma_{gs}}{\sigma_{gl}}. \tag{7.67}$$

Here σ_{ls} is the solid–liquid and σ_{gs} is the solid–gas interfacial tension, and σ_{gl} is the surface tension. θ is called the contact angle. It is the angle between the liquid-gas and liquid-solid interfaces. If there is complete wetting between the particle and melt, $\theta = 0$. The models give the critical We as a function of ρ^* and $\cos\theta$. If the particles wet the melt ($\theta = 0$) and $\rho^* \geqslant 1$, the critical We number is zero, that is, the particles are immersed in the melt whatever the velocity. From the definition of the critical We number it is found for non-wetting particles that if the particle size d_p is reduced, we must compensate by increasing the particle velocity, u. The critical We number is found to be around 100 for $\rho^* = 0.5$. The various equations show that the critical We number decreases with improved wetting, that is, with decreased θ and with increased density of the particles relative to the melt, that is, increased ρ^*.

Numerical work[20] on the penetration of single particles has taken the viscosity of the melt into account by introducing the dimensionless group:

$$M = \frac{\rho^3 g \nu^4}{\sigma^3}. \tag{7.68}$$

For penetration into a slag, the critical We number increases by several orders of magnitude.

Studies on the injection of particles using a carrier gas show that for a given set of operating conditions, some particles penetrate and some do not. The fraction that penetrates increases with We, ρ, $180° - \theta$, and $1/M$ and depends on the mass flow ratio between particles and carrier gas. Furthermore, the fraction depends on velocities and flow patterns in the melt. For instance, during injection, particles may accumulate or 'pile up' near the lance tip and finally choke up the lance.[21] Perhaps the ratio between the mass flow of particles and mass flow of carrier gas

$$r = \frac{\dot{M}_p}{\dot{M}_g} \tag{7.69}$$

is the most important factor in determining the fraction that penetrates. Model studies in the laboratory[20] indicate that if $r > 1$ then 80–90 per cent of the particles penetrate (for We > 200).

It should be pointed out that injection in industry is usually carried out for

$1 < r < 100$ and We > 1000 (see problem 7.3). This would indicate that recovery of particles is close to 100 per cent. However, even losses less than 1 per cent can be enonomically important. Furthermore, particles or dust in the exist gases may represent a serious environmental problem.

7.7.3 Size of bubbles and droplets in melts

In high-temperature melts, such as steel and copper, gases are fed into the melt only through lances, tuyeres or porous plugs. Due to the agressive environment impellers can not usually be employed. For these high-temperature processes convection and stirring is caused only by the momentum and buoyancy effects of the gas. The bubble size is then determined mainly by the gas flow rate and the geometry of the tuyeres or nozzles.

In the following some remarks are made concerning our state of knowledge regarding bubble size in high-temperature melts and in melts at lower temperatures.

In low-temperature melts, such as aluminium and magnesium, small bubbles can be produced using a variety of impeller systems and nozzles. In these, shear forces probably play the dominating role in fixing the bubble size. To some extent such mechanisms have been treated in the chemical engineering literature.[22] There are, however, problems in the transfer of the results to molten metals due to the high surface tensions encountered for metals and thermal expansion of the gas.

Bubble sizes from nozzles at low gas flow rates have been the object of extensive study in the literature.[5] In industry very high gas flow rates are commonly used, but such conditions are not easy to study in the laboratory for several reasons. One difficulty is the large tuyere size and gas volumes involved. Another problem is the unsteady and statistical nature of the bubbling or jetting that occurs at high gas flow rates. Finally, bubble size is different in various parts of the melt. For bubbling we may speak of at least three zones:[23]

(1) the primary (or large bubble) zone;

(2) the dispersed bubble zone;

(3) the surface or disengagement zone where slopping or splashing occurs.

The primary or large bubble zone is of special importance in processes where the melt viscosity is high. Examples of such processes are slag reduction or slag fuming units for the recovery of zinc and tin.[24] Often there is a low liquid seal depth and shallow tuyere submergence. As a consequence, the large primary bubbles formed at the tuyere tip may not have sufficient rise time to break into bubbles.

For highly viscous liquids such as slags it is found that viscosity increases bubble size. In addition to viscosity inertial and buoyancy effects must be taken into account at high gas flow rates. Bubbles are elongated in the axial direction of the nozzle axis due to the stagnation pressure of the gas jet inside the bubble. It has been possible to calculate bubble volumes based on a number of assumptions,[25] such as:

(1) constant flow of gas;

(2) no convective flow in the liquid;

(3) no effect of the preceding bubble;

(4) the shape of the bubble at an orifice is that of a prolate ellipsoid (where the eccentricity changes during growth of the bubble).

For low-viscosity liquids such as water and molten metals it is found theoretically and experimentally for a nozzle that points upwards:

$$N_v = 0.5\, N_I^{-\frac{1}{4}}. \tag{7.70}$$

N_v is a dimensionless volume:

$$N_v = \frac{V g^{\frac{3}{5}}}{\dot{Q}_g^{\frac{6}{5}}} \tag{7.71}$$

and N_I is the so-called injection number[26] that takes momentum of the gas into account:

$$N_I = \frac{\dot{Q}_g^{\frac{4}{5}}\, \rho_g/\rho}{g^{\frac{2}{5}}\, r_0^2} \tag{7.72}$$

where r_0 is the nozzle radius. Equation (7.70) is valid for the range $0.1 \leqslant N_I \leqslant 1$. The group N_v is recognized as that given by the well-known Davidson and Schüler equation[27] derived for gas flow rates so low that the bubble can be assumed to be spherical. Then

$$N_v = 1.378. \tag{7.73}$$

Davidson and Schüler considered the spherical bubble to detach when it is in tangential contact with the nozzle tip. The same criterion was employed for the elliptical bubble in the derivation of eqn (7.70). The meaning of N_I is perhaps best understood by taking the ratio between N_I and N_v:

$$\frac{N_I}{N_v \pi} = \frac{\rho_g \dot{Q}_g^2}{\rho \pi r_0^2 g V} = \rho_g \left(\frac{\dot{Q}_g}{\pi r_0^2}\right)^2 \frac{\pi r_0^2}{\rho g V}.$$

This is equal to the ratio between the momentum of the gas and the buoyancy force of the bubble.

For viscous liquids an additional dimensionless group E_0 must be taken into account.[25]

$$E_0 = 177\nu \left(\frac{\rho_g}{\rho \pi r_0^2 g \dot{Q}_g}\right)^{\frac{1}{3}}. \tag{7.74}$$

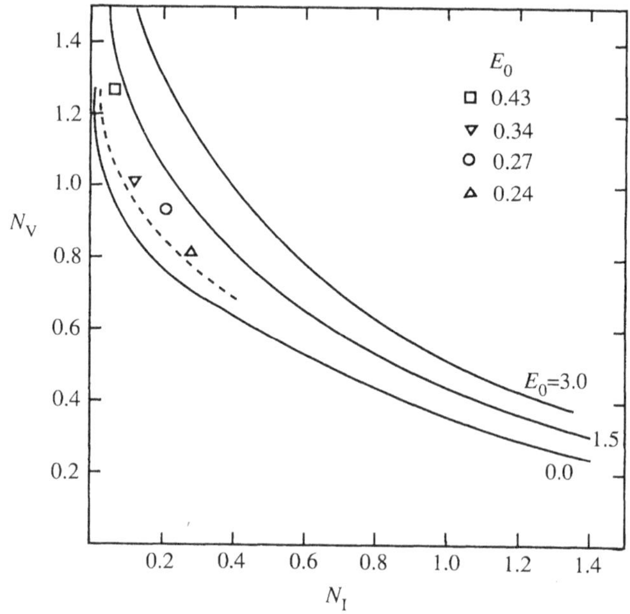

FIG. 7.4 Dimensionless bubble volume N_v as a function of the injection number N_I and a viscosity group E_0.[25]

Figure 7.4 gives N_v as a function of N_I and E_0.

The problem of the break-up and coalescence of bubbles or a dispersed liquid phase is difficult both from the experimental and theoretical points of view. Here only an indication can be given of the possible mechanisms and the resulting relations for bubble or droplet size. Pressure or shear forces can be the cause of break-up. For instance, drops can break up in the turbulent flow along a wall due to shear forces.[11] Similarly, shear forces should break up and disperse the gas forced through a rotating impeller. Very roughly a bubble or a droplet should fragment if the dynamic forces are greater than the surface tension forces holding it together:

$$\rho v^2 \pi \frac{d^2}{4} \sim \sigma \pi d$$

$$d \sim \frac{4\sigma}{\rho v^2} \tag{7.75}$$

where v is an eddy velocity. Equation (7.75) is equivalent to stating that the bubbles break up at a Weber number:

$$\mathrm{We} = \frac{\rho v^2 d}{\sigma} = 4.$$

where v_g is the mean gas velocity in the pipe and ϕ is the friction factor described in Chapter 5 here applied to the gas flow. From eqn (5.108) we have

$$\phi = 0.04\,\mathrm{Re}_g^{-\frac{1}{4}} \tag{7.83}$$

where $\mathrm{Re}_g = v_g 2R/\nu_g$ now refers to the gas phase and the velocity, v_g, is the difference between the gas and melt surface velocity.

The Schmidt number for gases is not exactly 1 but can vary between 0.5 and 2.5. Then, according to the Chilton–Colburn correlation[4] eqn (7.82) is replaced by

$$k_g \mathrm{Sc}_g^{\frac{2}{3}}/v_g = \phi. \tag{7.84}$$

Equation (7.83) inserted into eqn (7.84) then gives

$$k_g = 0.04\,\mathrm{Sc}_g^{-\frac{2}{3}}\,v_g\,\mathrm{Re}_g^{-\frac{1}{4}} \tag{7.85}$$

or

$$\frac{k_g 2R}{D_g} = 0.04\,\mathrm{Re}_g^{\frac{3}{4}}\,\mathrm{Sc}_g^{\frac{1}{3}}. \tag{7.86}$$

For engineering calculations with $3000 < \mathrm{Re}_g < 40000$ the following equation[29] is often applied for wetted-wall columns:

$$\frac{k_g 2R}{D_g} = 0.033\,\mathrm{Re}_g^{0.77}\,\mathrm{Sc}_g^{0.33}. \tag{7.87}$$

For the driving force the logarithmic mean at the two ends of the wetted-wall column is used.

For binary gas mixtures at low pressures, D_g is inversely proportional to the total pressure. D_g increases with increasing temperature, and is almost independent of composition for a given gas pair.[4] As an example, for Ar–O_2 at 2000 K and 0.1 atm pressure, $D_g = 5.17 \times 10^{-3}\,\mathrm{m^2/s}$. Then, if $\mathrm{Sc}_g = 0.5$, $\mathrm{Re}_g = 10^4$, and $2R = 0.1$, eqn (7.87) gives $k_g = 1.6\,\mathrm{m/s}$.

From eqns (7.85) and (7.17) we can now determine the relative resistance λ:

$$\lambda = \frac{0.015}{0.04}\left(\frac{\mathrm{Sc}_g}{\mathrm{Sc}}\right)^{\frac{2}{3}} \frac{v_m RT}{v_g K_j \gamma_j}\left(\frac{\mathrm{Re}_g}{\mathrm{Re}}\right)^{\frac{1}{4}} \mathrm{Re}^{\frac{1}{8}}. \tag{7.88}$$

Equation (7.88) is derived for pipe flow but should also be useful in the turbulent regime for other geometries, except that the factor $(\mathrm{Re}_g/\mathrm{Re})^{\frac{1}{4}}$ may have a slightly different exponent. To obtain the order of magnitude of the various terms we note that $\mathrm{Sc}_g \sim 1$, $\nu_g \sim 10^{-5}\,\mathrm{m^2/s}$, $\nu \sim 10^{-6}\,\mathrm{m^2/s}$, $\mathrm{Sc} \sim 100$, and $RT \sim 10^7$ J/kmol. $\mathrm{Re}^{\frac{1}{8}}$ is not much greater than 1. Then if the gas and melt velocities are roughly equal, the gas/liquid resistance ratio becomes

$$\lambda \sim \frac{10^5}{K_j \gamma_j}. \tag{7.89}$$

Notice that $K_j \gamma_j$ is a ratio between pressure given in units of Pa and concentration in $kmol/m^3$ (cf. Section 4.11).

Only if the vapour pressure is so high that $K_j \gamma_j$ is greater than 10^5, can resistance in the gas phase be neglected at one atmosphere pressure. We will see that this explains why it is a simple matter to remove Na from molten aluminium, while inert gas purging of Li and Ca proceeds much more slowly.[30,31] From Table 2.1 and Appendix 1 one obtains at 1000 K: $P_{Na}/[\% \, Na] = 1.06$ (with 1 bar and 1% as reference states). This gives $K_j \gamma_j = 1.06 \times 10^5 \times 23 \times 100/2350 = 1.04 \times 10^5$. The factor 10^5 converts pressure from bar to Pa. The denominator 2350 kg/m^3 is the density of molten Al and 23 kg/kmol is the atomic weight of Na. Thus the factor $2350/(23 \times 100)$ converts concentration from mass% to $kmol/m^3$.

For Li and Ca it is found that $K_j \gamma_j$ is far less than 10^5 so that resistance in the gas film dominates when we purge with inert gas. However, this resistance can be eliminated by using a gas such as Cl_2 that reacts with Li or Ca.

In Section 4.11 evacuation of Cu from molten iron was treated. Cu has a low vapour pressure and $K_{Cu} \gamma_{Cu} = 27.4$. For this low value, resistance in the gas dominates unless the pressure is very low. However, if a very high vacuum can be attained so that there are few or no collisions in the gas phase, resistance, k_g, in the gas is completely eliminated. To attain this the mean free path of the atoms in the gas must be of the same order of magnitude as the dimensions of the vacuum chamber. This requires that the total pressure be reduced to around 10^{-6} bar or 10^{-1} Pa.[32] In this case if the atoms collide with cold walls, they are removed by condensation.

7.8.2 Gas-side and interfacial resistance for diatomic gases

For monoatomic gases we found for low solubilities that the melt boundary layer determined the mass transfer resistance. A similar result is obtained for diatomic gases except that now the calculations become slightly more involved. Initially, we disregard any interfacial resistance and write in analogy with eqns (4.76) and (4.77) for the melt boundary layer:

$$\dot{n}_j = k(c_j - c_{ji}) \tag{7.90}$$

and for the boundary layer in the gas:

$$\dot{n}_j = \frac{2k_g}{RT}(p_{ji} - p_{bj}) \tag{7.91}$$

where now the equilibrium at the interface is given by

$$c_{ji}^2 K_j^2 \gamma_j^2 = p_{ji}. \tag{7.92}$$

The eqns (7.90) and (7.91) are altered to the form

$$c_{ji}^2 = \left(c_j - \frac{\dot{n}_j}{k}\right)^2$$

and

$$p_{ji} = p_{bj} + \frac{\dot{n}_j RT}{2k_g}.$$

Then the interfacial concentration and pressure c_{ji} and p_{ji} are eliminated from the two equations by employing eqn (7.92). One obtains from the resulting second-degree equation:

$$\frac{\dot{n}_j}{kc_j} = 1 + \frac{\lambda}{2c_j} - \left\{\left(\frac{\lambda}{2c_j}\right)^2 + \frac{p_{bj}}{K_j^2\gamma_j^2c_j^2} + \frac{\lambda}{c_j}\right\}^{\frac{1}{2}} \tag{7.93}$$

where

$$\frac{\lambda}{c_j} = \frac{kRT}{2k_g K_j^2 \gamma_j^2 c_j}. \tag{7.94}$$

If

$$\frac{\lambda}{c_j} = \frac{kRT}{2k_g K_j^2 \gamma_j^2 c_j} \ll 1 \tag{7.95}$$

eqn (7.93) reduces to

$$\dot{n}_j = k(c_j - c_{je}) \tag{7.96}$$

where c_{je} is the hypothetical concentration in the melt in equilibrium with the bulk concentration in the gas, $c_{je} = p_{bj}^{\frac{1}{2}}/(K_j\gamma_j)$. Equation (7.96) may be compared with the left-hand side of eqn (4.61). There hydrogen was taken as an example, $c_H = \rho[\%\,H]/(100\,m_H)$. If inequality (7.95) is fulfilled the total mass transfer coefficient k_t is equal to the mass transfer coefficient for the melt boundary layer, k. By comparison with eqns (7.80) and (7.89) we see that (7.95) is fulfilled if

$$\frac{10^5}{2K_j^2\gamma_j^2c_j} \ll 1.$$

It is realized that the inequality is not satisfied at sufficiently low values of c_j. At least in theory if not in practice, for sufficiently low values of c_j, the gas-side resistance (see eqn (7.99)) dominates for any two-atomic gas. The reason is that the partial pressure, p_{ji}, becomes very low because it is proportional to the second power of c_{ji}. Thus, p_{ji} becomes a bottleneck for the mass transfer. It is possible to reduce gas-side resistance by the use of a vacuum.

For hydrogen in pure aluminium at 1000 K one obtains from Table 2.1:

$p_{H_2}^{\frac{1}{2}}/[\% \text{ H}] = 9.43 \times 10^3$ (with 1 bar and 1 mass% as the reference states). Then

$$K_H^2 \gamma_H^2 c_H = \frac{(9.43 \times 10^3)^2 \times 10^5 \times 100}{2350} [\% \text{ H}] = 3.8 \times 10^{11} [\% \text{ H}]$$

(with 1 N/m^2 and 1 $kmol/m^3$ as the reference states). Resistance to transfer is greater on the metal side than on the gas side for practical values of [% H].

For nitrogen in iron, Table 2.2 gives, at 1800 K: $p_{N_2}^{\frac{1}{2}}/[\% \text{ N}] = 22.5$. Then

$$K_N^2 \gamma_N^2 c_N = \frac{22.5^2 \times 10^5 \times 100 \times 14}{7030} [\% \text{ N}] = 1.01 \times 10^7 [\% \text{ N}].$$

Thus, also, for nitrogen in steel, resistance in the melt boundary layer should dominate for [% N] > 0.01 = 100 p.p.m.

However, for nitrogen in steel a very strong effect of surface-active elements such as oxygen and sulphur is described in the literature[13, 14] where it is often assumed that a chemical reaction at the interface partially blocked by surface-active elements

$$N_2 = 2\underline{N} \tag{7.97}$$

slows down the mass transfer at higher oxygen and sulphur levels. Then eqn (7.90) is replaced by two equations:

$$\dot{n}_N = k(c_N - c_{N_i}) \tag{7.90}$$

where c_{N_i} is the concentration in the melt at the interface. The chemical reaction is presented as 'second order':

$$\dot{n}_N = k_c (c_{N_i}^2 - c_{Ne}^2) \tag{7.98}$$

where k_c is the associated chemical rate constant. k_c must be determined empirically; its value drops strongly with increasing activity of oxygen and sulphur.[12, 15] c_{Ne} is in equilibrium with p_{ji} so that $K_j^2 c_{Ne}^2 = p_{ji}$. Furthermore, there may be an interfacial resistance due to oxide films on surfaces exposed to oxidation (but not on pure argon bubbles). This complex behaviour of nitrogen is not pursued further here.

Until now we have discussed systems where the solubility of the diatomic gases is low. However, sometimes solubility is high. For instance, for O_2 dissolved in pure iron:

$$\tfrac{1}{2}O_2 = \underline{O} \text{ at } 1800 \text{ K}$$

we find $p_{O_2}^{\frac{1}{2}}/[\% \text{ O}] = 2.82 \times 10^{-4}$. This means that oxygen has a very high solubility in pure iron:

$$K_O^2 \gamma_O^2 c_O = \frac{(2.82 \times 10^{-4})^2 \times 10^5 \times 100 \times 16}{7030} [\% \text{ O}]$$

$$= 1.8 \times 10^{-3} [\% \text{ O}].$$

5. Hydrogen molecules diffuse back out of the boundary layer.
6. Hydrogen molecules dissociate and form atomic hydrogen on the surface.
7. Hydrogen atoms diffuse through the metal boundary layer.

The usual assumptions are made that the chemical reactions, dissociation, absorption, and desorption are rapid. This leaves steps 1, 5, and 7. In Section 7.8.2. we saw that step 7 is slow. However, steps 1 and 5 play an important role in determining p_{H_2} at the surface. Step 1 is described by

$$\dot{n}_{H_2O} = \frac{k_{H_2O}}{RT}\left(P_{H_2O} - p_{H_2Oi}\right). \tag{7.102}$$

Similarly, for step 5 we have for the flux of hydrogen molecules:

$$\dot{n}_{H_2} = \frac{k_{H_2}}{RT}\left(p_{H_2i} - p_{H_2}\right). \tag{7.103}$$

Here p_{H_2} is the partial pressure of hydrogen in the atmosphere above the melt. Unless there is a source of H_2 such as an oil or gas burner heating the melt, we can usually set $p_{H_2} = 0$. p_{H_2Oi} and p_{H_2i} are the partial pressures at the interface. Their ratio is determined by the equilibrium given by eqn (7.101). Finally step 7 gives

$$\dot{n}_H = k_H\left(c_{Hi} - c_H\right) \tag{7.104}$$

where

$$K_H^2 \gamma_H^2 c_{Hi}^2 = p_{H_2i}. \tag{7.105}$$

The mole balance for hydrogen for the reactions that take place at the surface is (noting that H_2O and H_2 both contain two hydrogen atoms)

$$2\dot{n}_{H_2O} = 2\dot{n}_{H_2} + \dot{n}_H. \tag{7.106}$$

By subtraction of eqn (7.103) from eqn (7.102) one obtains from eqn (7.106)

$$\frac{RT\dot{n}_H}{2} = k_{H_2O}p_{H_2O} - k_{H_2O}p_{H_2Oi} - k_{H_2}p_{H_2i}. \tag{7.107}$$

p_{H_2Oi} is now eliminated using the equilibrium constant for eqn (7.101):

$$\frac{p_{H_2Oi}}{p_{H_2i}} = K. \tag{7.108}$$

This gives

$$\frac{RT}{2}\dot{n}_H = k_{H_2O}p_{H_2O} - \left(k_{H_2O}K + k_{H_2}\right)p_{H_2i}. \tag{7.109}$$

Then c_{Hi} and p_{H_2i} are eliminated from eqns (7.104) and (7.109) by using eqn (7.105) in a similar way to the procedure employed for eqns (7.90) to (7.92). Solution of the second-degree equation leads to

$$\frac{\dot{n}_H}{k_H c_H} = -\left(1 + \frac{\lambda}{2c_H}\right) + \left\{\left(\frac{\lambda}{2c_H}\right)^2 + \frac{k_{H_2O} p_{H_2O}}{(k_{H_2O} K + k_{H_2}) K_H^2 \gamma_H^2 c_H^2} + \frac{\lambda}{c_H}\right\}^{\frac{1}{2}} \tag{7.110}$$

with

$$\frac{\lambda}{c_H} = \frac{k_H RT}{2(k_{H_2O} K + k_{H_2}) K_H^2 \gamma_H^2 c_H}. \tag{7.111}$$

Equation (7.110) may be compared with eqn (7.93). The signs are different since the flow is now into the melt. The equilibrium constant K for eqn (7.101) is very small so that the term $k_{H_2O}K$ may be neglected. As seen in Section 7.8.2, λ/c_H is very small. Then eqn (7.110) reduces to

$$\dot{n}_H = k_H \left\{ \frac{\left(\frac{k_{H_2O}}{k_{H_2}} p_{H_2O}\right)^{\frac{1}{2}}}{K_H \gamma_H} - c_H \right\}. \tag{7.112}$$

The result is the same as if we had assumed a partial pressure of hydrogen at the interface given by

$$p_{H_2} = \frac{k_{H_2O}}{k_{H_2}} p_{H_2O}. \tag{7.113}$$

A simple physical explanation is that water vapour is reduced to hydrogen at the surface. The factor k_{H_2O}/k_{H_2} takes into account diffusion of H_2 back out through the gas film.

p_{H_2O} may be obtained from measurements of the relative humidity and temperature. The water vapour saturation pressure $p_{H_2O}^{sat}(T)$ is tabulated, and p_{H_2O} is equal to this pressure multiplied by the relative humidity. Pick-up of hydrogen from water vapour is discussed further in problem 7.6.

7.11 FLUID DYNAMICS

7.11.1 Introduction

In this book we frequently refer to conditions in the 'bulk liquid'. It is often assumed that we have complete mixing so that concentrations and temperatures are the same throughout the bulk melt. In Chapter 5 and in Section 7.4 the words 'eddy velocity', 'eddy viscosity', 'eddy diffusivity', and 'eddy length' are mentioned repeatedly. No attempt will be made here to justify the use of these concepts in any rigorous manner. Rather, the aim is to give the reader some background of, and understanding for, the underlying ideas and theories.

To start out it must be emphasized that in the majority of cases our problems are turbulent, that is, flow varies in an irregular and random manner with time and position. The mathematical complexity of turbulence precludes any exact analysis at this time. We have no other recourse than the idea of Osborne Reynolds to work with the time-averaged equations of motion. Velocities, pressures, and temperatures have a mean and a fluctuating component:

$$v = \bar{v} + v'$$
$$p = \bar{p} + p' \tag{7.114}$$
$$T = \bar{T} + T'$$

The time average of the fluctuating component v', p', or T' is zero. The following rules of averaging apply for any two turbulent quantities f and g:[34]

$$\overline{f'} = 0, \qquad \bar{\bar{f}} = \bar{f}, \qquad \overline{f\bar{g}} = \bar{f}\bar{g}$$
$$\overline{f'\bar{g}} = 0, \qquad \overline{f+g} = \bar{f} + \bar{g}$$
$$\overline{\frac{\partial f}{\partial x}} = \frac{\partial \bar{f}}{\partial x}, \qquad \overline{fg} = \bar{f}\bar{g} + \overline{f'g'}.$$

For incompressible flow the continuity equation becomes

$$\frac{\partial \bar{v}_x}{\partial x} + \frac{\partial \bar{v}_x}{\partial y} + \frac{\partial \bar{v}_z}{\partial z} = 0 \tag{7.115}$$

$$\frac{\partial v_x'}{\partial x} + \frac{\partial v_x'}{\partial y} + \frac{\partial v_z'}{\partial z} = 0. \tag{7.116}$$

(In Chapter 4 we have discarded the dash for the fluctuating velocity in order to simplify the notation.) With these averaging rules the Navier–Stokes equation, for instance, for the x-component of the velocity, becomes

$$\rho \frac{D\bar{v}_x}{Dt} + \rho\left(\frac{\partial \overline{v_x'^2}}{\partial x} + \frac{\partial \overline{v_x'v_y'}}{\partial y} + \frac{\partial \overline{v_x'v_z'}}{\partial z}\right) = -\frac{\partial \bar{p}}{\partial x} + \mu \nabla^2 \bar{v}_x. \tag{7.117}$$

$\rho\overline{v_x'^2}$, $\rho\overline{v_x'v_y'}$, and $\rho\overline{v_x'v_z'}$ are called apparent, or Reynolds, stresses. These new terms are never negligible in any turbulent shear flow and are the source of our analytical difficulties, because their analytical forms are not known a priori. In essence, the time-averaging procedure has introduced new variables such as $\overline{v_x'v_y'}$ and $\overline{v_x'v_z'}$ which can be defined only through (unavailable) knowledge of the detailed turbulent structure. The variables are related not only to fluid physical properties but also to local flow conditions: velocity, geometry, surface roughness, and upstream history.

Equations (7.115), (7.116), and (7.117) (in three dimensions) are the basic relations for continuity and the mean momentum. These relations contain many unknown terms and do not give much information about the evolution of the turbulent structure. Other relations might be useful and Prandtl suggested formulating a mean mechanical energy equation (see discussion in[34]). He obtained a relation for the rate of change of what may be called the turbulent kinetic energy per unit mass:

$$k = \tfrac{1}{2}\,(v_x'^2 + v_y'^2 + v_z'^2). \tag{7.118}$$

The terms in the equation are rather complex.[35] In words:

The change in k per unit time
= Turbulent and viscous diffusion of turbulent kinetic energy
+ Production (redistribution) − $\dot{\varepsilon}$. (7.119)

Here $\dot{\varepsilon}$ is the dissipation rate of turbulence energy. $\dot{\varepsilon}$ has units of m^2/s^3.

Equations (7.117) and (7.119) contain unknown turbulence correlation terms. These terms must be modelled with semiempirical assumptions which will now be discussed.

7.11.2 Turbulence modelling assumptions

Perhaps the first move towards a model of turbulence can be attributed to Boussinesq.[36] More than a hundred years ago, he suggested that the effective shear stress, arising from the cross-correlation of fluctuating velocities, could be replaced by the product of the mean velocity gradient and a quantity termed the *turbulent viscosity*. $\tau = \rho\overline{v_z' v_x'}$ is replaced by

$$\tau = -\mu_t \frac{\partial \bar{v}_x}{\partial z}. \tag{7.120}$$

Unlike μ, the molecular viscosity, μ_t, is not a property of the fluid. It is determined by the structure of the turbulence and varies from point to point in the flow.

Pre-eminent among the models which employ algebraic relations for μ_t is Prandtl's proposal:[37, 10]

$$\mu_t = \rho l^2 \left| \frac{\partial \bar{v}_x}{\partial z} \right| \tag{7.121}$$

where l is the Prandtl eddy length. For near two-dimensional boundary layer flows the proposal has served very well. It is simple, is surprisingly accurate, and has a wide range of applicability. For example, the Prandtl eddy length concept has been employed in Section 5.14 concerning inclusion removal to walls. Prandtl took into account that near a wall the velocity gradient $d\bar{v}_x/dz$ is large for the fluid in turbulent flow. To explain the model proposed by Prandtl we exploit the analogy with molecular transport, go back to Fig. 2.16 and replace the molecule with a pulse (lump) of fluid. A pulse of fluid moving a distance l in the direction towards a wall neg. z-direction will be entering a slower moving part of the stream where it will increase locally the velocity in the x-direction. Thus a negative v_z' should correlate with a positive v_x'. The value of v_x' depends on the distance of travel, l, before the lumps lose their identity and are mixed into the fluid at the new position. v_x' is proportional to l and to the mean velocity gradient: $v_x' = l\mathrm{d}\bar{v}_x/\mathrm{d}z$, $v_z' = -l\mathrm{d}\bar{v}_x/\mathrm{d}z$, and $\tau = -\rho l^2|\mathrm{d}v_x/\mathrm{d}z|\mathrm{d}\bar{v}_x/\mathrm{d}z$. l is effectively the mean free path of a pulse of liquid, and is thus a measure of the scale of the turbulence eddies. In Section 5.14 it was also assumed that transfer of mass and momentum takes place by the same mechanisms. For the turbulent diffusion coefficient, $D_\mathrm{E} = v_z' l$. This relation was also used in Section 7.4 concerning a free surface.

One problem with the proposal by Prandtl was to determine the 'length scale' l. It can be specified fairly realistically for boundary layers along solid walls, $l = 0.4\,z$ where z is the distance from the wall. However, it varies in a complicated way in irregular channel sections, and in mixing and recirculating flows. For such flows it seems to be necessary to describe turbulence in terms of quantities such as k and $\dot{\varepsilon}$. Then eqn (7.119) must be included for k. In addition an equation for $\dot{\varepsilon}$ can be derived.[35] This leads to the so-called k–$\dot{\varepsilon}$ model for turbulence. The k–$\dot{\varepsilon}$ model is typical of the multi-equation models widely used in computer calculations of turbulent flow problems.[7,37,38] This also applies to turbulent flow in multiphase systems. The next section offers some comments concerning systems with several phases.

7.11.3 Multiphase flows

In process metallurgy multiphase flow arises when, for example, a liquid bath is stirred by injection of an inert gas. Injection of solid particles into a melt by a carrier gas is another example. Clogging of nozzles, transfer of inclusions to walls and top-slag, and filtration represent multiphase situations. In fact, in the field of refining metallurgy, all problems associated with liquid or solid inclusions or particles and their removal are examples of multiphase flow. In this book, with some exceptions, it is assumed that the particles or bubbles are so far apart that each behaves as a single particle or bubble surrounded by a liquid of infinite extent. Usually, the behaviour in this case is well known and allows us to describe the problem in a satisfactory manner. However, there are a number of cases when a more accurate analysis of two-phase flow would be desirable. For instance, in the description of gas-purging processes we would like to determine the residence time of the bubbles accurately. When we study the removal of inclusions by settling, we wish to characterize the behaviour of the slurry at the bottom of the settling unit. The inclusions in this part of the settling or holding furnace are so close together that they certainly interact. In filtration liquid inclusions may tend to coalesce to give a liquid film that escapes through the filter. This last example is perhaps particularly difficult in that the thermodynamics of surfaces is also involved.

For inclusions we are usually justified in assuming that the dispersed phase

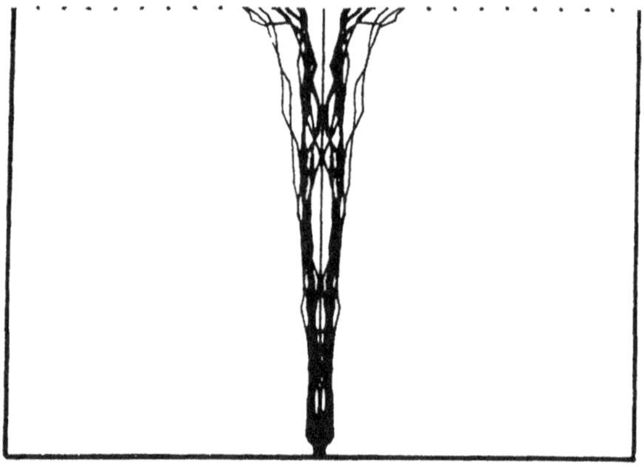

FIG. 7.6 Predicted bubble trajectories for the situation depicted in Fig. 7.7. Bubble diameter is 2.2×10^{-2} m and gas flow rate is 30 Nl/min.

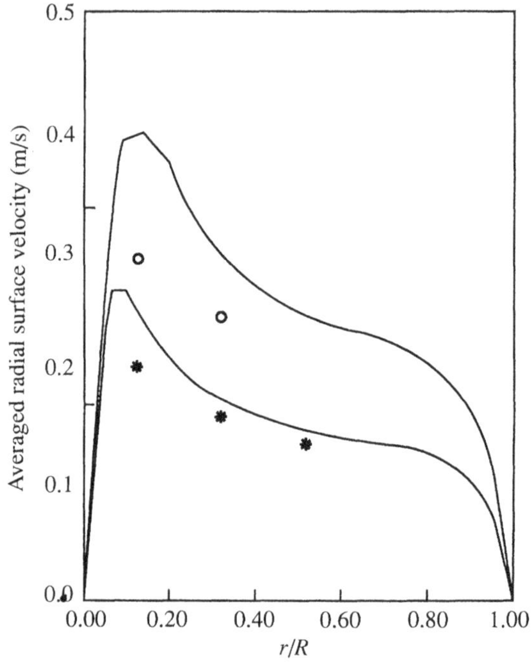

FIG. 7.7 A comparison between experimental surface velocities[39] and predicted velocities in steel. The results for a gas flow rate of 30 ml/min are given by stars (experiments) and a broken line (theory). The open rings and fully drawn line show the experimental and theoretical results, respectively, for 75 Nl/min.

does not affect the flow of the continuous phase. Obviously this is not the case for bubble stirring. Then the interaction between the bubble plume and the melt generates the circulating flow. The momentum acting on the liquid is equal but opposite in sign to the drag experienced by the bubbles. To model the properties of the bubble plume it has been postulated that the fluctuating liquid velocities which a bubble encounters are isotropic and possess a Gaussian probability distribution.[39] A bubble is assumed to interact with an eddy for a time equal to the eddy lifetime or bubble transit time, whichever is smaller.[39] These effects cause the bubbles to spread out and generate the plume as shown in Fig. 7.6. Figure 7.7 gives the velocities obtained from the model[39] and they agree quite well with measurements.[28,41] Such results indicate that 'turbulence models' can quite successfully represent the gross time-averaged aspects of many turbulent flow situations. This applies even to the multiphase case.

7.12 NUMERICAL SOLUTION

In the various chapters of this book we have seen that our problems are described in terms of conservation equations. Examples of such conservation equations are mass balances for impurities such as oxygen and hydrogen in metals. If the process is steady state, these mass balances can lead to simple algebraic equations. If time enters into the problem the mass balances contain time-dependent accumulation terms, and we enter the difficult field of differential equations. To allow us to handle such problems, we often assume that there is complete mixing in the bulk melt. Then the melt is described by one set of time-dependent variables for concentrations and temperature and an exact analytical solution is often possible. In many cases this simplistic approach to our problems is valuable especially from the point of view of understanding the main features of a process. Once the initial obstacles of a pedagogic nature have been surmounted, one may wish to obtain a more accurate description. For instance, perhaps there are zones in the melt that are near-stagnant so that the assumption of complete mixing is not valid. In other zones circulation of the melt may change the residence time of bubbles, inclusions, and so on. Then it becomes imperative to calculate the flow patterns and velocities throughout.

Unhappily for the refining metallurgist and perhaps especially for the engineer studying a stirred reactor, the numerical flow problems often belong to the most difficult class; the flow is two-way, recirculating.[35] Another name for this is that the problem is elliptic. Such problems require considerable computer storage and computer time.

A numerical solution of a flow problem consists of numbers that give the dependent variables such as velocities, concentrations, and temperatures in a set of points (grid points) placed throughout the melt. A numerical method includes the task of providing a set of algebraic equations for these unknown variables and of prescribing an algorithm for solving the equations. In the *control volume* formulation the melt is divided into a number of non-overlapping control volumes such that there is one control volume surrounding each grid point. Each of these control volumes behaves as a small (completely mixed) reactor. The behaviour of this small reactor is as described elsewhere in this book. Quantities such as mass, concentrations, momentum, and energy are exactly conserved for each control volume and of course over the whole calculation domain. This characteristic exists for any number of grid points, even when there is only one grid point corresponding to the assumption, often employed in this book, of complete mixing of the whole melt.

The algebraic equation that relate the various grid points are called *discretization equations* in the literature. Patankar[42] has formulated a number of rules for such equations in a readable and lucid manner.

The discretization equations are algebraic equations and are solved by the standard methods for linear algebraic equations. However, often one encounters non-linear situations even in heat conduction. For instance, the thermal conductivity k may depend on temperature, T, or a source term (chemical reaction term) in a heat balance equation may be a non-linear function of T. Then the coefficients in a discretization equation for T will themselves depend on T. Such situations are handled by iteration. This process, called an algorithm, involves the following steps.

1. Start with a guess or estimate of T at all grid points.
2. From these guessed values of T calculate tentative values of the coefficients in the discretization equation.
3. Solve the nominally linear set of algebraic equations to get new values of T.
4. Return to step 2 using the values from step 3 until further repetitions (called iterations) cease to produce any significant changes in T.

It is a simpler task to solve heat diffusion problems than to calculate the velocity field. The real difficulty here lies in the unknown pressure distribution. The pressure gradient forms a part of the source term for a momentum equation. Once the pressure is known, there is no particular problem in solving the momentum equations. When the correct pressure field is substituted into the momentum equations, the resulting velocity field satisfies the continuity equation. There is no obvious equation for obtaining pressure. However, the pressure field is indirectly, specified via the continuity equation. To obtain the correct pressure the main feature in the algorithm is a pressure correction equation based on the continuity equation. Again the procedure is iterative and goes on until the pressure no longer changes, that is, until a converged solution is obtained.

It should be mentioned that a staggered grid is employed; the grid points are different for velocity and pressure. The important advantage of this is that the pressure difference between two adjacent grid points now becomes the driving force for the velocity component located between these two points.

This very brief outline of numerical flow calculations concludes this chapter on mixing, mass transfer, and numerical models. Without doubt the reader will feel that this final section on fluid flow and numerical calculations was all too meagre and sketchy. Nevertheless, the author hopes the message has got across the numerical fluid flow calculation is a very useful field. Hopefully, the reader will proceed on his own to study the literature such as the books by Patankar[42] and Guthrie[7].

REFERENCES

1. Ueda, T., Furuyama, S., and Nakagawa R. (1981). Model experiment and its analysis on circulating continuous degassing. *Trans. ISIJ*, **21**, 716–25.
2. Seshadri, V. and DeSouza Costa, S. (1986). Cold model studies of R.H. degassing process. *Trans. ISIJ*, **26**, 133–8.
3. Kawabara, T, Umezawa, K., Mori, K., and Watanabe, H. (1988). Investigation of decarburization behaviour in RH-reactor. *Trans. ISIJ*, **28**, 305–14.
4. Bird, R.B., Steward, W.E., and Lightfoot, E.N. (1960). *Transport Phenomena*, Wiley, New York.
5. Brimacombe, J.K., Nikanishi, K., Anagbo, P.E., and Richards, G.G. (1990). Process dynamics: Gas-liquid. The Elliott Symposium, June 10–13, pp. 304–72, MIT, Cambridge, MA.
6. Ogawa, K. and Onoue, T. (1989). Mixing and mass transfer in ladle refining process. *ISIJ International*, **29**, 148–53.
7. Guthrie, R.I.L. (1992). *Engineering in process metallurgy* (rev. edn). Clarendon Press, Oxford.
8. Szekely, S. and Themelis, N.J. (1971). *Rate phenomena in process metallurgy*. Wiley-Interscience, New York.
9. Engh, T.A., Grip, C.-E., Hanson, L., and Worner, H.K. (1971). A one-dimensional model of the WOCRA steel refining launder. *Jernkont. Ann.*, **155**, 554–64.
10. Davies, J.T. (1972). *Turbulence phenomena*. Academic Press, New York.
11. Clift, R., Grace, J.R., and Weber, M.E. (1978). *Bubbles, drops and particles*. Academic Press, New York.
12. Ito, K., Amano, K., and Sakao, H. (1988). Kinetic study on nitrogen absorption and desorption of molten iron. *Trans. ISIJ*, **28**, 41–8.
13. Brückner, R. (1975). Interfacial convection and mass transfer. In *Kinetics of metallurgical processes in steelmaking* (ed. W. Dahl, W.K. Lange and D. Papamentillos), pp. 459–91. Verlag Stahleisen M.B.H. Düsseldorf.
14. Mori, K. and Sano, M. (1975). Effect of surface agents on convection and mass transfer. In *Kinetics of metallurgical process in steelmaking*, (ed. W. Dahl., K.W., Lange, and D. Papamantellos), pp. 492–513. Verlag Stahleisen M.B.H., Düsseldorf.
15. Belton, G.R. (1976). Langmuir adsorption, the Gibbs adsorption isotherm, and interfacial kinetics in liquid metal systems. *Met. Trans.*, **7B**, 35–42.
16. Carlsaw, H.S. and Jaeger, J.C. (1959). *Conduction of heat in solids*. Clarendon Press, Oxford.

17. Luikow, A. V. (1968). Analytical heat diffusion theory. *Academic Press*, New York.
18. Ida, T. and Guthrie, R. I. L. (1988). *The Physical properties of liquid metals*. Clarendon Press, Oxford.
19. Ozawa, Y. and Mori, K. (1983). Critical conditions for penetration of solid particles in liquid metal. *Trans. ISIJ.*, **23**, 769–74.
20. Lee, B. W. (1990). Single and multiple particle penetration across a gas-liquid interface relevant to injection metallurgy. Masters Thesis, Dep. of Chem. Engineering Science, Univ. of Melbourne, Australia.
21. Engh, T. A. and Johnston, A. (1988). Injection of non-wetting particles into melts. Proc. of the International Symposium on Reduction and Casting of Aluminium, Montreal, Pergamon Press, pp. 11–20.
22. Nagata, S. (1975). *Mixing, principles and applications*. Wiley, New York.
23. Nilmani, M., Engh, T. A., Gray, N. B., and Floyd, S. M. (1991). Plant and laboratory investigations on injection and related phenomena in some smelting and refining processes. *Scand. J. Metallurgy*, **19**, 278–87.
24. Waladan, M., Firkin, G. R., Bultitude-Paull. Lightfoot, B. W., Floyd, J. M. (1990), Top submerged lancing for recovery of zinc from ISF slag. In *Lead-zinc '90*, (ed. T. S. Mackey), pp. 607–28. TMS publication, Warrendale, Pa.
25. Engh, T. A. and Nilmani, M. (1988). Bubbling at high flow rates in inviscid and viscous liquids (slags). *Metall. Trans.*, **19B**, 83–94.
26. Wraith, A. E. (1971). Two stage bubble growth at a submerged orifice. *Chem. Engng-Sci.*, **26**, 1659–71.
27. Davidson, J. F. and Schüler, B. O. G. (1960). Bubble formation at an orifice in an inviscid liquid. *Trans. Instn. Chem. Engrs.*, **38**, 335–42.
28. Johansen, S. T., Robertson, D. G. C., Woje, K., and Engh, T. A. (1988). Fluid dynamics in bubble stirred ladles: Part I. Experiments. *Metall. Trans.*, **19B**, 74–54.
29. Sherwood, T. K., Pigford, R. L., and Wilke, C. R. (1975). *Mass transfer*. McGraw-Hill, New York.
30. Patak, F. (1983). Untersuchungen zur natrium- und lithium- -entfernung aus hüttenaluminium. Doktor-Ingenieurs dissertation, Fakultät für Bergbau und Hüttenwesen, Rheinisch-Westfälischen Technische Hochschule, Aachen.
31. Pedersen, T. (1984). Spylgassbehandling av aluminium for fjerning av hydrogen, løste alkalimetaller og inneslutninger, dr. ing. avhandling, Metallurgisk Institutt, Norges Tekniske, Høgskole, Trondheim.
32. Roth, A. (1986). *Vacuum technology*. North-Holland, Amsterdam.
33. Gariépy, B., Dubé, G., Simoneau, C., and Leblanc, G. (1984). The TAC process: A proven technology. In *Light Metals*, pp. 1267–79. Warrendale, Pa.
34. White, F. M. (1974). *Viscous fluid flow*. McGraw-Hill, New York.
35. Reynolds, A. J. (1974). *Turbulent flows in engineering*. Wiley, London.
36. Hinze, J. O. (1975). *Turbulence*, (2nd edn). McGraw-Hill, New York.
37. Launder, B. E. and Spalding, D. B. (1972). *Lectures in mathematical models of turbulence*. Academic Press, London.
38. Launder, B. E. (1989). Second-moment closure: present. . . . and future. *Int. J. Heat and Fluid Flow*, **10**, No. 4, 282–300.
39. Johansen, S. T. and Boysan, F. (1988). Fluid dynamics in bubble stirred ladles: Part 2, Mathematical modelling. *Metall. Trans.*, **19B**, 755–64.
40. Johansen, S. T., Boysan, F., and Engh, T. A. (1986). Numerical calculations of removal of inclusions and dissolution of refractory in a bubble stirred ladle. Proceedings of the fourth Japan-Nordic countries joint symposium on science and technology of process metallurgy, ISIJ, Nov. 17–18, Tokyo, pp. 182–215.

41. Hsiao, T.-C., Lehner, T., and Kjellberg, B. (1980). Fluid flow in ladles—experimental results. *Scand. J. Metallurgy*, **9**, 105–10.
42. Patankar, S. V. (1980). *Numerical heat transfer and fluid flow*. Hemisphere Publ. Corp., New York.
43. Liang, D. (1991). Flow generated by an impeller and mass transfer across gas-liquid interfaces in stirred vessels. Dr.ing.Thesis. The Norwegian Institute of Technology, Trondheim, Norway.
44. Bakke, P., Lauritzen, J.L., Engh, T.A., and Øymo, D. (1991). Hydrogen in magnesium, absorption, removal and measurement. In *Light metals* (ed. E.L. Rooy). Warrendale, Pa.

8

ADDITION OF ALLOY COMPONENTS

8.1 INTRODUCTION

There are innumerable ways in which the properties of a metal can be modified and improved by the addition of alloy components. This is especially the case for steel, and for aluminium and copper a vast array of alloys have also been developed. To achieve alloys with specific engineering characteristics with regard to strength, fatigue, ductility, and other physical properties, common practice is to start out with a 'pure' solvent metal and introduce alloying elements at very precise, predesigned levels. Thus addition of alloys, melting, and dissolution is a necessary operation in the handling of melts before casting. The thermal requirements and the rate of solution need to be understood precisely to develop good alloying practices. Also dissolution and melting rates may limit productivity in a number of steelmaking and other processes.[1]

Compared to solidification the literature concerning addition and melting of alloys is limited.[2-6] What happens to the additions proves to be surprisingly complex. For instance, a frozen shell may form for a transient period around the addition. Also the addition may melt inside the frozen shell if the melting-point of the addition is low compared to the melting-point of the surrounding fluid. Seemingly, each combination of melt and alloy addition behaves in a complex and special manner. For a given combination it is possible to describe and follow melting or dissolution with extensive numerical calculations.[7,8] However, a difficulty is always the lack of good data concerning physical quantities such as thermal conductivities and specific heats. Even worse is the problem of calculating the heat transfer coefficient between melt and addition or frozen shell. Figure 8.1 illustrates the various physical quantities needed to describe the problems associated with melting.

In this chapter an attempt is made to give the reader an understanding of the mechanisms involved. These mechanisms or *unit operations* are described as simply as possible in terms of analytical solutions. The specific heats, c, densities, ρ, thermal conductivities, k, and thermal diffusivities, α, are assumed to be constants independent of temperature. In order to describe conditions at the addition–shell interface it has been necessary to introduce a numerical solution (see Appendix 4). Hopefully, the background given in this chapter will help the reader to formulate in quantitative mathematical terms the behaviour of the special combination of addition and melt. Then one must also be aware of the approximations and assumptions involved. A good physical feeling for the problem, consideration of the dimensionless groups that enter, and analytical solutions obtained in special cases are valuable as a starting point and a check for a numerical solution of the problem at hand.

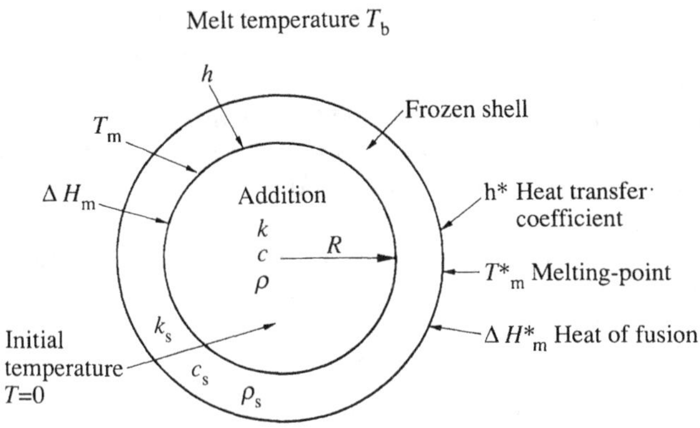

FIG. 8.1 The various physical quantities needed to describe melting

The most important methods for addition include:

(1) bulk additions to a furnace or to the ladle during tapping;

(2) powder injection by dense or dilute phase transport to the ladle (see also Chapter 4);

(3) Wire feeding to the ladle or the tundish and even in some cases to the casting mould.

Bulk additions of coarse material is widely practiced. Provided good recovery can be achieved, this method should also dominate in the future due to the favourable materials and equipment costs. Additions added to a tapping stream should be sized such that the material can be dissolved within the time of turbulent mixing.

Submerged powder injection has inherently good heat and mass transfer characteristics, and can be used for virtually all alloying elements.[9-12] The buoyancy of the carrier gas bubbles and the momentum of the particles provide stirring of the bath. Usually the stirring is an advantage, but it may become a problem if top-slag or dross is entrained. Other considerations are the relatively high capital costs for reliable and flexible injection systems.

Metal-encased alloy wires[13-15] are clearly more expensive than other forms of additions. Consequently, this method is only used for reactive elements such as calcium, titanium, or aluminium or for microalloying close to the casting operation. Wire feeding is attractive because of the simple equipment involved and the highly reproducible results. The metal sheath surrounding the alloys protects them from oxidation by atmosphere and slag and the wire penetrates some distance from the surface down into the metal before the sheath cover melts. This ensures a high yield of the alloy components.

Dissolution of an alloy addition into a melt may be divided into two main categories.

For class I alloys the melting range lies below the bath melting temperature. In this case dissolution is usually limited only by heat transfer from the melt to the lump or particle. For class II alloys the melting range lies above the bath temperature. Then dissolution is controlled by diffusion through a melt boundary layer. This process is slow compared to melting rates for class I alloys.

Steel is by far the most important metal both in terms of tonnage and value. Additions to steel are commonly in the form of ferroalloys which often have lower melting ranges than the pure elements and can be introduced more readily into the molten steel. Thus, ferroalloys often belong to class I. Most ferroalloys are produced in electric furnaces from the ores, usually with carbon as the reductant. Schematically, one may write

$$MO + \mathrm{Fe} + C = \mathrm{Fe}M + \mathrm{CO}. \tag{8.1}$$

The activity of the element, M, in the ferroalloy FeM is less than that for the pure element. This means in practice that FeM can be produced at lower temperatures and with less electrical energy consumption than the pure element, M. The main requirements of ferroalloys are uniformity in size and composition and these requirements are becoming ever more strict, reflecting steadily reduced tolerances on the composition of steels. Furthermore, the limits on impurities such as oxygen, nitrogen, carbon or carbides, and hydrogen are growing more severe.[16] Many ferroalloys have concentrations of oxygen and nitrogen in the range of 0.2 per cent[17] which is unacceptable when used for superclean steels with a total impurity level of 0.005 per cent. Particular attention needs to be paid to the reduction of nitrogen in the ferroalloys. The steelmaking approach is to produce the steel with a nitrogen concentration of 20–50 p.p.m., and then to shield the steel from the atmosphere during ladle treatment and continuous casting. Approximately 30 per cent of $\underline{\mathrm{N}}$ can be removed during vacuum treatment of low-carbon steels. However, in most steelmaking operations, any nitrogen introduced into the steels stays with the product. A problem with FeSi powder is that considerable amounts of nitrogen are located on or near the surface. Then the mean content of nitrogen becomes high.

Casting in the manufacture of ferroalloys must be given special consideration. Segregation during solidification may be severe in many alloying systems (see Chapter 9 concerning FeSi). A large and variable pick-up of oxygen and nitrogen may occur in flat-bed casting operations. Further oxidation as well as absorption of moisture takes place in storage and during the distribution of materials. Granulation of ferrosilicon has resulted in a more uniform product and a further advantage with granules compared to lumps is that the spherical geometry gives an increased penetration depth and residence time in the melt.[18] This indicates that similar methods should be developed for the whole range of ferroalloys.

Alloy additions to metals such as aluminium or magnesium often belong to the class II category. They dissolve into the melt by diffusion which is usually a slow process. Depending on their density relative to the melt, they may float up to the surface or settle to the bottom. In the first case there may be considerable losses due to oxidation or evaporation. In the second case dissolution rates may be prohibitively low, and the yields may vary from charge to charge. To solve such problems the methods mentioned previously can be supplemented by the use of ingots or waffles held in place in the melt with baskets or cages. However, such methods normally are time- and labour-consuming. Another approach is to use *master alloys* in analogy with ferroalloys in iron. For instance, for addition to aluminium there is available on the market aluminium alloyed with Sb, Bi, B, Cr, Co, Cu, Fe, Mg, Mn, Ni, Si, Sr, Ti, V, Zn, and Zr. These master alloys contain at least 50 per cent Al, and often much more. There seems to be at least three reasons for the use of master alloys.

1. Part of the mixing operation can be carried out in a separate unit.
2. Dissolution of the element from the master alloy is more rapid than for the alloy element addition alone. The main reason is probably that the master alloy has a density close to the melt density. Thus the tendency to float or settle is reduced.
3. For grain refining (see Chapter 9) the particles that promote the formation of a large number of small crystals can be produced in a separate operation.

Master alloys can also be produced directly at the plant and added in molten form. To stir the additions into the master alloy an impeller can sometimes be employed.

One stirring method developed for additions to aluminium is to create a vortex using an impeller.[19] The additions are then fed down into the hole created by the vortex. A problem is the solubility of the alloy additions. For instance, if the molten master alloy is to be mixed with three times the amount of metal, the alloy element concentration of the master alloy must be four times that of the final product. If the aluminium is to contain 7% Si, the master alloy must dissolve 28% Si. From the Al–Si temperature–composition diagram[20] it is seen that this requires a temperature of 800°C.

In the following sections we study the dissolution times for class I and class II alloy additions. Calculations are performed only for an assumed spherical geometry of the addition. Initially when a cold body is immersed in a molten metal bath, the surface is heated up and a temperature gradient is set up in the body. Heat is extracted by the addition, and a shell of solidified metal may be formed even though heat enters the shell from the surrounding melt. In this case heat conduction to the body initially dominates over heat transfer from the melt so that the shell grows. The temperature profile flattens out after a time; then heat transfer from the melt takes over and the frozen shell melts back.

In Table 2.14 data are given for the thermal conductivity of pure molten metals while some additional data concerning elements and ferroalloys are found in Table 8.1. Below, class II alloys, where mass transfer is rate limiting, are

TABLE 8.1 *Thermodynamic and physical data for some solid and liquid metals at the melting-point*[20,34,35,7]

	T_m (°C)	Sensible heat ΔH_m (kJ/kg)		ρ_s (kg/m^3)	ρ_l	c_s (J/kg K)	c_l	k_s (W/m K)	k_l
Fe	1536	1050	250	7270	7015	715	824	40	40
Al	660	672	397	2559	2385	1200	1180	210	95
Cu	1085	467	207	8320	8000	500	517	320	170
Mg	650	733	368	1640	1590	1280	1343	150	80
Cr	1857	1286	326	6950	6280	1040	756	36	
Si	1412	1280	1790	2270	2510	1025	970	20	55
Mn	1244	907	220		5730	830	840		95
	T_l								
FeCr (55% Cr)	1500	1100	325	7000		700		38	
FeSi (75% Si)	1320	1000	300	3500		700		10	
FeMn (80% Mn)	1266	900		7200		700			

treated. In a later section we review the literature concerning the heat transfer coefficients from melt to addition or solidified shell. Then some mathematical tools needed for solving the problems in melting and dissolution of alloys are introduced in Section 8.4. Students who are not very fond of mathematics may, at least initially, skip Section 8.4. From Section 8.6 onwards examples are given of calculations of the shell period and final melting down time.

8.2 CLASS II; DIFFUSION LIMITED DISSOLUTION OF ALLOYS

Alloys with a higher melting point than the bath temperature dissolve by mass transfer in the melt boundary layer. For metal systems mass transfer is a much slower process than heat transfer. Furthermore convection and diffusion in the melt boundary layer proceeds rapidly compared with solid state diffusion. Thus once the solid shell that initially forms has melted, dissolution is determined by the mass transfer coefficient k as described in the Chapter 7. Then the solid surface is heated up to the bath temperature while the surface composition may remain unchanged. Since viscosities and diffusion coefficients in molten metals are very similar to what we find in aqueous systems, chemical engineering theory and correlations can be applied. There are two main problems. One is to obtain reliable data, especially for the diffusion coefficients, D. (see also Section 2.2.9). The mass transfer coefficient k depends on D, for instance to the power $\frac{2}{3}$ (see also Chapter 7). The other problem is that dissolution sometimes does not proceed in a planar fashion.[21] However, in the following it is assumed that dissolution takes place on a well-defined planar front.

Once the mass transfer coefficient k is known one can calculate the dissolution time. A mass balance for the alloy component M gives:

Number of moles of M removed from the alloy addition per unit time
= Flow of moles of M transferred through the diffusion boundary layer in the melt

or

$$-\frac{\mathrm{d}r}{\mathrm{d}t}A\rho_M = kA(c_M - c_{M\mathrm{b}}) \tag{8.2}$$

where it may be noted that $(\mathrm{d}r/\mathrm{d}t)\,A$ is the volume of a spherical shell and A is the melt–shell contact area. r is the radius and ρ_M is the number of kmol/m^3 in the alloy addition. Here k is the mass transfer coefficient as described for instance in Chapter 7. It should not be confused with the thermal conductivity. c_M is the concentration of M in the melt in equilibrium with the concentration in the solid in units of kmol/m^3. $c_{M\mathrm{b}}$ is the concentration in the bulk melt. Thus $c_M - c_{M\mathrm{b}}$ is the driving force for diffusion through the melt boundary layer.

Integration of eqn (8.2) gives, disregarding the dependence of k on r,

$$t = \frac{\rho_M(R - r)}{k(c_M - c_{M\mathrm{b}})}. \tag{8.3}$$

Here R is the original radius of the addition. t in eqn (8.3) is the dissolution time from size R to r. The total dissolution time for the addition is

$$t_\mathrm{d} = \frac{\rho_M R}{k(c_M - c_{M\mathrm{b}})}. \tag{8.4}$$

To obtain the total time, the shell time, t_s, must be added to t_d. We find that usually t_s is much smaller than t_d.

As an example of the calculation of t_d we take the addition of niobium particles of 4 mm diameter to steel. The melting point of Nb is high, about 2500°C, so this should be a class II type of addition. For the diffusion coefficient of Nb in Fe at 1570°C we use eqn (2.132) and Table 2.13. Then

$$D = \frac{1.38\times10^{-23}\times1843}{2\pi0.8\times10^{-6}\times7030\times2\times1.43\times10^{-10}}\left(\frac{55.85+92.91}{2\times92.91}\right)^{\frac{1}{2}}$$
$$= 2.3\times10^{-9}\,\mathrm{m^2/s}.$$

Equation (5.44) with $\phi = 0.22$ for $\rho_M = 8570\,\mathrm{kg/m^3}$ now gives for the relative velocity between particle and melt:

$$u_r^2 = \left(\frac{8570}{7030} - 1\right) \times \frac{9.81 \times 4 \times 0.002}{0.22 \times 3}$$

or

$$u_r = 0.16\,\text{m/s}.$$

This gives

$$\text{Re} = \frac{0.16 \times 0.004}{0.8 \times 10^{-6}} = 800.$$

Then according to eqn (7.56)

$$\text{Sh} = \frac{kd}{D} = 1 + 0.724 \times 800^{0.48}\left(\frac{0.8 \times 10^{-6}}{2.3 \times 10^{-9}}\right)^{\frac{1}{3}}$$

or

$$k = 7.3 \times 10^{-5}\,\text{m/s}.$$

From eqn (8.4) we see that we must know c_M, the concentration of Nb in the melt in equilibrium with pure Nb. In fact from the temperature versus composition diagram[20] it is seen that a number of solid intermediate phases form. The result is that dissolution in fact does not proceed in a planar fashion. Rather, a hollow dendritic structure remains after most of the metal has diffused into the steel bath.[21] The melt in contact with the solid only contains about 20 mol% Nb. If c_{Mb} is negligible and eqn (8.4) is employed in spite of the non-planar dissolution:

$$t_d \sim \frac{R}{k0.20} = \frac{0.002}{7.3 \times 10^{-5} \times 0.20} \approx 137\,\text{s}.$$

In a stagnant bath the addition would settle to the bottom before dissolution was complete. Here the rate of dissolution would be very slow. Thus stirring (see Section 5.9) is important in order to keep the additions suspended and to increase k. Alternatively, particles smaller than 4 mm can be injected into the melt in order to reduce the relative velocity, or FeNb may be added. Then the density difference is less so that the relative velocity u_r is reduced.

For other high-melting alloys and ferroalloy additions to steel such as FeMo, FeB, and FeZr the problems are similar to those sketched here for FeNb. Dissolved oxygen slows down the dissolution rate.[4] For instance 700 p.p.m. oxygen reduces the dissolution rate for FeNb by a factor of 2.[4] Pure molybdenum dissolves very slowly;[21] there does not seem to be a simple explanation for this.

8.3 THE HEAT TRANSFER COEFFICIENT FOR MOLTEN METALS

In the calculation of the heat transfer coefficient for molten metals the high value of the thermal conductivity must be taken into account. As a result the thermal conductivity, k, and the thermal sublayer in the melt along the body play a much greater role than for normal liquids. Therefore one should be wary of applying heat transfer correlations derived for normal fluids to molten metals. For molten metals the ratio between kinematic viscosity and thermal diffusivity $\mathrm{Pr} = \nu/\alpha$ is very small, in the range 0.003–0.1. By comparison Pr lies between 0.65 and 1.1 for gases, between 0.9 and 50 for 'common' liquids such as water, and beyond 20 for viscous liquids such as oils and slags. Due to the importance of thermal diffusion, α, in the wall boundary layer for metals, the kinematic viscosity, ν, and the roughness of the walls play only a minor role, except that they have an effect on the velocity distribution. In part this is accounted for by including the variables v and d. This leaves only one dimensionless (Peclet) group to characterize heat transfer:

$$\begin{aligned}\mathrm{Pe} &= \frac{vd}{\alpha} = \frac{vd}{\nu}\frac{\nu}{\alpha}\\ &= \mathrm{Re}\,\mathrm{Pr}\end{aligned} \tag{8.5}$$

where v is the (mean) velocity of the metal past the solid surface. d is a characteristic dimension: for flow around a sphere, d is the diameter; for pipe flow, d is the diameter of the pipe. The Peclet number compares heat transport by convection to thermal conduction.

For turbulent pipe flow at very high Re numbers it is found empirically that the following equation represents quite well the data for many liquid metal systems:[22]

$$\mathrm{Nu} = 4 + 0.014\,\mathrm{Pe}^{0.8} \tag{8.6}$$

where $\mathrm{Nu} = hd/k$. The equation is valid when the length of the pipe is greater than $60\,d$. Also $\mathrm{Pe} > 100$.

For smooth plates laminar flow persists up to Re numbers of roughly 5×10^5. For laminar-forced convective flow over a plate, with $\mathrm{Pr} \ll 1^6$ theory gives a heat transfer coefficient:[23]

$$\mathrm{Nu} = 1.13\,\mathrm{Pe}^{\frac{1}{2}}. \tag{8.7}$$

This is the same result as for transfer to a free surface (cf. eqn (7.35)). Figure 8.2 compares the thermal and velocity boundary layers for $\mathrm{Pr} \ll 1$. It is seen that the thermal boundary layer is much greater than the velocity boundary layer. Then compared to the free surface case there is only a slight reduction in the flow of melt that supplies heat to the surface. For $\mathrm{Pr} < 0.4$ one may use a correction factor $(1 + 10\,\mathrm{Pr})^{-\frac{1}{6}}$.[24] For instance for $\mathrm{Pr} \approx 0.1$ this gives 0.89 so that

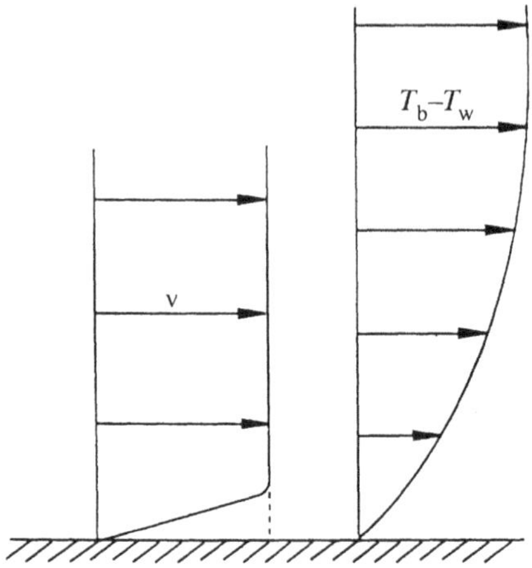

FIG. 8.2 Comparison of the velocity and thermal boundary layers for flow of a liquid metal along a solid surface.

$$\mathrm{Nu} \approx \mathrm{Pr}^{\frac{1}{2}}.$$

For flow past a sphere it is found experimentally that[25]

$$\mathrm{Nu} = 2 + 0.38\,\mathrm{Pe}^{\frac{1}{2}}. \tag{8.8}$$

If the velocity boundary layer that surrounds the sphere is disregarded, one would obtain (see also eqn (7.53))

$$\mathrm{Nu} = 2 + 1.13\,\mathrm{Pe}^{\frac{1}{2}}. \tag{8.9}$$

The discrepancy between eqns (8.9) and (8.8) may be partly explained by the effect of the velocity boundary layer. Probably this effect is somewhat greater for a sphere than for a plate. However, of greater importance is separation of the melt from the spherical surface at a polar angle of about 110 degrees.[23] In general the fluid particles near the surface, beyond the point of separation move in directions that differ from the course of the external stream. This 'backwater' fluid has a long residence time and is of less use in transferring heat. Other causes for the lower value are discussed at the end of this section.

The constant 2 on the right-hand side of eqn (8.9) takes pure conduction into account. This term is of the same order of magnitude as the convection term and even larger in some practical cases. The reason is the high thermal conductivity of metals. For instance, even if the Reynolds number for the addition is high, say 15 000, due to the low value of Pr ~ 0.1 the convection term is only:

$$0.38\,\mathrm{Pe}^{\frac{1}{2}} = 0.38 \times (0.1 \times 15\,000)^{\frac{1}{2}} = 14.7.$$

Thus even in this example with very high relative velocities convection-induced heat transfer is only around seven times greater than for pure heat conduction. A Reynolds numbers of 15 000 corresponds to a diameter of 0.02 m and relative velocity in steel of 0.6 m/s. For smaller dimensions and lower velocities pure heat conduction plays a relatively greater role. Initially when the addition is made to the melt the temperature gradients are high. This causes the melt to freeze around the addition. Only when the internal temperature gradients have been substantially reduced, and the temperatures increased correspondingly, does the shell start to remelt. The addition of FeCr to steel is taken as an example. The melting-point of the steel is assumed to be $T^*_{\mathrm{m}} = 1500°\mathrm{C}$. If the bath temperature is 1700°C this gives $T_{\mathrm{b}} - T^*_{\mathrm{m}} \approx 200°\mathrm{C}$. Then thermal conduction in the frozen shell and heat transfer in the melt are equal when

$$k_{\mathrm{s}}\frac{\mathrm{d}T}{\mathrm{d}r} = h(T_{\mathrm{b}} - T^*_{\mathrm{m}}).$$

Here k_{s} is the thermal conductivity of the shell and h is the heat transfer coefficient between melt and shell or, recalling that $\mathrm{Nu} = hd/k_{\mathrm{Fe}}$ (liquid),

$$\frac{\mathrm{d}T}{\mathrm{d}r}d = \mathrm{Nu}(T_{\mathrm{b}} - T^*_{\mathrm{m}})\frac{k_{\mathrm{Fe}}\,(\text{liquid})}{k_{\mathrm{Fe}}\,(\text{solid})}$$

or, disregarding the difference in the thermal conductivities,

$$\frac{\dfrac{\mathrm{d}T}{\mathrm{d}r}\dfrac{d}{2}}{T^*_{\mathrm{m}} - T(r=0, t=0)} = \frac{\mathrm{Nu}(T_{\mathrm{b}} - T^*_{\mathrm{m}})}{2\{T^*_{\mathrm{m}} - T(r=0, t=0)\}}.$$

If Nu = 16.7, as in the previous example,

$$\frac{\dfrac{\mathrm{d}T}{\mathrm{d}r}\dfrac{d}{2}}{T^*_{\mathrm{m}} - T(r=0, t=0)} = \frac{16.7 \times 200}{2 \times 1300} = 1.29$$

where $T(r = 0, t = 0)$ is taken as 200°C. Figure 8.3 sketches this situation for the temperature gradient and profile when the shell starts to remelt. Later when the shell has completely remelted, temperature gradients are even smaller and temperatures are higher. In this case the mathematical description of diffusion and melting of the spherical addition can be simplified. This is discussed in Section 8.4.

The local heat transfer coefficient is greatest at the front end of the sphere where it meets fresh fluid with the same temperature as the bulk, and lowest at the back side. This fact is not taken into account in the following sections. Only the mean heat transfer coefficient as presented in eqns (8.6)–(8.10) is employed.

There is considerable scatter in the experimental results for metals where

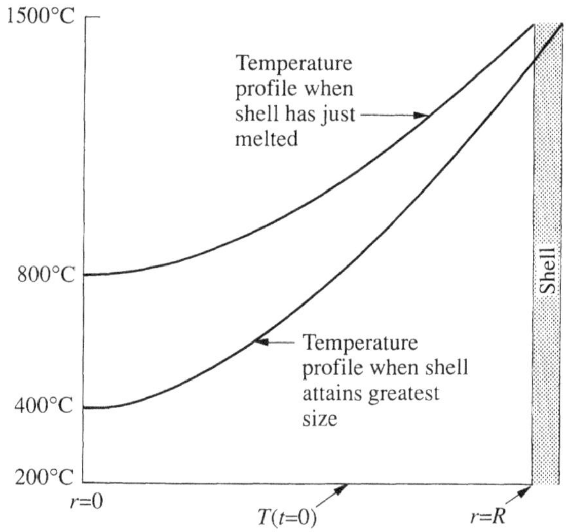

FIG. 8.3 Temperature profiles when the shell attains its greatest size and when the shell has just remelted. $Bi^* = 1.29$, $Bi^*b^* = 2.0$, $k/k_s = 1$, and $\alpha_s/\alpha = 1$.

$Pr \ll 1$. Experiments also tend to give Nu numbers lower than the values obtained from the theoretical formulas. There are a number of possible causes:

(1) oxide formation on the wall that introduces an extra thermal resistance;

(2) failure of some liquid metals to wet the wall;

(3) departures from a constant temperature along the wall.

Similar phenomena will also be present when additions are made to a molten bath, and this variability must be taken into consideration when calculations concerning melting are performed.

Before we get involved in the treatment of various problems concerning the class I addition of alloys to melts, some mathematical tools are presented in Section 8.4. They should be useful not only in handling problems concerning heat transfer when there is a change of state but also for solid state diffusion studies

8.4 SOLUTIONS OF HEAT DIFFUSION PROBLEMS

Problems in diffusion of heat or mass involving a change of state (melting) are often difficult to solve analytically[26] or even numerically. A further problem is that frequently the exact mechanisms and industrial conditions are not known. This makes it tempting to employ approximate solutions unless the situation requires the best possible description of the problem. In this last case extensive numerical calculations will usually be a necessary part of the work. As an introduction to the problems in the following, exact solutions are given for the case that the surface of an initially cold body ($T = 0$) is held at temperature T_m. We look at a plate, an infinite cylinder, and a sphere. The dimensions are indicated by the symbol R for radius. For the plate the thickness is $2R$. The equation for conservation of heat (enthalpy) is given by:

$$\frac{\partial\left(r^n k \dfrac{\partial T}{\partial r}\right)}{\partial r} = \frac{\partial (c\rho r^n T)}{\partial t} \tag{8.10}$$

where $n = 0$, 1, and 2 for a plate, cylinder, and sphere, respectively. If the thermal conductivity, k, specific heat, c, and density, ρ, can be regarded as independent of temperature, eqn (8.10) reduces to

$$\alpha \frac{\partial\left(r^n \dfrac{\partial T}{\partial r}\right)}{r^n \partial r} = \frac{\partial T}{\partial r} \tag{8.11}$$

where α is the thermal diffusivity, $\alpha = k/(c\rho)$. For a sphere it is found that

$$\frac{\partial}{r^2 \partial r}\left(r^2 \frac{\partial T}{\partial r}\right) = \frac{2}{r}\frac{\partial T}{\partial r} + \frac{\partial^2 T}{\partial r^2} = \frac{\partial^2 (rT)}{r \partial r^2}.$$

Thus for a sphere

$$\alpha \frac{\partial^2 (rT)}{\partial r^2} = \frac{\partial (rT)}{\partial t}. \tag{8.12}$$

rT for the sphere is given by the solution of the same problem for the plate. Due to the spherical symmetry, T is a symmetric function of r, $T(r) = T(-r)$. Thus rT is an antisymmetric function of r. To describe the spherical problems, we may therefore solve the one-dimensional plate problem

$$\alpha \frac{\partial^2 (rT)}{\partial r^2} = \frac{\partial (rT)}{\partial t}. \tag{8.13}$$

with boundary conditions:

$$\frac{rT}{RT_m} = 1 \quad \text{at} \quad r = +R \tag{8.14}$$

and

$$\frac{rT}{RT_m} = -1 \quad \text{at} \quad r = -R \tag{8.15}$$

for $t > 0$. It is then seen that

$$\frac{rT}{RT_{\mathrm{m}}} = \frac{r}{R} \tag{8.16}$$

and

$$\frac{rT}{RT_{\mathrm{m}}} = a_n \sin\left(\frac{n\pi r}{R}\right) \exp\left(-\alpha \frac{\pi^2}{R^2} n^2 t\right) \tag{8.17}$$

are solutions of eqn (8.11). Ultimately the whole sphere will reach the surface temperature T_{m}. This is the solution given by eqn (8.16). Equation (8.17) gives the harmonic of degree n where n is an integer. It is one of the transient terms that describe the change from the initial temperature to the final temperature. Equation (8.17) is zero at $r = \pm R$ in accordance with the condition that the temperature remains constant and equal to T_{m} on the surface as given by eqn (8.16). The amplitudes a_n are determined by conditions at time $t = 0$. If the initial temperature is chosen to be zero, then from eqn (8.16) and all possible solutions such as eqn (8.17), the following condition is obtained:

$$\frac{r}{R} + \sum_{n=1}^{\infty} a_n \sin\left(\frac{n\pi r}{R}\right) = 0.$$

Fourier analysis tells us that for $-1 < r/R < +1$

$$\frac{r}{R} = \frac{2}{\pi}\left\{\sin\left(\frac{\pi r}{R}\right) - \frac{\sin}{2}\left(\frac{2\pi r}{R}\right) + \frac{\sin}{3}\left(\frac{3\pi r}{R}\right) - \cdots\right\}. \tag{8.18}$$

Thus $a_n = (-1)^n 2/(n\pi)$ so that finally

$$\frac{rT}{RT_{\mathrm{m}}} = \frac{r}{R} + \frac{2}{\pi}\sum_{n=1}^{\infty} \frac{(-1)^n \sin}{n}\left(\frac{n\pi r}{R}\right) \exp\left(\frac{-\alpha\pi^2 n^2 t}{R^2}\right). \tag{8.19}$$

If the initial temperature, $T(t = 0)$, is not zero, then T_{m} is replaced by $T_{\mathrm{m}}^{*} = T_{\mathrm{m}} - T(t = 0)$. It is seen that the factors $\exp(-\alpha\pi n^2 t/R^2)$ suppress the terms with a high n (higher harmonics) after a very short time. Physically this means that the higher harmonics that change over a small distance are rapidly smoothed out by the diffusion process. After a time $t > R^2/(\alpha\pi^2)$ all the harmonics except the first have died out and the temperature is described by

$$\frac{T}{T_{\mathrm{m}}} \approx 1 - 2\,\frac{\sin(\pi r/R)}{\pi r/R} \exp\left(\frac{-\alpha\pi^2 t}{R^2}\right). \tag{8.20}$$

It should be recalled that the amplitude $a_1 = -2/\pi$ in eqn (8.18) depended on the temperature profile at time $t = 0$. a_1 was determined from the Fourier series, eqn (8.18). In the approximation represented by eqn (8.20) only the main feature (the lowest harmonic) of the diffusion process has been retained. This type of approximation is often useful in the study of diffusion and dissolution problems. It may be noted that after a time $t > 4R^2/(\alpha\pi^2)$ even the lowest harmonic fades away and the temperature is equal to T_m throughout.

The boundary condition at $r = R$, may be that the heat flux, $\dot{q} = -k\partial T/\partial r$ instead of temperature, is constant. For the case of zero initial temperature, one then obtains[27]

$$T = \dot{q}\left(\frac{3t}{\rho c R} + \frac{5r^2 - 3R^2}{10kR} - \frac{2R^2}{kr}\sum_{n=1}^{\infty}\frac{\sin(r\beta_n/R)\exp(-\alpha\beta_n^2 t/R^2)}{\beta_n^2 \sin\beta_n}\right) \tag{8.21}$$

where β_n for $n = 1, 2, 3, \ldots$ are the positive roots of

$$\tan\beta = \beta. \tag{8.22}$$

The solution of this equation is[27]:

$$\beta_1 = 4.493,\ \beta_2 = 7.725,\ \beta_3 = 10.904,\ \beta_4 = 17.221.$$

After a time $t > R^2/(\alpha\beta_1^2)$ all the harmonics except the first have died out.

In the following, in addition to the previous solutions, we will also make use of the case when the surface temperature (at $r = R$) changes at a constant rate $T = \kappa t$. The initial temperature is zero. Then[27]

$$T = \kappa\left(t - \frac{R^2 - r^2}{6\alpha} - \frac{2R^3}{\alpha\pi^3 r}\sum_{n=1}^{\infty}\frac{(-1)^n}{n^3}\sin\left(\frac{n\pi r}{R}\right)\exp(-\alpha\pi^2 n^2 t/R^2)\right). \tag{8.23}$$

As for the case of constant surface temperature, only the first harmonic is left after a time $t > R^2/(\alpha\pi^2)$. Furthermore, it may be noted that as in the case of constant flux, the first term in eqn (8.23) changes linearly with time, and T increases with r^2.

Often in problems concerning change of phase such as melting and dissolution the times involved are short so that $t \ll R^2/(\alpha\pi^2)$. Then a large number of terms must be included in, for instance, eqn (8.19) to describe temperatures and heat fluxes. It is fortuitous that often only the temperature profile close to the surface is needed and that the method of sources and sinks is often practical to use when $t < R^2/(\alpha\pi^2)$. We imagine the solid to be continued in all directions and by the process of taking images in the bounding planes, we determine the distribution of sources and sinks which fulfils the boundary conditions for temperature or heat flux. The solution for the diffusion equation presented in Chapter 2 when discussing diffusion coefficients is an example of an instantaneous source. The case treated was in one dimension. The tracer was added at time $t = 0$ and spread out in a body of infinite extent.

Using the method of sources and sinks[27,28] eqn (8.19) is replaced by

$$\frac{T}{T_m} = \frac{R}{r}\sum_{n=0}^{\infty}\left\{\operatorname{erfc}\left(\frac{(2n+1)R - r}{2(\alpha t)^{\frac{1}{2}}}\right) - \operatorname{erfc}\left(\frac{(2n+1)R + r}{2(\alpha t)^{\frac{1}{2}}}\right)\right\}. \tag{8.24}$$

Here

$$\mathrm{erfc}(x) = \frac{2}{\pi^{\frac{1}{2}}}\int_x^\infty \exp(-v^2)\mathrm{d}v.$$

In particular, erfc (2) = 0.0047, and erfc (0) = 1. It is seen for $r = R$ that $T/T_\mathrm{m} = 1$ when $t > 0$ since all the terms cancel except the first, which is equal to erfc (0). erfc $(((2n+1)R - r)/2(\alpha t)^{\frac{1}{2}})$ has the instantaneous source located at position $r = (2n+1)\,R$. To fulfil the boundary conditions given in (8.14) and (8.15), the solution requires an infinite number of such instantaneous sources at positions $r = R$, $r = 3R$, $r = 5R$, ... and instantaneous sinks at positions $r = -R$, $r = -3R$, $r = -5R$, ... Near $r = R$, the remote instantaneous sources and sinks at positions $r = \pm(2n+1)R$ will only be noticeable at long times, that is, for times greater than n^2R^2/α. This is essentially the time it takes the heat pulse at $r = \pm(2n+1)R$ to diffuse to $r = R$. If we require that $t < R^2/(4\alpha)$ and $r \sim R$, then $(2n+1)R \mp R > 4(\alpha t)^{\frac{1}{2}}$ for $n > 0$ and it is only necessary to take the instantaneous heat source at $r = R$ into account in eqn (8.24):

$$\frac{Tr}{T_\mathrm{m}R} \approx \mathrm{erfc}\left(\frac{R-r}{2(\alpha t)^{\frac{1}{2}}}\right). \tag{8.25}$$

This solution is the same as if the surface of the sphere had been replaced by a semi-infinite solid held at surface temperature T_m. In other words for small values of the time $t < R^2/(4\alpha)$, the solution near the surface has not yet been affected by the spherical geometry.

Until now we have studied the situation when the boundary conditions at the surface $r = R$ are known. In problems involving a change of state this is often not the case. For instance, if a frozen shell surrounds the granule, both the temperature and the heat flux change with time. Both must match at the sphere and at the shell interface.

For the shell the heat diffusion equation is

$$\alpha_\mathrm{s}\frac{\partial^2(rT_\mathrm{s})}{\partial r^2} = \frac{\partial(rT_\mathrm{s})}{\partial t}$$

or, in dimensionless variables, $x = r/R$ and $\tau = 4\alpha t/R^2$,

$$\frac{\alpha_\mathrm{s}\partial^2(xT_\mathrm{s}/T_\mathrm{m})}{4\alpha\partial x^2} = \frac{\partial(xT_\mathrm{s}/T_\mathrm{m})}{\partial\tau} \tag{8.26}$$

where α_s is the thermal diffusivity of solidified metal.

At the interface between the sphere and shell the temperature is the same in the two phases. Therefore at $x = 1$

$$\frac{T}{T_\mathrm{m}}(x=1) = \frac{T_\mathrm{s}}{T_\mathrm{m}}(x=1) = \phi(\tau). \tag{8.27}$$

Furthermore, the heat fluxes are the same at the interfaces

$$\dot{q}(x=1) = \dot{q}_s(x=1) = \dot{q}(\tau). \tag{8.28}$$

The equation assumes that there is no reaction heat at the interface.

The solution of eqn (8.26) must satisfy these two time-dependent functions. Such a solution[27] is

$$\frac{T_s x}{T_m} = \phi(\tau) + \frac{(x-1)^2 4\alpha\dot{\phi}(\tau)}{\alpha_s 2!} + \frac{(x-1)^4 (4\alpha)^2}{\alpha_s^2 4!}\ddot{\phi}(\tau)$$
$$+ \cdots (x-1)\psi(\tau) + \frac{(x-1)^3 4\alpha\dot{\psi}(\tau)}{\alpha_s 3!}$$
$$+ \frac{(x-1)^5 (4\alpha)^2}{\alpha_s^2 5!}\ddot{\psi}(\tau) + \cdots. \tag{8.29}$$

The dots here denote differentiation with respect to τ. The function $\psi(\tau)$ is equal to

$$\psi(\tau) = \left\{ \frac{\partial \left(\dfrac{T_s x}{T_m} \right)}{\partial x} \right\}_{x=1}. \tag{8.30}$$

$\psi(\tau)$ may be determined from the heat flux at the surface $\dot{q}(\tau)$:

$$\dot{q}(\tau) = -\frac{k_s T_m}{R}\left\{ \frac{\mathrm{d}(T_s/T_m)}{\mathrm{d}x} \right\}_{x=1}$$
$$= -\frac{k_s T_m}{Rx}\left\{ \frac{\mathrm{d}\left(\dfrac{T_s x}{T_m}\right)}{\mathrm{d}x} - \frac{T_s}{T_m} \right\}_{x=1}$$
$$= -\frac{k_s T_m}{R}\{\psi(\tau) - \phi(\tau)\}. \tag{8.31}$$

Once the shell has formed and remelted or partly remelted it is time for the addition to melt. This is a typical moving boundary problem[27] and the solutions of the heat diffusion equation mentioned previously may not be practical to use. In this case the following solution of eqn (8.12) useful for short times, t, can be applied instead of the solution given earlier:

$$\frac{Tr}{T_m R} = b_0 + \frac{b_1}{R}r + \frac{b_2}{R^2}(r^2 + 2\alpha t)$$
$$+ \frac{b_3}{R^3}(r^3 + 6r\alpha t) + \frac{b_4}{R^4}(r^4 + 12r^2\alpha t + 12\alpha^2 t^2)$$
$$+ \frac{b_5}{R^5}(r^5 + 20r^3\alpha t + 60r\alpha^2 t^2) + \cdots. \tag{8.32}$$

or in dimensionless units:

$$\frac{T}{T_{\text{m}}}x = b_0 + b_1 x + b_2(x^2 + \tau/2) + b_3\left(x^3 + \frac{3x\tau}{2}\right) + b_4(x^4 + 3x^2\tau + \tfrac{3}{4}\tau^2) + b_5\left(x^5 + 5x^3\tau + \frac{15x\tau^2}{4}\right) + \cdots. \quad (8.33)$$

Since T must be an even function of x as mentioned previously, $b_0 = b_2 = b_4 = b_6 = 0$, and so on, and eqn (8.33) reduces to

$$\frac{T}{T_{\text{m}}} = b_1 + b_3\left(x^2 + \tfrac{3}{2}\tau\right) + b_5\left(x^4 + 5x^2\tau + \tfrac{15}{4}\tau^2\right) + \cdots. \quad (8.34)$$

Equation (8.34) is useful when the temperature profiles in the addition have flattened out after the shell period (see eqn (8.50)). In this case it may suffice to include the first two or three terms to describe the temperature. τ is the dimensionless time taken from the start of melting of the addition and $T/T_{\text{m}} = b_1 + b_3x^2 + b_5x^4 + \cdots$ describes the temperature profile when melting starts.

8.5 BOUNDARY CONDITIONS FOR THE SHELL SOLIDIFICATION AND ADDITION MELTING FRONTS

Initially a frozen shell of the melt forms, grows to its maximum size, and then remelts. For class I additions the addition itself begins to melt before the shell has disappeared. In the following, boundary conditions are given for this case. Strictly, they apply only when:

(1) melt and addition have a well-defined melting-point, T_{m};

(2) the thermal conductivities and diffusivities do not change significantly with temperature;

(3) the effect of a difference in densities between melt and addition on heat transfer may be neglected and melting fronts are planar (not dendritic);

(4) there is no heat of reaction or heat of dissolution between addition and melt.

In many cases the last assumption is not valid and then eqn (8.28) does not apply. This is the case for FeSi additions to steel. The FeSi reacts exothermically with the solid iron to form liquid Fe_2Si. Thus melting of the addition starts prematurely and triggers dissolution and melting of the outer iron shell.[29,30] This gives short melting times for FeSi alloys. Sometimes silicon is added to other ferroalloys to form powder alloy compacts. This can reduce the dissolution time and increase the yield of class II alloys.[31,32]

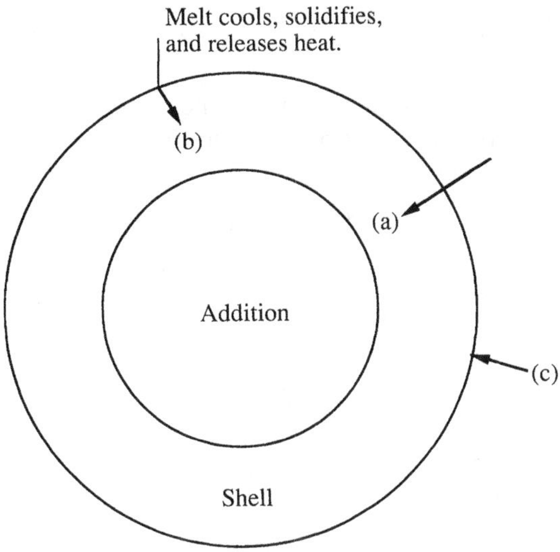

(a) Heat flux diffusing into the shell (and addition) = (b) Heat flux released by cooling and solidification of the melt + (c) Heat flux supplied by heat transfer from the melt

FIG. 8.4 Heat balance for the surface of a solidfying shell.

For the solidifying shell layer, the heat flux released by the cooling and solidification of the melt, and the heat supplied by heat transfer from the melt, is equal to the flux of heat diffusing into the previously solidified shell (see Fig. 8.4):

$$k_{\mathrm{s}}\left(\frac{\mathrm{d}T_{\mathrm{s}}}{\mathrm{d}r}\right)_{r=sR} = h^*(T_{\mathrm{b}} - T^*_{\mathrm{m}}) + (c_{\mathrm{l}}\rho_{\mathrm{l}}(T_{\mathrm{b}} - T^*_{\mathrm{m}}) + \Delta H^*_{\mathrm{m}}\rho_{\mathrm{s}})\frac{\mathrm{d}(sR)}{\mathrm{d}t}. \tag{8.35}$$

s is the radial distance divided by R, the index s refers to the solidified shell, and T^*_{m} is the melting-point of the melt. Notice that $\mathrm{d}T_{\mathrm{s}}/\mathrm{d}r$ will be positive. T_{b} is the (constant) temperature in the bulk melt, ΔH^*_{m} is the heat of melting of the melt in J/kg, and $c_{\mathrm{l}}\rho_{\mathrm{l}}(T_{\mathrm{b}} - T^*_{\mathrm{m}})$ is the heat removed from the melt when cooling from T_{b} to T^*_{m}. In dimensionless units eqn (8.35) simplifies to

$$b^*\frac{\mathrm{d}s}{\mathrm{d}\tau} = \frac{1}{\mathrm{Bi}^*}\left\{\frac{\mathrm{d}(T_{\mathrm{s}}/T^*_{\mathrm{m}})}{\mathrm{d}x}\right\}_{x=s} - 1 \tag{8.36}$$

where Bi* is a Biot number that gives the ratio between convective heat transfer and thermal conduction in the shell:

$$\mathrm{Bi}^* = \frac{h^*(T_\mathrm{b} - T^*_\mathrm{m})}{k_\mathrm{s} T^*_\mathrm{m}/R} = \frac{\mathrm{Nu}\,k_\mathrm{l}(T_\mathrm{b} - T^*_\mathrm{m})}{2k_\mathrm{s} T^*_\mathrm{m}} \tag{8.37}$$

(cf. Fig. 8.3 and the discussion concerning the figure). b^*/α is a ratio between the heat extracted from the melt during solidification and cooling of the melt and the heat transferred from the melt:

$$b^* = \frac{(\Delta H^*_\mathrm{m}\rho_\mathrm{s} + c_\mathrm{l}\rho_\mathrm{l}(T_\mathrm{b} - T^*_\mathrm{m}))4\alpha}{h^*(T_\mathrm{b} - T^*_\mathrm{m})R}. \tag{8.38}$$

It proves to be useful to write eqn (8.36) in the form:

$$b^* s^2 \frac{\mathrm{d}s}{\mathrm{d}\tau} = \frac{1}{\mathrm{Bi}^*}\left(s\left\{\frac{\mathrm{d}(T_\mathrm{s}x/T^*_\mathrm{m})}{\mathrm{d}x}\right\}_{x=s} - (T_\mathrm{s}s/T^*_\mathrm{m})\right) - s^2. \tag{8.39}$$

For melting of the addition ΔH^*_m, ρ_s, k_s, T^*_m, and h^* are substituted by symbols that refer to the addition: ΔH_m, ρ, k, T_m, h. Then

$$\mathrm{Bi} = \frac{h(T_\mathrm{b} - T_\mathrm{m})}{kT_\mathrm{m}/R} \tag{8.40}$$

$$b = \frac{\{\Delta H_\mathrm{m}\rho + c_\mathrm{l}\rho_\mathrm{l}(T_\mathrm{b} - T_\mathrm{m})\}4\alpha}{h(T_\mathrm{b} - T_\mathrm{m})R}. \tag{8.41}$$

Notice that here c_l and ρ_l refer to the molten addition and not to the melt as in eqn (8.38). Here b/α is a ratio between the heat consumed when the addition melts and the heat transferred from the melt. Equation (8.36) is replaced by

$$b\frac{\mathrm{d}s}{\mathrm{d}\tau} = \frac{1}{\mathrm{Bi}}\left\{\frac{\mathrm{d}(T/T_\mathrm{m})}{\mathrm{d}x}\right\}_{x=s} - 1. \tag{8.42}$$

In eqn (8.42) we have assumed that the addition is removed as it melts. The mechanism for removal is then either gravity or shear forces generated by the relative movement between addition and melt or by turbulence.

8.6 THE FROZEN SHELL PERIOD

In the following we calculate the time it takes the frozen shell around the cold, spherical addition to grow to its greatest size and then melt away again. During this period heat has been supplied to the addition, and it is important to calculate the temperature profile in the sphere when melting (for class I additions) finally takes place. The total time for melting is the sum of the shell period and the final melting down time. Often the shell period is short compared with the total time. However, the shell period is important because it gives the initial conditions for the melting down period. As mentioned towards the end of Section 8.3, for metallic additions the shell period may be sufficiently long to allow the temperature gradients to flatten out. Then simple relations such as eqns (8.20) and (8.33) can be employed to describe the temperature profiles.

The shell period represents a complex state of affairs with a composite solid composed of addition and shell. Also the shell dimensions change with time. The key to the solution lies in the description of the temperature and heat flux at the addition–shell interface. We assume that during the shell period the temperature at the interface is given by

$$\frac{T}{T_\mathrm{m}} = \phi(\tau) = a_1 + a_2\left(\frac{\tau}{\tau_0}\right)^{\frac{1}{2}} + a_3\left(\frac{\tau}{\tau_0}\right) + \cdots \tag{8.43}$$

where τ_0 is the shell period (the dimensionless time it takes the shell to form and remelt). The constant coefficients a_1 and a_3 are for a constant temperature and a temperature that varies linearly with time, respectively. It will be seen that a_2 takes into account the effect of a constant heat flux. Further terms are not included here. The temperature profile close to the surface according to eqn (8.25) is given by:[27]

$$\begin{aligned}\frac{T}{T_\mathrm{m}}x \approx a_1 \operatorname{erfc}\left(\frac{1-x}{\tau^{\frac{1}{2}}}\right) &+ a_3\left(\left\{\left(\frac{\tau}{\tau_0}\right) + 2\frac{(1-x)^2}{\tau_0}\right\}\operatorname{erfc}\left(\frac{1-x}{\tau^{\frac{1}{2}}}\right)\right.\\ &\left. - \frac{2\tau^{\frac{1}{2}}(1-x)}{\pi^{\frac{1}{2}}\tau_0}\exp\left\{-\frac{(1-x)^2}{\tau}\right\}\right)\\ &+ a_2\left(\left(\frac{\tau}{\tau_0}\right)^{\frac{1}{2}}\exp\left\{-\frac{(1-x)^2}{\tau}\right\}\right.\\ &\left. - \frac{(1-x)}{\tau_0^{\frac{1}{2}}}\pi^{\frac{1}{2}}\operatorname{erfc}\left(\frac{1-x}{\tau^{\frac{1}{2}}}\right)\right)\end{aligned} \tag{8.44}$$

where $x = r/R$.

As mentioned previously in connection with eqn (8.25) this solution can only be applied when τ and $\tau_0 \leqslant 1$.

From eqn (8.44) the fluxes $\dot{q}$ at $x = 1$ may be obtained:

$$\dot{q} = -k\frac{\mathrm{d}T}{\mathrm{d}r} = -\frac{k}{R}\frac{\mathrm{d}T}{\mathrm{d}x} = -\frac{kT_\mathrm{m}}{R}\frac{\mathrm{d}(T/T_\mathrm{m})}{\mathrm{d}x}.$$

Now since

$$\frac{\mathrm{d}(T/T_\mathrm{m})}{\mathrm{d}x} = \frac{\mathrm{d}(Tx/T_\mathrm{m})}{x\mathrm{d}x} - \frac{T}{xT_\mathrm{m}}$$

one obtains at $x = 1$

$$\dot{q} = -\frac{2kT_\mathrm{m}}{R}\left[\frac{1}{\tau^{\frac{1}{2}}}\left(\frac{a_1}{\pi^{\frac{1}{2}}} + 2a_3\left(\frac{\tau}{\tau_0}\right)\frac{1}{\pi^{\frac{1}{2}}} + a_2\left(\frac{\tau}{\tau_0}\right)^{\frac{1}{2}}\frac{\pi^{\frac{1}{2}}}{2}\right) - \frac{T}{T_\mathrm{m}2}\right]. \tag{8.45}$$

This gives

$$\dot{q}(\tau) = -\frac{2kT_\mathrm{m}}{\pi^{\frac{1}{2}}R}\left(\frac{a_1}{\tau^{\frac{1}{2}}} + 2a_3\frac{\tau^{\frac{1}{2}}}{\tau_0} + \frac{a_2\pi}{2\tau_0^{\frac{1}{2}}}\right) + \frac{kT_\mathrm{m}}{R}\phi(\tau). \tag{8.46}$$

It is seen that a_2 refers to a constant heat flux. If we have determined a_1, a_2, and a_3, the heat flux into the addition can be calculated. In an example in the following, eqn (8.46) will be used to determine the heat supplied to the addition during the shell period. In Appendix 4 eqns (8.27)–(8.31), (8.39), (8.43), and (8.46) are solved to give the shell time τ_0 and the maximum radius $s_\mathrm{m}R$ of addition and shell. Figures 8.5, 8.6, and 8.7 give τ_0, s_m, and a_1 and a_3 in eqn (8.43) as a function of a Biot number, Bi*, and a number $\mathrm{B}_\mathrm{i}^* b^*$. Bi* is here essentially a ratio between heat transfer and thermal conduction, while Bi* b^*/α gives a ratio between heat extracted from the melt during solidification of the shell and thermal conduction in the shell. In the calculations the ratio of the thermal conductivities of the shell and addition have been chosen to be equal to one. Also the ratio of the thermal diffusivities is set equal to one. Figure 8.5 is applied to the addition of FeCr to steel in Section 8.8. Figure 8.7 is used to calculate the heat transfer to the addition during the shell period.

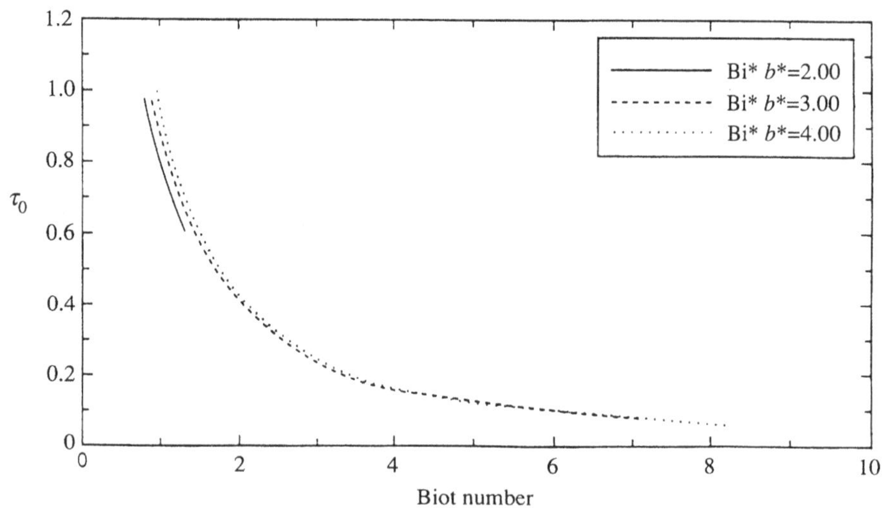

FIG. 8.5 The shell time τ_0 as a function of Bi* and Bi*b^*. In the calculation $k_\mathrm{s}/k = 1$ and $\alpha_\mathrm{s}/\alpha = 1$.

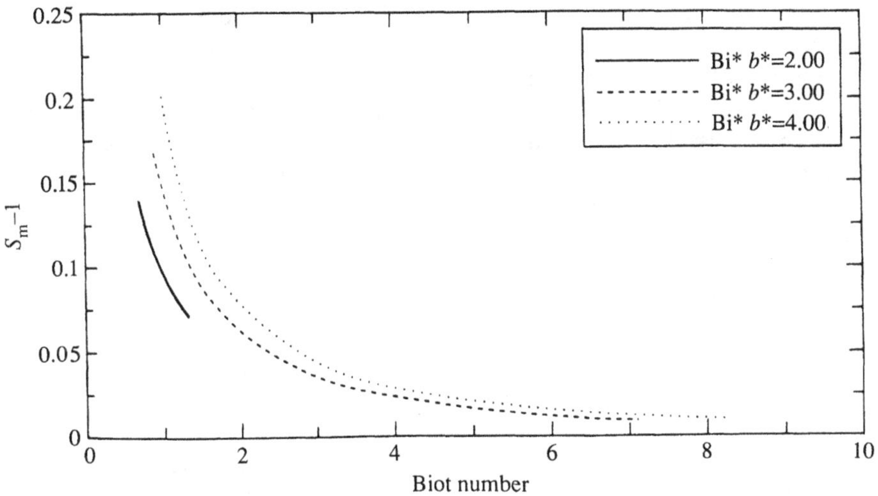

FIG. 8.6 The maximal extent of the shell $(s_m - 1)\,R$ as a function of Bi* and Bi*b*. In the calculation $k_s/k = 1$ and $\alpha_s/\alpha = 1$.

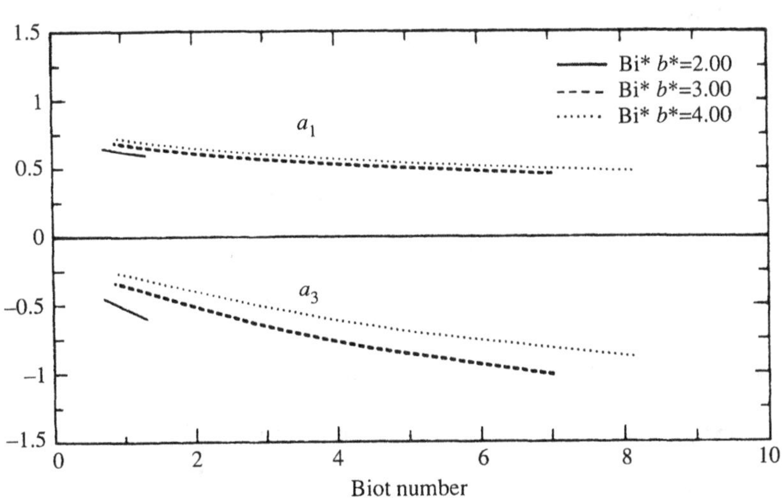

FIG. 8.7 The coefficients a_1 and a_3 for the temperature at the interface between addition and shell as a function of Bi* and Bi*b*. In the calculation $k_s/k = 1$ and $\alpha_s/\alpha = 1$.

It is recalled from eqns (8.37) and (8.7) that Bi* is proportional to $R^{\frac{1}{2}}$ and that $\tau_0 = 4\alpha t_0/R^2$. From Fig. 8.5 it is seen that τ_0 decreases with increasing Bi* roughly as 1/Bi*. Thus the shell time t_0 depends on R to a power $R^{\frac{3}{2}}$. For class II (diffusion-limited) alloy additions, the dissolution time, t_d, is proportional to R. Also for class I alloys the melting down time, t_a, is proportional to the ratio between volume and surface, that is, proportional to R (cf. (8.62b)). Thus for alloy additions where the dimensions are sufficiently small, the shell time becomes insignificant compared to the dissolution or melting down times.

8.7 MELTING DOWN TIME FOR CLASS I; HEAT TRANSFER LIMITED ADDITION OF ALLOYS

Often in practical applications it is important to determine the time, t_a, required to melt down the addition. If the addition had reached the melting-point temperature, such a calculation would have been very simple. Then $\mathrm{d}(T/T_m)/\mathrm{d}x = 0$ and eqn (8.42) reduces to

$$b\frac{\mathrm{d}s}{\mathrm{d}\tau} = -1 \tag{8.47}$$

or, integrated since $s = 1$ for $\tau = 0$,

$$s = 1 - \tau/b. \tag{8.48}$$

In terms of dimensional quantities this becomes

$$r = R - \frac{th(T_b - T_m)}{\Delta H_m \rho + c_l \rho_l (T_b - T_m)}. \tag{8.49}$$

Notice that c_l and ρ_l are specific heat and density of the molten addition, respectively. In the following we set $\rho_l = \rho$. The melting front moves inwards with a constant velocity given by the ratio between the heat transfer $h(T_b - T_m)$ and the heat requirements of melting.

In general the addition is not at the melting-point throughout. The temperature gradient in eqn (8.42) must be taken into account even though, as mentioned previously in Section 8.4, it is reduced to low values for shell times $\tau_0 > 4/\pi^2$ (cf. eqn 8.20)). An approximate solution may be obtained when $\tau_0 > 4/\pi^2$ by retaining only the first two terms in eqn (8.34). Then

$$\left\{\frac{\mathrm{d}(T/T_m)}{\mathrm{d}x}\right\}_{x=s} \approx 2b_3 s. \tag{8.50}$$

The temperature gradient is proportional to the diameter $2s$ of the addition. The proportionality constant b_3 is determined from a heat balance for the sensible heat in the addition:

Total heat conducted into the solid during the time, t_a, for melting of the addition
= The remaining sensible heat demand when melting of the addition starts

or

$$4\pi \int_0^{t_a} kr^2 \frac{dT}{dr} dt = H - H_s \tag{8.51}$$

where $r = Rs$, H is the initial heat demand for bringing the addition up to just below the melting point, and H_s is the heat supplied before melting of the addition starts.

If equation (8.50) is inserted into eqn (8.51), and $\tau = 4\alpha t/R^2$ and $\alpha = k/(c\rho)$, then

$$2b_3 \int_0^{\tau_a} s^3 d\tau = \frac{H - H_s}{c\rho T_m \pi R^3}. \tag{8.52}$$

Insertion of eqn (8.50) into eqn (8.42) results in

$$b \frac{ds}{d\tau} = \frac{2b_3 s}{\text{Bi}} - 1. \tag{8.53}$$

Integration of this equation with $s = 1$ for $\tau = 0$ gives

$$s = \left(1 - \frac{\text{Bi}}{2b_3}\right) \exp(X) + \frac{\text{Bi}}{2b_3} \tag{8.54}$$

where

$$X = \frac{2b_3\tau}{\text{Bi}\, b}. \tag{8.55}$$

When melting is completed at $\tau = \tau_a$, then $s = 0$ and

$$\exp(X_a) = 1/\beta \tag{8.56}$$

where

$$\beta = 1 - 2b_3/\text{Bi} \tag{8.57}$$

$$X_a = \frac{2b_3\tau_a}{\text{Bi}\, b}. \tag{8.58}$$

Equations (8.56), (8.57), and (8.58) give

$$\tau_a = -\frac{b \ln \beta}{1 - \beta}. \tag{8.59}$$

This gives $X_a = -\ln \beta$. Insertion of eqns (8.54)–(8.57) into eqn (8.52) results in

$$\int_0^{X_a} (-\beta \exp(X) + 1)^3 \, dX = \frac{(1-\beta)^3 (H - H_s)}{(\Delta H_m + c_l(T_b - T_m))4\pi\rho R^3}. \tag{8.60}$$

If the integration is carried out and eqn (8.57) is employed, one obtains

$$-\frac{(1-\beta^3)}{3} + \frac{3(1-\beta^2)}{2} - 3(1-\beta) - \ln\beta = \frac{(1-\beta)^3 (H - H_s)}{(\Delta H_m + c_l(T_b - T_m))4\pi\rho R^3}. \tag{8.61}$$

Notice that

$$\frac{H - H_s}{(\Delta H_m + c_l(T_b - T_m))4\pi\rho R^3}$$

on the right-hand side of eqn (8.60) is $\frac{1}{3}$ of the ratio between the remaining sensible heat per unit volume $(H - H_s)/4\pi R^3/3)$ and the heat consumed per unit volume as the addition melts. With eqns (8.61) and (8.59) one can obtain τ_a/b as a function of the ratio:

$$\frac{H - H_s}{(\Delta H_m + c_l(T_b - T_m))4\pi\rho R^3/3}.$$

τ_a/b is the ratio between the actual melting down time, τ_a, and the melting down time, b, if there was no sensible heat. Figure 8.8 presents τ_a/b as a function of $(H - H_s)/\{(\Delta H_m + c_l(T_b - T_m))4\pi\rho R^3/3\}$. It is seen that τ_a/b increases with this ratio. If the remaining sensible heat is very high compared to the heat consumed as the addition melts, $H - H_s > 6(\Delta H_m + c_l(T_b - T_m))\rho 4\pi R^3/3$, then eqns (8.59) and (8.61) reduce to the straight line:

$$\tau_a = b\left(\frac{1}{3}\,\frac{(H - H_s)}{(\Delta H_m + c_l(T_b - T_m))4\pi\rho R^3/3} + \frac{11}{6}\right) \tag{8.62a}$$

$$t_a = \frac{R}{h(T_b - T_m)}\left(\frac{(H - H_s)}{3\times 4\pi R^3/3} + \tfrac{11}{6}(\Delta H_m + c_l(T_b - T_m))\rho\right) \tag{8.62b}$$

For small values of the relative sensible heat demand $(H - H_s)/(\Delta H_m + c_l(T_b - T_m)4\pi\rho R^3/3)$, the sensible heat demand may simply be added to the heat of melting and melting is assumed to proceed at a constant rate. For large values of the relative sensible heat demand the first term in eqn (8.62b) corresponds to the assumption that the surface area remains unchanged and equal to $4\pi R^2$.

A problem with this procedure is that the heat transferred during the shell period, H_s, must be known. In Figs 8.4, 8.5, and 8.6 only the special case is presented where the thermal conductivities and diffusivities for addition and shell are equal, respectively. We are attempting to employ a more direct numerical approach instead of using Appendix 4 to calculate H_s.

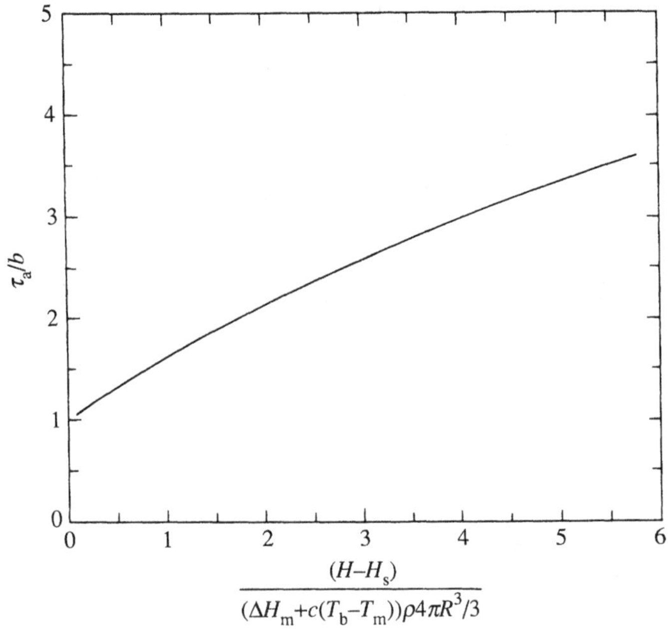

FIG. 8.8 The melting down time τ_a as a function of b (the melting time if the addition was at the melting-point) and the ratio between the remaining sensible heat and the heat requirements of melting.

8.8 EXAMPLE CALCULATION OF THE TOTAL TIME FOR MELTING OF A FeCr ADDITION

The example discussed in Section 8.3 is further pursued here. From eqn (8.37) it is found that Bi* = 1.29, and from eqns (8.37) and (8.38) one finds that

$$\text{Bi}^*\, b^* = \frac{(\Delta H_m^* \rho_s + c_l \rho_l (T_b - T_m^*))4\alpha}{c_s \rho_s (T_m^* - T(t=0))\alpha_s} \tag{8.63}$$

where α_s refers to the thermal diffusivity of the solidified shell. We see that Bi* b^* is a ratio between the heat released by solidification and cooling of the melt, and a sensible heat of the solidified shell. The factors α/α_s also enters. From Table 8.1 it is found that

$$\text{Bi}^*\, b^* = \left(\frac{250000 \times 7270 + 824 \times 7015 \times 200}{715 \times 7270 \times 1300}\right) \frac{4 \times 7.8 \times 10^{-6}}{7.7 \times 10^{-6}} = 1.78$$

From Fig. 8.5 one then obtains for the dimensionless shell time, $\tau_0 = 0.55$. This gives for the shell time:

$$t_0 = \frac{\tau_0 R^2}{4\alpha} = \frac{0.55 \times (0.5 \times 10^{-1})^2}{4 \times 7.8 \times 10^{-6}}$$

$$= 44 \text{ seconds.}$$

Here we have taken the diameter $2R$ equal to 10^{-1} m.

To obtain the final melting down time we must first calculate how much heat, H_s, has entered the addition during the shell period. The heat flux is obtained from eqns (8.43) and (8.46):

$$\dot{q}(\tau) = -\frac{(T_m - T(t=0))k}{R}\left\{a_1\left(\frac{2}{\pi^{\frac{1}{2}}\tau^{\frac{1}{2}}} - 1\right) + a_2\left(\frac{\pi^{\frac{1}{2}}}{\tau_0^{\frac{1}{2}}} - \frac{\tau^{\frac{1}{2}}}{\tau_0^{\frac{1}{2}}}\right) + a_3\left(\frac{4\tau^{\frac{1}{2}}}{\pi^{\frac{1}{2}}\tau_0} - \frac{\tau}{\tau_0}\right)\right\} \tag{8.64}$$

$$H_s = -4\pi R^2 \int_0^{t_0} \dot{q}(\tau)\mathrm{d}t = -\frac{4\pi R^2 R^2}{4\alpha}\int_0^{\tau_0} \dot{q}(\tau)\mathrm{d}\tau$$

$$= (T_m - T(t=0))\,\pi R^3 c\rho\left\{a_1\left(\frac{4\tau_0^{\frac{1}{2}}}{\pi^{\frac{1}{2}}} - \tau_0\right) + a_2\left(\pi^{\frac{1}{2}}\tau_0^{\frac{1}{2}} - \tfrac{2}{3}\tau_0\right) + a_3\left(\frac{8\tau_0^{\frac{1}{2}}}{3\pi^{\frac{1}{2}}} - \frac{\tau_0}{2}\right)\right\}. \tag{8.65}$$

From Fig. 8.7 it is found that $a_1 = 0.60$ and $a_3 = -0.60$. From eqn (8.43) it is seen that

$$a_2 = 1 - a_1 - a_3. \tag{8.66}$$

Thus $a_2 = 1.00$. Equation (8.65) gives $H_s = 2775$ kJ. Using the sensible heat per unit mass given in the FeCr row of Table 8.1, the sensible heat of the addition on taking into account the initial temperature of 200°C becomes

$$H = (1100\text{–}122)\frac{4\pi}{3}(0.05)^3 \times 7000 = 3580 \text{ kJ}.$$

Finally

$$\frac{H - H_s}{(\Delta H_m + c_1(T_b - T_m))4\pi\rho R^3/3} = \frac{(3580 - 2775) \times 10^3}{(325 \times 10^3 + 700 \times 200)\,7000\,\frac{4\pi}{3}\,0.05^3}$$

$$= 0.48$$

From Fig. 8.8 one finds $\tau_\alpha/b = 1.25$. b is obtained from eqn (8.41) and $\mathrm{Nu} = 16.7$. Thus $b = 1.53$. Then

$$\tau_a = 1.25 \times 1.53 = 1.91$$

or

$$t_a = \frac{1.91 \times (0.5 \times 10^{-1})^2}{4 \times 7.8 \times 10^{-6}} = 153.$$

Thus the total time is

$$t_{\text{tot}} = t_0 + t_a = 44 + 153 = 227\ \text{s}$$

or roughly four minutes. This value is relatively low due to the high value of the heat transfer coefficient employed. Roughly it corresponds to the 'homogenization' time for FeCr additions to an electric arc furnace equipped with electromagnetic stirrers.[33] At temperatures below the liquidus temperature we have a class II alloy, and mass transfer is limiting. Then the dissolution time for the FeCr may be as high as forty minutes.[4]

It is instructive to perform the following calculation. We disregard the shell period and simply add the sensible heat, H, to the heat of melting in eqn (8.49). This then gives for the total time

$$t_{\text{tot}} = \frac{((\Delta H_m \rho + c_l(T_b - T_m))\rho_l + H)R}{h(T_b - T_m)} \tag{8.67}$$

$$= \frac{(325 \times 10^3 + 700 \times 200 + 978 \times 10^3) \times 7000 \times 0.05}{16.7 \times \frac{40}{0.1} \times 200}$$

$$= 378\ \text{s}$$

This calculation is nearly a factor of two too high for the total time. The main reason is the assumption that the melting front moves inwards at a constant rate. In fact the addition initially actually expands during the shell period. Immediately after the shell period ends melting proceeds relatively slowly because heat transfer first covers mainly the latent heat demand. Therefore the addition has a larger surface area for heat transfer than in the simple model that leads to eqn (8.67). This is the reason that eqn (8.67) gives too long a total time. Nevertheless, eqn (8.67) is useful in practical calculations where we only wish to obtain a rough upper limit for the total time.

A more accurate calculation of the melting down time than that given by eqn (8.67) may be required. On the other hand one may not wish to get involved in calculations concerning the shell period. The simpler procedure is to describe the process in terms of two periods.

1. The addition retains its original volume and shape until the whole sensible heat demand has been satisfied. The surface is at the melting-point T_m. There is no shell.
2. When the addition finally melts, only the latent heat ΔH_m and the superheat $c(T_b - T_m)$ must be supplied to the shrinking sphere.

If we employ this model in our example with $\mathrm{Nu} = h2R/k_{\mathrm{Fe}} = 16.7$, this gives for the first period:

$$\frac{H}{h(T_b - T_m)4\pi R^2} = \frac{978 \times 10^3 \times 7000 \times 0.05}{16.7 \times \dfrac{40}{0.1} \times 200 \times 3}$$

$$= 85\ \mathrm{s}$$

Note that there is a 3 in the denominator reflecting the volume to surface area of a sphere. For the final melting down period:

$$\frac{(\Delta H_m + c_l(T_b - T_m))\rho R}{h(T_b - T_m)} = \frac{(325 \times 10^3 + 700 \times 200)\, 7000 \times 0.05}{16.7 \times \dfrac{40}{0.1} \times 200}$$

$$= 122\ \mathrm{s}.$$

Then the total time is $85 + 122 = 207$ s. This happens to be close to the value 227 s obtained when the shell period is taken into account.

8.9 SUMMARY

It is difficult to describe alloy addition to a melt in a general manner. Special problems are the lack of good data for the thermal conductivity and difficulties in the determination of the heat transfer coefficient. In the simplified treatment given here the temperature dependence of the various physical and thermodynamic quantities is neglected. Also systems have not been considered where the heat of solution is important. For instance, the addition of FeSi to molten steel is not discussed. Ferrosilicon has an exothermic heat of solution in a steel melt and thus dissolves rapidly.[29]

Calculation of the shell time was illustrated using FeCr as an example. Here it is assumed that the addition is removed as it melts.

In Section 8.8 dissolution limited by heat transfer (class I additions) was discussed. Dissolution limited by diffusion (class II additions) was treated in Section 8.2.

REFERENCES

1. Wright, J.K. (1989). Steel dissolution in quiescent and gas stirred Fe/C melts. *Met. Trans.*, **20B**, 363–74.
2. Guthrie, R.I.L., Clift, R., and Henein, H. (1975). Contacting problems associated with aluminium and ferro-alloy additions in steelmaking-hydrodynamic aspects. *Met. Trans.*, **6B**, 321–9.

3. Guthrie, R.I.L. (1977). Addition kinetics in steelmaking. *Electric Furnace Proceedings*, pp. 30–41.
4. Argyropoulos, S.A. (1990). On the recovery and solution rate of ferroalloys. *Iron and Steelmaker*, May, 77–86.
5. Joo, S. and Guthrie, R.I.L. (1988). The role of physical and mathematical models for ladle metallurgy operations. Iron and Steel section, C.I.M. Montreal, Quebec. 1–28.
6. Mair, H. and Marsh, D.B. (1989). Advancement of cored wire applications within the steel and cast metal industries. *Proceedings INFACON'89*, April, New Orleans, Louisiana. **Vol 1**, 132–50.
7. Argyropoulus, S.A. and Guthrie, R.I.L. (1982). Dissolution kinetics of ferro-alloys in steelmaking. *Proceedings Steelmaking Conference*, Pittsburgh, pp. 156–67.
8. Argyropoulus, S.A. and Guthrie, R.I.L. (1982). The dissolution of titanium in liquid steel. *Metall. Trans.* **15B**, 47–58.
9. Lehner, T. (1977). Reactor models for powder injection. Scaninject, Luleå, Sweden, pp. 11.1–11.48.
10. Engh, T.A., Larsen, K., and Venås, K. (1979). Penetration of particle-gas jets into liquids. *Ironmaking and Steelmaking*, **No. 6**, 268–73.
11. Pedersen, T. and Myrbostad, E. (1985). Alloying in aluminium by injection. International Seminar on refining and alloying of liquid aluminium and ferro-alloys, Trondheim, Norway, pp. 333–42. Aluminium-Verlag, Düsseldorf.
12. Irons, G.A. (1980). Fundamental and practical aspects of lance design for powder injection processes. Scaninject IV, Part 1, Luleå, Sweden, pp. 3.1–3.25.
13. Mucciardi, F.A. and Guthrie, R.I.L. (1983). Aluminium wire feeding in steelmaking. *Iron and Steelmaker Trans.* **Vol 3**, 53–9.
14. Herbert, A., Notman, G.K., Morris, D., Knowles, S., and Jemson, C. (1986). Experiences with powder and wire injection at British Steel Corp., Lackenby works, basic oxygen steelmaking plant. Scaninject IV, Part 2, Luleå, Sweden, pp. 27.1–27.36.
15. Külünk, B. and Guthrie, R.I.L. (1988). Optimisation studies for aluminium wire feeding operations in steelmaking ladles. *Ironmaking and Steelmaking*, **15**, No. 6, 293–304.
16. Aukrust, E. (1985). Trends in steel developments and the impact on ferro-alloy production. International Seminar on refining and alloying of liquid aluminium and ferro-alloys, Trondheim, Norway, pp. 37–48. Aluminium-Verlag, Düsseldorf.
17. Wada, H., Lee, S.W., and Pehlke, R.D. (1986). Nitrogen solubility in liquid Fe and Fe–Mn alloys. *Metall. Trans.*, **17B**, 238–9.
18. Guthrie, R.I.L. (1990). Modelling of Gransil additions to baths of liquid steel. McGill University, report.
19. Gariépy, B., Dubé, G., Paré, J. and Weaver, C. (1988). Aluminium ladle metallurgy (ALM): A new process for more efficient alloy preparation. In *Light Metals*, (ed. L.G. Boxall), pp. 457–66. Warrendale, Pa.
20. Brandes, E.A. (1983). *Smithells metals reference book*. Butterworths, London.
21. Gourtsoyannis, L., Guthrie, R.I.L., and Ratz, G.A. (1984). The dissolution of ferromolybdenum, ferroniobium and rare earth (lanthanide) silicide in cast iron and steel melts. *Proceedings of the electric furnace conference*, *AIME*, **42**, 119–32.
22. Davies, J.T. (1972). *Turbulence phenomena*. Academic Press, New York.
23. Schlichting, H. (1968). *Boundary-layer theory*. McGraw-Hill, New York.
24. Bennett, C.O. and Meyers, J.E. (1974). *Momentum, heat, and mass transfer*. McGraw-Hill, New York.

25. Guthrie, R.I.L. (1989). *Engineering in process metallurgy*. Clarendon Press, Oxford.
26. Crank, J. (1984). *Free and moving boundary problems*. Clarendon Press, Oxford.
27. Carslaw, H.S. and Jaeger, J.C. (1971). *Conduction of heat in solids*. Clarendon Press, Oxford.
28. Luikov, A.V. (1968). *Analytical heat diffusion theory*. Academic Press, New York.
29. Argyropoulos, S.A. and Guthrie, R.I.L. (1979). The exothermic dissolution of 50 wt. % ferro-silicon in molten steel. *Canadian Metallurgical Quarterly*, **Vol. 18**, 267–81.
30. Woolley, F. and Elliott, J.F. (1967). Heats of solution of aluminium, copper, and silicon in liquid iron. *Trans-TMS-AIME*, **Vol. 239**, 1872–83.
31. Argyropoulos, S.A. (1984). The effect of microexerthermicity and macroexothermicity on the dissolution of ferroalloys in steel. *Proceedings of the electric furnace conference, AIME*. **42**, 133–48.
32. Schade, J., Argyropoulus, S.A., and McLean, A. (1990). Cored-wire microexothermic alloys for tundish metallurgy. *Iron and Steelmaker*, Jan., 53–65.
33. Plöckinger, E. and Straube, H. (1965). *Die Edelstahlerzeugung*. Springer Verlag, Vienna.
34. Hultgren, R., Desai, P.D., Hawkins, D.T. Gleiser, M., Kelley, K.K. and Wagman, D. D. (1973). *Selected values of the thermodynamic properties of binary alloys*. American Soc. for metals, Metals Park, Ohio.
35. Barin, I. (1989). *Thermochemical data of pure substances*, Parts I and II, VCH Verlagsgesellschaft mbH, Weinheim, Germany.

9

SOLIDIFICATION AND REFINING

9.1 INTRODUCTION

A worker in refining metallurgy is not only concerned with attaining the specified mean composition of his melt; he must also take into consideration what happens during solidification. Problems with the accumulation of impurities may force him to reduce impurities down to very low levels. For example, to keep $\underline{P}$ below 50 p.p.m. in the grain boundaries in steel, the mean $\underline{P}$ levels may have to be less than 15 p.p.m. With such problems in mind microsegregation is treated briefly here. Microsegregation describes the variation of concentration over distances of the order of the dendrite size. Variation in the concentration of elements over much longer distances may also occur. Such distances may be part of, or may constitute the full length of the cast. This effect is called macrosegregation.

Macrosegregation is a problem not only for the final product, it also creates difficulties in attaining a known and constant composition of the alloy added to the melt before casting. For instance, ferrosilicon is usually cast in blocks and then broken up into lumps. Macrosegregation in the block may lead to a varying composition of the FeSi lump additions to the steel melt.

Small amounts of additions and stirring during casting may profoundly influence the properties of the metal. This is briefly described towards the end of this chapter.

9.2 MICROSEGREGATION

In the solidification of an alloy over a temperature range, there is a variation of solute in the growing solid dendrite from that first deposited until the transformation into solid of the final liquid. Within a scale determined by the dendrite arms spacing, this variation of composition is termed microsegregation. To describe segregation in this case one very often assumes that there is no diffusion in the solid. Then the assumption that there is complete mixing in the interdendritic liquid and no mass flow in or out of the volume element leads to a simple conservation equation for the solute.

The amount of solute in the liquid is fc_l where f is the fraction liquid and c_l is the liquid concentration of the solute. Later when the fraction liquid is $f + \Delta f$ and the solute concentration is $c_l + \Delta c_l$, the amount of solute that remains in the liquid is (see Fig. 9.1)

$$(f + \Delta f)(c_l + \Delta c_l).$$

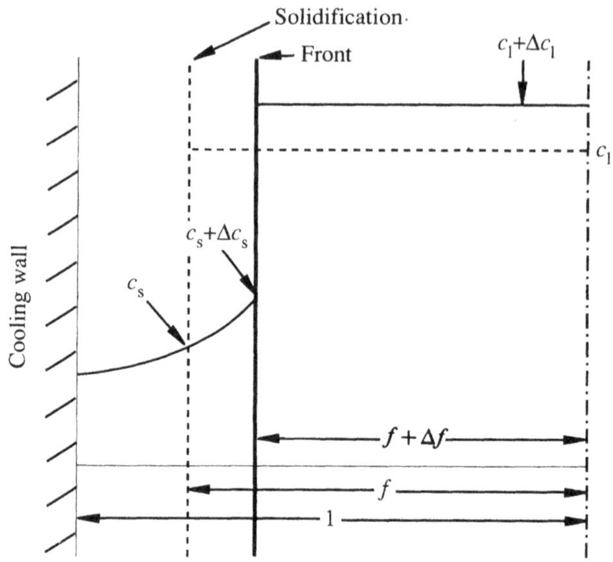

FIG. 9.1 Concentration of solute in the melt, c_l, and in the solid, c_s, as a function of fraction melt, f.

The concentration of solute in the solidified material is c_s so that the amount trapped in the solidified material is $-\Delta f c_s$. This is equal to the solute rejected by the liquid. Thus:

$$fc_l - (f + \Delta f)(c_l + \Delta c_l) = -\Delta f c_s \tag{9.1}$$

or on simplifying and dividing by $\Delta c_l f$,

$$\frac{\Delta f (c_l - c_s)}{\Delta c_l f} + \frac{\Delta f}{f} = -1.$$

If Δc_l and Δf are infinitesimal this becomes

$$\frac{\mathrm{d}f}{f} = -\frac{\mathrm{d}c_l}{c_l - c_s}. \tag{9.2}$$

The ration, c_s/c_l, called the partition ratio or coefficient, k, (see Fig. 9.12) can be obtained from a temperature versus concentration diagram (assuming no undercooling). For dilute solutions the temperature versus concentration lines may often be regarded as straight lines so that k is a constant. Then eqn (9.2) becomes

$$(k-1)\frac{\mathrm{d}f}{f} = \frac{\mathrm{d}c_l}{c_l} \tag{9.3}$$

or, integrated

$$(k-1)\int_1^f \frac{\mathrm{d}f}{f} = \int_{c_{\mathrm{in}}}^{c} \frac{\mathrm{d}c_\mathrm{l}}{c_\mathrm{l}}. \tag{9.3a}$$

This gives the Scheil equation

$$\frac{c_\mathrm{l}}{c_\mathrm{in}} = f^{k-1} \tag{9.4}$$

where c_{in} is the concentration of the solute when solidification begins. For the solid

$$\frac{c_\mathrm{s}}{c_\mathrm{in}} = kf^{k-1}. \tag{9.5}$$

Figure 9.2 shows the effect of a decreasing k, that is, the effect of the liquid rejecting the solute component. If $k = 1$, there is no segregation. If $k = 0.1$ the solute component is mainly found in the last fraction to solidify.

Equation (9.5) must be regarded as an extreme case of segregation. In fact mixing of the liquid is only partial. There is a local variation of solute from the centre of the exterior of the dendrite arms. There may also be rapid solid state diffusion during solidification, particularly for carbon in cast iron and steel. These various effects tend to give a reduced 'effective' value of k, k_{eff}, often employed in an empirical version of the Scheil equation:

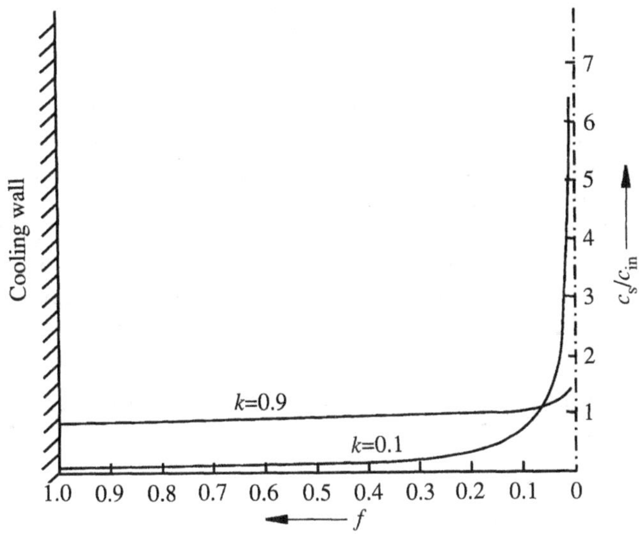

FIG. 9.2 Ratio $c_\mathrm{s}/c_\mathrm{in}$ between the concentration of solute in solidified material, c_s, and the concentration before solidification, c_{in}, as a function of fraction melt, f.

$$\frac{c_s}{c_{in}} = k_{eff} f^{k_{eff}-1}. \tag{9.6}$$

We will find that the Scheil eqn (9.6) can also be used to describe macrosegregation. Then convection effects and solidification shrinkage is also taken into account.

9.3 MACROSEGREGATION

During solidification, segregation of the different elements occurs very often. The variation in concentration of an element over a distance of the order of the dendrite size, called microsegregation, was discussed in the previous section. Variation in the concentration of an element over much longer distances, called macrosegregation, is treated here.

There may be several reasons for macrosegregation in a cast. It is also important to realize that macrosegregation may or may not be beneficial to the quality of the cast. Sometimes, when some macrosegregation is unavoidable a clever treatment of the cast may reduce the harm due to the segregation.

Many authors have discussed various mechanisms for macrosegregation.[1-3] Macrosegregation of an element in a cast is caused by:

(1) a different solute content of the solid and the liquid phase, combined with
(2) separation of the solid and the liquid phase.

A difference in the solubility of an element in the liquid and the solid phase is the normal situation during solidification so that the equilibrium partition ratio (or the effective partition ratio k_{eff}) is different from 1.

Table 9.1 gives a list of mechanisms that may separate liquid and solid. Normally in a cast the liquid phase may move while the solid phase stays fixed.

In the following a study of casting of FeSi is mentioned as an example of segregation during casting. Ferrosilicon is used as an alloy addition in steelmaking (see Chapters 6 and 8) in which case it is important that the content of Si is the same throughout the charge material. The cast is a square block of iron–silicon with a content of 75 wt% Si.[4] When a liquid of 75 wt% Si–25 wt% Fe is cooled down to 1315°C the silicon starts to precipitate. The

TABLE 9.1 *Mechanisms that give separation of solid and liquid*

Liquid motion in mushy zone may be:
(1) due to temperatures gradients;
(2) caused by concentration gradients;
(3) induced by thermal contractions;
(4) caused by convection in the bulk melt which in turn may be induced by temperature or concentration gradients, thermal contractions, or by electromagnetic stirring.

liquid will be enriched in iron until the eutectic composition is reached. Normally, with rapid solidification, the cast consists of fine grains of silicon and $FeSi_2$ (actually non-stoichiometric $FeSi_{2.5}$). The melt has been cooled both from the bottom and from the top. When cooled from the bottom little vertical segregation seems to occur. The concentration in the solidified material is very nearly the same as in the melt, about 75% Si. This is shown in Fig. 9.3. Figure 9.4 shows that when the block is unidirectionally cooled from the top with approximately the same solidification conditions, more segregation occurs and the concentration of Si is higher. To ensure that only a planar solidification front is considered, only data from the first part of the solidification is included in Figs 9.3 and 9.4.

The reasons for the segregation with cooling from the top can easily be explained. As mentioned above the silicon will precipitate at the solidification front. The rejected melt will therefore be enriched in iron. This gives a heavier liquid compared to the bulk melt. This heavy liquid sinks downwards and induces a return convective flow of bulk liquid. Finally, the iron-enriched liquid will solidify in the bottom of the block. The result is a segregation profile as shown in Fig. 9.4.

When the block is cast with unidirectional cooling from the bottom a different segregation pattern occurs. The solidification front will consist of solid precipitated silicon and liquid iron–silicon. The rejected iron-rich liquid will be of higher density than the bulk density and will allow little bath movement. This results in less segregation as shown on Fig. 9.3.

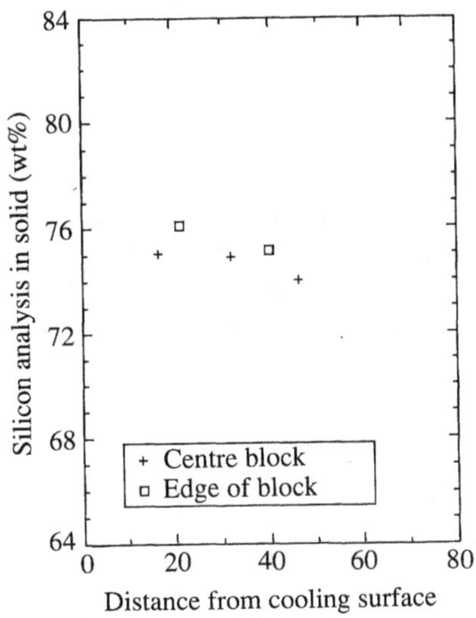

FIG. 9.3 Si analysis in the solid as a function of the distance from the cooling surface. Cooling of the melt is from the bottom.

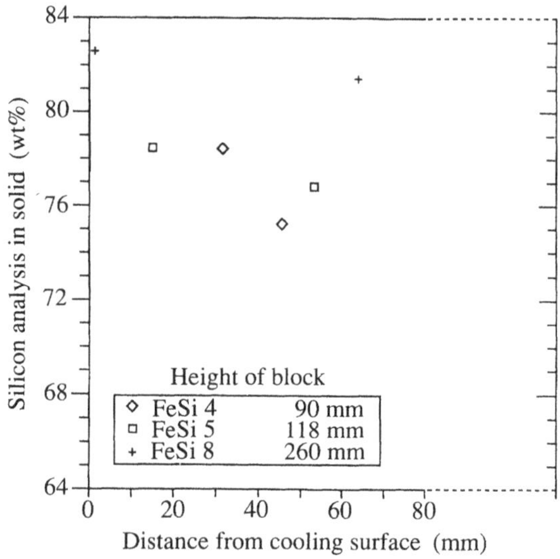

FIG. 9.4 Si analysis in the solid as a function of the distance from the cooling surface. Cooling of the melt is from the top.

In summary the following two important points should be taken into account for FeSi.

1. The temperature–composition diagram (the partition ratio between solid and liquid) during solidification. Under thermodynamic equilibrium conditions only silicon should precipitate at temperatures above 1207°C. If iron is regarded as the solute, $k_{Fe} \approx 0$.
2. The effect of liquid–solid separation during the solidification. The density of the melt depends on temperature and composition. Due to density differences, gravity creates bath movement and liquid–solid separation when cooled from the top. Gravity may prevent the bath movement (natural convection) when cooled from below since the heavier iron-rich liquid phase tends to remain at the bottom. Thus, the direction of the solidification front is important.

9.3.1 Modelling of macrosegregation

In general, macrosegregation is caused by a number of mechanisms and is a complex process. The Scheil eqn (9.6) simply describes macrosegregation as complete mixing in the liquid zone and no diffusion in the solid.

A number of studies have been performed to describe the macrosegregation which results from mass flow of solute-rich liquid to feed solidification and thermal contractions. Flemings[1] gives an equation for the macrosegregation as a result of interdendritic flows:

$$\frac{\partial f}{\partial c_1} = -\frac{(1-\beta)}{(1-k)}\left(1 - \frac{v}{\mathrm{d}z/\mathrm{d}t}\right)\frac{f}{c_1}. \tag{9.7}$$

Equation (9.7) is for the one-dimensional case when the solidification front moves with velocity $\mathrm{d}z/\mathrm{d}t$ in the z-direction. v is the local velocity of the interdendritic liquid relative to the solid.

$$\beta = (\rho_s - \rho)/\rho$$

is the solidification shrinkage and

$$1 - \beta = \rho/\rho_s.$$

If v is zero, eqn (9.7) reduces to eqn (9.3), expect that now the difference in liquid and solid densities is taken into account. Often the solidification velocity $\mathrm{d}z/\mathrm{d}t$ in eqn (9.7) is given in terms of the rate of temperature change $\partial T/\partial t$ just after solidification very close to the solidification front and the corresponding temperature gradient $\partial T/\partial t$. Now since temperature is a function of position z and time:

$$T = T(z, t)$$

and

$$\mathrm{d}T = \left(\frac{\partial T}{\partial z}\right)_t \mathrm{d}z + \left(\frac{\partial T}{\partial t}\right)_z \mathrm{d}t. \tag{9.8}$$

On the solidification front the temperature is constant so that

$$\mathrm{d}T = 0.$$

Then

$$\frac{\mathrm{d}z}{\mathrm{d}t} = -\frac{\dfrac{\partial T}{\partial t}}{\dfrac{\partial T}{\partial z}}. \tag{9.9}$$

Now eqn (9.7) becomes

$$\frac{\partial f}{\partial c_1} = -\frac{(1-\beta)}{(1-k)}\left(1 + \frac{v\,\partial T/\partial z}{\partial T/\partial t}\right)\frac{f}{c_1}. \tag{9.10}$$

Recently, several attempts have been made to calculate numerically the liquid fluid flow in the mushy zone during solidification. The procedure requires a calculation of the interdendritic fluid flow. Often this is performed by using Darcy's law.[5,6] To describe macrosegregation this equation, together with mass balances, momentum, and energy balances, has been solved numerically in recent years.[1] However, this procedure will not be examined here.

In the following it will be assumed, as in the previous section, that there is complete mixing in the bulk liquid with concentration c_b. Furthermore, we usually assume that there is no diffusion in the solid so that eqn (9.6) applies. Then c_l is replaced by c_b and the excess solute in the mushy zone is neglected. A total mass transfer resistance $1/k_t$ is introduced which sums resistance to mass transfer in the mushy two-phase zone and in the boundary layer between solid and bulk/melt. k_t must be regarded as empirical.

9.4 THE MASS TRANSFER COEFFICIENT k_t FROM SOLID TO BULK LIQUID

Figure 9.5 sketches the solute,c, concentration profile from the solid to the bulk liquid. The solid is moving with velocity $v = \mathrm{d}z/\mathrm{d}t$ to the right. Relative to the solid the liquid moves to the left with velocity $-v$. To take mass transfer into account a turbulent diffusion coefficient, D_E, is introduced that is much greater than the atomic diffusion coefficient. Then the flux of solute is given by

$$\dot{n} = -cv - D_E \frac{\mathrm{d}c}{\mathrm{d}z} \tag{9.11}$$

and in the solid

$$\dot{n}_s = -c_s v. \tag{9.12}$$

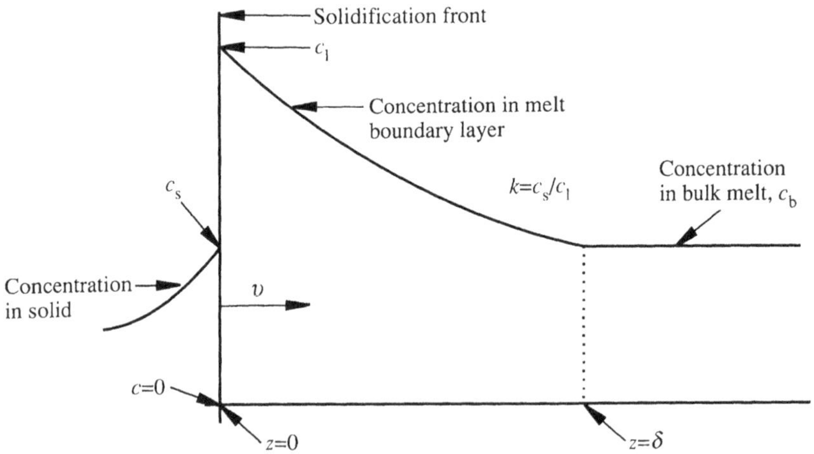

FIG. 9.5 Concentration of solute in the melt boundary layer at a solidification front moving with velocity, v.

At $z = 0$, $\dot{n} = \dot{n}_s$. Since no solute is consumed in the boundary layer, the flux and v may depend on time but not on z, and the solution of the differential equation (9.11) becomes

$$c = A \exp\left(-\frac{vz}{D_E}\right) + B \tag{9.13}$$

where A and B are integration constants. They are determined by $\dot{n} = \dot{n}_s$, eqn (9.12), and the condition that $c = c_b$ at $z = \delta$ where δ is the thickness of the diffusion boundary layer. This gives

$$\frac{c_b - c_s}{c_l - c_s} = \exp\left(-\frac{v\delta}{D_E}\right)$$

or, if we introduce the mass transfer coefficient $k_t = D_E/\delta$ (see Chapter 7) to replace the quantities D_E and δ,

$$\frac{c_b - c_s}{c_l - c_s} = \exp\left(-\frac{v}{k_t}\right). \tag{9.14}$$

From this equation we see that if the mass transfer coefficient is small, that is $k_t \ll v$, the solid concentration, c_s, is the same as the bulk melt concentration, c_b. If $k_t \gg v$ the liquid concentration at the interface, c_l, is the same as c_b.

To integrate the Scheil equation in the form

$$\frac{df}{f} = -\frac{dc_b}{c_b - c_s} \tag{9.2a}$$

it is seen that we must know the ratio

$$k_{eff} = \frac{c_s}{c_b}. \tag{9.15}$$

This is obtained from eqn (9.14) on inserting $c_l = c_s/k$. Then

$$k_{eff} = \frac{k}{k + (1 - k)\exp\left(-\frac{v}{k_t}\right)}. \tag{9.16}$$

v/k_t may in general depend on time or on fraction liquid, f. If $v/k_t \ll 1$, then $k_{eff} = k$. If $v/k_t \gg 1$, then $k_{eff} = 1$. If v/k_t does not depend on f and we insert eqn (9.15) into (9.2a) and integrate, we obtain eqn (9.6).

9.5 SEGREGATION OF ALLOYS DISPLACED FROM THE EUTECTIC COMPOSITION

It will be seen that if the mass transfer coefficient, k_t, is too large, considerable segregation may occur during solidification from alloys removed from the eutectic composition. The results are of a somewhat general nature, although FeSi is used as an example.

If k_t is small and the ratio v/k_t is large, the iron that is rejected during precipitation of Si from the FeSi melt with about 25% Fe accumulates in the mushy zone. Then after a short transient period the mushy zone reaches the eutectic composition, c_E, about 42% Fe. Then eqn (9.14) becomes

$$\frac{c_b - c_s}{c_E - c_s} = a \tag{9.14a}$$

where, to simplify matters, $\exp(-v/k_t)$ has been replaced by a. From eqn (9.14) c_b can be given in terms of c_s and eqn (9.2a) can be integrated. The initial condition is $c_b = c_{in}$ for $f = 1$. Equation (9.14a) is employed to make use of this condition. Then rearranging slightly

$$c_E - c_s = \frac{c_E - c_b}{1 - a}. \tag{9.14b}$$

If a is constant for all f, we obtain from eqns (9.2*a*) and (9.14b)[2]

$$\frac{c_E - c_s(f)}{c_E - c_{in}} = \frac{f^{a/(1-a)}}{(1 - a)}. \tag{9.17}$$

It is seen that if k_t, and therefore a, is small there is little or no segregation. This is the situation in Fig 9.3. Figure 9.4 shows the segregation encountered when solidifying from the top. In experiments 4, 5, and 8 the solidification velocity v was estimated to be 4×10^{-5} m/s, 3.4×10^{-5} m/s, and 7×10^{-5} m/s, respectively.[4] From eqn (9.17) it is found from Fig. 9.4 that $v/k_t = 1.3$, 1.9, and 2.7, respectively. This gives $k_t = 3 \times 10^{-5}$ m/s, 1.8×10^{-5} m/s, and 2.6×10^{-5} m/s. Mass transfer in the melt boundary layer is caused by natural convection induced by temperature and concentration gradients in the melt. The effect of concentration gradients probably dominates. Problem 9.1 attempts to calculate the k_t caused by the concentration gradients in the experiments and it is found that $k_t \sim 4.5 \times 10^{-5}$ m/s. This value is roughly a factor of two too high. However, in the calculation, resistance to mass transfer in the mushy zone has not been taken into account. The admittedly rough correspondence with the calculated value for k_t indicates that k_t can indeed be obtained from computer calculations that should include both the temperature and the concentration effects that cause the natural convection in the mushy zone and in the melt.

If the ratio v/k_t is very small, there is no accumulation of iron in the mushy zone. The iron is instead transferred directly to the bulk melt. Silicon only is precipitated. Then the bulk melt should reach the eutectic composition c_E before precipitation of the eutectic can come about. In problem 9.2 we find the ratio between the amounts of pure silicon and of eutectic that is precipitated from FeSi with 75% Si in this simple, idealized case.

9.6 REFINING OF ALLOYS BY PARTIAL SOLIDIFICATION

Refining by partial solidification or melting is based on the fact that when a melt containing one or more solutes is cooled, or conversely a solid is melted, the crystals have a different impurity content from the liquid. An example of industrial applications is the elimination of high melting-point impurities and dross from molten lead (see Section 3.5.3). The methods conventionally employed are skimming or sedimentation. At somewhat greater costs sedimentation can be enhanced by centrifuging. Filtration is employed to remove dross and intermetallics from molten metals such as aluminium and magnesium. The main practical difficulty in all cases is usually an efficient and complete physical separation of solid and liquid. Traditionally, this has limited refining by fractional solidification to processes where only a small fraction solid was removed. Techniques such as filtration under high pressures have improved the separation of solid and liquid.[7] This is particularly the case if a non-dendritic structure is formed. This development allows refining by fractional solidification for higher solid fractions, for instance, up to $\frac{1}{2}$.

Here we look at the application where a solidified fraction with a low content

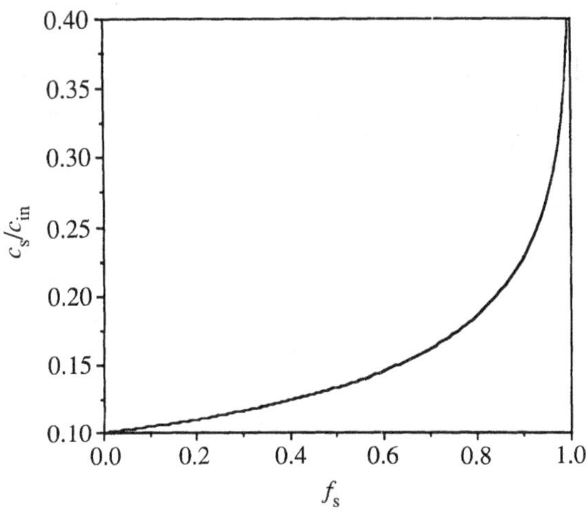

FIG. 9.6 The refining ratio as a function of fraction solidified, $f_s = 1 - f$, for a partition ratio $k = 0.1$.

of solutes (impurities) is extracted from a molten alloy. This is possible in many cases since very often the partition coefficient (see Fig. 9.2) for the solute is less than 1, for instance 0.1. Again the Scheil equation, eqn (9.5), may be employed. It is integrated from liquid fraction $f = 1$ to f. Then the mean relative impurity content, $\bar{c}$, called the refining ratio in the solidified part, $1 - f$, is

$$\bar{c}_s = \frac{1}{1 - f} \int_f^1 c_s \, df$$

or

$$\frac{\bar{c}_s}{c_{in}} = \frac{k}{1 - f} \int_f^1 f^{k-1} \, df = \frac{1 - f^k}{1 - f}. \tag{9.18}$$

Equation 9.18 is shown in Fig. 9.6 for $k = 0.1$. If there is diffusion in the solid the ratio $\bar{c}_s/c_{in}$ is greater. Furthermore, the low refining ratios illustrated in the figure are achieved only if the liquid is completely removed from the solid.

9.7 REFINING OF ALLOYS BY CONTINUOUS DRAINING OF LIQUID

This process[8] involves heating an alloy within its liquid–solid region while simultaneously draining off the (interdendritic) liquid. The liquid must be continuously expelled during heating, and heating must proceed at such a slow rate that diffusion in the solid has time to proceed. Filtration under high pressures is the method of removal. As shown theoretically in the following, remarkably low refining ratios may be attained. However, residual liquid increases the refining ratios significantly. Here we look at the ideal case that there is no residual liquid.

To describe the process we imagine that the temperature increases in small steps and that the temperature remains constant for a time sufficient for the mean solid composition $\bar{c}_s$ to come to equilibrium with the liquid c_l:

$$\bar{c}_s = kc_l. \tag{9.19}$$

Notice that now $\bar{c}_s$ is representative of the solute composition throughout the solid particles. The liquid with composition c_l is removed in each temperature interval and can no longer influence what happens later when the temperature is again increased slightly. Therefore we have

$$\frac{dc_l}{dT} = 0 \tag{9.20}$$

for the change of solute concentration with temperature T.

The solute is either in the solid or liquid state so that

$$\bar{c}_s f_S + c_l f = \text{constant} \tag{9.21}$$

or, in terms of temperature increase,

$$\frac{\mathrm{d}}{\mathrm{d}T}(\bar{c}_s f_S + c_l f) = 0 \tag{9.22}$$

where f_S is the solid fraction so that

$$f_S + f = 1. \tag{9.23}$$

Now eqns (9.20) and (9.22) give the following when the temperature is increased:

$$\mathrm{d}(\bar{c}_s f_S) + c_l \mathrm{d}f = 0 \tag{9.24}$$

and employing eqns (9.19) and (9.23),

$$\frac{\mathrm{d}\bar{c}_s}{\bar{c}_s} = \left(\frac{1}{k} - 1\right)\frac{\mathrm{d}f_S}{f_s}.$$

Integration from $\bar{c}_s = c_{in}$ for $f_S = 1$ results in

$$\frac{\bar{c}_s}{c_{in}} = f_S^{1/k-1}. \tag{9.25}$$

For instance, if half the material is removed by partial melting, the refining ratio for the remainder is, for $k = 0.1$

$$\frac{\bar{c}_s}{c_{in}} = 0.5^9 = 0.002.$$

This startling result depends, as previously pointed out, on an efficient continuous removal of interdendritic liquid (see also problem 9.3) and on heating so slowly that the solid is always in equilibrium with the liquid that is being melted off.

9.8 ZONE REFINING

Zone melting, illustrated in Fig. 9.7, can avoid contamination from a container (crucible). Its potential for ultrapurification of materials was first recognized when it was applied to germanium. The priciple can also be used to grow single crystals and to obtain a controlled uniform or non-uniform composition within the crystal.

Only a small portion of the bar is melted and this molten zone is moved slowly from one end of the bar to the other. Movement is always in the same direction, for instance from bottom to top of the vertical bar. Figure 9.8 shows the solute distribution during and after one pass. The concentration in the solidified material is $c_s = c_l k$ or if the heater is moving too fast only $c_l k_{eff}$ where k_{eff} is

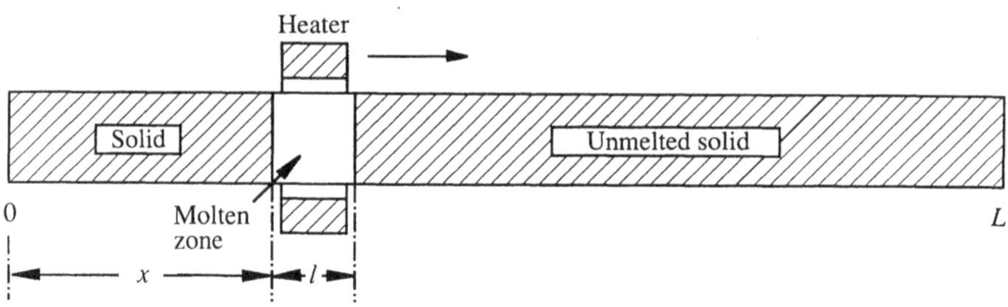

FIG. 9.7 Molten zone of length l traversing a cylindrical ingot of length L.[9]

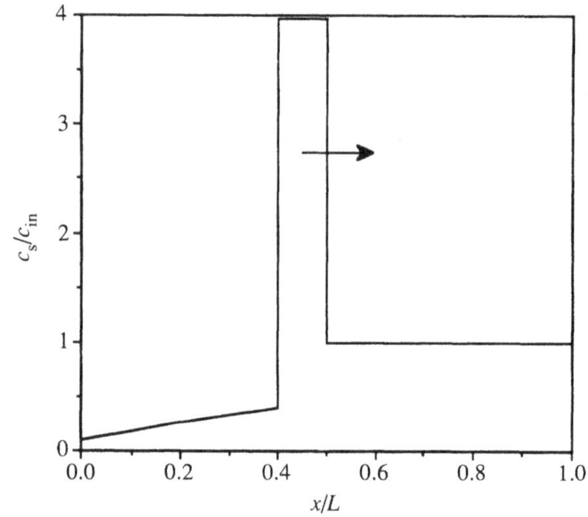

FIG. 9.8 Solute distribution in zone refining during one pass. $l/L = 0.1$ and $k = 0.1$.

given by eqn (9.16). The rejected solute accumulates in the molten zone. When the heater has moved a distance dx, the accumulated solute is

$$l\mathrm{d}c_l = \mathrm{d}x(c_{in} - c_s) \tag{9.26}$$

where c_{in} is the constant original composition of the bar. l is the length of the molten zone. At $x = 0$ there has been no accumulation of solute so that here $c_l = c_{in}$. Integration of eqn (9.26), assuming that the zone length l and k or k_{eff} are constants, gives

$$\frac{c_s}{c_{in}} = 1 - (1 - k)\exp(-xk/l). \tag{9.27}$$

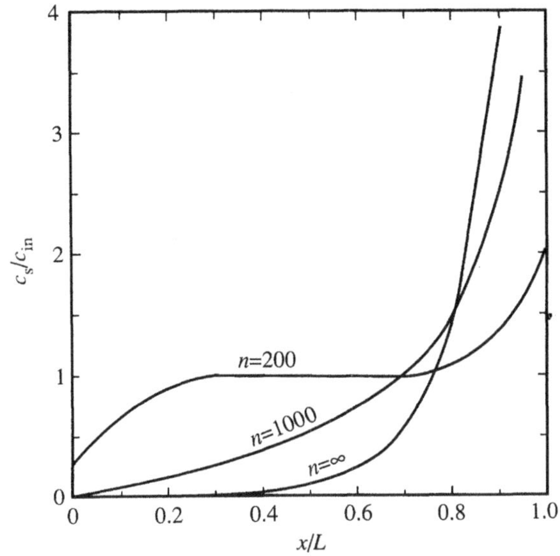

FIG. 9.9 The refining ratio versus distance x/L with $l/L = 0.01$ and a large, k, $k = 0.9524$. n is the number of passes of the molten zone.[9]

There is no simple way to calculate the solute distribution after multipass zone melting, but a large number of computed curves are given by the inventor of this process W.G. Pfann.[9] One set of these curves is shown in Fig. 9.9 for a single-phase alloy with a very high effective partition ratio, $k_{\text{eff}} = 0.9524$ and a ratio of zone length to ingot length of 1:100. n is the number of passes by the heater. It is seen that the solute is slowly displaced to the right. For an infinite number of passes there is a limiting curve which may be calculated in a rather simple manner since it does not depend on time. When the heater moves a distances dx, the difference between the solute that leaves the molten zone and the solute that enters is $\mathrm{d}x(c(x+l) - c(x))$. This is equal to the accumulation

$$l\mathrm{d}c_{\mathrm{l}}.$$

Furthermore

$$c_{\mathrm{l}} k_{\text{eff}} = c_{\mathrm{s}}(x)$$

so that

$$\frac{l}{k_{\text{eff}}} \frac{\mathrm{d}c_{\mathrm{s}}(x)}{\mathrm{d}x} = c_{\mathrm{s}}(x+l) - c_{\mathrm{s}}(x). \tag{9.28}$$

A solution of this equation is

$$\frac{c_s}{c_{in}} = A \exp(bx) \tag{9.29}$$

where b is found by inserting eqn (9.29) into eqn (9.28):

$$\frac{lb}{k_{eff}} = \exp(lb) - 1. \tag{9.30}$$

A is obtained from the condition that the solute remains in the ingot:

$$\int_0^L c_s(x)\mathrm{d}x = c_{in}L \tag{9.31}$$

where L is the ingot length. Thus

$$\frac{A}{b}(\exp(bL) - 1) = L. \tag{9.32}$$

Equation (9.30) gives, for $k_{eff} = 0.9524$

$$\frac{bl}{0.9524} = \exp(bl) - 1$$

with the solution $bl = 9.68/100$. Furthermore, from eqn (9.32):

$$A = \frac{bL}{\exp(bL) - 1} = \frac{9.68}{\exp(9.68) - 1} = 0.0006.$$

This gives for eqn (9.29)

$$\frac{c_s}{c_{in}} = 0.0006 \exp(9.68\, x/L).$$

At $x/L = 0.77$, c_s is equal to the original concentration c_{in}. At the mid-point of the ingot, $x/L = 0.5$,

$$\frac{c_s}{c_{in}} = 0.076.$$

In problem 9.4 we determine the limiting curve for $k = 0.9$ and $l/L = 1/100$.

9.9 NUCLEATION AND GRAIN REFINING

Most of the properties of metals and alloys are affected by the as-cast structure, even after extensive deformation, heat treatment, and recrystallization have taken place. The reason is that the treatment of the solid usually cannot substantially change the distribution of constituents and inclusions. Hence, the most important step in the control of metallic properties is to achieve the best possible as-cast structure. This is often the responsibility of a process metallurgist working in the cast house.

Generally, the objective has been to create a structure of fine, equiaxed grains with random orientation. However, a fine grain is not always the most important condition for optimal properties. Especially in multiphase alloys and in materials to be extensively worked, a small interdendritic spacing that produces a good distribution of phases and inclusions may be more important than fine grain. Fortunately grain refining often reduces the variations in concentrations in addition to producing fine grain.

Grain refining in casting can be produced only by increasing the number of nucleating centres, so that a large number of crystals is formed which soon impinge on each other and prevent further growth. Hence enhanced nucleation is the key to grain refining. These nucleation centres must be solid in the melt. It is important that they survive in the solid state in contact with the surrounding melt at its temperature. They are thus derived from solid nucleants already present in the melt or are added purposely as centres for nucleation. The nucleants are covered by solidified layers of a phase that acts as nuclei for solidification.

Many authors have given requirements for a good nucleant, but none of the listings seem to be generally applicable.[10] The most widely accepted relationship between nucleant and nucleus is the similarity of lattice. It is an advantage that the nucleant has a crystal structure and lattice parameters very close to those of the nucleus. However, some nucleants are effective even though only a few atoms in the lattices coincide.[10]

Furthermore, the nucleus must 'wet' the nucleant. Perhaps the best approach for the study of wetting is to look at the three-phase system; nucleant, solidified melt (nucleus), and melt. Such three-phase systems were discussed in Section 5.5. The interfacial tension between solid and liquid metal was examined in Section 2.2.8. If the nucleant is wetted by the nucleus, the growing nucleus can spread out over the nucleant. With a good fit between the lattice of the nucleant (phase 2) and nucleus (phase 3) the interfacial tension σ_{23} may be small enough to be negligible. Then according to the Young–Dupré equation, eqn (5.10), the nucleus spreads out if $\sigma_{12} > \sigma_{13}$. Thus a comparison of the nucleant–melt interfacial tension, σ_{12}, with the nucleus–melt interfacial tension, σ_{13}, is needed to decide if a crystal is a good nucleant. The problem is that very little reliable data is available for solid–liquid[11] and solid–solid interfacial tensions.

Practical methods to attain grain refining are discussed in the following.

9.9.1 Nucleating additions

The grain refinement method most used industrially is the addition of nucleating agents, known as grain refiners or inoculants. This is especially true for the Al, Mg, and Cu alloys where the additions may be so small (<0.10 per cent) that they do not appreciably affect other properties. For Ni alloys a substantial amount (>1 per cent) must be added to get the refining effect.

Grain refining of Al by Ti addition was discovered more than half a century ago. Nevertheless, two theories are still presented to explain the mechanism of grain refinement. One theory contends that TiB_2 or TiC (particles formed by the reaction of Ti with carbon impurities) are the nucleating agents. TiB_2 and TiC as well as other carbides, borides, and nitrides have crystal structures with planes that closely match some of the aluminium planes. Then growth of aluminium on the nucleant can take place with little distortion of the lattice.

The second theory stresses the importance of Al_3Ti particles created by he addition of Ti. On cooling a melt containing over 0.15 wt% Ti (see Fig. 9.10),[12] primary Al_3Ti solid particles first separate, and then the following peritectic reaction (solid + liquid = solid) takes place at the crystal-melt interface:

$$Al_3Ti(s) + Al(l) = \underline{Ti} \text{ (solid solution in alpha-aluminium).} \quad (9.33)$$

However, this explanation based on equilibrium considerations seems to break down when additions of Al–Ti–B equivalent to 0.005% Ti prove to be effective

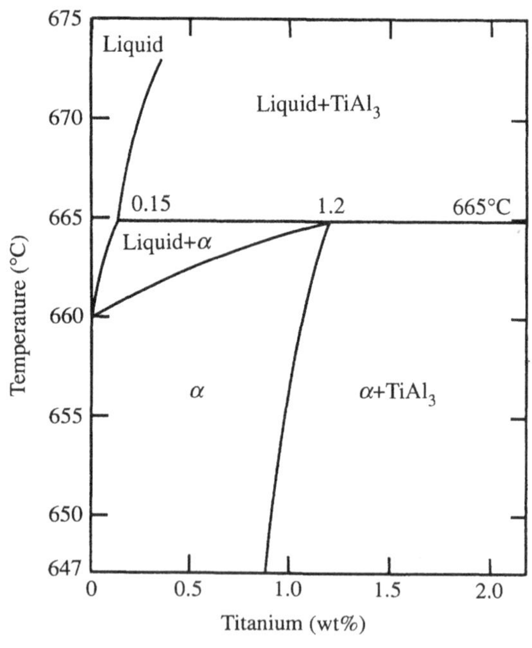

FIG. 9.10 Aluminium-rich end of the Al–Ti composition versus temperature diagram[12] showing the peritectic at 665°C and 0.15% Ti.

grain refiners. To explain this situation arguments of a kinetic nature have been mobilized;[13] some of the Ti in the TiB_2 inoculant is replaced by Al to form a $(TiAl)B_2$ phase. This phase in turn reacts with more Al to form a layer of Al_3Ti on the TiB_2 particle. Finally, this Al_3Ti nucleates the alpha-Al according to eqn (9.33). At the addition level of a few hundreds of a per cent of Ti, normal in grain refining practice, the equilibrium liquidus temperature of the alpha-Al is some tenths of a degree above the melting-point. It is assumed that these alpha crystals are the actual nuclei already present when the freezing temperature is reached, so that crystalization can start without need for under-cooling for nucleation.

The behaviour of grain refiners changes with holding time in the melt. Figure 9.11 shows the dependence of grain size on holding time for some typical grain refiners in aluminium. It is seen that there is an optimal holding time; beyond this point in time the effect fades.

In order to combat the problem of fade and to make more economic use of the grain refiner addition, the addition is usually made continuously in the form of a 9 mm master alloy rod, containing typically 5% Ti and 1% B.[14] The rod is best added where turbulent mixing occurs in order to speed up the dissolution of the aluminium matrix and disperse the molten borides in the molten metal. Often the refiner is added to the gas purging unit (see Chapter 4). Then the time between adding the grain refiner and the nucleant reaching the casting unit

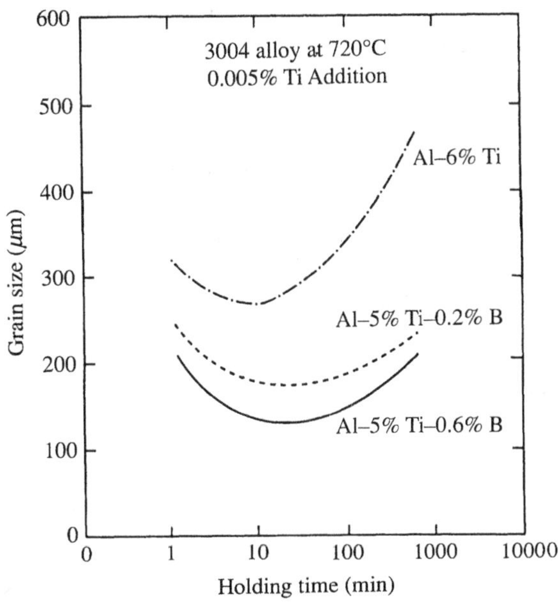

FIG. 9.11 Grain size as a function of holding time in molten metal for some commonly used master alloys for aluminium alloys.[14]

roughly corresponds to the optimal holding time (see Fig. 9.11). Probably loss of grain refiner due to flotation is not a serious problem since the TiB_2 does not seem to wet the bubbles. However, clustering of the TiB_2 particles caused by the turbulent stirring may be a greater problem.

Zirconium is extensively used to grain-refine magnesium alloy castings.[15] The effect is usually explained in terms of a peritectic reaction in which Zr particles exist in the melt above the melting-point. The peritectic temperature is 654°C compared to the melting-point of Mg at 650°C. The Zr particles can react with the melt acquiring a layer of Zr-enriched solid solution. This layer serves as nuclei during solidification. However, according to the equilibrium temperature versus composition diagram for the Mg–Zr system, Zr particles should only form when the zirconium content exceeds 0.55 per cent. In fact, grain refining occurs with contents around 0.1 per cent. Kinetics arguments have been introduced to clear up this problem. However, some workers have explained the effect in terms of nucleation by Zr or a zirconium compound. Evidently, the mechanisms suggested for Zr grain refining of Mg are very similar to those proposed for Ti-based grain refining of Al alloys.

9.10 GRAIN REFINING, CONSTITUTIONAL SUPERCOOLING, AND STIRRING

In the preceding chapter we have seen how a solute-rich layer normally builds up in front of a solidifying planar interface. For temperature versus composition systems as illustrated in Fig. 9.12 the liquidus temperature drops with an increased content of solute. Therefore there is a zone in front of the solidification front where the equilibrium freezing temperature is lower than in the bulk melt. Since heat is being drawn away from the melt, the temperature is lower than in the bulk melt. However, there may be a zone in front of the solid–liquid interface where the actual temperature is lower than the local equilibrium freezing temperature yet where the liquid metal persists. In this case we say that the liquid is constitutionally supercooled. The term 'constitutional' reflects the importance of the solute composition change (see Fig. 9.5). In a supercooled zone nucleation may be enhanced giving a smaller grain size. A criterion for the existence and extent of the supercooled zone is given in the following.

The concentration gradient dc/dz, of the solute at the interface is found from eqns (9.11) and (9.12):

$$\frac{dc_l}{dz} = -\frac{v}{D_E} c_l(1-k). \tag{9.34}$$

The relationship between temperature and composition is obtained from the slope of the liquidus line in Fig. 9.12:

$$\frac{\Delta T_l}{\Delta c} = -b. \tag{9.35}$$

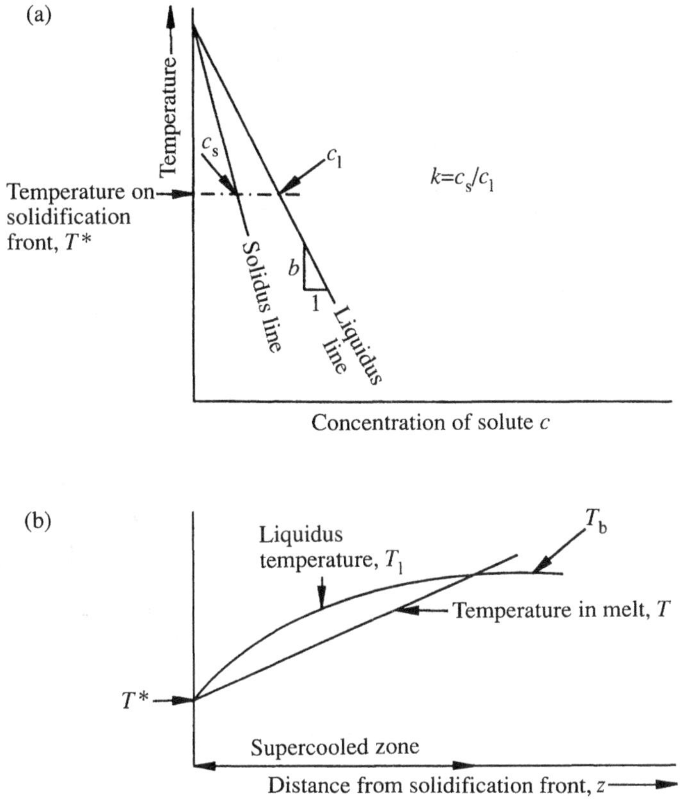

FIG. 9.12 Constitutional supercooling in alloy solidification: (a) temperature versus composition diagram; (b) comparison of the liquidus temperature profile with the melt temperature.

Equations (9.34) and (9.35) then give, for the slope of the liquidus temperature at the solidification interface,

$$\left(\frac{dT_l}{dz}\right)_{z=0} = \frac{v}{D_E} b(1-k)c_l. \tag{9.36}$$

If there is to be a supercooled zone, dT_l/dz must be greater than the temperature gradient $(dT/dz_{z=0})$:

$$\left(\frac{dT_l}{dz}\right)_{z=0} > \left(\frac{dT}{dz}\right)_{z=0}$$

or

$$\frac{vb}{D_E}(1-k)c_l > \left(\frac{dT}{dz}\right)_{z=0} \tag{9.37}$$

where c_l is obtained from eqn (9.14). If there is little convection, that is, if the mass transfer coefficient is small, then $c_l = c_b/k$ and eqn (9.37) becomes

$$\frac{vb(1-k)c_b}{D_E k} > \left(\frac{dT}{dz}\right)_{z=0}. \tag{9.38}$$

When stirring is vigorous $c_l \approx c_b$ and the criterion for supercooling becomes

$$\frac{vb(1-k)c_b}{D_E} > \left(\frac{dT}{dz}\right)_{z=0}. \tag{9.39}$$

It is seen that the conditions for supercooling are favoured by a high velocity of solidification, v, and a large value of b (a steep equilibrium liquidus temperature versus composition curve). Furthermore, it is an advantage to minimize the temperature gradient dT/dz. In practice this may be accomplished by reducing the pouring temperature and by electromagnetic stirring.[16]

Copper alloys are usually quite adequately grain refined by alloying alone.[15] Aluminium alloys exhibit similar behaviour when solute is added. However, in practice, for Al alloys it is also necessary to resort to nucleants as described in the previous section. Constitutional undercooling then provides favourable conditions for the nuclei to expand on the nucleants and for the resulting large number of nuclei to grow into fine grains.

Finally, it should be pointed out that the conditions for creating a supercooled zone as given by eqns (9.38) and (9.39) are identical to the criteria that give the transition from a planar to a cellular interface.[2] Supercooling results in instability of the solidification front since any protruberance forming on the interface finds itself in supercooled liquid and therefore will not disappear. Therefore when inequalities (9.38) or (9.39) are fulfilled, the structure should be cellular. With vigorous stirring of the melt it may be possible to melt off the tips of the cells and thus create additional nuclei. This increase of nuclei in the supercooled melt results in solidification structures where the grains are smaller and randomly oriented (equiaxed) instead of columnar. The multiplication process occurs naturally due to the shear from convection during solidification but can be strongly enhanced by the shear induced by electromagnetic stirring.[16,14] Supercooling is important also in that it allows the nuclei swept into the melt to grow.

REFERENCES

1. Flemings, M.C. (1990). Segregation in castings and ingots. Elliott Symposium, June 10–13, Cambridge, Mass., pp. 253–72. AIME, Warrendale, Pa.

2. Flemings, M.C. (1974). *Solidification processing*. McGraw-Hill, New York.
3. Minkoff, I. (1986). Solidification and cast structure. Wiley, New York.
4. Tveit, H. (1988). Størkning av 75% ferrosilisium, forløp, struktur og styrke. Dr. ing. Thesis, Metallurgisk Inst., NTH, Trondheim, Norway.
5. Mehrabian, R., Keane, M., and Flemings, M.C. (1970). Interdendritic fluid flow and macrosegregation; influence of gravity. *Metall. Trans.*, **1**, 1209–20.
6. Ganesan, S. and Poirier, D. R. (1990). Conservation of mass and momentum for the flow of interdendritic liquid during solidification. *Metall. Trans. B.*, **21B**, 173–81.
7. Lux, A. L. and Flemings, M. C. (1979). Refining by fractional solidification. *Metall. Trans. B.*, **10B**, 71–8.
8. Lux, A.L. and Flemings, M.C. (1979). Refining by fractional melting. *Metall. Trans. B.*, **10B**, 79–84.
9. Pfann, W.G. (1966). Zone melting. Wiley, New York.
10. Mondolfo, L.F. (1983). Grain refinement in the casting of non-ferrous alloys. In *Grain refinement in castings and welds*, (ed. G.J. Abbaschian and S.A. David), pp. 3–50. AIME, Warrendale, Pa.
11. Delannay, F., Froyen, L., and Deruyttere, A. (1987). Review the wetting of solids by molten metals and its relation to the preparation of metal-matrix composites. *J. of Materials Science*, **22**, 1–16.
12. Hansen, M., and Anderko, K. (1958). Constitution of binary alloys. McGraw-Hill, New York.
13. Arnberg, L. Bäckerud, L., and Klang, H. (1982). Possible grain refining mechanisms in aluminium as a result of addition of master alloys of the Al–Ti–B type. In *Grain refinement in castings and welds*, (ed. G. J. Abbaschian and S. A. David), pp. 165–182. AIME Warrendale, Pa.
14. Granger, D.A. (1985). Practical aspects of grain refining aluminium alloy melts. In *Refining and alloying of liquid aluminium and ferro-alloys*, (ed. T.A. Engh, S. Lyng, and H.A. Øye, NTH, Trondheim, Norway), pp. 229–46. Aluminium-Verlag, Düsseldorf.
15. Stefanescu, D.M., (1988) *Metals Handbook*, (9th edn) Vol. 15, Casting. Metals Park, Ohio.
16. Tzavaras, A.A. (1984). Solidification control by electromagnetic stirring—state of the art. In *Continuous Casting*, Vol. 3 (ed. J.J. Moore), pp. 47–67. Iron and Steel Soc. of AIME, Warrendale, Pa.

10

REFINING AND RECYCLING METALLURGY IN THE FUTURE

10.1 INTRODUCTION

Refining metallurgy deals with the treatment of the extracted or recycled metal by purification and alloying. Traditionally, the refining metallurgist has had a good knowledge of extractive metallurgy and of the impurities present in the raw metal. Often there are problems with oxidation or evaporation of alloy additions that result in a low or variable yield. These problems must be solved and taken into account in the melt treatment.

The refining metallurgist must continue to play a role in producing metals with improved strength properties. For instance in the production of steels for liquified natural gas and steels for use at high temperatures it is very important that the content of impurities such as sulphur and phosphorus is very low.[1,2] Thus it seems inevitable that impurity levels will be pushed ever lower.

It is important to have an understanding of physical and mechanical metallurgy. This insight is required because alloying not only covers the addition of metal components and carbon to the solvent metal, but is also concerned with additions that influence the nucleation and solidification behaviour, precipitation as a finely dispersed second phase, and so on. In addition to the treatment of molten metals the refining metallurgist should take an active part in materials processing such as continuous casting, near-net shape casting,[3] and production of (molten) metal-based composite materials.[4] For such materials, inclusions, impurity levels, and interfacial properties are very important.[5]

The expanding use of recycled metal presents a further challenge. Therefore the refining metallurgist must also have a knowledge of commercial alloys, their composition, and impurity content. In fact a new field in metallurgy is now starting to develop. It is concerned with the processing and refining of recycled metals for reuse. For many reasons this area should become of vital importance in the near future. In a few illustrative examples this will be discussed in the following.

10.2 INTRODUCTION TO RECYCLING

By the year 2050, 10 billion people are likely to live on this planet. Ideally, all would enjoy standards of living equivalent to developed countries like the US and Japan. However, if they consume natural resources such as copper, cobalt, molybdenum, nickel, and petroleum at current US rates, the known resources would not last for many decades.[6] On the waste side of the ledger, 10 billion people would generate at least 4 billion tonnes of solid waste every year, enough to bury Los Angeles 100 metres deep. Furthermore the rate of increase of the CO_2 levels in the atmosphere would lead to global warming.[7]

These forecasts are not meant to scare and paralyse us. Instead they emphasize the incentives for recycling, conservation, and a switch to alternative materials. The metallurgist has an important role to play in bringing about the necessary change. Traditionally individual manufacturing processes take in raw materials and generate products to be sold plus waste to be disposed of. Ideally, industrial processes should be transformed into a more integrated model: an industrial ecosystem. Wastes are minimized and the effluents of one process, whether they are spent catalysts from petroleum refining, fly and bottom ash from electric-power generation, or slag from a metallurgical process, should serve as the raw material for another process. The industrial ecosystem should function as an analogue of biological ecosystems. Characteristic of a (tropical) ecosystem is the great diversity of species and the specialization of the various species. Although an industrial system can never approach such a degree of variation and specialization, the automotive aluminium recycling industry in the US is an encouraging example that it is possible to develop a more closed industrial ecosystem. In this industry there are more than 20 000 independent enterprises, mostly specializing in one or two facets of the recycling loop: salvaging parts, dismantling wrecks, baling hulks for steelmaking, shredding for metal scrap values, and metal upgrading. There is a remarkably high and consistent recovery of 70–80 per cent of the available aluminium in spite of the fact that the automotive scrap is 'recycle unfriendly'. Unfortunately, this industrial ecosystem has not yet grown to include the car manufacturing end of the loop.

Materials in an ideal industrial ecosystem are not depleted any more than in a biological one. A chunk of steel could potentially show up one year in a tin can, the next year in an automobile, and ten years later in the skeleton of a building. Manufacturing processes in an industrial ecosystem transform circulating stocks of materials from one shape to another. The circulating stock decreases when some material is unavoidably lost, and it increases to meet the needs if the population increases. Such recycling still requires the expenditure of energy and the unavoidable generation of wastes and harmful by-products, but at much lower levels than are typical today. In the industrialized countries such as the US, Japan, and Western Europe the population more or less remains constant. Ideally, in these areas, consumption of metals such as copper, aluminium, and steel should to a large extent be taken care of by recycling. It will be seen, however, that recycling is intimately related to factors such as energy and labour costs, process technology, and the infrastructure for collection and transport. The following discussion concerning steelmaking illustrates a situation where there is little economic incentive for recycling.

This does not mean that recycling is an uninteresting concept but instead that adjustments are required in the other factors. Thus the iron- and steelmaking technology should be changed so as to utilize the large amount of iron scrap now accumulating in the US. This is discussed in the following section not only because of the process metallurgy and refining aspects, but also to illustrate the importance of economic and infrastructure factors. Clearly, the recycling and refining metallurgist of the future must have an understanding of technology and society that goes far beyond thermodynamics!

10.3 THE IRON AND STEEL INDUSTRY; RECYCLING AND REFINING

In the iron and steel sector the infrastructure and technology for recycling has existed for a long time. In fact the open hearth furnace was developed to utilize steel scrap more than a hundred years ago. The economy of the open hearth depended on the supply of cheap fuel in the form of gas or oil and on low labour costs. Competition from iron-making blast furnaces combined with oxygen converter processes for steel has practically driven the open hearth out of business. However, oxygen converters can handle only about 30 tonnes of steel scrap to be mixed with 100 tonnes of pig iron from the blast furnace. In the US today the ratio between the output of recoverable steel scrap and consumption of iron and steel is about 1 to 2. Thus the blast furnace–oxygen converter route can only accept part of the available scrap. More recently mini-mills have been built that melt down local scrap in electric arc furnaces. These low-volume mills have increased their share of US steel production, but not enough to compensate for the lost demand for scrap to feed the open hearth furnaces. Furthermore, mini-mills produce only a limited range of steel products, and many of these products must be made from scrap containing very low levels of impurities. There are two reasons for this. Levels of unwanted impurities such as copper cannot be lowered by dilution with virgin pig iron as is the case in an oxygen converter. Also, the refining capacity of the arc furnace unit is very limited. If electric arc furnaces are to make significant inroads into the US stock of scrap iron, they must be coupled with a separate refining ladle or converter so that a wider range of products can be manufactured.

The scrap problem could partly, be reduced if one could encourage manufacturers to generate 'recycle friendly' scrap. For instance, perhaps a method could be developed for removing the copper wiring from automotive scrap, or alternatively aluminium wiring could replace copper in cars. The main point is, however, that as long as steel scrap has little or no market value such measures will probably not be taken. Conversely, if steel scrap had a market value, scrap that contained impurities such as copper would fetch a lower price.[10] Then the market mechanisms and the ingenuity of inventors and engineers would solve the difficulty caused by copper. So now the metallurgist is back again to the question of whether the iron and steelmaking process route can be modified so as to make use of the vast amounts of available scrap (750 million tonnes in 1987 in the US)[6] in an economically competitive manner.

To answer the question posed above we take a new look at the existing blast furnace and oxygen converter route. The oxygen converter has become an extremely efficient steel making unit from the point of view of productivity and efficient use of energy. The possibilities for refining (removal of P and S) are good and post-combustion of CO to CO_2 in the converter has improved the capacity to accept scrap. Ultimately however, this capacity has been limited by the carbon content of the pig iron from the blast furnace. However, if carbon as fuel is added to the converter, its scrap capacity is improved.

A different approach would be to add iron carbide, Fe_3C in an arc furnace-refining ladle (converter) route. Fe_3C would be an alternative to the use of sponge iron or could be employed together with sponge iron additions. Thus the use of Fe_3C would increase the capacity of an arc furnace–converter route to accept impure scrap. However, a practical and economically viable process[8] for producing Fe_3C has not yet been developed. To conclude, scrap will go on accumulating in the US until a solution such as that sketched above is found.[10] In 1982 recoverable scrap amounted to 610 million tonnes. In 1987 the figure had risen to 750 million.

10.4 RECYCLING IN THE ALUMINIUM INDUSTRY

Some years ago beverage cans in parks and streets and in overflowing rubbish dumps were such an environmental nuisance and problem that some countries seriously considered prohibiting their production and use. This threat was one of the reasons why the primary aluminium industry in the US in the mid-1960s entered the recycling business. In competition with the secondary industry, mentioned in the introduction, advanced scrap collection, preparation, remelting, and refining methods were developed. The primary industry in the US is 'closing the loop' by recycling more than half the beverage cans back into container sheet. This recycling is described briefly in the following.

Collection direct from the customer has been simplified by the use of reverse vending machines that pay cash on the spot for used beverage containers (UBCs). In the recycling plant the material is shredded and compacted in a hammermill.[9] Then delacquering[11] takes place. This is illustrated in Fig. 10.1. There is a unit for magnetic separation of heavy metals and an air knife to remove loose lacquer particles. In a process developed by Alcoa[9] the lid alloy (0.2% Si, 0.35% Fe, 0.15% Cu, 0.2–0.5% Mn, 4–5% Mg, 0.25% Zn) is separated from the can body stock alloy (0.3% Si, 0.7% Fe, 0.25% Cu, 1–1.5% Mn, 0.8–1.3% Mg, 0.25% Zn). These alloys are transferred to melting furnaces, the composition is adjusted, and standard refining operations such as hydrogen removal by inert gas purging are performed. Finally the alloys are

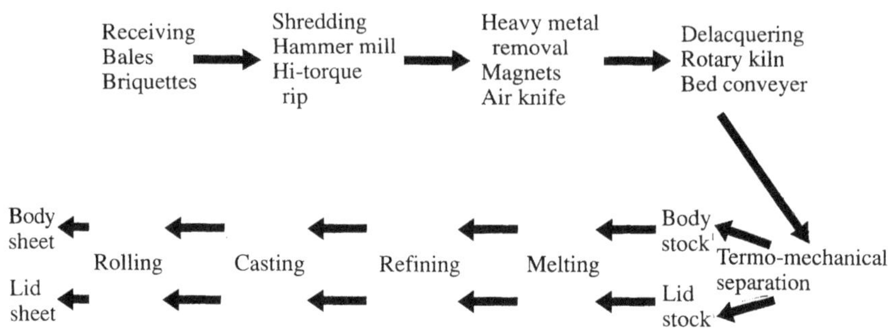

FIG. 10.1 Used beverage containers recycled back to body sheet and lid sheet for new containers.[9]

cast into ingots and rolled into sheet. Then the lid and body sheet is shipped to the can manufacturer.

A recycling plant is seen to be rather complex. Not only does it include melting down, refining, casting, and rolling, but also operations reminiscent of mineral beneficiation techniques.

10.5 RECYCLING OF LEAD, ZINC, AND TIN

Since the 1970s and 1980s strict workplace and ambient air quality standards have been mandatory for lead plants. In the US it is required that discardible slags contain less than 0.3–0.5% Pb and less than 2.0–2.5% Zn. Throughout much of the 1980s, the lead, zinc, and tin industries were in a state of crisis as demand stagnated, pricess fell, and environmental pressures intensified.[12] Simultaneously, secondary raw materials from lead batteries, zinc plant leach, and steel plant dusts have been accumulating. The problem that faces us is the recirculation of these waste materials while complying with the environmental regulations both for the recycling and primary industries. One approach is to integrate primary and secondary smelting. The rationale for this is to attain an economy of scale and combine smelting and refining processes using new technologies.[13–15] A different approach is to build specialized plants based on only secondary materials; in the case of lead, principally lead accumulator scrap.[15, 16] Figure 10.2 gives a flowsheet for such a lead recycling plant. There are three stages: mechanical separation of the scrap, reduction and smelting of the lead-containing fractions, and finally refining and alloying of the lead bullion into a commercial product. It should be noted that the propylene fraction has also been upgraded into a product that can be sold. However, the hard rubber separators are stored on the plant site. The slag that contains 3–5% Pb + Sb and 10–15% S is deposited on a dump within the plant area. In

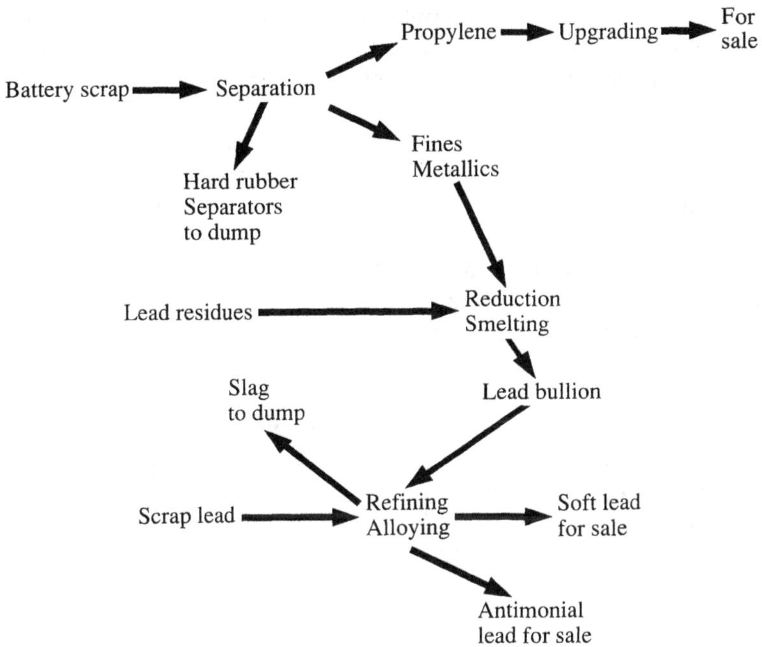

FIG. 10.2 Lead recycling flowsheet.[16]

a mineral beneficiation type of operation the batteries are broken up in a double roll crusher, and a pick-up magnet removes non-lead metal particles before the feeding bolt goes to the hammermill. The refining operations are roughly as indicated in Section 3.5.3.

As is the case for recycling of automotive scrap, the plant contains a mineral beneficiation and a refining section. The metallic part of the output is sold commercially. One hopes that the Pb, Sb, and S content of the slag can be reduced to such low levels that it can be discarded as landfill. Ideally, some method of separating and reusing the battery acid should be developed, but the expanding use of gel batteries makes the situation worse. Another problem is the dusts, drosses, and slags generated in the various stages of the refining processes (see Section 3.5.3 and Fig. 3.5). To transform the wastes into saleable by-products with present technology will require substantial recycling and the labour-intensive operation of small units. Further research and market studies will be necessary.

Until now zinc production has been dominated by the electrolytic process. Three processes have been employed commercially for the precipitation of iron. The processes are named for the mineral type of residue produced: the jarosite, the goethite, and the hematite processes. Of the three residue products only hematite has been marketed with any success (to the cement and tile industries). Both jarosite and goethite are regarded as hazardous solid wastes due to the content of leachable lead, cadmium, and at times arsenic and other impurities. At most electrolytic zinc plants the iron residue has been impounded since start-up. With waste volumes of 0.3–0.45 m^3/tonne of zinc produced, the storage and pond management costs have become significant.

The impoundment of residues will probably prevent the construction of any new electrolytic zinc plants.[12] Problems with the electrolytic zinc process are augmented by the high consumption of Zn for refining the electrolyte by cementation. This consumes 5–7% of a plant's zinc capacity. Due to the presence of Zn in the precipitates, the Cu, Ni, and Co fetch only a low by-product credit. Furthermore, recycled zinc materials such as dust from electric arc furnaces and dust from fuming lead slags cannot currently be used in electrolytic zinc plants. Today we do not have a solution to the problems outlined above, but an interesting partial solution may be a proposal to leach waste Zn dusts in H_2SO_4 and use the electrolyte to electrogalvanize steel.[17]

Tin refining involves the removal of many of the same impurities as in the lead industry: bismuth, silver, indium, copper, and antimony. These cannot be discarded. As in lead refining the challenge is to transform these wastes into products that the market is willing to buy.

10.6 CONCLUSION

Metals and alloys play a vital role in human existence and growing populations will require an increasing supply of metals. Metals have an enormous range of properties and uses; they are durable or can be rendered durable (with a protective oxide surface layer); and they may be used over and over again. The metallurgical industry and the metallurgist must cover the needs of society with the least possible damage to the environment.

Metals are usually found in nature as oxides and sulphides, and extraction requires energy and usually carbon as reductant. Thus production of primary metal contributes to an increase of CO_2 in the atmosphere and to global warming.[7] Also for this reason the best possible use must be made of existing resources already in metallic form. The free enterprise system should be the best tool to accomplish this. An important challenge to the worker in the recycling industry is to convert the waste materials into products that the market is willing to pay for. The wastes that remain after extracting the valuable parts should ideally be so harmless and inert that they can be discarded as landfill or returned to a depleted mining site.

REFERENCES

1. Furukimi, O. and Saito, Y. (1990). The effects of grain boundary phosphorus segregation and heat treatment on toughness of 9% Ni steel and its welded joint. *Trans. ISIJ*, **30**, 390–6.

2. Mineura, K., Takahashi, I., and Tanaka, K. (1990). Deoxidation and desulfurization of pressured liquid high nitrogen stainless steels with calcium. *Trans. ISIJ*, **30**, 192–8.
3. Reichelt, W., Kapellner, W., and Steffen, R. (1988). Endabmessungsnahe Herstellung vom Flachprodukten. *Stahl u. Eisen*, **108**, 409–17.
4. Maxwell, P.B., Martins, G.P., Olson, D.L., and Edwards, G.R. (1990). The infiltration of aluminium into silicon carbide compacts. *Metall. Trans. B*, **21B**, 475–85.
5. Delanny, F., Froyen, L., and Deruyttere, A. (1987). Review, the wetting of solids by molten metals and its relation to the preparation of metal-matrix composites. *J. of Materials Sci.*, **22**, 1–16.
6. Frosch, R.A. and Gallopoulos, N.E. (1990). Strategies for manufacturing, in managing planet earth. W.H. Freeman and Co, New York.
7. Schneider, S.H. (1990). The changing climate, in managing planet earth. W.H. Freeman and Co, New York.
8. Stephens, Jr. (1977). Process for the direct production of steel. US patent 4053301.
9. Van Linden, J.H.L. (1990), Aluminium recycling—everybody's business, technological challenges and opportunities. In *Light Metals*, (ed. C.M. Bickert), pp. 675–82. Warrendale, Pa.
10. Spoel, H. (1990). The current status of scrap metal recycling. *J. of Metals*, April, 38–41.
11. McCabbin, J.C. (1986), Decoating of used beverage containers in a thin down-flow bed. In *Light Metals*, (ed. R.E. Miller), pp. 917–22. Warrendale, Pa.
12. Bounds, C.O. (1990). Innovative processes: lead–zinc–tin. The Elliott Symp., June 10–13, pp. 73–91. Cambridge, Ma.
13. Pusateri, J.F., Bounds, C.O., and Lherbier, L.W. (1988). Zinc recovery via the flame reactor process. *J. of Metals*, Aug., 31–5.
14. Waladan, M., Firkin, G.R., Bultitude-Paull, J., Lightfoot, B.W., and Floyd, J.M. (1990), Top submerged lancing for recovery of zinc from ISF slag. In *Lead–Zinc' 90*, (ed. T.S. Mackey and R.D. Prengemon) pp. 607–28. Warrendale, Pa.
15. Pedersen, T., Aune, J.A., and Cundall, R.H. (1990). The Elkem multi-purpose furnace for the recovery of zinc and lead from waste materials. In *Lead–Zinc' 90*, (ed. T.S. Mackey and R.D. Prengamon) pp. 857–879. Warrendale, Pa.
16. Schenker, G. (1990). Lead recycling from battery scrap and other raw materials in Metaleurop's lead smelting plant in oker. In *Lead–Zinc' 90*, (eds. T. S. Mackey and R. D. Prengamon) pp. 979–99. Warrendale, Pa.
17. Cole, E.R., Dattilo, M., and O'Keefe, T.J. (1987). *Nonferrous waste as a source of zinc for electrogalvanizing*. Bureau of Mines, 9125, US Dept. of Interior.

APPENDIX 1 SELECTED STANDARD GIBBS ENERGIES

Standard Gibbs energies, in joule/mole, are given in the table below as

$$\Delta G^0 = \overline{\Delta H^0} - T\,\overline{\Delta S^0}$$

where $\overline{\Delta H^0}$ and $\overline{\Delta S^0}$ are average values for the stated range of temperatures. The error arising from the equation is usually less than the error in the tabulated data. This allows a very compact presentation of thermodynamic data. The references (which are listed at the end of Chapter 2) and the tables are to be regarded only as a general guide to the reader. For guidance graphs of ΔG^0 available in the literature[107] may also be useful. When working on a specialized project the original literature should be consulted and a database employed.

The data given here is primarily based on a compilation by Turkdogan,[24] data from the JANAF tables,[8] and thermodynamic data by Kubaschewski and Alcock,[3] Hultgren *et al.*,[10] and Barin and Knacke.[70,71]

Compounds and elements are given in the solid state, except for O_2, N_2, CO, CO_2, Cl_2, F_2, H_2, S_2, P_2, and SO_2 which are given in the gaseous state. If a compound that is normally in the solid state is liquid or gaseous, an l or g indicates this. In the temperature range column, m, b, or s indicate the melting-point, boiling-point, or sublimation point, respectively; d signifies decomposition. Estimates of errors in the data are not presented here, but are usually found in the references.

Nowadays one can calculate thermodynamic functions using a computer program and a database. Estimation and prediction of thermodynamic values have been simplified even though the system is described by very sophisticated equations and has uncertain factors based on the choice of the solution model. Typical thermodynamic databases are SOLGASMIX,[108,109] FACT,[110] CSIRO, Thermo-Calc,[111] and so on.[112]

Reaction	$\overline{\Delta H^0}$ (joule)	$\overline{\Delta S^0}$ (joule)	Range (°C)	Reference
Ag = Ag(l)	11 300	9.16	961 m	10
Ag(l) = Ag(g)	258 300	106.25	961–2163 b	10
AgBr = AgBr(l)	9200	13.10	430 m	70
AgBr(l) = AgBr(g)	192 000	104.6	1560 b	70
AgBr(l) = Ag(l) + $\frac{1}{2}$ Br_2	99 010	27.8	961–1560 b	70
AgCl = AgCl(l)	13 000	17.83	455 m	70
AgCl(l) = AgCl(g)	177 800	96.65	1564 b	70
AgCl(l) = Ag(l) + $\frac{1}{2}$ Cl_2	104 470	23.0	961–1564 b	70
AgF = Ag + $\frac{1}{2}$ F_2	200 800	54.8	25– 435 m	70

Reaction	$\overline{\Delta H^0}$ (joule)	$\overline{\Delta S^0}$ (joule)	Range (°C)	Reference
$AgI = AgI(l)$	9410	11.3	558 m	70
$AgI(l) = AgI(g)$	144 300	81.2	1505 b	70
$AgI(l) = Ag(l) + \frac{1}{2} I_2(g)$	77 100	22.8	961–1505 b	70
$Ag_2O = 2\ Ag + \frac{1}{2} O_2$	30 540	66.1	25	3
$Ag_2S = Ag_2S(l)$	7870	7.1	830 m	71
$Ag_2S = 2\ Ag(s) + \frac{1}{2} S_2$	161 300	168.6	25–830 m	71
$Al = Al(l)$	10 800	11.55	660 m	10
$Al(l) = Al(g)$	304 600	109.5	660–2520 b	10
$AlCl(g) = Al(l) + \frac{1}{2} Cl_2$	77 500	−58.16	660–2000	8
$Al_2Cl_6(g) = 2\ Al + 3\ Cl_2$	1 292 000	242.2	25– 660	8
$AlF(g) = Al(l) + \frac{1}{2} F_2$	292 000	−54.2	660–2000	8
$AlF_3 = Al(l) + \frac{3}{2} F_2$	1 508 000	257.9	660–1276 s	8
$AlF_3(g) = Al(l) + \frac{3}{2} F_2$	123 000	79.37	660–2000	8
$AlB_{12} = Al(l) + 12\ B$	220 000	7.45	660–2200 m	71
$Al_4C_3 = 4\ Al(l) + 3\ C$	265 000	95.1	660–2200 m	8
$AlN = Al(l) + \frac{1}{2} N_2$	327 000	115.5	660–2000	8
$AlP = Al(l) + \frac{1}{2} P_2$	249 500	104.4	660–1700	71
$Al_2O_3 = 2\ Al(l) + \frac{3}{2} O_2$	1 687 300	327	660–2054 m	8
$Al_2O(g) = 2\ Al(l) + \frac{1}{2} O_2$	170 700	−49.4	660–2000	8
$Al_2O_2(g) = 2\ Al(l) + O_2$	470 700	28.8	660–2000	8
$AlPO_4 = Al + \frac{1}{2} P_2 + 2\ O_2$	1 802 000	450	25– 660	71
$3\ Al_2O_3 \cdot 2\ SiO_2 = 3\ Al_2O_3 + 2\ SiO_2$	−8600	−17.4	25–1750 m	8
$Al_2O_3 \cdot SiO_2 = Al_2O_3 + SiO_2$	8815	3.89	25–1700	8
$Al_2O_3 \cdot TiO_2 = Al_2O_3 + TiO_2$	25 300	3.93	25–1860 m	71
$As = \frac{1}{4} As_4(g)$	36 570	41.9	25– 630 s	10
$Au = Au(l)$	12 550	9.41	1064 m	10
$B = B(l)$	50 210	21.8	2030 m	10
$B = B(g)$	570 150	147.8	25–2030 m	10
$BCl_3(g) = B + \frac{3}{2} Cl_2$	404 000	51.7	25–2030	8
$BF_3(g) = B + \frac{3}{2} F_2$	1 140 600	65.1	25–2030	8
$B_4C = 4\ B + C$	41 500	5.56	25–2030	8
$BN = B + \frac{1}{2} N_2$	250 000	87.61	25–2030	8
$B_2O_3 = B_2O_3(l)$	24 060	33.3	450 m	8
$B_2O_3(l) = 2\ B(s) + \frac{3}{2} O_2$	1 229 000	210	450–2043 b	8
$BO(g) = B(s) + \frac{1}{2} O_2$	3800	−88.8	25–2030	8
$Ba = Ba(l)$	7740	7.74	727 m	10, 114
$Ba(l) = Ba(g)$	152 000	70.4	729–1894 b	10, 114
$BaCl_2 = BaCl_2(l)$	15 980	12.93	962 m	8
$BaCl_2(l) = Ba(l) + Cl_2$	811 300	121.5	962–1622	8
$BaF_2 = BaF_2(l)$	23 350	14.23	1368 m	8
$BaF_2(l) = Ba(l) + F_2$	1 154 000	129	1368–1622	8
$BaO = Ba(l) + \frac{1}{2} O_2$	557 000	103	729–1622	8
$BaS = Ba(l) + \frac{1}{2} S_2$	544 000	123.4	729–1622	71

Reaction	$\overline{\Delta H^0}$ (joule)	$\overline{\Delta S^0}$ (joule)	Range (°C)	Reference
$BaCO_3 = BaO(s) + CO_2$	250 800	147.0	800–1060 m	72, 106
$Be = Be(l)$	11 715	7.53	1287 m	8
$Be(l) = Be(g)$	300 000	108.6	1287–2468 b	8
$BeCl_2(g) = Be(l) + Cl_2$	380 000	−5.86	1287–2000	8
$BeF_2(g) = Be(l) + F_2$	819 000	2.34	1287–2000	8
$Be_2C = 2\ Be + C$	93 300	13.8	25–1287	8
$Be_2C = 2\ Be(l) + C$	115 000	28.5	1287–2000	8
$Be_3N_2 = 3\ Be(s) + N_2$	587 000	185.4	25–1287	8
$Be_3N_2 = 3\ Be(l) + N_2$	616 000	203	1287–2200 m	8
$BeO = Be(s) + \frac{1}{2}\ O_2$	608 000	97.7	25–1287	8
$BeO = Be(l) + \frac{1}{2}\ O_2$	613 600	100.9	1287–2000	8
$BeS = Be(s) + \frac{1}{2}\ S_2$	300 000	86.6	25–1287	71
$BeO \cdot Al_2O_3 = BeO + Al_2O_3$	4180	−3	25–1870 m	8
$3\ BeO \cdot B_2O_3 = 3\ BeO + B_2O_3(l)$	58 400	25.5	450–1495 m	8
$2\ BeO \cdot SiO_2 = 2\ BeO + SiO_2$	10 900	7.1	25–1560 m	8
$BeSO_{4_\beta} = BeO + SO_2 + \frac{1}{2}\ O_2$	254 400	232	25–1287	8
$Bi = Bi(l)$	11 300	20.75	271 m	10
$Bi(l) = Bi(g)$	192 000	99.4	272–1661 b	10
$BiCl_3 = BiCl_3(l)$	23 640	46.6	234 m	71
$BiCl_3(l) = Bi(l) + \frac{3}{2}\ Cl_2$	351 000	160.5	272– 439 b	71
$BiF_3 = BiF_3(l)$	21 590	23.4	649 m	71
$BiF_3(l) = BiF_3(g)$	125 500	104.6	927 b	71
$BiF_3(l) = Bi(l) + \frac{3}{2}\ F_2$	853 500	182.8	649– 927 b	71
$Bi_2O_3 = Bi_2O_3(l)$	59 850	54.6	824 m	71
$Bi_2O_3(l) = 2\ Bi(l) + \frac{3}{2}\ O_2$	445 000	159.6	824–1500	71
$Bi_2S_3 = Bi_2S_3(l)$	78 200	74.5	777 m	71
$Bi_2S_3 = 2\ Bi(l) + \frac{3}{2}\ S_2$	360 000	274	272– 777 m	71
Graphite = Diamond	1440	−4.48	29– 900	8
$C = C(g)$	713 500	155.5	1750–3800 s	8
$CH_4(g) = C + 2\ H_2$	91 040	110.7	500–2000	8
$CCl_4(g) = C + 2\ Cl_2$	89 100	129.2	25–2000	8
$CF_4(g) = C + 2\ F_2$	933 200	151.5	25–2000	8
$CN(g) = C + \frac{1}{2}\ N_2$	−433 500	−99.6	25–2000	8
$CO = C + \frac{1}{2}\ O_2$	110 540	−89.35	−150– 500	114
$CO = C + \frac{1}{2}\ O_2$	114 400	−85.77	500–2000	8, 114
$CO_2 = C + O_2$	395 350	−0.54	500–2000	8, 114
$CO_2 = C + O_2$	393 500	−2.88	−50– 500	114
$COS(g) = C + \frac{1}{2}\ O_2 + \frac{1}{2}\ S_2$	202 800	−9.96	500–2000	8, 114
$CS(g) = C + \frac{1}{2}\ S_2$	−163 000	−88	25–2000	8
$CS_2(g) = C + S_2$	11 400	−6.5	25–2000	8
$Ca = Ca(l)$	8536	7.7	842 m	8, 73
$Ca(l) = Ca(g)$	156 800	89.5	842–1482 b	8, 114
$CaBr_2 = CaBr_2(l)$	28 870	28.45	742 m	8
$CaBr_2(l) = Ca(l) + Br_2(g)$	670 000	105	842–1482	8

Reaction	$\overline{\Delta H^0}$ (joule)	$\overline{\Delta S^0}$ (joule)	Range (°C)	Reference
$CaCl_2 = CaCl_2(l)$	28 535	27.3	772 m	8
$CaCl_2(l) = CaCl_2(g)$	235 000	106.4	1936 b	8
$CaCl_2(l) = Ca(l) + Cl_2$	800 000	146	842–1482	8
$CaF_2 = CaF_2(l)$	29 700	17.6	1418 m	8
$CaF_2(l) = CaF_2(g)$	308 700	110	2533 b	8
$CaF_2 = Ca(l) + F_2$	1 220 000	162	842–1482	8
$CaC_2 = Ca(l) + 2\ C$	60 250	−26.3	842–1482	70
$Ca_3N_2 = 3\ Ca + N_2$	43 514	200	25– 842	71
$CaAl_2 = CaAl_2(l)$	63 200	46.9	1080 m	71
$CaAl_2(l) = Ca(l) + 2\ Al(l)$	189 000	3.35	1080–1482	71
$CaAl_4 = CaAl_4(l)$	78 700	75.0	777 m	71
$CaAl_4(l) = Ca(l) + 4\ Al(l)$	200 000	3.0	842–1482	71
$Ca_2Si = 2\ Ca + Si$	210 000	22.0	25– 842	71
$CaSi = Ca + Si$	150 000	15.5	25– 842	71
$CaSi_2 = Ca + 2\ Si$	150 000	28.5	25– 842	71
$Ca_3P_2 = 3\ Ca + P_2$	650 000	216	25– 842	71
$CaO = CaO(l)$	79 500	24.7	2927 m	8
$CaO = Ca(l) + \frac{1}{2}\ O_2$	640 000	108.6	842–1482	8
$CaS = Ca(l) + \frac{1}{2}\ S_2$	548 000	104	842–1482	8
$3\ CaO \cdot Al_2O_3 = 3CaO + Al_2O_3$	12 600	−24.7	500–1535 m	74
$CaO \cdot Al_2O_3 = CaO + Al_2O_3$	18 000	−19.0	500–1605 m	74
$CaO \cdot 2\ Al_2O_3 = CaO + 2\ Al_2O_3$	17 000	−25.5	500–1750 m	74
$CaCO_3 = CaO + CO_2$	161 300	137.2	700–1200	75
$2\ CaO \cdot Fe_2O_3 = 2\ CaO + Fe_2O_3$	53 100	−2.5	700–1450 m	74
$CaO \cdot Fe_2O_3 = CaO + Fe_2O_3$	30 000	−4.8	700–1216 m	74
$3\ CaO \cdot P_2O_5 = 3\ CaO + P_2 + \frac{5}{2}\ O_2(g)$	2 314 000	556	25–1730 m	71
$2\ CaO \cdot P_2O_5 = 2\ CaO + P_2 + \frac{5}{2}\ O_2$	2 190 000	586	25–1353 m	71
$CaSO_{4_\beta} = CaO + SO_2 + \frac{1}{2}\ O_2$	460 000	238	950–1195	76
$CaSO_{4_\alpha} = CaO + SO_2 + \frac{1}{2}\ O_2$	454 000	232.0	1195–1365 m	76
$3\ CaO \cdot SiO_2 = 3\ CaO + SiO_2$	120 000	−6.7	25–1500	71
$2\ CaO \cdot SiO_2 = 2\ CaO + SiO_2$	120 000	−11.3	25–2130 m	71
$3\ CaO \cdot 2\ SiO_2 = 3\ CaO + 2\ SiO_2$	237 000	9.6	25–1500	71
$CaO \cdot SiO_2 = CaO \cdot SiO_2(l)$	56 000	33.0	1540 m	71
$CaO \cdot SiO_2 = CaO + SiO_2$	92 500	2.5	25–1540 m	71
$3\ CaO \cdot 2\ TiO_2 = 3\ CaO + 2\ TiO_2$	207 000	−11.5	25–1400	74
$4\ CaO \cdot 3\ TiO_2 = 4\ CaO + 3\ TiO_2$	293 000	−17.6	25–1400	74
$CaO \cdot TiO_2 = CaO + TiO_2$	80 000	3.35	25–1400	74
$3\ CaO \cdot V_2O_5 = 3\ CaO + V_2O_5$	332 000	0.0	25– 670	74
$2\ CaO \cdot V_2O_5 = 2\ CaO + V_2O_5$	265 000	0.0	25– 670	74
$CaO \cdot V_2O_5 = CaO + V_2O_5$	146 000	0.0	25– 670	74

Reaction	$\overline{\Delta H^0}$ (joule)	$\overline{\Delta S^0}$ (joule)	Range (°C)	Reference
$CaO \cdot ZrO_2 = CaO + ZrO_2$	39 300	0.4	25–2000	71
$CaO \cdot MgO \cdot SiO_2 = CaO + MgO + SiO_2$	125 000	+4.0	25–1200	71
$2\ CaO \cdot MgO \cdot 2\ SiO_2 = 2\ CaO + MgO + 2\ SiO_2$	205 000	0.0	25–1454 m	71
$CaO \cdot MgO \cdot 2\ SiO_2 = CaO \cdot MgO \cdot 2\ SiO_2(l)$	128 500	77.1	1392 m	71
$CaO \cdot MgO \cdot 2\ SiO_2 = CaO + MgO(s) + 2\ SiO_2$	163 000	18.8	25–1392 m	71
$3\ CaO \cdot Al_2O_3 \cdot 3\ SiO_2 = 3\ CaO + Al_2O_3 + 3\ SiO_2$	389 000	100	25–1400	71
$2\ CaO \cdot Al_2O_3 \cdot SiO_2 = 2\ CaO + Al_2O_3 + SiO_2$	170 000	8.8	25–1500	71
$CaO \cdot Al_2O_3 \cdot SiO_2 = CaO + Al_2O_3 + SiO_2$	106 000	14.2	25–1400	71
$CaO \cdot Al_2O_3 \cdot 2\ SiO_2 = CaO \cdot Al_2O_3 \cdot 2\ SiO_2(l)$	167 000	91.4	1553	71
$CaO \cdot Al_2O_3 \cdot 2\ SiO_2 = CaO + Al_2O_3 + 2\ SiO_2$	139 000	17.2	25–1553	71
$CaO \cdot TiO_2 \cdot SiO_2 = CaO + TiO_2 + SiO_2$	122 600	10.9	25–1400	71
$Cd = Cd(l)$	6190	10.4	321 m	10, 114
$Cd(l) = Cd(g)$	101 700	97.9	321– 766 b	10, 114
$CdBr_2 = CdBr_2(l)$	33 300	39.7	568 m	71
$CdBr_2 = Cd(l) + Br_2(g)$	344 000	151.3	321– 568 m	71
$CdCl_2 = CdCl_2(l)$	30 100	35.8	568 m	71
$CdCl_2 = Cd(l) + Cl_2$	390 000	153	321– 568 m	71
$CdI_2 = CdI_2(l)$	20 700	31.2	388 m	71
$CdI_2(l) = Cd(l) + I_2(g)$	238 000	107.0	388– 744 b	71
$CdSO_4 = CdO + SO_2 + \frac{1}{2}\ O_2$	372 000	277	25–1000	70
$CdO \cdot SiO_2 = CdO + SiO_2$	24 600	4.7	25–1241 m	71
$CdO = Cd(l) + \frac{1}{2}\ O_2$	263 000	105	321– 767	71
$CdS = Cd(l) + \frac{1}{2}\ S_2$	215 500	97.2	321– 767	70
$CdO \cdot Al_2O_3 = CdO + Al_2O_3$	−16 700	−23.0	25– 900	71
$Ce = Ce(l)$	5480	5.105	798 m	10
$Ce(l) = Ce(g)$	413 700	111.8	798–3422 b	10
$CeH_2 = Ce + H_2$	208 400	154	25– 798	71

Reaction	$\overline{\Delta H^0}$ (joule)	$\overline{\Delta S^0}$ (joule)	Range (°C)	Reference
$CeCl_3 = CeCl_3(l)$	53 550	49.1	817 m	71
$CeCl_3 = Ce + \frac{3}{2} Cl_2$	1 050 000	240	25– 798	71
$CeF_3 = CeF_3(l)$	56 500	33.05	1437 m	71
$CeF_3 = Ce + \frac{3}{2} F_2$	1 770 000	252	25– 798	71
$CeB_6 = Ce(l) + 6 B$	376 000	55.2	798–2190 m	71
$Ce_2C_3 = 2 Ce(l) + 3 C$	188 000	−14.5	798–1200	71
$CeC_2 = Ce(l) + 2 C$	85 230	−27.0	798–2250 m	77
$CeN = Ce(l) + \frac{1}{2} N_2$	489 000	177.1	2000–2575	78
$CeAl_2 = CeAl_2(l)$	81 600	47.0	1465 m	71
$CeAl_2 = Ce(l) + 2 Al(l)$	213 000	50.6	798–1465 m	71
$CeAl_4 = Ce + 4 Al$	176 000	21.3	25– 660	71
$CeSi_2 = Ce(l) + 2 Si$	200 000	16.3	798–1427 m	71
$Ce_2O_3 = 2 Ce + \frac{3}{2} O_2$	1 788 000	286.6	25– 798	79
$CeO_{1.72} = Ce + 0.86 O_2$	990 000	170	25– 798	71
$CeO_{1.83} = Ce + 0.915 O_2$	1 030 000	182	25– 798	71
$CeO_2 = Ce + O_2$	1 084 000	212	25– 798	79
$CeS = Ce(l) + \frac{1}{2} S_2$	535 000	90.96	798–2450 m	77
$Ce_2O_2S = 2 Ce(l) + O_2 + \frac{1}{2} S_2$	1 770 000	333	798–1400	80
$Ce_2O_3 \cdot Al_2O_3 = Ce_2O_3 + Al_2O_3$	79 500	−20.9	1100–1500	74
$Co = Co(l)$	16 200	9.16	1495 m	8, 114
$Co(l) = Co(g)$	387 000	121.3	1495–2925 b	8, 114
$CoCl_2 = CoCl_2(l)$	44 800	44.2	740 m	8
$CoCl_2 = Co + Cl_2$	307 000	130	25– 740 m	8
$CoF_2 = Co + F_2$	686 000	141.4	25–1127 m	8
$Co_2B = 2 Co + B$	130 000	10.9	25–1200	71
$CoB = Co + B$	96 200	7.5	25–1460 m	71
$CoAl = CoAl(l)$	62 760	32.7	1645 m	71
$CoAl = Co + Al(l)$	133 900	29.7	660–1495	71
$Co_2Al_5 = 2 Co + 5 Al(l)$	364 000	92	660–1170 m	71
$CoAl_3 = Co + 3 Al$	155 000	8.4	25– 660	71
$Co_2Al_9 = 2 Co + 9 Al$	318 000	15.0	25– 660	71
$Co_2P = 2 Co + \frac{1}{2} P_2$	264 000	98.0	25–1100	71
$CoP = Co + \frac{1}{2} P_2$	197 000	86.6	25–1100	71
$CoP_3 = Co + \frac{3}{2} P_2$	406 000	228	25–1100	71
$CoO = Co + \frac{1}{2} O_2$	246 000	78.6	25–1495	8
$Co_3O_4 = 3 Co + 2 O_2$	957 000	467	25– 700	8
$Co_9S_8 = 9 Co + 4 S_2$	1 326 000	666	25– 788	81
$CO_4S_3 = 4 Co + \frac{3}{2} S_2$	388 500	146	788– 877	81
$CoS_2 = Co + S_2$	280 000	182.0	25– 600	71
$CoO \cdot Cr_2O_3 = CoO + Cr_2O_3$	59 400	8.37	700–1400	74
$CoO \cdot Fe_2O_3 = CoO + Fe_2O_3$	24 700	−6.5	500–1200	74
$CoSO_4 = CoO + SO_2 + \frac{1}{2} O_2$	335 000	264	25– 600	71
$2 CoO \cdot SiO_2 = 2 CoO \cdot SiO_2(l)$	100 400	59.4	1417 m	71
$2 CoO \cdot SiO_2 = 2 CoO + SiO_2$	28 000	4.2	25–1417 m	71

Reaction	$\overline{\Delta H^0}$ (joule)	$\overline{\Delta S^0}$ (joule)	Range (°C)	Reference
$2\ CoO \cdot TiO_2 = 2\ CoO + TiO_2$	22 260	−1.1	25–1575 m	71
$CoO \cdot TiO_2 = CoO + TiO_2$	24 700	6.3	500–1400	74
$Cr = Cr(l)$	16 950	7.95	1857 m	8, 114
$Cr(l) = Cr(g)$	348 500	118.4	1857–2669 b	8
$CrCl_2 = CrCl_2(l)$	32 200	29.6	815 m	71
$CrCl_2 = Cr + Cl_2$	389 000	119.7	25– 815 m	71
$CrCl_3 = Cr + \frac{3}{2}\ Cl_2$	548 000	215.5	25– 945 s	71
$CrF_2 = Cr + F_2$	774 000	134	25–1100 m	71
$CrF_3 = Cr + \frac{3}{2}\ F_2$	1 110 000	225.5	25–1100	71
$CrB = Cr + B$	75 300	6.7	25–1000	71
$CrB_2 = Cr + 2\ B$	92 050	8.4	25–1000 m	71
$Cr_4C = 4\ Cr + C$	96 200	−11.7	25–1520 m	70
$Cr_{23}C_6 = 23\ Cr + 6\ C$	310 000	−77.4	25–1500	8
$Cr_7C_3 = 7\ Cr + 3\ C$	153 500	−37.2	25–1857	8
$Cr_3C_2 = 3\ Cr + 3\ C$	79 100	−17.6	25–1857	8
$Cr_2N = 2\ Cr + \frac{1}{2}\ N_2$	99 200	47.0	1000–1450	8, 82
$CrN = Cr + \frac{1}{2}\ N_2$	113 400	73.2	25– 500	8, 82
$Cr_3Si = 3\ Cr + Si$	106 300	3.4	25–1412	71
$Cr_5Si_3 = 5\ Cr + 3\ Si$	219 200	−15.6	25–1412	71
$CrSi = Cr + Si$	53 560	−3.9	25–1412	71
$CrSi_2 = Cr + 2\ Si$	79 500	1.97	25–1412	71
$Cr_2O_3 = 2\ Cr + \frac{3}{2}\ O_2$	1 110 000	247	900–1650	8, 83
$Cr_3O_4 = 3\ Cr + 2\ O_2$	1 355 000	265	1650–1665 m	84
$CrO(l) = Cr + \frac{1}{2}\ O_2$	334 000	63.8	1665–1750	84
$CrS = Cr + \frac{1}{2}\ S_2$	202 500	56.0	1100–1300	85
$Cr_2(SO_4)_3 = Cr_2O_3 + 3\ SO_2 + \frac{3}{2}\ O_2$	892 000	820	600– 800	86
$Cs = Cs(l)$	2092	6.95	28.5 m	8
$Cs(l) = Cs(g)$	71 700	76.2	285– 675 b	8
$Cu = Cu(l)$	13 050	9.62	1085 m	8
$Cu(l) = Cu(g)$	308 200	108.9	1085–2570 b	8
$CuCl(g) = Cu(l) + \frac{1}{2}\ Cl_2$	−63 300	−66.8	1085–2000	8
$Cu_3Cl_3(g) = 3\ Cu(l) + \frac{3}{2}\ Cl_2$	314 000	47.0	1085–2000	8
$CuF(g) = Cu(l) + \frac{1}{2}\ F_2$	−20 900	−66.1	1085–2000	8
$Cu_3P = 3\ Cu + \frac{1}{2}\ P_2$	188 300	84.1	25– 800	71
$CuP_2 = Cu + P_2$	238 500	155	25–1085	71
$Cu_2O = Cu_2O(l)$	56 820	37.66	1244 m	8
$Cu_2O = 2\ Cu + \frac{1}{2}\ O_2$	168 400	71.25	25–1085	8
$Cu_2O(l) = 2\ Cu(l) + \frac{1}{2}\ O_2$	119 000	39.5	1236–2000	8
$CuO = Cu + \frac{1}{2}\ O_2$	152 300	85.4	25–1085	8
$Cu_2S_\gamma = 2\ Cu + \frac{1}{2}\ S_2$	140 700	43.3	25– 435	87
$Cu_2S_\mu = 2\ Cu + \frac{1}{2}\ S_2$	132 000	30.8	435–1129 m	87
$Cu_2S = Cu_2S(l)$	9000	6.40	1129 m	87
$Cu_{1.738}S_\mu = 1.738\ Cu + \frac{1}{2}\ S_2$	113 600	26.6	435– 620	87
$CuS = Cu + \frac{1}{2}\ S_2$	115 600	76.0	25– 430	88

Reaction	$\overline{\Delta H^0}$ (joule)	$\overline{\Delta S^0}$ (joule)	Range (°C)	Reference
$CuFeS_2 = Cu + Fe + S_2$	278 600	115	557– 700	89
$Cu_2O \cdot Fe_2O_3 = Cu_2O + Fe_2O_3$	37 700	19.0	25–1100	71
$CuSO_4 = \frac{1}{2}\ CuO \cdot CuSO_4 + \frac{1}{2}\ SO_2 + \frac{1}{4}\ O_2$	152 600	136	400– 800	90
$CuO \cdot CuSO_4 = 2\ CuO + SO_2 + \frac{1}{2}\ O_2$	297 000	250	500– 900	90
$Fe_\delta = Fe(l)$	13 800	7.61	1536 m	8
$Fe(l) = Fe(g)$	363 600	116.0	1536–2860 b	8
$FeCl_2 = FeCl_2(l)$	43 100	45.3	677 m	8
$FeCl_2(l) = FeCl_2(g)$	110 000	84.7	1074 b	8
$FeCl_2(g) = Fe + Cl_2$	167 000	−25.1	1074–2000	8
$FeCl_3(g) = Fe + \frac{3}{2}\ Cl_2$	260 000	25.4	332–2000	8
$Fe_2Cl_6(g) = 2\ Fe + 3\ Cl_2$	650 000	181	332–2000	8
$FeF_2 = Fe + F_2$	700 000	134	25–1100 m	8
$Fe_2B = 2\ Fe + B$	87 900	18.4	25–1390 m	71
$FeB = Fe + B$	79 500	10.5	25–1650 m	71
$Fe_3C = 3\ Fe_\alpha + C$	−29 000	−28.0	200– 727	91
$Fe_3C = 3\ Fe_\gamma + C$	−11 230	−11.0	727–1137	91
$Fe_4N = 4\ Fe + \frac{1}{2}\ N_2$	33 500	70.0	400– 680	92
$Fe_{0.947}O = Fe_{0.947}O(l)$	31 340	19.0	1370 m	8
$Fe_{0.947}O = 0.947\ Fe + \frac{1}{2}\ O_2$	263 700	64.4	25–1370 m	8
$FeO(l) = Fe(l) + \frac{1}{2}\ O_2$	256 000	54.7	1371–2000	8
$Fe_3O_4 = 3\ Fe + 2\ O_2$	1 100 000	307	25–1597 m	8
$Fe_2O_3 = 2\ Fe + \frac{3}{2}\ O_2$	814 000	251	25–1500	8
$FeS = Fe_\gamma + \frac{1}{2}\ S_2$	164 000	61.0	988–1190 m	93
$FeS_2 = FeS + \frac{1}{2}\ S_2$	182 000	188	630– 760	93
$FeO \cdot Cr_2O_3 = Fe + \frac{1}{2}\ O_2 + Cr_2O_3$	317 000	72.6	750–1536	94
$FeO \cdot Cr_2O_3 = Fe(l) + \frac{1}{2}\ O_2 + Cr_2O_3$	331 000	80.3	1536–1700	94
$Fe_2(SO_4)_3 = Fe_2O_3 + 3\ SO_2 + \frac{3}{2}\ O_2$	772 000	724	400– 800	95
$Fe_2(SO_4)_3 = 2\ FeSO_4 + SO_2 + O_2$	396 000	352	430– 630	95
$FeSO_4 = \frac{1}{2}\ Fe_2O_3 + SO_2 + \frac{1}{2}\ O_2$	203 500	202.3	500– 630	95
$2\ FeO \cdot SiO_2 = 2\ FeO \cdot SiO_2(l)$	92 050	61.7	1220 m	71
$2\ FeO \cdot SiO_2 = 2\ FeO + SiO_2$	36 200	21.0	25–1220 m	71
$2\ FeO \cdot TiO_2 = 2\ FeO + TiO_2$	33 900	5.86	25–1100	74
$FeO \cdot TiO_2 = FeO + TiO_2$	33 500	12.1	25–1300	74
$FeO \cdot V_2O_3 = Fe + V_2O_3 + \frac{1}{2}\ O_2$	289 000	62.3	750–1536	94
$FeO \cdot V_2O_3 = Fe(l) + V_2O_3 + \frac{1}{2}\ O_2$	301 000	70.0	1536–1700	94
$FeO \cdot WO_3 = FeO + WO_3$	55 000	−10.0	700–1150	74

Reaction	$\overline{\Delta H^0}$ (joule)	$\overline{\Delta S^0}$ (joule)	Range (°C)	Reference
$Ga = Ga(l)$	5610	17.2	30 m	8
$Ga(l) = Ga(g)$	264 000	106.9	30–2204 b	8
$Ga_2O_3 = 2\ Ga(l) + \frac{3}{2}\ O_2$	1 090 000	324	30–1795 m	71
$GaS = Ga(l) + \frac{1}{2}\ S_2$	276 000	111	30– 960 m	71
$Ga_2S_3 = 2\ Ga(l) + \frac{3}{2}\ S_2$	720 000	318	30–1090 m	71
$Ge = Ge(l)$	37 000	30.5	937 m	10
$Ge(l) = Ge(g)$	335 000	108	937–2834 b	10
$GeO_2 = Ge + O_2$	575 000	188	25– 937	71
$HCl(g) = \frac{1}{2}\ H_2 + \frac{1}{2}\ Cl_2$	94 100	−6.40	25–2000	8
$HF(g) = \frac{1}{2}\ H_2 + \frac{1}{2}\ F_2$	274 500	−3.47	25–2000	8
$H_2O(l) = H_2O(g)$	41 090	110.1	100 b	8
$H_2O(g) = H_2 + \frac{1}{2}\ O_2$	247 500	55.86	25–2000	8
$H_2S(g) = H_2 + \frac{1}{2}\ S_2$	91 630	50.6	25–2000	8
$Hf = Hf(l)$	24 300	9.6	2227 m	8
$Hf(l) = Hf(g)$	574 200	117.7	2227–4598 b	8
$HfB_2 = Hf + 2\ B$	335 000	11.1	25–1200	71
$HfC = Hf + C$	230 000	7.5	25–2000	70
$HfO_{2_\alpha} = Hf + O_2$	1 060 000	174	25–1700	70
$HfO_{2_\alpha} = HfO_{2_\beta}$	10 500	5.3	1700	70
$Hg(l) = Hg(g)$	60 750	96.78	−39– 356 b	8
$HgO_{red} = Hg(l) + \frac{1}{2}\ O_2$	90 800	70.3	25– 500	3, 8
$HgS_{red} = Hg(l) + \frac{1}{2}\ S_2$	53 000	82.0	25– 345	3
$HgS_{red} = HgS_{black}$	4000	6.3	345	3
$In = In(l)$	3260	7.62	157 m	10
$In(l) = In(g)$	235 000	100.3	157–2073 b	10
$In_2O_3 = 2\ In(l) + \frac{3}{2}\ O_2$	919 000	309.4	157–1910 m	71
$InS = InS(l)$	36 000	37.3	692 m	71
$InS(l) = In(l) + \frac{1}{2}\ S_2$	155 000	60.7	692–1500	71
$In_2S_3 = 2\ In(l) + \frac{3}{2}\ S_2$	544 000	286	157– 900	71
$Ir = Ir(l)$	26 150	9.62	2443 m	10, 114
$Ir(l) = Ir(g)$	615 000	131	2443–4424 b	10, 114
$IrO_3(g) = Ir + \frac{3}{2}\ O_2$	−16 700	45.2	25–1500	71
$IrO_2 = Ir + O_2$	234 000	170	25–1000	70
$Ir_2S_3 = 2\ Ir + \frac{3}{2}\ S_2$	430 000	300	25–1000	71
$IrS_2 = Ir + S_2$	268 000	190	25–1000	71
$K = K(l)$	2335	6.95	63 m	8
$K(l) = K(g)$	84 500	82.0	63– 766 b	8
$KCl = KCl(l)$	26 300	25.2	771 m	8
$KCl(l) = K(l) + \frac{1}{2}\ Cl_2$	474 000	132	771–1437 b	8
$KCl(g) = K(g) + \frac{1}{2}\ Cl_2$	306 000	35.5	1437–2000	8
$KF = KF(l)$	27 200	24.06	858 m	8

Reaction	$\overline{\Delta H^0}$ (joule)	$\overline{\Delta S^0}$ (joule)	Range (°C)	Reference
$KF(l) = K(g) + \frac{1}{2} F_2$	607 500	142	858–1517 b	8
$KF(g) = K(g) + \frac{1}{2} F_2$	418 400	37.7	1517–2000	8
$La = La(l)$	6200	5.19	920 m	10
$La(l) = La(g)$	416 000	113	920–3460 b	10
$LaCl_3 = LaCl_3(l)$	54 400	48.2	855 m	70
$LaCl_3 = La + \frac{3}{2} Cl_2$	1 060 000	225.0	25– 855	70
$LaF_3 = LaF_3(l)$	50 200	28.5	1493 m	70
$LaF_3 = La(l) + \frac{1}{2} F_2$	1 780 000	241	25– 920	71
$LaN = La + \frac{1}{2} N_2$	300 000	106	25– 920	71
$La_2O_3 = 2\ La + \frac{3}{2} O_2$	1 790 000	278	25– 920	71
$LaS = La(l) + \frac{1}{2} S_2$	527 000	104	920–1500	71
$La_2S_3 = 2\ La(l) + \frac{3}{2} S_2$	1 420 000	286	920–1500	71
$Li = Li(l)$	3000	6.61	181 m	8
$Li(l) = Li(g)$	151 300	93.85	181–1347 b	8
$LiCl = LiCl(l)$	19 800	23.0	610 m	8
$LiCl(l) = Li(l) + \frac{1}{2} Cl_2$	382 000	52.4	610–1383 b	8
$LiCl(g) = Li(g) + \frac{1}{2} Cl_2$	360 000	42.2	1383–2000	8
$LiF = LiF(l)$	27 070	24.14	848 m	8
$LiF(l) = Li(l) + \frac{1}{2} F_2$	583 400	66.8	848–1342	8
$LiF(g) = Li(g) + \frac{1}{2} F_2$	500 000	43.0	1717b–2000	8
$Li_3AlF_6 = Li_3AlF_6(l)$	86 200	81.4	785 m	8
$Li_3AlF_6(l) = 3\ Li(l) + Al(l) + 3\ F_2$	3 240 000	400	785–1342 m	8
$Li_3N = 3\ Li(l) + \frac{1}{2} N_2$	200 000	160	181–1000	8
$Li_2O = Li_2O(l)$	58 600	31.8	1570 m	8
$Li_2O = 2\ Li(l) + \frac{1}{2} O_2$	603 000	135	181–1342	8
$Li_2S = 2\ Li(l) + \frac{1}{2} S_2$	515 000	121	181–1000	71
$LiOH = LiOH(l)$	21 000	28.2	471 m	8
$LiOH(l) = \frac{1}{2} Li_2O + \frac{1}{2} H_2O(g)$	28 000	21.8	471–1007 d	8
$Li_2O \cdot Al_2O_3 = Li_2O \cdot Al_2O_3(l)$	25 000	13.3	1610 m	74
$Li_2 \cdot Al_2O_3 = Li_2O + Al_2O_3$	107 000	−10.6	500–1300	74
$Li_2CO_3 = Li_2CO_3(l)$	44 800	45.1	720 m	8
$Li_2CO_3(l) = Li_2O + CO_2$	148 000	78.7	720–1570	8
$Li_2O \cdot Fe_2O_3 = Li_2O + Fe_2O_3$	40 000	−25.3	25– 700	74
$Li_2O \cdot SiO_2 = Li_2O \cdot SiO_2(l)$	28 000	19.0	1201 m	8
$Li_2O \cdot SiO_2 = Li_2O + SiO_2$	144 000	3.8	25–1201 m	8
$Li_2O \cdot 2\ SiO_2 = Li_2O \cdot 2\ SiO_2(l)$	54 000	41.2	1034 m	8
$Li_2O \cdot 2\ SiO_2 = Li_2O + 2\ SiO_2$	146 400	0.0	25–1034 m	8
$Li_2O \cdot TiO_2 = Li_2O + TiO_2$	130 000	−3.3	25– 900	74
$Mg = Mg(l)$	8960	9.71	650 m	8, 114
$Mg(l) = Mg(g)$	129 600	95.1	650–1093 b	8
$MgCl_2 = MgCl_2(l)$	43 100	43.7	714 m	8
$MgCl_2 = Mg + Cl_2$	637 000	155.4	25– 650	8

Reaction	$\overline{\Delta H^0}$ (joule)	$\overline{\Delta S^0}$ (joule)	Range (°C)	Reference
$MgCl_2(l) = Mg(g) + Cl_2$	649 000	157.7	714–1437 b	8
$MgF_2 = MgF_2(l)$	58 700	38.2	1263 m	8
$MgF_2 = Mg + F_2$	1 120 000	171.2	25– 650	8
$MgF_2(l) = Mg(g) + F_2$	1 172 000	215.6	1263–2263 b	8
$MgB_2 = Mg + 2\ B$	92 000	10.5	25– 650	8
$MgB_4 = Mg + 4\ B$	108 000	11.2	25– 650	8
$Mg_3N_2 = 3\ Mg + N_2$	464 000	203	25– 650	8
$Mg_2Si = Mg_2Si(l)$	85 800	62.5	1102 m	8
$Mg_2Si = 2\ Mg(l) + Si$	100 000	39.3	650–1090	8
$MgO = Mg + \frac{1}{2}\ O_2$	601 000	107.6	25– 650	8
$MgO = Mg(g) + \frac{1}{2}\ O_2$	730 000	204	1090–2000	8
$MgS = Mg + \frac{1}{2}\ S_2$	410 000	94.4	25– 650	8
$MgS = Mg(g) + \frac{1}{2}\ S_2$	540 000	193	1090–1700	8
$MgO \cdot Al_2O_3 = MgO + Al_2O_3$	35 600	−2.09	25–1400	8, 74
$MgCO_3 = MgO + CO_2$	116 000	173.4	25– 402 d	8
$MgO \cdot Cr_2O_3 = MgO + Cr_2O_3$	42 900	7.1	25–1500	74
$MgO \cdot Fe_2O_3 = MgO + Fe_2O_3$	19 250	−2.1	700–1400	74
$MgO \cdot MoO_2 = MgO + MoO_2$	−13 800	−22.6	900–1200	74
$MgO \cdot MoO_3 = MgO + MoO_3$	54 000	−13.6	25– 795	74
$MgSO_4 = MgO + SO_2 + \frac{1}{2}\ O_2$	371 000	261	670–1050	8, 95
$2\ MgO \cdot SiO_2 = 2\ MgO \cdot SiO_2(l)$	71 100	32.6	1898 m	8
$2\ MgO \cdot SiO_2 = 2\ MgO + SiO_2$	67 200	4.31	25–1898 m	8
$MgO \cdot SiO_2 = MgO \cdot SiO_2(l)$	75 300	40.6	1577 m	8
$MgO \cdot SiO_2 = MgO + SiO_2$	41 100	6.1	25–1577 m	8
$3\ MgO \cdot P_2O_5 = 3\ MgO \cdot P_2O_5(l)$	121 300	75	1348 m	8
$3\ MgO \cdot P_2O_5 = 3\ MgO + P_2 + \frac{5}{2}\ O_2$	2 000 000	510	25–1348 m	8
$2\ MgO \cdot TiO_2 = 2\ MgO + TiO_2$	26 360	1.26	25–1500	8
$MgO \cdot TiO_2 = MgO + TiO_2$	27 600	3.14	25–1500	8
$2\ MgO \cdot V_2O_5 = 2\ MgO + V_2O_5$	87 500	0	25– 670	74
$MgO \cdot V_2O_5 = MgO + V_2O_5$	41 800	8.4	25– 670	74
$MgO \cdot WO_3 = MgO + WO_3$	74 000	10.9	25–1200	8, 74
$Mn = Mn(l)$	12 100	7.95	1246 m	8
$Mn(l) = Mn(g)$	235 800	101.2	1246–2061 b	8
$MnCl_2 = MnCl_2(l)$	37 700	40.8	650 m	70
$MnCl_2 = Mn + Cl_2$	478 000	128	25– 650 m	70
$MnCl_2(g) = Mn(l) + Cl_2$	302 000	9.8	1244–2062	70
$Mn_3C = 3\ Mn(l) + C$	14 000	−1.1	25–1037 m	71
$Mn_7C_3 = 7\ Mn + 3\ C$	128 000	21.0	25–1200	71
$Mn_5Si_3 = Mn_5Si_3(l)$	173 000	110	1300 m	71
$MnSi = Mn + Si$	61 500	6.3	25–1246	71
$MnP = Mn + \frac{1}{2}\ P_2$	167 000	87.0	25–1147 m	71
$MnP_3 = Mn + \frac{3}{2}\ P_2$	381 000	240	25–1246	71
$MnO = Mn + \frac{1}{2}\ O_2$	390 000	76.3	25–1246	96
$Mn_3O_4 = 3\ Mn + 2\ O_2$	1 385 000	344	25–1246	70

Reaction	$\overline{\Delta H^0}$ (joule)	$\overline{\Delta S^0}$ (joule)	Range (°C)	Reference
$Mn_3O_4 = 3\ MnO + \frac{1}{2}\ O_2$	232 000	117	925–1540 m	97
$Mn_2O_3 = 2\ Mn + \frac{3}{2}\ O_2$	954 000	255	25–1246	70
$Mn_2O_3 = \frac{2}{3}\ (Mn_3O_4) + \frac{1}{6}\ O_2$	35 000	28.1	800–1000	98
$MnO_2 = Mn + O_2$	519 000	181	25– 510 d	96
$MnS = MnS(l)$	26 100	14.5	1530 m	71
$MnS = Mn + \frac{1}{2}\ S_2$	296 500	76.74	700–1200	99
$MnO\cdot Al_2O_3 = Mn + Al_2O_3$	48 100	7.32	500–1200	74
$MnSO_{4_\alpha} = \frac{1}{3}\ Mn_3O_4 + SO_2 + \frac{1}{3}\ O_2$	295 600	227.4	700– 849	95
$MnSO_{4_\beta} = \frac{1}{3}\ Mn_3O_4 + SO_2 + \frac{1}{3}\ O_2$	250 200	187.2	849–1030	95
$2\ MnO\cdot SiO_2 = 2\ MnO\cdot SiO_2(l)$	89 600	55.4	1345 m	71
$2\ MnO\cdot SiO_2 = 2\ MnO + SiO_2$	53 600	25.0	25–1345 m	71
$MnO\cdot SiO_2 = MnO\cdot SiO_2(l)$	67 000	42.8	1291 m	71
$MnO\cdot SiO_2 = MnO + SiO_2$	28 000	2.76	25–1291 m	71
$Mo = Mo(l)$	27 800	9.6	2624 m	8, 114
$Mo(l) = Mo(g)$	591 000	120	2624–4705 b	8
$Mo_2C = 2\ Mo + C$	45 600	−4.2	25–1100	70
$MoC = Mo + C$	7530	−5.4	25– 700	70
$Mo_2N = 2\ Mo + \frac{1}{2}\ N_2$	60 700	14.6	25– 500	71
$Mo_3Si = 3\ Mo + Si$	119 000	2.5	25–1412	71
$Mo_5Si_3 = 5\ Mo + 3\ Si$	311 000	4.1	25–1412	71
$MoSi_2 = Mo + 2\ Si$	132 600	2.8	25–1412	71
$MoO_2 = Mo + O_2$	578 000	166.5	25–2000	8
$MoO_2(g) = Mo + O_2$	18 400	33.9	25–2000	8
$MoO_3 = MoO_3(l)$	47 700	45.2	801 m	8
$MoO_3 = Mo + \frac{3}{2}\ O_2$	740 000	247	25– 801 m	8
$MoO_3(g) = Mo + \frac{3}{2}\ O_2$	360 000	59.4	25–2000	8
$MoS_2 = Mo + S_2$	397 000	182	25–1185	8, 71
$Mo_2S_3 = 2\ Mo + \frac{3}{2}\ S_2$	594 000	265	25–1200	8
$NH_3(g) = \frac{1}{2}\ N_2 + \frac{3}{2}\ H_2$	53 700	116.5	25–2000	8
$NO(g) = \frac{1}{2}\ N_2 + \frac{1}{2}\ O_2$	−90 400	−12.68	25–2000	8
$NO_2(g) = \frac{1}{2}\ N_2 + O_2$	−32 300	63.3	25–2000	8
$Na = Na(l)$	2594	6.99	98 m	8
$Na(l) = Na(g)$	101 300	87.9	98– 897 b	8, 114
$NaCl = NaCl(l)$	28 160	26.2	801 m	8
$NaCl = Na(l) + \frac{1}{2}\ Cl_2$	411 600	93.1	98– 801 m	8
$NaCl(l) = Na(g) + \frac{1}{2}\ Cl_2$	464 000	134	801–1465 b	8
$NaF = NaF(l)$	33 350	26.3	996 m	8
$NaF = Na(l) + \frac{1}{2}\ F_2$	577 000	105.5	98– 996 m	8
$NaF(l) = Na(g) + \frac{1}{2}\ F_2$	624 000	148	996–1787 b	8
$Na_3AlF_6 = Na_3AlF_6(l)$	107 300	83.5	1012 m	113
$Na_3AlF_6 = 3\ Na(l) + Al + 3\ F_2$	3 310 000	554	98– 660	113

Reaction	$\overline{\Delta H^0}$ (joule)	$\overline{\Delta S^0}$ (joule)	Range (°C)	Reference
$Na_3AlF_6(l) = 3\ Na(g) + Al(l) + 3F_2$	3 380 000	623	1012–2000	113
$NaCN = NaCN(l)$	8790	10.5	562 m	8
$NaCN = Na(l) + C + \frac{1}{2}\ N_2$	90 500	34.1	98– 562 m	8
$NaCN(l) = Na(g) + C + \frac{1}{2}\ N_2$	152 000	83.7	883–1530 b	8
$Na_2O = Na_2O(l)$	47 700	34.0	1132 m	8
$Na_2O = 2\ Na(l) + \frac{1}{2}\ O_2$	422 000	141	98–1132 m	8
$Na_2O(l) = 2\ Na(g) + \frac{1}{2}\ O_2$	519 000	235	1132–1950 d	8
$Na_2O \cdot Al_2O_3 = Na_2O + Al_2O_3$	185 000	−2.9	500–1131	74
$Na_2CO_3 = Na_2O + CO_2$	297 000	118	25– 850 m	8
$Na_2CO_3(l) = Na_2O(l) + CO_2$	316 400	131	1132–2000	8
$Na_2O \cdot Cr_2O_3 = Na_2O + Cr_2O_3$	203 000	−5.8	25–1132	71
$Na_2O \cdot Fe_2O_3 = Na_2O + Fe_2O_3$	88 800	−14.6	25–1132	70
$Na_2O \cdot SiO_2 = Na_2O \cdot SiO_2(l)$	51 800	38.0	1089 m	8
$Na_2O \cdot SiO_2 = Na_2O + SiO_2$	238 000	8.8	25–1089 m	8
$Na_2O \cdot 2\ SiO_2 = Na_2O \cdot 2\ SiO_2(l)$	35 600	31.0	874 m	8
$Na_2O \cdot 2\ SiO_2 = Na_2O + 2\ SiO_2$	233 000	−3.8	25– 874 m	8
$Nb = Nb(l)$	26 900	9.8	2467 m	8
$Nb(l) = Nb(g)$	690 000	134.35	2467–4740 b	8
$NbB_2 = Nb + 2\ B$	251 000	9.2	25–1100	71
$Nb_2C = 2\ Nb + C$	194 000	11.7	25–1500	70
$NbC_{0.98} = Nb + 0.98\ C$	137 000	2.4	25–1500	8
$Nb_2N = 2\ Nb + \frac{1}{2}\ N_2$	251 000	83.3	25–2400 m	70
$NbN = Nb + \frac{1}{2}\ N_2$	230 000	77.8	25–2050 m	70
$NbO = NbO(l)$	84 000	38.5	1937 m	8
$NbO = Nb + \frac{1}{2}\ O_2$	415 000	87.0	25–1937	8
$Nb_2O_5 = Nb_2O_5(l)$	104 300	58.4	1512 m	8
$Nb_2O_5 = 2\ Nb + \frac{5}{2}\ O_2$	1 888 000	420	25–1512 m	8
$NbO_2 = NbO_2(l)$	92 000	42.3	1902 m	8
$NbO_2 = Nb + O_2$	784 000	167	25–1902 m	8
$Ni = Ni(l)$	17 500	10.1	1453 m	8, 114
$Ni(l) = Ni(g)$	386 000	121.4	1453–2911 b	8, 114
$NiCl_2 = NiCl_2(g)$	225 000	179	972 s	8
$NiCl_2 = Ni + Cl_2$	305 400	146	25– 972 s	8
$NiF_2 = Ni + F_2$	653 000	149	25–1474 m	71
$NiB = Ni + B$	176 000	8.4	25–1000	71
$Ni_4B_3 = 4\ Ni + 3\ B$	318 000	33.0	25–1580 m	71
$Ni_3C = 3\ Ni + C$	−40 000	−17.0	25– 500	71
$Ni_3Al = 3\ Ni + Al(l)$	176 000	32.2	660–1395 m	71
$NiAl = Ni + Al(l)$	138 000	28.0	660–1453 m	71
$NiSi = NiSi(l)$	43 100	34.0	992 m	71
$NiSi = Ni + Si$	91 000	7.9	25– 992 m	71
$Ni_7Si_{13} = 7\ Ni + 13\ Si$	632 000	20.0	25– 972 m	71
$Ni_3Ti = 3\ Ni + Ti$	146 000	26.0	25–1378 m	71
$NiTi = Ni + Ti$	67 000	12.0	25–1240 m	71

Reaction	$\overline{\Delta H^0}$ (joule)	$\overline{\Delta S^0}$ (joule)	Range (°C)	Reference
$NiO = Ni + \frac{1}{2} O_2$	235 600	86.0	25–1984 m	100
$Ni_3S_2 = 3 Ni + S_2$	330 000	163	25– 789 m	71
$NiS = Ni + \frac{1}{2} S_2$	146 000	72.0	25– 500	8
$NiO \cdot Al_2O_3 = NiO + Al_2O_3$	4200	−12.5	700–1300	71
$Ni(CO)_4(g) = Ni + 4 CO$	160 000	410	25– 500	8, 114
$NiO \cdot Cr_2O_3 = NiO + Cr_2O_3$	53 600	8.4	700–1200	74
$NiO \cdot Fe_2O_3 = NiO + Fe_2O_3$	19 900	−3.8	582–1400	74
$NiSO_4 = NiO + SO_2 + \frac{1}{2} O_2$	347 500	293.2	600– 860	95
$2 NiO \cdot SiO_2 = 2 NiO + SiO_2$	15 500	9.2	25–1545 m	71
$NiO \cdot TiO_2 = NiO + TiO_2$	18 000	8.4	500–1400	74
$NiO \cdot WO_3 = NiO + WO_3$	46 000	0.0	500–1100	74
$Os = Os(l)$	31 800	9.62	3027 m	10, 114
$Os(l) = Os(g)$	745 000	141	3027–5008 b	10, 114
$OsO_2 = Os + O_2$	290 000	175	25– 900	71
$OsO_4(g) = Os + 2 O_2$	334 000	145	131b– 70	71
$OsS_2 = Os + S_2$	270 000	195	25–1100	71
$P_{white} = P(l)$	657	2.05	44 m	8
$P_{red} = \frac{1}{4} P_4(g)$	32 100	45.6	25– 431 s	8
$P_4(g) = 2 P_2$	217 000	139	25–1700	8
$PH_3(g) = \frac{1}{2} P_2 + \frac{3}{2} H_2$	71 500	108	25–1700	8
$PCl_3(g) = \frac{1}{2} P_2 + \frac{3}{2} Cl_2$	474 000	209	25–1300	8
$PCl_5(g) = \frac{1}{2} P_2 + \frac{5}{2} Cl_2$	420 000	280	25–1700	8
$PF_3(g) = \frac{1}{2} P_2 + \frac{3}{2} F_2$	1 030 000	137	25–1700	8
$PF_5(g) = \frac{1}{2} P_2 + \frac{5}{2} F_2$	1 660 000	306	25–1700	8
$PO(g) = \frac{1}{2} P_2 + \frac{1}{2} O_2$	77 800	−11.6	25–1700	8
$PO_2(g) = \frac{1}{2} P_2 + O_2$	386 000	60.3	25–1700	8
$P_4O_{10}(g) = 2 P_2 + 5 O_2$	3 155 000	1011	358 s–1700	8
$Pb = Pb(l)$	4810	8.0	327 m	8
$Pb(l) = Pb(g)$	182 000	90.1	327–1746 b	8
$PbCl_2 = PbCl_2(l)$	21 900	28.2	501 m	8
$PbCl_2(l) = Pb(l) + Cl_2$	325 000	103.5	501– 953 b	8
$PbCl_2(g) = Pb(l) + Cl_2$	190 000	−7.5	953–1750	8
$PbF_2(l) = PbF_2(l)$	14 730	13.35	830 m	8
$PbF_2 = Pb(l) + F_2$	628 000	360	830–1200	8
$PbO = PbO(l)$	27 500	23.7	886 m	8
$PbO(l) = Pb(l) + \frac{1}{2} O_2$	181 000	68.0	886–1535 b	8
$Pb_3O_4 = 3 Pb(l) + 2 O_2$	702 500	370	328–1200	71
$PbO_2 = Pb(l) + O_2$	272 000	194	328– 900	71
$PbS = Pb(l) + \frac{1}{2} S_2$	163 000	88.0	328–1113 m	8
$PbO \cdot B_2O_3 = PbO(l) + B_2O_3(l)$	102 500	33.0	885–1535	8
$PbO \cdot 2 B_2O_3 = PbO(l) + 2 B_2O_3(l)$	166 500	79.5	885–1535	8
$PbSO_4 = PbO + SO_2 + \frac{1}{2} O_2$	401 000	262	25–1090 m	70

Reaction	$\overline{\Delta H^0}$ (joule)	$\overline{\Delta S^0}$ (joule)	Range (°C)	Reference
$2\ PbO \cdot SiO_2 = 2\ PbO \cdot SiO_2(l)$	51 050	50.3	743 m	8
$2\ PbO \cdot SiO_2(l) = 2\ PbO(l) + SiO_2$	33 500	−6.7	885–1500	8
$PbO \cdot SiO_2 = PbO \cdot SiO_2(l)$	26 000	25.1	764 m	8
$PbO \cdot SiO_2 = PbO(l) + SiO_2$	25 100	1.26	885–1500	8
$Pd = Pd(l)$	17 600	9.6	1552 m	10, 114
$Pd(l) = Pd(g)$	352 000	108.6	1552–2961 b	10, 114
$PdO = Pd + \frac{1}{2}\ O_2$	111 000	96.7	25– 870 m	71
$PdS = Pd + \frac{1}{2}\ S_2$	134 000	91.2	25– 970 m	71
$PdS_2 = Pd + S_2$	200 000	167	25– 972 m	71
$Pd_4S = 4\ Pd + \frac{1}{2}\ S_2$	134 000	86.0	25– 761 m	71
$Pt = Pt(l)$	19 700	9.63	1772 m	10, 114
$Pt(l) = Pt(g)$	521 000	127	1772–3823 b	10, 114
$PtO_2(g) = Pt + O_2$	164 000	0.0	25–1700	71
$PtS = Pt + \frac{1}{2}\ S_2$	149 000	101.5	25–1200	71
$PtS_2 = Pt + S_2$	233 500	182	25–1200	71
$Pu = Pu(l)$	2845	3.1	640 m	10
$Pu(l) = Pu(g)$	335 000	96.0	640–3230 b	10
$PuF_3 = PuF_3(l)$	54 400	32.0	1426 m	71
$PuF_3 = Pu(l) + \frac{3}{2}\ F_2$	1 540 000	228.0	640–1426 m	71
$PuN = Pu(l) + \frac{1}{2}\ N_2$	303 000	84.8	640–2000	71
$Pu_2C_3 = 2\ Pu(l) + 3\ C$	132 000	−16.3	640–2000	71
$PuC_2 = Pu(l) + 2\ C$	44 400	−22.6	640–2252 m	71
$PuO_2 = Pu(l) + O_2$	1 046 000	180	640–2390 m	71
$Rb = Rb(l)$	2190	7.0	39.5 m	8
$Rb(l) = Rb(g)$	76 320	79.7	39.5– 697 b	8
$Rb_2O = 2\ Rb(l) + \frac{1}{2}\ O_2$	335 000	154	25– 688	71
$Re = Re(l)$	33 200	9.62	3180 m	10
$Re(l) = Re(g)$	705 000	120	3180–5600 b	10
$ReO_2 = Re + O_2$	428 000	170	25–1200	71
$ReS_2 = Re + S_2$	300 000	191	25–1200	71
$Rh = Rh(l)$	21 500	9.62	1960 m	10, 114
$Rh(l) = Rh(g)$	505 000	127.4	1960–3694 b	10, 114
$Rh_2O_3 = 2\ Rh + \frac{3}{2}\ O_2$	377 000	266	25–1000	71
$RhO_2 = Rh + O_2$	200 000	−19.6	25–1200	71
$Ru = Ru(l)$	24 270	9.62	2250 m	10
$Ru(l) = Ru(g)$	605 000	137	2250–4146 b	10
$RuO_2 = Ru + O_2$	300 000	162	25–1200	71
$RuO_3(g) = Ru + \frac{3}{2}\ O_2$	79 500	61.0	25–1600	71
$RuO_4(g) = Ru + 2\ O_2$	180 000	143	25–1700	71
$RuS_2 = Ru + S_2$	330 000	191	25–1200	71
$S = S(l)$	1715	4.44	115 m	8
$S(l) = \frac{1}{2}\ S_2$	58 600	68.3	115– 445 b	8
$S_2 = 2\ S(g)$	469 300	161.3	25–1700	71

Reaction	$\overline{\Delta H^0}$ (joule)	$\overline{\Delta S^0}$ (joule)	Range (°C)	Reference
$S_4(g) = 2\ S_2$	62 800	115.5	25–1700	71
$S_6(g) = 3\ S_2$	276 000	305	25–1700	71
$S_8(g) = 4\ S_2$	400 000	450	25–1700	71
$SO(g) = \frac{1}{2}\ S_2 + \frac{1}{2}\ O_2$	57 800	−5.0	445–2000	8
$SO_2 = \frac{1}{2}\ S_2 + O_2$	362 000	72.7	445–2000	8
$SO_3(g) = \frac{1}{2}\ S_2 + \frac{3}{2}\ O_2$	458 000	163.3	445–2000	8
$Sb = Sb(l)$	19 900	22	631 m	10
$Sb(l) = Sb(g)$	85 300	45.6	631–1587 b	10
$Sb_2O_3 = Sb_2O_3(l)$	55 000	59.2	656 m	71
$Sb_2O_3(l) = \frac{1}{2}\ Sb(l) + \frac{3}{2}\ O_2$	661 000	198	656–1456 b	71
$Sc = Sc(l)$	14 100	7.78	1539 m	10
$Sc(l) = Sc(g)$	327 200	105.7	1539–2828 b	10
$ScCl_3 = ScCl_3(l)$	67 360	54.3	967 m	71
$ScCl_3 = Sc + \frac{3}{2}\ Cl_2$	891 200	230	25– 967 m	71
$ScF_3 = Sc + \frac{3}{2}\ F_2$	1 648 500	230	25–1000	71
$ScN = Sc + \frac{1}{2}\ N_2$	314 000	98.3	25–1539	71
$Sc_2O_3 = 2\ Sc + \frac{3}{2}\ O_2$	1 900 000	291	25–1539	70
$Se = Se(l)$	5440	11.0	220 m	70, 114
$Se(l) = \frac{1}{2}\ Se(g)$	59 300	61.1	221– 685 b	70, 114
$SeO(g) = \frac{1}{2}\ Se_2(g) + \frac{1}{2}\ O_2$	9200	−4.2	685–1700	71
$SeO_2(g) = \frac{1}{2}\ Se_2(g) + O_2$	178 000	66.1	685–1700	71
$Si = Si(l)$	50 540	30.0	1412 m	8
$Si(l) = Si(g)$	395 000	111	1412–3232 b	8
$SiCl_4(g) = Si + 2\ Cl_2$	660 000	129	62 b–1412	113
$SiI_4(g) = Si + 2\ I_2(g)$	231 000	115	301 b–1412	113
$Si_3N_{4_\alpha} = 3\ Si + 2\ N_2$	724 000	315	25–1412	8, 101
$Si_3N_{4_\alpha} = 3\ Si(l) + 2\ N_2$	874 500	405	1412–1700	8, 101
$SiC_\beta = Si + C$	73 000	7.7	25–1412	8
$SiC_\beta = Si(l) + C$	122 600	37.0	1412–2000	8
$SiP = Si + \frac{1}{2}\ P_2$	130 000	−89.0	25–1410 d	70
$SiO(g) = Si + \frac{1}{2}\ O_2$	104 000	−82.5	25–1412	8
$SiO_{2_{quartz}} = SiO_2(l)$	7700	4.52	1423 m	8
$SiO_{2_{quartz}} = Si + O_2$	907 000	176	25–1412	8
$SiO_{2_{cristobalite}} = SiO_2(l)$	9600	4.81	1723 m	8
$SiO_{2_{cristobalite}} = Si + O_2$	906 000	176	25–1412	8
$SiS(g) = Si + \frac{1}{2}\ S_2$	−51 800	−81.6	700–1412	8, 102
$SiS_2 = SiS_2(l)$	8400	6.15	1090 m	8
$SiS_2 = Si + S_2$	326 000	139	25–1090	8
$Sn = Sn(l)$	7030	13.9	232 m	10
$Sn(l) = Sn(g)$	296 000	103	232–2603 b	10
$SnCl_2(l) = SnCl_2(g)$	81 600	88.2	652 b	71
$SnCl_2(g) = Sn(l) + Cl_2$	226 000	13.0	652–1200	71
$SnO(g) = Sn(l) + \frac{1}{2}\ O_2$	−6300	−50.9	232–1700	71
$SnO_2 = Sn(l) + O_2$	575 000	198	232–1630 m	70
$SnS(g) = Sn(l) + \frac{1}{2}\ S_2$	−26 000	−49.4	232–1700	71

Reaction	$\overline{\Delta H^0}$ (joule)	$\overline{\Delta S^0}$ (joule)	Range (°C)	Reference
$SnS_2 = Sn(l) + S_2$	285 000	196	232– 765 m	71
$Sr = Sr(l)$	8370	7.95	777 m	8, 114
$Sr(l) = Sr(g)$	141 500	85.7	768–1412 b	8
$SrCl_2 = SrCl_2(l)$	16 200	14.1	874 m	8
$SrCl_2 = Sr + Cl_2$	824 000	148	25– 768	8
$SrF_2 = SrF_2(l)$	29 700	16.95	1477 m	8
$SrF_2 = Sr + F_2$	1 213 000	167	25– 768	8
$SrO = Sr(l) + \frac{1}{2} O_2$	597 000	102.4	768–1377	8
$SrS = Sr(l) + \frac{1}{2} S_2$	519 000	96	768–1377	8, 71
$SrCO_3 = SrO + CO_2$	215 000	142	700–1243 d	103
$SrO \cdot Al_2O_3 = SrO + Al_2O_3$	71 000	−4.2	25–1300	71
$SrSO_4 = SrO + SO_2 + \frac{1}{2} O_2$	548 000	273	25–1600 m	71
$2\ SrO \cdot SiO_2 = 2\ SrO + SiO_2$	213 000	5.86	25– 900	71
$SrO \cdot SiO_2 = SrO + SiO_2$	134 000	4.2	25– 900	71
$SrO \cdot TiO_2 = SrO + TiO_2$	137 000	2.1	25– 900	71
$Ta = Ta(l)$	31 600	9.62	3014 m	8
$Ta(l) = Ta(g)$	739 800	127.9	3014–5453 b	8
$TaCl_5(l) = TaCl_5(g)$	53 000	104.6	234 b	8
$TaCl_5(g) = Ta + \frac{5}{2} Cl_2$	754 000	169.4	234–2000	8
$TaF_5(l) = TaF_5(g)$	50 600	101	224 b	71
$TaF_5(g) = Ta + \frac{5}{2} F_2$	1 814 000	183	224–1700	71
$TaB_2 = Ta + 2\ B$	209 000	9.6	25–1700	70
$Ta_2C = 2\ Ta + C$	201 000	2.1	25–1700	71
$TaC = Ta + C$	142 000	1.2	25–1700	8
$Ta_2N = 2\ Ta + \frac{1}{2} N_2$	264 000	91	25–1700	71
$TaN = Ta + \frac{1}{2} N_2$	247 000	81.2	25–1700	71
$Ta_2Si = 2\ Ta + Si$	126 000	−3.3	25–1412	70
$Ta_5Si_3 = 5\ Ta + 3\ Si$	335 000	−12.6	25–1412	70
$TaSi_2 = Ta + 2\ Si$	117 000	22.6	25–1412	70
$TaO(g) = Ta + \frac{1}{2} O_2$	−188 000	−87	25–2000	8
$TaO_2(g) = Ta + O_2$	209 000	−20.5	25–2000	8
$Ta_2O_5 = Ta_2O_5(l)$	151 000	70.3	1785 m	8
$Ta_2O_5 = 2\ Ta + \frac{5}{2} O_2$	2 025 000	413	25–1785 m	8
$Te = Te(l)$	17 500	24.2	450 m	10
$Te(l) = \frac{1}{2} Te_2(g)$	58 000	45.9	450– 988 b	10
$TeO(l) = \frac{1}{2} Te_2(g) + \frac{1}{2} O_2$	7100	−6.0	1009–1700	71
$Th = Th(l)$	16 300	7.95	1755 m	10
$Th(l) = Th(g)$	527 000	104.4	1755–4790 b	10
$ThI_4 = ThI_4(l)$	48 000	57.0	570 m	71
$ThI_4 = Th + 2\ I_2(g)$	782 000	290	25– 570 m	71
$ThN = Th + \frac{1}{2} N_2$	377 000	86.2	25–1755	71
$Th_3N_4 = 3\ Th + 2\ N_2$	1 314 000	351	25–1755	71
$ThO(g) = Th + \frac{1}{2} O_2$	67 000	−52.7	1600–2000	104
$ThO_2 = Th + O_2$	1 220 000	183	25–2000	71

Reaction	$\overline{\Delta H^0}$ (joule)	$\overline{\Delta S^0}$ (joule)	Range (°C)	Reference
$Ti = Ti(l)$	15 500	7.95	1666 m	8
$Ti(l) = Ti(g)$	427 000	120	1666–3358 m	8
$TiCl_4(g) = Ti + 2\ Cl_2$	764 000	121	25–1666	8
$TiF_4(g) = Ti + 2\ F_2$	1 554 000	124	25–1666	8
$TiI_4(g) = Ti + 2\ I_2(g)$	402 000	117.6	25–1666	8
$TiB = Ti + B$	163 000	5.9	25–1666	8
$TiB_2 = Ti + B_2$	285 000	20.5	25–1666	8
$TiC = Ti + C$	185 000	12.6	25–1666	8
$TiN = Ti + \frac{1}{2}\ N_2$	336 000	93.3	25–1666	8
$TiO_\beta = Ti + \frac{1}{2}\ O_2$	515 000	74.1	25–1666	8
$TiO_{2\ rutile} = Ti + O_2$	941 000	177.6	25–1666	8
$Ti_2O_3 = 2\ Ti + \frac{3}{2}\ O_2$	1 502 000	258	25–1666	8
$Ti_3O_5 = 3\ Ti + \frac{5}{2}\ O_2$	2 435 000	421	25–1666	8
$Tl = Tl(l)$	4140	7.2	304 m	10
$Tl(l) = Tl(g)$	170 000	97.6	304–1473 b	10
$U = U(l)$	8540	6.07	1132 m	10
$U(l) = U(g)$	477 000	105.4	1132–4130 b	10
$UC = U(l) + C$	110 000	1.84	1132–2520 m	70
$U_2C_3 = 2\ U(l) + 3\ C$	222 000	2.5	1132–1700	71
$UC_{1.93} = U(l) + 1.93\ C$	90 400	9.2	1132–1700	71
$UN = U + \frac{1}{2}\ N_2$	293 000	81	25–1132	70
$UO_2 = U(l) + O_2$	1 087 000	172	1132–2000	70
$U_4O_9 = 4\ U + \frac{9}{2}\ O_2$	4 492 000	740	25– 600	70
$U_3O_8 = 3\ U + 4\ O_2$	3 560 000	653	25– 600	70
$UO_3 = U + \frac{3}{2}\ O_2$	1 227 000	251	25– 600	70
$V = V(l)$	22 850	10.4	1902 m	8
$V(l) = V(g)$	463 000	126	1902–3406 b	8
$VB = V + B$	138 000	5.86	25–2000	71
$V_2C = 2\ V + B$	146 400	3.35	25–1700	70
$VC = V + C$	102 000	9.6	25–2000	71
$VN_{0.46} = V + 0.23\ N_2$	130 000	44.4	25–1700	8
$VN = V + \frac{1}{2}\ N_2$	215 000	82.4	25–2346 d	8
$VO = V + \frac{1}{2}\ O_2$	425 000	80	25–1800	8
$V_2O_3 = 2\ V + \frac{3}{2}\ O_2$	1 203 000	238	25–2070 m	8
$VO_2 = V + O_2$	706 000	155	25–1360 m	8
$V_2O_5 = V_2O_5(l)$	64 400	68.3	670 m	8
$V_2O_5(l) = 2\ V + \frac{5}{2}\ O_2$	1 447 000	321.6	670–2000	8
$W = W(l)$	35 560	9.62	3407 m	8, 114
$W(l) = W(g)$	821 000	141	3407–5658 b	8
$W_2C = 2\ W + C$	30 540	−2.34	1302–1400	105
$WC = W + C$	42 300	4.98	900–1302	105
$WO_2 = W + O_2$	581 000	172	25–1724 d	8

Reaction	$\overline{\Delta H^0}$ (joule)	$\overline{\Delta S^0}$ (joule)	Range (°C)	Reference
$WO_3 = WO_3(l)$	73 400	42.1	1472 m	8
$WO_3 = W + \frac{3}{2}\ O_2$	833 500	245	25–1472 m	8
$Y = Y(l)$	11 380	6.32	1526 m	10
$Y(l) = Y(g)$	379 000	105.4	1526–3334 b	10
$YCl_3 = YCl_3(l)$	31 400	31.6	721 m	71
$YCl_3 = Y + \frac{3}{2}\ Cl_2$	968 000	227	25– 721	71
$YF_3 = YF_3(l)$	28 000	19.6	1155 m	71
$YF_3 = Y + \frac{3}{2}\ F_2$	1 711 000	219	25–1155 m	71
$YN = Y + \frac{1}{2}\ N_2$	300 000	99.6	25–1526	71
$Y_2O_3 = 2\ Y + \frac{3}{2}\ O_2$	1 900 000	282	25–1526	70
$Y_2O_3 \cdot Zr_2O_3 = Y_2O_3 + Zr_2O_3$	21 000	0.0	25–1200	71
$Zn = Zn(l)$	7320	10.6	420 m	8
$Zn(l) = Zn(g)$	118 000	100.3	420– 907 b	8
$ZnCl_2(l) = ZnCl_2(g)$	119 200	118.7	732 b	70
$ZnCl_2(l) = Zn(l) + Cl_2$	402 500	131	420– 732 b	70
$ZnCl_2(g) = Zn(g) + Cl_2$	393 000	105	907–1700	70
$ZnF_2 = ZnF_2(l)$	41 800	36.44	875 m	71
$ZnF_2 = Zn(l) + F_2$	769 000	175	420– 875 m	71
$ZnF_2(l) = Zn(g) + F_2$	832 000	227	907–1505 b	71
$ZnO = Zn(g) + \frac{1}{2}\ O_2$	460 240	190.4	907–1700	70
$ZnS = Zn(l) + \frac{1}{2}\ S_2$	278 000	108	420– 907	71
$ZnS(g) = Zn(g) + \frac{1}{2}\ S_2$	−5020	30.5	1182 b–1700	71
$ZnO \cdot Fe_2O_3 = ZnO + Fe_2O_3$	9620	3.8	25– 700	71
$ZnSO_4 = ZnO + SO_2 + \frac{1}{2}\ O_2$	328 000	267	25– 700	8
$Zr = Zr(l)$	20 900	9.83	1852 m	8
$Zr(l) = Zr(g)$	580 000	124	1852–4429 b	8
$ZrCl_4(g) = ZrCl_4(g)$	110 500	181.5	336 s	8
$ZrCl_4(g) = Zr + 2\ Cl_2$	871 000	116	336–2000	8
$ZrF_4 = ZrF_4(g)$	237 700	202	906 s	8
$ZrF_4(g) = Zr + 2\ F_2$	1 678 000	128.4	903–2000	8
$ZrB_2 = ZrB_2(l)$	104 600	31.5	3050 m	8
$ZrB_2 = Zr + 2\ B$	328 000	23.4	25–1852	8
$ZrC = Zr + C$	197 000	9.2	25–1852	8
$ZrN = Zr + \frac{1}{2}\ N_2$	364 000	92	25–1852	8
$ZrSi = Zr + Si$	160 000	4.2	25–1412	8
$ZrO(g) = Zr + \frac{1}{2}\ O_2$	−54 600	−67.0	1300–1852	8
$ZrO_2 = ZrO_2(l)$	87 000	29.5	2677 m	8
$ZrO_2 = Zr + O_2$	1 092 000	184	25–1852	8
$ZrS(g) = Zr + \frac{1}{2}\ S_2$	−237 000	−78.2	25–1852	8
$ZrS_2 = Zr + S_2$	700 000	178	25–1552 m	71
$ZrO \cdot SiO_2 = ZrO + SiO_2$	26 800	12.6	25–1707 m	71

APPENDIX 2 EFFECT OF COLLISION ON THE SIZE DISTRIBUTION OF INCLUSIONS

At high levels of inclusions in molten steel, say greater than 100 p.p.m., collision of particles may become important. By collision small particles may grow into large ones. The collision mechanism is obviously a binary mechanism, that is, the rate is proportional to the square of the concentration of inclusions,[1-3] whereas the removal of inclusions to the walls is proportional to the concentration. Therefore, at sufficiently low levels of inclusions, collisions become unimportant compared with separation to the walls and top-slag. Here an estimate is derived of the critical concentration in steel where the effect of collisions may be neglected.

In general, particles suspended in turbulent flow may approach one another by several mechanisms. Collisions by Brownian diffusion should not be important for particles larger than 1 μm. Furthermore, inertial and gravitational mechanisms do not seem to play a significant role for particles less than 100 μm.[2,4] If we assume that there are few or no inclusions greater than 100 μm, we are left with only one mechanism for collision, namely shear. Smoluchowski[1] considered a suspension of particles moving under uniform shear, that is, laminar flow. He assumed that the particles were randomly distributed throughout the fluid and that they followed the fluid streamlines as if no other particles were present (up to the moment of impact). This gave for the number of collisions $\dot{N}_{12}$ per unit volume between particles of type 1 and 2:

$$\dot{N}_{12} = \tfrac{4}{3}\, c_1 c_2 d_{12}^3 \frac{\mathrm{d}v}{\mathrm{d}z}. \tag{A2.1}$$

Here c_1 and c_2 are the number of particles per unit volume of type 1 and 2, respectively, $\mathrm{d}v/\mathrm{d}z$ is the velocity gradient normal to the direction of particle travel, and d_{12} is the sum of the particle radii, $d_{12} = a_1 + a_2$.

Equation (A2.1) has been applied to colloid collisions in turbulent flow.[3,5] Then $\mathrm{d}v/\mathrm{d}z$ is replaced by the mean velocity gradient. For isotropic turbulence and assuming that $|\partial v/\partial z|$ is normally distributed, we obtain

$$\left|\frac{\partial v}{\partial z}\right| = \left(\frac{2\dot{\varepsilon}}{15\nu}\right)^{1/2}. \tag{A2.2}$$

Then

$$\dot{N}_{12} = c_1 c_2 d_{12}^3 \left(\frac{8\pi\dot{\varepsilon}}{15\nu}\right)^{1/2} \tag{A2.3}$$

where $\dot{\varepsilon}$ is the average power input per unit mass of suspension.

Equation (A2.3) is valid when the particles experience the same conditions as in Smoluchowski's original calculation for laminar flow. This means that d_{12} is small compared with the smallest eddy size. Furthermore, the particles are to follow the fluid motion completely. The smallest eddy size λ is roughly[6]

$$\lambda = \left(\frac{\nu^3}{\dot{\varepsilon}}\right)^{1/4}.$$

Typically, for steel $\dot{\varepsilon} \sim 0.01\,\mathrm{m^2/s^3}$. Then $\nu = 0.8 \times 10^{-6}\,\mathrm{m^2/s}$ gives $\lambda = 84.5\,\mu\mathrm{m}$. It should be noted that due to the $\frac{1}{4}$ power dependence, λ is not very sensitive to $\dot{\varepsilon}$. Thus, it is seen that the first condition is fulfilled for a stirred bath except for the largest inclusions.

As shown in Section 5.8 inclusions less than 100 μm in size obey Stokes' law. Particles regain their velocity after a collision in a time interval, the *relaxation time*, of

$$\tau_p = \frac{(\rho_p + \frac{1}{2}\rho)\, d_p^2}{\rho\, 18\nu}. \qquad \text{(A2.4)}$$

For instance $\rho_p/\rho = \frac{1}{2}$ and $d_p = 100\,\mu\mathrm{m}$ gives for steel

$$\tau_p = \frac{1.5 \times 10^{-8}}{18 \times 0.8 \times 10^{-6}} = 10^{-3}\,\mathrm{s}.$$

This time is short compared to the lifetime of the smallest eddy:

$$\tau_e = \left(\frac{\nu}{\dot{\varepsilon}}\right)^{1/2} = \left(\frac{0.8 \times 10^{-6}}{10^{-2}}\right)^{1/2} = 0.9 \times 10^{-2}\,\mathrm{s}.$$

Thus the inclusions should follow the fluid motion so that the second condition for the validity of equ (A2.3) is realized.

As particles suspended in a fluid approach each other to within a diameter or so, they begin to influence each others' motion by disturbance of the intervening fluid. This part of a particle–particle approach is treated as a separate problem and can be described by a *collision efficiency*, η. If N is the number of particles in the volume swept out by the target (whose cross-section is a circle with radius d_{12}) the collision efficiency η is given by

$$\eta = \frac{\text{Number of collisions made}}{N}. \qquad \text{(A2.5)}$$

η is defined in the same manner as in the sections on flotation and filtration in Chapter 5. Considerable effort has been expended to calculate η.[4,7] However, experimental data for η seems to be limited.[2,8] For hydrosols in the size range 1–30 μm it seems that very roughly η may be described by a relation of the form[4]

$$\eta \sim \frac{10^{-7} a_1 a_2}{(a_1 + a_2)^3}. \tag{A2.6}$$

This expression is employed in the following discussion.

We determine the lowest concentration of inclusions where collision effects significantly change the size distribution eqn (5.37) given in Section 5.7:

$$f(a) = b_1 \exp(-a/b_2). \tag{5.37}$$

For simplicity the index N to indicate that the distribution is number size has been discarded. To start with it should be useful to relate b_1 and b_2 to various characteristics of the inclusions. The constant b_2 was determined from the mass balance (differential equation (5.24)). b_1 is fixed by the total number of inclusions N_i per unit volume melt:

$$N_i = \int_o^\infty f(a)\,\mathrm{d}a = b_1 b_2. \tag{A2.7}$$

The total volume of inclusions, V_i, per unit volume melt is determined by b_2

$$V_i = \int_o^\infty f(a)\,\frac{4\pi a^3}{3}\,\mathrm{d}a = \frac{b_1\,4\pi \times 6\,b_2^4}{3}. \tag{A2.8}$$

Then the mean inclusion volume is

$$\frac{4\pi\overline{a^3}}{3} = \frac{V_i}{N_i} = 8\pi b_2^3 \tag{A2.9}$$

where $\overline{a^3}$ is the mean value of a^3.

The mean value of a^3 is equal to $6b_2^3$. Therefore the most interesting inclusion size is around $2b_2$. This value is employed in the discussions which follow.

As previously, we assume that the inclusions are spherical not only before but also after a collision takes place.[9] This is certainly the case if the inclusions are molten. For instance, SiO_2 inclusions melt at 1722°C and at lower temperatures if they contain small amounts of CaO and Al_2O_3. Furthermore, we assume that all collisions result in coalescence.

Following Lindborg and Torsell[10] we take the annihilation as well as the formation of particles into account. Then eqn (5.23) for a steady state situation is expanded to give

$$V\frac{\mathrm{d}f(a)}{\mathrm{d}a}\frac{\mathrm{d}a}{\mathrm{d}t} = -f(a)k_t A - V\int_0^\infty f(a_2)\,W(a_1, a_2)\,f(a)\,\mathrm{d}a_2$$
$$+\frac{V}{2}\int_0^\infty f(a_1)\,W(a_1, a_2)\,f(a_2)\,\frac{a^2}{a_2^2}\,\mathrm{d}a_1. \tag{A2.10}$$

Here $\mathrm{d}a/\mathrm{d}t$ is given by eqn (5.21) and

$$W(a_1, a_2) = \left(\frac{8\pi\dot{\varepsilon}}{15\nu}\right)^{1/2} \frac{(a_1 + a_2)^3\, 10^{-7} a_1 a_2}{(a_1 + a_2)^3}$$

$$= \left(\frac{8\pi\dot{\varepsilon}}{15\nu}\right)^{1/2} \times 10^{-7} a_1 a_2. \qquad \text{(A2.11)}$$

To estimate the effect of collisions eqn (5.37) is used to obtain approximate values of the two integrals I_a and I_f, respectively, in eqn (A2.10). The annihilation term I_a gives

$$I_a = f(a)\, Vk_1 \int_0^\infty a a_2 f(a_2)\, da_2$$

$$= f(a)\, Vk_1 a b_1 b_2^2 \qquad \text{(A2.12)}$$

where

$$k_1 = \left(\frac{8\pi\dot{\varepsilon}}{15\nu}\right)^{1/2} 10^{-7} \qquad \text{(A2.13)}$$

has the dimensions of velocity (m/s). By comparison of eqn (A2.12) with the $f(a)k_t A$ term in eqn (A2.10), it is found that annihilation is unimportant if

$$Vk_1 a b_1 b_2^2 \ll k_t A$$

or, from eqns (A2.8) and (A2.9),

$$k_1 V_i \ll \frac{(\overline{a^3})^{1/3}}{\tau} \qquad \text{(A2.14)}$$

where τ is the time constant for removal of inclusions ($= V/k_t A$). Here $(\overline{a^3})^{1/3}$ may be determined metallographically. $\tau = k_t A/V$ must be obtained from fluid dynamic calculations or from batch experiments where a pulse addition of inclusions is carried out. In simple terms eqn (A2.14) says that the removal by collision must be less than removal to walls and top-slag. For instance, if

$$\tau = 300\,\text{s}, \qquad (\overline{a^3})^{1/3} = 10\,\mu\text{m},$$

$$V_i = 100\,\text{p.p.m. (by volume), and}$$

$$k_1 = 10^{-7} \left(\frac{8\pi \times 0.01}{15 \times 0.8 \times 10^{-6}}\right)^{1/2} = 1.45 \times 10^{-5}\,\text{m/s},$$

the left-hand side of the inequality is 1.45×10^{-9} while the right-hand side is 33×10^{-9}.

The formation term I_f in eqn (A2.10) gives

$$I_f = \frac{Vk_1 b_1^2 a^2}{2} \int_a^{a2^{2/3}} \frac{\exp(-v/b_2) a_1 a_2\, dv}{(1 - a_1^2/a_2^2) a_2^2} \qquad \text{(A2.15)}$$

where $v = a_1 + a_2 = a_1 + (a^3 - a_1^3)^{1/3}$ and $dv = da_1\,(1 - a_1^2/a_2^2)$. By solving a second-degree equation in a_1 one finds that

$$\left(1 - \frac{a_1^2}{a_2}\right) a_2^2 = a_2^2 - a_1^2 = v(v - 2a_1) = 2v^{1/2}\left(\frac{a^3}{3} - \frac{v^3}{12}\right)^{1/2}.$$

Furthermore $a_1 a_2 = (v^3 - a^3)/(3v)$. Then eqn (A2.15) becomes

$$I_f = \frac{Vk_1 b_1^2 a^2}{4 \times 3^{1/2}} \int_a^{a2^{2/3}} \frac{\exp(-v/b_2)(v^3 - a^3)\,dv}{(a^3 - v^3/4)^{1/2} v^{3/2}}.$$

A substitution, $p = v/a$, leads to

$$I_f = \frac{Vk_1 b_1^2 a^3}{4 \times 3^{1/2}} \int_1^{2^{2/3}} \frac{\exp(-pa/b_2)(p^3 - 1)\,dp}{(1 - p^3/4)^{1/2} p^{3/2}}.$$

The value of the integral is less than $\exp(-a/b_2)$ so that

$$I_f < \frac{Vk_1 b_1^2 a^3}{4 \times 3^{1/2}} \exp(-a/b_2). \qquad \text{(A2.16)}$$

To take formation into account (disregarding annihilation) eqn (A2.10) may be written in the form

$$\frac{df(a)}{da} = -\frac{f(a)}{b_2} + \frac{I_f}{k_t A b_2} \qquad \text{(A2.17)}$$

or, integrated,

$$f(a) = b_1 \exp(-a/b_2) + \frac{\exp(-a/b_2)}{k_t A b_2} \int_0^a I_f \exp(a/b_2)\,da. \qquad \text{(A2.18)}$$

From inequality (A2.16) it is seen that the formation term is less than $\exp(-a/b_2) V k_1 b_1^2 a^4/(k_f A b_2 4 \times 3^{1/2} \times 4)$.

If formation is to be unimportant then

$$\frac{Vk_1 b_1^2 a^4}{k_t A b_2 4 \times 3^{1/2} \times 4} \ll b_1. \qquad \text{(A2.19)}$$

On the assumption that the values of a that are interesting in practice are roughly equal to $2b_2$, we again derive inequality (A2.14).

Equations (A2.14) and (A2.19) for annihilation and formation cannot be applied to very large inclusions. The reason for this is partly the approximations employed in the solution of eqns (A2.10) and (A2.15) and partly the assumed exponential size distribution, eqn (5.37). In fact, large inclusions have such a high k_t value that $f(a)$ for them drops at a steeper rate than indicated by eqn (5.37).

It is concluded that inequality (A2.14) gives a condition for when collision can be neglected. The smaller the time constant τ, the less important collision effects become. k_1 is given by eqn (A2.13) containing the factor 10^{-7} m from eqn (A2.6). Even if eqn (A2.6) only gives an approximate functional dependence on a_1 and a_2 and only employs a rough value of the factor 10^{-7},[11] below around 100 p.p.m. inclusions collisions should become unimportant. According to eqn (A2.13) k_1 increases with the specific stirring power $\dot{\varepsilon}$ as $\dot{\varepsilon}^{1/2}$. However, $1/\tau$ also increases with $\dot{\varepsilon}$, usually as $\dot{\varepsilon}$. It is possible to produce locally high values of $\dot{\varepsilon}$ electromagnetically, by mechanical stirring, or by employing a gas jet. In this case collisions may become important at inclusion levels substantially below 100 p.p.m.

REFERENCES

1. Smoluchowski, M. (1917). Versuch einer mathematischen teori der koagulasjonskinetic kolloider lösungen. *Z. Phys. Chem.*, **92**, 129.
2. De Boer, C. B. J., Hoedermakers, C. F. M., and Thoenes, D. (1989). Coagulation in turbulent flow. *Chem. Eng. Res. Des.*, **67**, Parts I and II, 301–15.
3. Abrahamsen, J. (1975). Collision rates of small particles in a vigorously turbulent fluid. *Chem. Eng. Sci.*, **30**, 1372–9.
4. Davis, R. H. (1984). The rate of coagulation of a dilute polydisperse system of sedimenting spheres. *J. Fluid Mech.*, **145**, 179–99.
5. Saffman, P. G. and Turner, S. S. (1956). On the collision of drops in turbulent clouds. *J. Fluid Mech.*, **1**, 16–30.
6. Levich, V. G. (1962). *Physicochemical Hydrodynamics*. Prentice-Hall Englewood Cliffs, New Jersey.
7. Pearson, H. J. Valioulis, I. A., and List, E. J. (1984). Monte Carlo simulation of coagulation in discrete particle-size distributions. *J. Fluid Mech.*, **143**, 367–411.
8. Higashitani, K., Yamauchi, K., Matsuno, Y., and Hosokawa, G. (1983). Turbulent coagulation of particles dispersed in turbulent flow. *J. Chem. Eng. Japan*, **16**, 299–304.
9. Torsell, K. (1967). Removal of inclusions resulting from deoxidation of iron with silicon. *Jernkontorets Annaler*, **151**, 890–949.
10. Lindborg, U. and Torsell, K. (1968). A collision model for the growth and separation of deoxidation products. *Trans. AIME*, **242**, 94–102.
11. De Boer, G. B. J., De Weerd, C., and Thoenes, D. (1989). Coagulation in turbulent flow: Part II. *Chem. Eng. Res. Des.*, **67**, 308–15.

APPENDIX 3 EFFECT OF PROTRUSIONS ON THE FLOW AND ON THE BOUNDARY CONDITIONS FOR THE CONCENTRATION OF INCLUSIONS

When the protrusion height, $e - a$, is less than the boundary layer thickness, δ_l, the protrusions have no effect on drag between wall and fluid.[1] Then the velocity and flow patterns outside δ_l should not be significantly affected. We have this situation if the protrusion Reynolds number is

$$\frac{v_{\mathrm{m}}(e - a)}{v} < 100. \tag{A3.1}$$

The turbulent diffusivity which describes the flow within δ_l in fact probably depends on $e - a$ and on the number of protrusions N/A per unit area. However, with the present state of our knowledge, we do not attempt to determine this dependence and there is no alternative other than to employ eqn (5.97).

It is important to estimate the number of protrusions N/A required in order to justify using the boundary condition eqn (5.101). Then we make use of the fact that fluid and particles essentially move parallel to the wall since the velocity fluctuations are large parallel to the wall and much smaller in the normal direction. Roughly the distance between protrusions is $(A/N)^{1/2}$. If an inclusion is regarded as a point, then a protrusion with diameter $e - a$ effectively captures inclusions over a distance $e - a + 2a = e + a$. The number 2 is present because an inclusion can be captured on either side of a protrusion. The probability p that an inclusion moving towards the protrusions actually hits one is

$$p = \frac{e + a}{(A/N)^{1/2}}. \tag{A3.2}$$

If the inclusion travels a distance L along the wall, this process is repeated $L/(A/N)^{1/2}$ times. Then finally the probability that the inclusion is not removed is given by

$$(1 - p)^{L/(A/N)^{1/2}} \sim 1 - \frac{Lp}{(A/N)^{1/2}}.$$

This gives the following approximate condition for removal:

$$\frac{L}{(A/N)^{1/2}} \frac{(e + a)}{(A/N)^{1/2}} = \frac{L(e + a)N}{A} = 1. \tag{A3.3}$$

For instance, if $L = 1$ m and $e + a = 30\,\mu$m,

$$\left(\frac{A}{N}\right)^{1/2} = 5.5 \times 10^{-3}\,\mathrm{m}.$$

Thus in the example the average distance between protrusions should be less than 5.5 mm in order to justify the boundary condition, eqn (5.101). The protrusions may be created by inclusions already precipitated or by the original roughness of the walls.

REFERENCE

1. Schlichting, H. (1968). *Boundary-layer theory*. McGraw-Hill, New York.

APPENDIX 4 THE NUMERICAL SOLUTION OF THE EQUATIONS FOR A SOLIDIFIED SHELL

Due to the presence of terms such as $a_2\tau^{1/2}$ in the relation for the surface temperature, eqn (8.43), the solution of the heat diffusion equation, eqn (8.29), will contain an infinite number of terms. It will be seen that eqn (8.29) can be transformed to a more amenable form.

Let us start by looking at the terms that contain $\phi(\tau)$. Then we split T_x/T_{m} into two parts:

$$\frac{T_x}{T_{\mathrm{m}}} = S_1(x) + S_2(x). \tag{A4.1}$$

Here $S_1(x)$ refers to $\phi(\tau)$:

$$S_1(x) = \phi(\tau) + \sum_{n=1}^{\infty} \left((x-1)^2 \frac{4\alpha}{\alpha_{\mathrm{s}}}\right)^n \frac{\phi^{(n)}(\tau)}{(2n)!} \tag{A4.2}$$

a_1 and $a_3\tau$ in eqn (8.43) only contribute to the two first terms. The term $a_2\tau^{1/2}$ in $\phi^{(n)}(\tau)$ gives

$$a_2\tfrac{1}{2}\left(\tfrac{1}{2}-1\right)\left(\tfrac{1}{2}-2\right)\cdots\left(\tfrac{1}{2}-(n-1)\right)\tau^{1/2-n} = -a_2\tfrac{1}{2}\left(n-\tfrac{3}{2}\right)\left(n-\tfrac{5}{2}\right)\cdots \tfrac{1}{2}(-1)^n\tau^{1/2-n}.$$

Now

$$(2n)! = 2n(2n-2)(2n-4)\cdots 2(2n-1)(2n-3)\cdots 1$$

so that

$$(2n)! = 2^{2n}n!\left(n-\tfrac{1}{2}\right)\left(n-\tfrac{3}{2}\right)\cdots\tfrac{1}{2}. \tag{A4.3}$$

Thus

$$S_1(x) = a_1 + \frac{(x-1)^2\,2\alpha a_3}{\tau_0\alpha_{\mathrm{s}}} + a_3\left(\frac{\tau}{\tau_0}\right) - a_2\left(\frac{\tau}{\tau_0}\right)^{1/2}\sum_{n=0}^{\infty}\left(\frac{-(x-1)^2\alpha}{\alpha_{\mathrm{s}}\tau}\right)^n\frac{1}{n!(2n-1)} \tag{A4.4}$$

or since

$$\frac{1}{n(2n-1)} = \frac{2}{2n-1} - \frac{1}{n}$$

the infinite series is equal to

$$\sum_{n=1}^{\infty}\left(-\frac{(x-1)^2\alpha}{\alpha_s\tau}\right)^n\frac{1}{(n-1)!\,(n-\frac{1}{2})}-\sum_{n=0}^{\infty}\left(-\frac{(x-1)^2\alpha}{\alpha_s\tau}\right)^n\frac{1}{n!}.$$

Here the last series is simply the exponential function $\exp(-(x-1)^2\alpha/(\alpha_s\tau))$. Then eqn (A4.4) becomes

$$S_1(x)=a_1+\frac{(x-1)^2\,2\alpha a_3}{\tau_o\alpha_s}+a_3\left(\frac{\tau}{\tau_o}\right)+a_2\left(\frac{\tau}{\tau_o}\right)^{1/2}\exp\left(-\frac{(x-1)^2\alpha}{\alpha_s\tau}\right)$$
$$-a_2\left(\frac{\tau}{\tau_o}\right)^{1/2}\sum_{n=1}^{\infty}\left(-\frac{(x-1)^2\alpha}{\alpha_s\tau}\right)^n\frac{1}{(n-1)!\,(n-\frac{1}{2})}.$$

Finally, this last series can be transformed to the error integral so that

$$S_1(x)=a_1+\frac{(x-1)^2 2\alpha a_3}{\tau_o\alpha_s}+a_3\frac{\tau}{\tau_o}+a_2\left(\frac{\tau}{\tau_o}\right)^{1/2}\exp\left(-\frac{(x-1)^2\alpha}{\alpha_s\tau}\right)$$
$$-a_2\left(\frac{(x-1)^2\alpha\pi}{\alpha_s\tau_o}\right)^{1/2}\operatorname{erf}\left\{\left(\frac{(x-1)^2\alpha}{\alpha_s\tau}\right)^{1/2}\right\} \qquad \text{(A4.5)}$$

where $\operatorname{erf}(x)=2\pi^{-1/2}\int_0^x\exp(-v^2)\,dv$. Equation (A4.5) can be checked by expanding in a Maclaurin series, carrying out the integration, and comparing term by term with the infinite series.

Similarly, for the terms with $\psi(\tau)$ in eqn (8.29) one obtains from eqns (8.31) and (8.46):

$$S_2(x)=(x-1)\left\{\psi(\tau)+\frac{(x-1)^2 2\alpha\left(1-\frac{k}{k_s}\right)a_3}{\tau_o\alpha_s^3}\right.$$
$$-\left(\frac{\tau}{\tau_o}\right)^{1/2}\left(a_2\left(1-\frac{k}{k_s}\right)+\frac{4ka_3}{\pi^{1/2}k_s\tau_o^{1/2}}\right)$$
$$\times\sum_{n=1}^{\infty}\left(-\frac{(x-1)^2\alpha}{\alpha_s\tau}\right)^n\frac{1}{n!\,(2n+1)\,(2n-1)}+\frac{2ka_1}{\pi^{1/2}k_s\tau_o^{1/2}\left(\frac{\tau}{\tau_o}\right)^{1/2}}$$
$$\left.\times\sum_{n=1}^{\infty}\left(-\frac{(x-1)^2\alpha}{\alpha_s\tau}\right)^n\frac{1}{n!\,(2n+1)}\right\}. \qquad \text{(A4.6)}$$

Now

$$\frac{2}{(2n+1)(2n-1)}=\frac{1}{2n-1}-\frac{1}{2n+1}$$

so that eqn (A4.6) simplifies to

$$S_2(x) = (x-1)\left[\psi(\tau) + \frac{(x-1)^2 2\alpha\left(1-\frac{k}{k_s}\right)a_3}{\tau_o\alpha_s^3}\right.$$

$$-\left(\frac{\tau}{\tau_o}\right)^{1/2}\left(\frac{a_2}{2}\left(1-\frac{k}{k_s}\right)+\frac{2ka_3}{\pi^{1/2}k_s\tau_o}\right)$$

$$\times\sum_{n=1}^{\infty}\left(-\frac{(x-1)^2\alpha}{\alpha_s\tau}\right)^n\frac{1}{n!(2n-1)}$$

$$+\left\{\left(\frac{\tau}{\tau_o}\right)^{1/2}\left(\frac{a_2}{2}\left(1-\frac{k}{k_s}\right)+\frac{2ka_3}{\pi^{1/2}k_s\tau_o^{1/2}}\right)+\frac{2ka_1}{\pi^{1/2}k_s\tau_o^{1/2}\left(\frac{\tau}{\tau_o}\right)^{1/2}}\right\}$$

$$\left.\times\sum_{n=1}^{\infty}\left(-\frac{(x-1)^2\alpha}{\alpha_s\tau}\right)^n\frac{1}{n!(2n+1)}\right].$$

This becomes

$$S_2(x) = (x-1)\left[\psi(\tau) + \frac{(x-1)^2 2\alpha(1-k/k_s)a_3}{\tau_o\alpha_s^3}\right.$$

$$-\left(\frac{\tau}{\tau_o}\right)^{1/2}\left(\frac{a_2}{2}\left(1-\frac{k}{k_s}\right)+\frac{2ka_3}{\pi^{1/2}k_s\tau_o}\right)\left(1-\exp\left(-\frac{(x-1)^2\alpha}{\alpha_s\tau}\right)\right.$$

$$\left.-\left(\frac{(x-1)^2\alpha\pi}{\alpha_s\tau}\right)^{1/2}\operatorname{erf}\left(\left(\frac{(x-1)^2\alpha}{\alpha_s\tau}\right)^{1/2}\right)\right)$$

$$-\left\{\left(\frac{\tau}{\tau_o}\right)^{1/2}\left(\frac{a_2}{2}\left(1-\frac{k}{k_s}\right)+\frac{2ka_3}{\pi^{1/2}k_s\tau_o^{1/2}}\right)+\frac{2ka_1}{\pi^{1/2}k_s\tau_o^{1/2}\left(\frac{\tau}{\tau_o}\right)^{1/2}}\right\}$$

$$\left.\times\left\{1-\frac{1}{2}\left(\frac{\alpha_s\tau\pi}{\alpha(x-1)^2}\right)^{1/2}\operatorname{erf}\left(\left(\frac{(x-1)^2\alpha}{\alpha_s\tau}\right)^{1/2}\right)\right\}\right]$$

or

$$S_2(x) = (x-1)\left\{\left(1-\frac{k}{k_s}\right)\left(a_1+a_3\frac{\tau}{\tau_o}\right)+\frac{k\pi^{1/2}a_2}{k_s\tau_o^{1/2}}\right.$$

$$\left.+\frac{(x-1)^2}{\tau_o}\left(1-\frac{k}{k_s}\right)\frac{2\alpha a_3}{3\alpha_s}\right\}$$

$$+ (x-1)\left(\frac{\tau}{\tau_o}\right)^{1/2}\left\{\left(\left(1-\frac{k}{k_s}\right)\frac{a_2}{2}+\frac{2ka_3}{k_s\pi^{1/2}\tau_o^{1/2}}\right\}\left\{\exp\left(-\frac{(x-1)^2\alpha}{\tau\alpha_s}\right)\right.$$

$$+ (x-1)\left(\frac{\alpha\pi}{\tau\alpha_s}\right)^{1/2}\operatorname{erf}\left(\frac{(x-1)^2\alpha}{\tau\alpha_s}\right)^{1/2}\right\}$$

$$+\frac{1}{2}\left(\frac{\alpha_s\pi}{\alpha\tau_o}\right)^{1/2}\left[\left\{\left(1-\frac{k}{k_s}\right)\frac{a_2}{2}+\frac{2ka_3}{k_s\pi^{1/2}\tau_o^{1/2}}\right\}\tau+\frac{2ka_1\tau_o^{1/2}}{k_s\pi^{1/2}}\right]$$

$$\times \operatorname{erf}\left(\frac{(x-1)^2\alpha}{\tau a_s}\right)^{1/2}. \tag{A4.7}$$

Together with eqn (A4.1), eqns (A4.5) and (A4.7) describe the temperature profile in the solidified shell. However, the coefficients a_1, a_2, and a_3 must be obtained first. They are determined by the boundary condition, eqn (8.27). To introduce this boundary condition the position of the solidification front as a function of time is described by a series expansion:

$$s = 1 + A_1\left\{\left(\frac{\tau}{\tau_o}\right)^{1/2} + b_1\left(\frac{\tau}{\tau_o}\right) + b_2\left(\frac{\tau}{\tau_o}\right)^{3/2}\right\} + \ldots \tag{A4.8}$$

where again the coefficients A_1, b_1, and b_2 are unknown. To determine a_1, a_2, a_3, A_1, b_1, and b_2 we have found it useful to introduce the dimensionless time, τ_m, when the shell attains its greatest extent $s = s_m$, and the time when the shell has completely remelted (disappeared), $\tau = \tau_o$. Then for $\tau = \tau_o$ and $s = 1$ from eqn (A4.8):

$$1 + b_1 + b_2 = 0 \tag{A4.9}$$

and since $ds/d\tau = 0$ at $\tau = \tau_m$, eqn (A4.8) gives

$$1 + 2b_1\left(\frac{\tau_m}{\tau_o}\right)^{1/2} + 3b_2\left(\frac{\tau_m}{\tau_o}\right) = 0. \tag{A4.10}$$

Again $s = s_m$ at $\tau = \tau_m$ is obtained from eqn (A4.8)

$$s_m = 1 + A_1\left(\left(\frac{\tau_m}{\tau_o}\right)^{1/2} + b_1\left(\frac{\tau_m}{\tau_o}\right) + b_2\left(\frac{\tau_m}{\tau_o}\right)^{3/2}\right). \tag{A4.11}$$

The condition that the temperature T_m is constant on the solidification front gives

$$\frac{T(s)\,s}{T_m} = s. \tag{A4.12}$$

To determine the remaining six constants, boundary condition eqns (A4.12) and (8.39) are employed for $\tau = 0$, $\tau = \tau_m$, and $\tau = \tau_o$. Equations (A4.12) and (A4.1) for $\tau = 0$ at $s = 1$ give

$$a_1 + \frac{a_1 k}{k_s}\left(\frac{\alpha_s}{\alpha\tau_o}\right)^{1/2} \operatorname{erf}\left\{\frac{A_1}{\tau_o^{1/2}}\left(\frac{\alpha}{\alpha_s}\right)^{1/2}\right\} = 1. \tag{A4.13}$$

From eqns (A4.1), (8.39), and (A4.8) for the term $\tau^{-1/2}$ one obtains

$$\frac{\mathrm{Bi}^* b^* A_1}{2\tau_o^{1/2}} = \frac{2ka_1}{\pi^{1/2} k_s} \exp\left(-\frac{A_1^2}{\tau_o}\frac{\alpha}{\alpha_s}\right). \tag{A4.14}$$

For $\tau = \tau_m$ from eqns (A4.12) and (A4.1):

$$S_1(x = s_m, \tau = \tau_m) + S_2(x = s_m, \tau = \tau_m) = s_m \tag{A4.15}$$

and from eqn (8.39), since $\mathrm{d}s/\mathrm{d}\tau = 0$,

$$\left(\frac{\mathrm{d}(T_s x/T_m)}{\mathrm{d}x_{x=s_m}}\right) = 1 + \mathrm{Bi}^* s_m. \tag{A4.16}$$

Then

$$\begin{aligned}
&\left(1 - \frac{k}{k_s}\right)\left(a_1 + a_3\frac{\tau_m}{\tau_o}\right) + \frac{a_2 k\pi^{1/2}}{k_s\tau_o^{1/2}} + \frac{(s_m - 1)4\alpha a_3}{\tau_o\alpha_s}\\
&+ (s_m - 1)^2\left(1 - \frac{k}{k_s}\right)\frac{2\alpha a_3}{\alpha_s\tau_o} + \exp\left(-\frac{(s_m - 1)^2\alpha}{\tau_m\alpha_s}\right)\left(\frac{\tau_m}{\tau_o}\right)^{1/2}\\
&\times\left\{\frac{a_2}{2}\left(1 - \frac{k}{k_s}\right) + \frac{2ka_3}{k_s\pi^{1/2}\tau_o^{1/2}} - \frac{a_2 4(s_m - 1)\alpha}{\alpha_s\tau_m} + \frac{2ka_1\tau_o^{1/2}}{k_s\pi^{1/2}\tau_m}\right\}\\
&+ \left(\frac{\alpha\pi}{\alpha_s\tau_o}\right)^{1/2} \operatorname{erf}\left\{\left(\frac{(s_m - 1)^2\alpha}{\tau_m\alpha_s}\right)^{1/2}\right\}\\
&\times\left\{\left(a_2\left(1 - \frac{k}{k_s}\right) + \frac{4ka_3}{k_s\pi^{1/2}\tau_o^{1/2}}\right)(s_m - 1) - a_2\right\} = 1 + \mathrm{Bi}^* s_m.
\end{aligned} \tag{A4.17}$$

Finally for $\tau = \tau_o$

$$\phi(\tau_o) = 1$$

or

$$a_1 + a_2 + a_3 = 1. \tag{A4.18}$$

Equation (8.39) becomes

$$\mathrm{Bi}^* b^* \left(\frac{\mathrm{d}s}{\mathrm{d}\tau}\right)_{\tau=\tau_o} = \left(\frac{\mathrm{d}(T_s x/T_m)}{\mathrm{d}x}\right)_{x=1} - 1 - \mathrm{Bi}^*$$

or

$$\frac{\text{Bi}^* b^* A_1}{\tau_o}\left(\tfrac{1}{2} + b_1 + \tfrac{3}{2} b_2\right) = \left(1 - \frac{k}{k_s}\right)(a_1 + a_3) + \frac{a_2 k \pi^{1/2}}{k_s \tau_o^{1/2}} + \frac{a_2}{2}\left(1 - \frac{k}{k_s}\right) + \frac{2k(a_1 + a_3)}{k_s \pi^{1/2} \tau_o^{1/2}} - 1 - \text{Bi}^*. \tag{A4.19}$$

The equations for the solidified shell are rather complicated to solve even numerically. We choose to work backward through the equations, as follows.

1. We start by guessing initial values of τ_o, τ_m, and s_m.
2. By using the initial values of τ_o, τ_m, and s_m in equations (A4.9), (A4.10), and (A4.11) we can calculate b_1, b_2, and A_1. These values are inserted into equation (A4.13) to give a_1, which enables us to solve equations (A4.18) and (A4.15) for a_2 and a_3. The next step is to solve equation (A4.17) to give Bi*. Then equation (A4.14) gives the final parameter b^*.
3. All parameters calculated in 2 are then used in equation (A4.19) to give a new value for τ_o. The new and old value of τ_o are averaged (in order to make the iteration more stable) to give a value of τ_o to be used as a new input value in 1.
4. Now the whole process of 2 and 3 is repeated until τ_o converges within a limit of $\Delta\tau_o < 0.000\,001$.
5. So far we are able to determine τ_o, but we cannot tell if our initial guess of τ_m and s_m has any resemblance to a real situation. Fortunately we do have information not yet used. Parameters Bi* and b^* are physical quantities, that is, they depend on our actual system and we can calculate the product Bi* b^* independently (see Chapter 8, equations (8.37) and (8.38)).
6. Then we go through step 1–4 for a set of incrementally increased values of τ_m and s_m, and calculate the value of Bi* b^* for each combination of τ_m and s_m. By comparing it against Bi* b^* from 5 we can sort out the combinations of τ_m and s_m that give the right value of Bi* b^*.
7. We are now left with a set of τ_m and s_m which gives the correct value of Bi* b^*, and each of the combinations also give a τ_o and a Bi*. Then we are able to plot τ_o versus Bi*, τ_m versus Bi*, and s_m versus Bi*. Since Bi* b^* depends on quantities such as forced convection, preheating of the alloy, and so on, we have chosen to present the plots for some values of Bi* b^*.

PROBLEMS

PROBLEMS FOR CHAPTER 2

2.1 *Configurational Entropy* A crystal containing N_A atoms of A is placed in contact with a crystal that contains N_B atoms of B. In how many ways can the atoms mix? What is the entropy increase ΔS? According to Stirling's formula:

$$\ln N! \approx N \ln N - N + \tfrac{1}{2} \ln (2\pi N).$$

Neglect the last term and show that ΔS is equal to the entropy of mixing of an ideal solution:

$$-\frac{\Delta S}{k} = N_A \ln x_A + N_B \ln x_B$$

where x_A and x_B are the mole fractions of A and B, respectively.

2.2 *Effect of pressure on solids and liquids* Using eqn (2.9) show that $(\partial \mu_i / \partial p)_T = V_i$ is the volume (of the solid or liquid) per unit mole of component i. The pressure dependence of V_i is given by

$$V_i = V_i^0 (1 - \kappa (p - 10^5))$$

where V_i^0 is the volume at a pressure of 10^5 N/m^2 and κ is the compressibility. Find for steel:

$$\mu_i (10^6\ N/m^2) - \mu_i (10^5\,\mathrm{N/m^2}).$$

The density is $\rho_m = 7800\,\mathrm{kg/m^3}$. The atomic weight is 55.85. $\kappa = 0.6 \times 10^{-11}\,\mathrm{m^2/N}$.

What comment do you have on the effect of pressure at modest pressures for solids and liquids?

2.3 *The solubility product of MnS in steel* Find the product $a_{Mn}\, a_S$ at 1550°C. We assume that the activity of $a_{MnS} = 0.5$.

2.4 *Concentration of dissolved oxygen at equilibrium in stainless steel* The composition of the steel is 18% Cr, 10% Ni, and 0.75% Si. Assume that silicon determines the oxygen activity ($a_{SiO_{2\,cristobalite}} = 1$). The temperature is 1600°C. Determine the oxygen concentration.

2.5 *Oxygen in equilibrium with carbon in steel* Find the equilibrium oxygen content in steel at 1600°C when the Fe–C–O alloy contains 1% $\underline{C}$. $p_{CO} =$ 1 bar. Can you ignore the interaction coefficients? Why are only e_C^C and e_O^C important?

2.6 Conversion between interaction coefficients Derive eqn (2.39).

2.7 Gas (H_2O pores in solid Cu) in copper melts Assume that dissolved oxygen in molten copper has an activity of $a_o = 0.01\%$. The temperature is 1100°C. Assume that hydrogen in the melt is in equilibrium with H_2, $p_{H2} = 0.04$ bar. Find a_H. Find p_{H_2O} given by the equilibria with p_{H_2} and $\underline{O}$.

2.8 Removal of Ca from molten FeSi Calcium is removed from FeSi (75% Si by mass) by the exchange reaction

$$2\,\underline{Ca} + SiO_2 = Si + 2\,CaO.$$

The slag used for the extraction contains equal mole fractions CaO, Al_2O_3, and SiO_2. The Raoultian activity coefficient of calcium in FeSi is taken to be $\gamma^0_{Ca} = 0.002$. Find the concentration of calcium in the melt at equilibrium with the slag a 1550°C. Assume that Raoult's law applies to Si.

2.9 Surface tension of Fe–S The radius of S^{2+} ions is 1.84 Å. What is the surface tension at 1650°C when [% S] = 0.1? The surface tension of pure iron at 1650°C is taken to be 1.90 N/m. Also $f_2 = f_1 = 1$.

2.10 The CO–$\underline{C}$ equilibrium in stainless steel In an AOD converter oxygen diluted with argon reacts with carbon in the molten steel:

$$\tfrac{1}{2}\,O_2 + \underline{C} = CO\ (\text{gas}). \qquad \text{(I)}$$

The following unwanted reaction may take place:

$$\tfrac{3}{2}\,O_2 + 2\,\underline{Cr} = Cr_2O_3\ (\text{slag}). \qquad \text{(II)}$$

(a) Calculate the ratio (at equilibrium):

$$K = \frac{p_{CO}}{a_C}$$

as a function of $a_{Cr_2O_3}$ and a_{Cr} in the bath, and the absolute temperature T. Use the data in Appendix 1.

(b) Assume that the bath has composition [% C] = 0.2, [% Ni] = 10, and [% Cr] = 18. Neglect the effect of other components and calculate p_{CO}/[% C] at 1615°C. $a_{Cr_2O_3}$ is taken to be 1.

PROBLEMS FOR CHAPTER 3

3.1 Gas (SO_2 pores in solid Cu) in copper melts Assume that $a_s = 0.03\%$ and that oxygen is determined by the equilibrium with Cu_2O. What is the partial pressure of SO_2 at 1100°C?

3.2 Removal of $\underline{S}$ by gas purging We are blowing N_2 through a Cu melt that contains sulphur and oxygen. Assume that SO_2 in the bubbles is in equilibrium with [% S] (that resistance to mass transfer is negligible). Indicate how [% S] will change with time if the flow of N_2 is kept constant.

3.3 Reduction of CdO by C Determine the temperature where CdO is reduced by C in equilibrium with CdO, 0.5 atm CO, and 0.1 bar Cd vapour.

3.4 Reduction of CeF_3 by Ca Find ΔG^0 for the reaction

$$3\,Ca(g) + 2\,CeF_3(s) = 2\,Ce(l) + 3\,CaF_2(s)$$

and show that it is negative.

3.5 Decomposition of $Ni(CO)_4$ Gas to Ni(s) and CO Assume that $p_{Ni(CO)_4} = 0.01$ bar and $p_{CO} = 1$ bar. Find the temperature.

PROBLEMS FOR CHAPTER 4

4.1 A steel melt of mass M kg contains $[\%\,S]_{in}$. Also $(\%\,S)_{in} = 0$. This sulphur is to be extracted to M_s kg of slag. The distribution coefficient is $K = (\%\,S)/[\%\,S] = 10.3$. Determine the final content of S in the steel. Assume that equilibrium is attained and that there is complete mixing in the metal and slag phases.

4.2 In problem 4.1 how much slag must be added to reduce sulphur levels from 400 parts per million (p.p.m.) to 100 p.p.m.?

4.3 Determine [% S] in problem 4.2 as a function of time when $M = 1000$ kg and the total metal-side mass transfer coefficient between metal and slag $k_t = 5 \times 10^{-4}$ m/s. The contact area between metal and slag is $A_s = 0.3$ m^2. The steel density is $\rho = 7000$ kg/m^3.

4.4 We are removing sodium from Al in a batch reactor purged with inert gas. Sodium is eliminated by burning with air at the top surface. The mass transfer coefficient for the top surface is $k_{ts} = 3 \times 10^{-4}$ m/s and the area $A_s = 1$ m^2. Sodium is also removed by the bubbles, and the amount of inert gas is 2 Nm^3/hour. We assume that these bubbles are saturated with sodium. The ratio between the partial pressure of sodium p_{Na} and [% Na] is ($f_{Na} = 1$):

$$K = \frac{p_{Na}}{[\%\,Na]} = 1.$$

One bar pressure and 1 mass% are the reference states. The density of the aluminium is $\rho = 2350$ kg/m^3, the molar weight of sodium is

23 kg/kmol, the volume of metal is 0.4 m^3 and $\underline{Na}$ is the same everywhere throughout the melt; [% Na] $\ll$ 1. The volume of 1 kmol is 22.4 Nm^3. How long does it take to reduce the levels of sodium to a tenth of the original value?

4.5 In problem 4.4 we assumed that the bubbles were saturated with sodium. How far are they from saturation if the bubble surface area $A = 0.5\ m^2$ and the mass transfer coefficient for the bubbles is $k_t = 2 \times 10^{-4}$ m/s? Neglect the effect of pressure change up through the melt.

4.6 Sodium is removed in two continuous back-mix tanks in series. Each gas-purged tank has the same size and behaves as described in problem 4.4. The flow of aluminium is 4 m^3 per hour. What is the content of sodium leaving the second unit compared with the sodium level in the metal that enters the first unit?

4.7 Suppose that the conditions are the same as in problem 4.6 except that the purge gas now contains sufficient Cl_2 to react with the sodium that diffuses to the bubbles. The bubble surface area is 0.5 m^2 and the mass transfer coefficient for the bubbles is 2×10^{-4} m/s. What is the sodium level in the output divided by the input sodium content in this case?

4.8 We are bubbling argon into 20 tonnes of molten steel in order to remove $\underline{H}$. It is assumed that equilibrium is attained between H_2 and $\underline{H}$. The equilibrium constant is $K_H^2 = p_{H_2}/[\%\ H]^2 = 1.37 \times 10^5$ and $f_H = 1$. Find the amount of argon, Q_g, required to reduce $\underline{H}$ from 4 p.p.m. to 2 p.p.m. 1 kmol of gas equals 22.4 Nm^3 and the pressure above the bath is 1 atm. You may assume that $p_{H_2} \ll p_{argon}$. $\underline{H}$ is the same everywhere throughout the melt and the melt is covered by a lid that retains the gas from the bubbles and prevents pick-up of hydrogen.

4.9 The problem is the same as in 4.8 except that the total pressure is only 0.5 bar. It is also assumed that in this case equilibrium is attained between hydrogen in bubbles and in the melt. How much inert gas is now required?

4.10 For problem 4.8 we wish to calculate to what extent the bubbles have reached equilibrium when they reach the bath surface for [% H] = 0.001. The contact area for the bubbles is $A = 2\ m^2$ and the mass transfer coefficient is $k_t = 10^{-4}$ m/s. Neglect the effect of pressure change up through the melt. $\dot{Q}_g = 2\ Nm^3$/hour, the molar volume is 22.4 Nm^3 per kmol and $\rho = 7000\ kg/m^3$.

4.11 We wish to remove $\underline{Mn}$ from a Fe batch melt by vacuum treatment. The mass transfer coefficient for the melt boundary layer is $k = 10^{-4}$ m/s. For the evaporation:

$$k_e = \frac{\gamma_{Mn} p^{\circ}_{Mn} M_{Fe}}{\rho(2\pi M_{Mn} RT)^{1/2}}$$

where $p^{o}_{Mn} = 46.65 \times 10^2$ Pa, $M_{Mn} = 54.9$ kg/kmol, $R = 8.314$ kJ/kmol K, $\gamma_{Mn} = 1$, $T = 1823$ K, $\rho = 7000$ kg/m^3, and $M_{Fe} = 55.85$ kg/kmol.

Neglecting resistance and pressure on the gas side of the interface, determine the total mass transfer coefficient k_t. Assume that the melt is completely mixed with volume $V = 0.5$ m^3 and that the surface area of the bath is $A_s = 0.5$ m^2. Find the time required to reduce the $\underline{Mn}$ concentration down to a tenth of the original level.

4.12 In an AOD converter oxygen diluted with argon reacts with carbon

$$\tfrac{1}{2}\,O_2 + \underline{C} = CO\ \text{(gas)}. \qquad \text{(I)}$$

Also we have the undesired reaction

$$\tfrac{3}{2}\,O_2 + 2\,\underline{Cr} = Cr_2O_3.$$

We assume that oxygen first reacts with Cr to cover the bubbles with a layer of Cr_2O_3. Then reaction (I) proceeds. Assume that the contact area between the bubbles and melt is sufficient to attain the equilibrium given by eqn (I). Also assume that the equilibrium $3\,\underline{C} + Cr_2O_3 = 2\,\underline{Cr} + 3\,CO$ is described by $k = p_{CO}/[\%\ C] = 2$.

Where pressure is in bar. We assume that $a_{Cr_2O_3} = 1$ (constant) and that [% Cr] should not change noticeably. We have $M = 65$ tonnes of melt and $\dot{G} = 45$ Nm^3 of inert gas per minute. Set up a dynamic balance (differential equation) for carbon removal. 1 kmol equals 22.4 Nm^3 and the atomic weight of carbon is 12.

Integrate the differential equation. How many minutes does it take to remove carbon from 0.2% down to 0.04%?

PROBLEMS FOR CHAPTER 5

5.1 Nucleation of SiO_2 inclusions in steel Silicon is dissolved in steel to give [% Si] = 0.3, and the steel contains [% O] = 0.035 when the silicon has dissolved.

(a) Calculate the critical radius for nucleation r^*! The interfacial tension is $\sigma = 0.7$ J/m^2, $V_m = 28 \times 10^{-6}$ m^3/mol, and the steel temperature is 1600°C.

(b) Calculate the number of nuclei.

(c) Find the growth rate (in μm per second) da/dt of the inclusions if the oxygen concentration in equilibrium with SiO_2 is 0.010%. The density of the steel is 7030 kg/m^3 and $k_o \approx 10^{-4}$ m/s. How much does an inclusion grow per minute?

5.2 Removal of inclusions by settling

(a) Find the velocity of rise, u, of inclusions in aluminium. The density of the bath is $\rho = 2.35 \times 10^3$ kg/m^3 and of the inclusions is $\rho_p = 2000$ kg/m^3; the acceleration of gravity is $g = 9.81$ m/s^2. The radius of the (spherical) inclusions is $a = 3 \times 10^{-5}$ m and the kinematic viscosity of molten aluminium is $\nu = 0.4 \times 10^{-6}$ m^2/s.

(b) Assume that there is complete mixing in the bulk bath. Supersaturation Δ[% O] is very small so that the growth of inclusions is not due to supersaturation and the inclusions are removed by rising into the top-slag according to problem (a). The bath volume is V and the contact area between the slag and melt is A_s.

Find the concentration of inclusions, c_p, as a function of time. At time $t = 0$, $c_p = c_p\ (t = 0)$. Use u from problem (a) and assume $V = 1$ m^3 and $A_s = 0.5$ m^2.

What is $c_p(t)/c_p(t = 0)$ at time $t = 75$ min?

(c) Assume that the number–size distribution of inclusions, $f_N(t, a)$, initially (at $t = 0$) is $f_N(a) = B \exp(-ba)$. The velocity of rise u, and V and A_s are known. As in problem (b), the inclusions do not grow and do not collide. Find $f_N(t, a)$.

5.3 *Removal of inclusions by stirring* We have evacuated the atmosphere above the melt and very stable refractories have been used. We assume that we may neglect dissolved oxygen and that pick-up of oxygen from the refractories and atmosphere is small so the effect of supersaturation Δ[% O] is negligible. The coefficient for transfer of the inclusions to the walls and top-slag is k_t, the wall and top-slag contact area is A, and the melt volume is V. What is the mass of inclusions in per cent, [% O]$_i$, at time t when the concentration initially was [% O]$_i$ $(t = 0)$?

Give the [% O]$_i$/[% O]$_i$ $(t = 0)$ ratio for $t = 15$ min. What is [% O]$_i$ when [% O]$_i$ $(t = 0) = 2 \times 10^{-2}$. $V = 1$ m^3, $A = 2$ m^2, and $k_t = 10^{-3}$ m/s.

5.4 *Effect of oxygen pick-up on removal of oxide inclusions* Assume in problem 5.3 that there is a constant pick-up of oxygen from the walls and atmosphere. This pick-up is described by a mass transfer coefficient $k = 10^{-4}$ m/s and a constant driving force term for pick-up from the walls and atmosphere Δ[% O] $= 10^{-2}$. Neglecting dissolved oxygen, and assuming that [% O]$_i$ $(t = 0) = 2 \times 10^{-2}$, find [% O]$_i$ at $t = 15$ min.

5.5 *Removal of particles (inclusions) from melts by flotation* The fraction of the melt that the bubbles effectively sweep through per second is given by

$$\dot{\alpha}_e = \frac{\eta \pi R^2 N u_{br}}{V}$$

where R is the bubble radius, u_{br} is the relative velocity between bubbles and melt, N is the number of bubbles, and V is the melt volume. η is the collection efficiency.

$$N = \frac{3\dot{Q}H}{4\pi R^3 v_{bw}}$$

where $\dot{Q}$ is the volume flow of gas, H is the bath depth, and v_{bw} is the bubble velocity relative to the wall.

(a) Find $\dot{\alpha}_e$ when $\dot{Q} = 1\ m^3/hr$, $H = 1\ m$, $2R = 0.01\ m$, $u_{br} = v_{bw}$, $\eta = 0.1$, and $V = 1\ m^3$.

(b) The inclusions are to be removed in a completely mixed batch flotation unit. What is the number of inclusions per unit volume, c, at time t when the initial concentration was c_i?

(c) If the conditions are as given in (a) and (b), how much time does it take to reduce the content of inclusions to a tenth?

(d) List mechanisms that determine the collision efficiency.

5.6 Filtration of aluminium We have a bed of Al_2O_3 balls, and Al_4C_3 inclusions are to be filtered out. The balls have a diameter of $d = 10^{-2}$ m and the collection efficiency is $\eta = 0.05$; the void fraction in the filter is $\varepsilon = 0.6$.

(a) Give the ratio c_o/c_{in} for the inclusions as a function of d, ε, η, and filter depth H. c_i is the concentration in and c_o is the concentration out of the inclusions.

(b) In problem (a) insert $z = H = 0.2$ m.

(c) It may be argued that one should not use a continuous model as in problem (a) but instead take into account what happens around each (spherical) collector. Assuming that the concentration of inclusions in front of a spherical collector i is c_i, find the concentration behind the collector, c_{i+1}. The volume flow of metal towards the collector is $\dot{q}$ (m^3/s) and the collection efficiency, η, is known.

(d) By repeated use of the result from (c) find the concentration out of the filter, c_o, when the input to the filter is c_{in}. Assume that the metal passes n collectors.

(e) Show that the result given in problem (d) may, as an approximation, be replaced by an exponential expression when η is small.

(f) For a filter of height H, cross-section A, void fraction ε, and collectors of diameter d, calculate the number of collectors encountered by the melt in passing through the filter.

(g) Use the results from problems (e) and (f) and show that the same answer is obtained as in problem (a).

5.7 Removal of oxide inclusions from steel by electromagnetic stirring Determine the time constant, τ, for removal of inclusions. We assume that resistance is in the restricted turbulence boundary layer, the specific stirring power input is taken to be $\dot{\varepsilon} = 0.015\,m^2/s^3$, and $Re^{1/8} = 5.6$. The roughness of the walls is $e = 20\,\mu m$, the kinematic viscosity of the steel is $\nu = 0.8 \times 10^{-6}\,m^2/s$, and there is no pick-up of oxygen by the melt. Growth of inclusions can be neglected. The ladle is cylindrical with height-to-diameter ratio 1, and inclusions are removed to bottom, wall, and top-slag.

PROBLEMS FOR CHAPTER 7

7.1 Short-circuiting of the flow in a continuous reactor Sodium in aluminium is removed from a continuous-stirred reactor by burning at the surface of the melt. The mass transfer coefficient for the surface is k_s, the contact area is A, and the volume flow of the melt is constant and equal to $\dot{q}$. A volume flow $\dot{q}_s$ short-circuits the surface (the concentration of sodium remains equal to the input concentration c_i). $\dot{q} - \dot{q}_s$ of the flow is completely mixed and contacts the melt surface. What is the ratio between the mean outlet concentration $\bar{c}_o$ and c_i? What is this ratio if $kA/\dot{q} \gg 1$?

7.2 Velocity of bubbles Find the rise velocity of a bubble with diameter 4.5 mm in aluminium.

7.3 Injection of particles Particles are injected into a melt through a lance using N_2 as carrier gas. The inner lance diameter is 2.54 cm and 1 Nm^3 of gas is blown into the melt per min. The temperature at the lance tip is 300°C and the (static) pressure is 1.5 atm. The velocity of the particles is half the velocity of the gas and 1 kg of powder is injected every second; the mean particle diameter is 10^{-3} m. The surface tension of the melt is 1 N/m and the density of the melt is 7000 kg/m^3. Find the We number as defined in eqn (7.65) and the mass flow ratio, r. The molar volume under normal conditions is 22.4 l and the mole weight of N_2 is 28. Do we expect the particles to enter the melt?

7.4 Bubble size at nozzle tip 1 m^3 of gas (at the bath temperature) is blown upwards into a melt per minute. The ratio $\rho_g/\rho = 10^{-3}$ and the nozzle, flush with the bottom wall, has $r_0 = 4.5 \times 10^{-3}$ m. Find the volume V of the primary bubble.

7.5 Removal of Ca in Al with Cl_2 from a continuous reactor Suppose that the vapour pressure of Ca and the transfer of a Ca through the top surface can be neglected, and suppose that there is complete mixing in the melt. 80% of the Ca is to be removed. Find the required bubble surface contact area if the volume flow of the melt is $\dot{q}$ and k_{Ca} is known.

What volume flow of inert gas plus Cl_2 must be used if the bubble diameter is 0.5×10^{-2} m and the residence time, τ, is 1 s? (see Section 4.5). Assume that $\dot{q} = 10^{-3}\,\text{m}^3/\text{s}$ and $k_{Ca} = 5 \times 10^{-4}\,\text{m/s}$.

Make a comment on the gas consumption. How can we reducte the gas consumption? What about three identical tanks in series?

7.6 Hydrogen pick-up from H_2O vapour The partial pressure of water vapour above an aluminium melt is 10^{-2} atm. Assume that

$$\frac{k_{H_2O}}{k_{H_2}} = \left(\frac{D_{H_2O}}{H_{H_s}}\right)^{2/3}$$

At 0°C. For H_2O in air

$$D_{H_2O} = 0.24 \times 10^{-4}\,\text{m}^2/\text{s}$$

and for H_2 in air

$$D_{H_2} = 0.63 \times 10^{-4}\,\text{m}^2/\text{s}.$$

Assuming that the temperature dependences of H_{H_2O} and D_{H_2} are the same, find the hydrogen concentration in the melt when it has been in contact with the water vapour for a long time at 700°C. Use eqn (4.56) and assume that $f_H = 1$.

PROBLEMS FOR CHAPTER 9

9.1 Calculation of the mass transfer coefficient k_t for FeSi (75%).

For natural convection assume that

$$\text{Sh} = 0.15\,\text{Ra}^{1/3}$$ [1]

where

$$\text{Ra} = \frac{g\Delta\rho L^3}{\nu\rho D} \quad \text{and} \quad \text{Sh} = \frac{k_t L}{D}.$$

L is a characteristic dimension of the melt and $\Delta\rho$ is the density difference. On the basis of the three experiments, we take as a rough mean value that there is a 5 per cent difference in iron content between the melt and the solidified material. For 25% Fe the density of the melt is 3200 kg/m³. For 42% Fe the melt density is 4400 kg/m³.[2] By interpolation this gives

$$\Delta\rho = (4400 - 3200)\,\frac{5}{42 - 25} = 353\,\text{kg/m}^3$$

$$D_{Fe} \sim 5 \times 10^{-9}\,\text{m}^2/\text{s}.$$

There is little or no experimental data for the kinematic viscosity ν, but the result is not very sensitive to ν. Taking $\nu = 10^{-6}\,\text{m}^2/\text{s}$, find k_t.

9.2 If we solidify FeSi (75% Si) very slowly with good mixing in the melt, find the relative amounts of pure Si and eutectic (58% Si).

9.3 For partial melting assume that there is a constant cake wetness

$$\beta = f'/f_c$$

where the prime refers to the fluid part of the cake, and f_c is the fraction cake: $f_c + f = 1$, $f_c = f_s + f'$. Derive an equation for $\bar{c}_s$ and insert $k = 0.1$ and $\beta = 0.1$ in the answer. The fraction cake is $\frac{1}{2}$. Note that

$$\bar{c}_c f_c = \bar{c}_s f_s + c_1 f'. \tag{I}$$

9.4 What is the limiting curve in zone refining when $k = 0.9$ and $l/L = 0.01$? Find c_s/c_{in} at $x/L = 0.8$.

REFERENCES

1. Guthrie, R. I. L. (1989). *Engineering in process metallurgy*. Clarendon Press, Oxford.
2. Tveit, H. (1988). Størkning av 75% ferrosilisium, forløp, struktur og styrke. Dr. ing. Thesis, Metallurgisk Inst., NTH, Trondheim, Norway.
3. Iida, T. and Guthrie, R. I. L. (1988). *The physical properties of liquid metals*. Clarendon Press, Oxford.

ANSWERS TO PROBLEMS

ANSWERS FOR CHAPTER 2

2.1 $$W = \frac{(N_A + N_B)!}{N_A! N_B!}$$

$$\frac{\Delta S}{k} = \ln W.$$

2.2 $$\Delta\mu_i = \mu_i\,(10^6\,\mathrm{N/m^2}) - \mu_i\,(10^5\,\mathrm{N/m^2})$$

$$= V_i^0 \left(p - 10^5 - \frac{\kappa}{2}\,(p - 10^5)^2 \right)$$

$$V_i^0 = \frac{55.85 \times 10^{-6}}{7.8} = 7.16 \times 10^{-6}\,\mathrm{m^3/mol}$$

$$\Delta\mu_i = 7.16 \times 10^{-6}\,(9 \times 10^5) \left(1 - \frac{0.6}{2} \times 10^{-11} \times 9 \times 10^5 \right) = 6.4\ \mathrm{J/mol}.$$

Small effect.

2.3 From Appendix 1 and Table 2.2,

$$\Delta G^0 = 151520 - 84.92\,T$$

for MnS(l) = $\underline{\mathrm{Mn}}$ + $\underline{\mathrm{S}}$

$$a_{\mathrm{Mn}}\,a_{\mathrm{S}} = 0.5 \times 1.24 = 0.62.$$

2.4 From Table 2.7, $e_{\mathrm{O}}^{\mathrm{Cr}} = -0.04$, $e_{\mathrm{O}}^{\mathrm{Ni}} = 0.006$, and $e_{\mathrm{O}}^{\mathrm{Si}} = -0.131$, $e_{\mathrm{Si}}^{\mathrm{Si}} = 0.11$, $e_{\mathrm{Si}}^{\mathrm{Cr}} = -0.0003$, and $e_{\mathrm{Si}}^{\mathrm{Ni}} = 0.005$. Then $f_{\mathrm{O}} = 0.17$, $f_{\mathrm{Si}} = 1.34$. From Appendix 1 and Table 2.2, $\Delta G^0 = 590740 - 229.02\,T$ J/mol for the reaction $SiO_2 = \underline{\mathrm{Si}} + 2\underline{\mathrm{O}}$. This gives $a_{\mathrm{SiO_2}}/a_{\mathrm{Si}}a_{\mathrm{O}}^2 = 32\,498$. Then

$$[\%\,\mathrm{O}] = (0.75 \times 1.34 \times 32\,498)^{-1/2}/0.17 = 325 \times 10^{-4}\% = 325\ \mathrm{p.p.m.}$$

2.5 From Table 2.7, $e_{\mathrm{C}}^{\mathrm{C}} = 0.124$, $e_{\mathrm{O}}^{\mathrm{O}} = -0.2$, $e_{\mathrm{C}}^{\mathrm{O}} = -0.34$, and $e_{\mathrm{O}}^{\mathrm{C}} = -0.45$. From Appendix 1 and Table 2.2, ΔG^0 for CO (gas) = $\underline{\mathrm{C}}$ + $\underline{\mathrm{O}}$ is

$$\Delta G^0 = 19\,840 + 40.62\,T.$$

Then

$$\frac{a_{\mathrm{C}}\,a_{\mathrm{O}}}{p_{\mathrm{CO}}} = 2.1 \times 10^{-3}.$$

Then taking the logarithm

$$\log f_{\mathrm{C}} + \log f_{\mathrm{O}} + \log [\%\,\mathrm{O}] = -2.68$$

and

$$e_{\mathrm{C}}^{\mathrm{C}}[\%\,\mathrm{C}] + e_{\mathrm{C}}^{\mathrm{O}}[\%\,\mathrm{O}] + e_{\mathrm{O}}^{\mathrm{O}}[\%\,\mathrm{O}] + e_{\mathrm{O}}^{\mathrm{C}}[\%\,\mathrm{C}] + \log[\%\,\mathrm{O}] = -2.68.$$

The solution is $[\%\,\mathrm{O}] = 4.4 \times 10^{-3}$.

No, if the interaction coefficients had been ignored $[\%\,\mathrm{O}] = 2.1 \times 10^{-3}$. The terms $e_{\mathrm{C}}^{\mathrm{O}}[\%\,\mathrm{O}] + e_{\mathrm{O}}^{\mathrm{O}}[\%\,\mathrm{O}]$ are not important due to the relatively low value of $[\%\,\mathrm{O}]$.

2.6 From eqn (2.38) one obtains:

$$2.3 \log f_i = \ln \frac{\gamma_i}{\gamma_i^0} + \ln\left(1 + \sum_{j=2}^{k} x_j\,(M_j/M_1 - 1)\right)$$

$$\ln\left(1 + \sum_{j=2}^{k} x_j\,(M_j/M_1 - 1)\right) \approx \sum_{j=2}^{k} x_j\,(M_j/M_i - 1).$$

If first-order terms in x_j are compared, eqn (2.39) is obtained.

2.7 From Table 2.3 ΔG^0 for $\frac{1}{2}\,\mathrm{H_2} = \underline{\mathrm{H}}$ is 43510 + 31.38 T J/mol. Thus

$$\frac{a_{\mathrm{H}}}{p_{\mathrm{H_2}}^{1/2}} = 5.1 \times 10^{-4}.$$

For $p_{\mathrm{H_2}} = 0.04$ bar, $a_{\mathrm{H}} = 1$ p.p.m.

$$\mathrm{H_2O} = 2\underline{\mathrm{H}} + \underline{\mathrm{O}}$$

$$\Delta G^0 = 249170 + 25.44\,T\ \mathrm{J/mol}.$$

This gives $p_{\mathrm{H_2O}} = a_{\mathrm{O}} a_{\mathrm{H}}^2 \exp(\Delta G^0/RT) = 0.01 \times 10^{-8} \times 6.43 \times 10^{10} =$ 6.4 bar.

2.8 From Figs. 2.14 and 2.15 one finds $a_{\mathrm{CaO}} = 0.008$ and $a_{\mathrm{SiO_2}} = 0.2$. From Appendix 1, ΔG^0 for the reaction 2 Ca(l) + $\mathrm{SiO_2}$ = Si(l) + 2 CaO is $-323\,460 + 11.2\,T$. Then

$$\frac{a_{\mathrm{Ca}}^2 a_{\mathrm{SiO_2}}}{a_{\mathrm{CaO}}^2 \times a_{\mathrm{Si}}} = 2.1 \times 10^{-9}.$$

Insertion of the activity values for $\mathrm{SiO_2}$ and CaO give

$$\frac{a_{\mathrm{Ca}}^2}{a_{\mathrm{Si}}} = 0.66 \times 10^{-12}.$$

For a_{Si} we employ Raoults law

$$x_{Si} = \frac{75/28}{75/28 + 25/55.85} = 0.86.$$

Then $a_{Ca} = 0.75 \times 10^{-6}$ and

$$x_{Ca} = \frac{a_{Ca}}{\gamma^0_{Ca}} = 3.8 \times 10^{-4}$$

or, in mass%

$$[\%\,Ca] = \frac{100 \times 3.8 \times 10^{-4} \times 40}{0.14 \times 55.85 + 0.86 \times 28} = 0.05.$$

2.9 The area occupied by one sulphur atom is $\pi \times 1.84^2 = 10.63$ Å^2. From Fig. 2.23 $-\Delta G^0/RT = 7.39$,

$$b_2 = \frac{10.63 \times 55.85}{4.83 \times 32} - 0.0006 = 3.84$$

$$\sigma = 1.9 - \frac{55.85 \times 8.3145 \times 1923}{32 \times 3.84 \times 4.83 \times 10^{-20} \times 6.022 \times 10^{23}}$$

$$\times \ln\left(\frac{3.84 \times 1620 \times 0.1}{100} + 1\right) = 1.41\ \text{N/m}.$$

2.10

(a) eqn (I):

	ΔG^0
$\frac{1}{2}\,O_2 + C = CO$ (gas)	$-114400 - 85.77\,T$
$\underline{C} = C$	$-\ \ 22590 + 42.26\,T$
$\frac{1}{2}\,O_2 + \underline{C} = CO$ (gas)	$-136990 - 43.5\ \ T.$

eqn (II):

$\frac{3}{2}\,O_2 + 2\,Cr = Cr_2O_3$	$-1110000 + 247.0\,T$
$2\,\underline{Cr} = 2\,Cr$	$-\ \ 38500 + 93.7\,T$
$\frac{3}{2}\,O_2 + 2\,Cr = Cr_2O_3$	$-1148500 + 340.7\,T$

3 × eqn (I) − eqn (II):

	ΔG^0
$3\,\underline{C} + Cr_2O_3 = 3\,CO + 2\,\underline{Cr}$	$737530 - 471.2\,T$

$$\ln\left(\frac{p^3_{CO}\,a^2_{Cr}}{a^3_{C}\,a_{Cr_2O_3}}\right) = -\frac{88704}{T} + 56.67$$

$$K = \frac{p_{CO}}{a_C} = \frac{a_{Cr_2O_3}^{1/3}}{a_{Cr}^{2/3}} \exp\left(-\frac{29568}{T} + 18.89\right).$$

(b) $T = 1615 + 273 = 1888$, $e_C^C = 0.124$, $e_C^{Cr} = -0.024$, $e_C^{Ni} = 0.012$, $e_{Cr}^C = -0.12$, and $e_{Cr}^{Ni} = 0.0002$.

$$\log f_C = 0.124 \times 0.2 - 0.024 \times 18 + 0.012 \times 10 = -0.2872$$

$$f_C = 0.52$$

$$\log f_{Cr} = -0.12 \times 0.2 + 0.0002 \times 10 = -0.022$$

$$f_{Cr} = 0.95$$

$$\frac{p_{CO}}{[\%\,C]} = \frac{0.52}{(0.95 \times 18)^{2/3}} \exp\left(-\frac{29568}{1888} + 18.89\right) = 2.0.$$

ANSWERS FOR CHAPTER 3

3.1

	ΔG^0
$SO_2 = \frac{1}{2} S_2 + O_2$	$362000 - 72.7 \times 1373$
$4\,Cu(l) + O_2 = 2\,Cu_2O$	$-389000 + 181.0 \times 1373$
From Table 2.3:	
$\frac{1}{2} S_2 = \underline{S}$	$-119660 + 25.23 \times 1373$
$SO_2 + 4\,Cu(l) = 2\,Cu_2O + \underline{S}$	$-146660 + 133.53 \times 1373$

$$p_{SO_2}/a_S = 24.8 \text{ and } p_{SO_2} = 24.8 \times 0.03 = 0.75 \text{ atm}.$$

3.2 The amount of S removed by gas purging is proportional to [% S]. Thus the rate of change of sulphur in the bath is proportional to [% S] or in mathematical terms

$$-\mathrm{d}[\%\,S]/\mathrm{d}t \propto [\%\,S]$$

or

$$\frac{\mathrm{d}[\%\,S]}{[\%\,S]} = -\text{constant} \times t,$$

or integrating

$$\frac{[\%\,S]}{[\%\,S]_0} = \exp(-\text{constant} \times t).$$

The concentration of sulphur declines exponentially (compare Fig. 3.8).

3.3 According to Appendix 1:

	ΔG^0
$CdO = Cd(g) + \frac{1}{2} O_2$	$364700 - 202.9\,T$
$\frac{1}{2} O_2 + C = CO$	$-114400 - 85.77\,T$
$CdO + C = Cd(g) + CO$	$250300 - 288.7\,T$

$$\frac{p_{Cd}p_{CO}}{a_{CdO}a_C} = \exp\left\{\left(-\frac{250300}{T} + 288.7\right)\Big/ 8.31451\right\}$$

or inserting $p_{Cd} = 0.1$, $p_{CO} = 0.5$

$$\ln(0.05) = -\frac{30104}{T} + 34.72.$$

$T = 798$ K or 525°C.

3.4 From Appendix 1:

	ΔG^0
$2\,CeF_3 = 2\,Ce(l) + 3\,F_2$	$3551000 - 514.21\,T$
$3\,Ca(l) + 3\,F_2 = 3\,CaF_2$	$-3660000 + 486.00\,T$
$3\,Ca(l) + 2\,CeF_3 = 3\,CaF_2 + 2\,Ce(l)$	$-109000 - 28.21\,T$

3.5 From Appendix 1, for

$$Ni(CO)_4 = Ni + 4\,CO$$

$$\Delta G^0 = 160000 - 410\,T.$$

Then

$$\frac{p_{Ni(CO)_4}}{p^4_{CO}} = \exp\left\{\left(\frac{160000}{T} - 410\right)\Big/ 8.31451\right\}$$

or

$$\ln(0.01) = \frac{19243}{T} - 49.3.$$

$T = 431$ K or 158°C.

ANSWERS FOR CHAPTER 4

4.1 $\dfrac{[\%\,S]}{[\%\,S]_{in}} = 1/(1 + 10.3\,M_s/M)$

4.2 $\dfrac{M_s}{M} = 0.29.$

4.3 $\dfrac{[\% S]}{[\% S]_{in}} = 0.25 + 0.75 \exp(-1.4 \times 10^{-3} t).$

4.4 $t = 47.3$ min.

4.5 $Z = 98\%$ of saturation.

4.6 $\dfrac{[\% Na]_2}{[\% Na]_{in}} = 0.6.$

4.7 $\dfrac{[\% Na]_2}{[\% Na]_{in}} = 0.54.$

4.8 $Q_g = 41\ Nm^3.$

4.9 $Q_g = 20.5\ Nm^3.$

4.10 $\psi/[\% H] = 1$. This gives an efficiency (see fig. 4.16) $Z^2 = 0.71$ (or 71%).

4.11 It takes 457 min.

4.12 $$-\frac{M \times 1000}{100 \times 12}\frac{d[\% C]}{dt} = \frac{\dot{G}K[\% C]}{22.4(1 - K[\% C]}$$

$$-\frac{1}{K}\ln\frac{[\% C]}{[\% C]_{in}} + ([\% C] - [\% C]_{in}) = at$$

where $a = \dot{G}\ 100 \times 12/(22.4 M\ 1000)$. Therefore $t = 17.4$ min.

ANSWERS FOR CHAPTER 5

5.1 (a) $r^* = 10.0 \times 10^{-10}$ m.

(b) $$c^* = 70300\left(\frac{0.3}{28}\right)^{1/3} \times \left(\frac{0.035}{16}\right)^{2/3} \times \frac{6.022 \times 10^{23}}{(1 \times 4)^{2/3}} \times 3$$

$$\times \exp\left(\frac{-4\pi \times 0.7 \times 10.0^2 \times 10^{-20}}{3 \times 3 \times 1.38 \times 1873 \times 10^{-23}}\right)$$

$$= 7 \times 10^9\ \text{nuclei/m}^3.$$

(c) $$\rho_O = \frac{2}{V_m} = \frac{2}{28 \times 10^{-3}} = 71.4\ \frac{\text{kmol O}}{\text{m}^3\ \text{inclusion}}$$

$$\frac{da}{dt} = \frac{10^{-4} \times 7030 \times (0.035 - 0.010)}{71.4 \times 100 \times 16}$$

$$= 0.15 \times 10^{-6}\,\text{m/s} = 0.15\,\mu\text{m/s}.$$

An inclusion grows 9.2 μm in a minute.

5.2 (a) 7.3×10^{-4} m/s.

(b) $$\frac{c_p(t)}{c_p} = \exp\left(-\frac{A_s u}{V} t\right)$$

$$\frac{c_p(t = 75\,\text{min})}{c_p(t = 0)} = \exp\left(-\frac{0.5 \times 7.3 \times 10^{-4} \times 75 \times 60}{1.0}\right)$$

$$= 0.19.$$

(c) Mass balance:

$$-\frac{V\,\mathrm{d}f_N(a)}{\mathrm{d}t} = f_N(a)\,u\,A_s.$$

Integration gives

$$\frac{f_N(a, t)}{f_N(a, t = 0)} = \exp\left\{-\frac{A_s u t}{V}\right\}$$

$$f_N(a, t) = B \exp\left(-ba - \frac{A_s u t}{V}\right).$$

5.3 $$\frac{[\%\,O]_i}{[\%\,O]_i\,(t = 0)} = \exp\left(-\frac{k_t A t}{V}\right)$$

$t = 15$ min.

$$\frac{[\%\,O]_i}{[\%\,O]_i\,(t = 0)} = \exp\left(-\frac{10^{-3} \times 2 \times 15 \times 60}{1}\right) = 0.165.$$

$[\%\,O]_i = 0.33 \times 10^{-2}$.

5.4 A kmole balance for oxygen in inclusion is

Reduction of oxygen in the melt = Oxygen (as inclusions) transferred to the walls and top slag − Oxygen in inclusions due to pick-up

or

$$-\frac{\rho V \mathrm{d}[\%\,O]_i}{M_O\,100\,\mathrm{d}t} = \frac{\rho k_t A\,[\%\,O]_i}{M_O\,100} - \frac{\rho k A\,\Delta[\%\,O]}{M_O\,100}$$

where M_O is the atomic weight of oxygen in kg/kmol. Therefore,

$$-\frac{V\mathrm{d}[\%\,\mathrm{O}]_\mathrm{i}}{\mathrm{d}t} = k_\mathrm{t} A\,[\%\,\mathrm{O}]_\mathrm{i} - kA\,\Delta[\%\,\mathrm{O}].$$

Integration gives:

$$\int_{[\%\,\mathrm{O}]_\mathrm{i}\,(t=0)}^{[\%\,\mathrm{O}]_\mathrm{i}\,(t)} \frac{\mathrm{d}[\%\,\mathrm{O}]_\mathrm{i}}{[\%\,\mathrm{O}]_\mathrm{i} - \dfrac{k}{k_\mathrm{t}}\Delta[\%\,\mathrm{O}]} = -\int_0^t \frac{k_\mathrm{t} A\,\mathrm{d}t}{V}$$

$$\ln \frac{[\%\,\mathrm{O}]_\mathrm{i}\,(t) - \dfrac{k}{k_\mathrm{t}}\Delta[\%\,\mathrm{O}]}{[\%\,\mathrm{O}]_\mathrm{i}\,(t=0) - \dfrac{k}{k_\mathrm{t}}\Delta[\%\,\mathrm{O}]} = -\frac{k_\mathrm{t} A}{V}t$$

$t = 15$ min gives

$$[\%\,\mathrm{O}]_\mathrm{i} = 0.165 \times (1.9 \times 10^{-2}) + 1 \times 10^{-3}$$
$$= 0.41 \times 10^{-2}.$$

5.5 (a) $\dot{\alpha}_\mathrm{e} = \dfrac{3\dot{Q}H\eta}{4RV} = 4.17 \times 10^{-3}$ 1/s.

(b) $\dfrac{c}{c_\mathrm{O}} = \exp(-\dot{\alpha}_\mathrm{e} t)$.

(c) $t = \dfrac{1}{\dot{\alpha}_\mathrm{e}} \ln \dfrac{c_\mathrm{O}}{c} = 552$ seconds.

(d) Interception, gravity, inertial forces, hydrodynamic effects, and Brownian diffusion.

5.6 (a) $\dfrac{c}{c_\mathrm{i}} = \exp\left(-\dfrac{3(1-\varepsilon)\eta}{2d\varepsilon}H\right)$.

(b) $\dfrac{c}{c_\mathrm{i}} = 0.37$.

(c) $\dfrac{c_{i+1}}{c_\mathrm{i}} = 1 - \eta$.

(d) $c_n = (1-\eta)^n c_\mathrm{in}$.

(e) $\dfrac{c}{c_\mathrm{in}} = (1-\eta)^n$.

$$\ln \frac{c}{c_\mathrm{in}} = n \ln(1-\eta) \approx -n\eta.$$

$$\frac{c}{c_{\text{in}}} \approx \exp(-n\eta).$$

(f) Volume of filter $= AH$

Volume of collectors $= AH(1-\varepsilon) = N\pi d^3/6$

Number of collectors $= N = 6AH(1-\varepsilon)/(\pi d^3)$

Melt cross-sectional area $= A\varepsilon$

Total cross-section of the collectors $= N\pi d^2/4$

The number of collectors encountered by the melt is:

$$n = \frac{N\pi d^2/4}{A\varepsilon} = \frac{3H(1-\varepsilon)}{2d\varepsilon}.$$

(g) $\dfrac{c}{c_{\text{in}}} = \exp(-n\eta)$ from (e);

$n = \dfrac{3H(1-\varepsilon)}{2d\varepsilon}$ from (f);

$$\frac{c}{c_{\text{in}}} = \exp\left(\frac{3H(1-\varepsilon)}{2d\varepsilon}\eta\right).$$

This is the same result as in eqn (5.85).

5.7 $\dfrac{1}{\tau} = \dfrac{0.015 \times (20 \times 10^{-6})^2 \times 0.85 \times 10^{-3}}{5.6 \times 0.64 \times 10^{-12}}$

$\tau = 176$ s or 2.9 minutes.

ANSWERS FOR CHAPTER 7

7.1 c_o is the sodium concentration in the part of the melt that contacts the surface. Then a mass balance for sodium gives

$$c_i\dot{q} = c_i\dot{q}_s + kA\,c_o + c_o(\dot{q} - \dot{q}_s).$$

Then

$$\frac{c_o}{c_i} = \frac{1}{1 + \dfrac{kA}{\dot{q} - \dot{q}_s}} \qquad \dot{q}\bar{c}_o = c_o(\dot{q} - \dot{q}_s) + c_i q_s.$$

Then

$$\frac{\bar{c}_o}{c_i} = \frac{\left(1 + \dfrac{kA\,\dot{q}_s}{(\dot{q} - \dot{q}_s)\,\dot{q}}\right)}{1 + \dfrac{kA}{\dot{q} - \dot{q}_s}}$$

If $kA/\dot{q} \to \infty$

$$\frac{\bar{c}_o}{c_i} = \frac{\dot{q}_s}{\dot{q}}.$$

7.2 $Eo = 0.52$, $H = 94.6$. From eqn (7.62), $\mathrm{Re} = 3374$, so $v = 0.30\,\mathrm{m/s}$.

7.3 The gas velocity is

$$v = \frac{(300 + 273) \times 1.5 \times 4 \times 10^4}{273 \times \pi \times 2.54^2 \times 60} = 103.6\,\mathrm{m/s}.$$

The particle velocity is $u = 51.8\,\mathrm{m/s}$, so

$$\mathrm{We} = \frac{7000 \times 51.8^2 \times 10^{-3}}{2} = 9393.$$

$$r = \frac{60}{28/22.4} = 48.$$

Yes, the particles should enter the melt (unless they have a melting-point lower than 300°C, stick to the walls, and choke up the lance, or the lance may clog up due to oscillations in the transport line).[1,2]

7.4 $$N_l = \frac{(\frac{1}{60})^{0.8} \times 10^{-3} \times 10^6}{9.81^{0.4} \times 4.5^2} = 0.75$$

$$N_V = 0.5 \times 0.75^{-0.25} = 0.54$$

$$V = \frac{0.54}{9.81^{0.6}}\left(\frac{1}{60}\right)^{6/5} = 10^{-3}\,\mathrm{m}^3.$$

7.5 $\dot{q}[\%\,\mathrm{Ca}] = \dot{q}[\%\,\mathrm{Ca}]\,0.2 + k_{\mathrm{Ca}}A[\%\,\mathrm{Ca}]\,0.2$. Thus

$$A = \frac{\dot{q}\,4}{k_{\mathrm{Ca}}}$$

$A = \dot{Q}\,6\,\tau/d$, and

$$\dot{Q} = \frac{\dot{q}\,4\,d}{6\,k_{\mathrm{Ca}}} = \frac{10^{-3} \times 4 \times 10^{-2} \times 0.5}{6 \times 5 \times 10^{-4}} = 0.0067\,\mathrm{m}^3/\mathrm{s}.$$

We are using a lot of gas! We can reduce the gas consumption by having several tanks in series. Three identical tanks in series would require that

$(0.2)^{1/3} = 0.585$ of the sodium be removed in each unit. Then for each unit

$$A = \frac{\dot{q}\,0.415}{k_{Ca}\,0.585}$$

and

$$3A = \frac{\dot{Q}\,6\,\tau}{d}$$

gives

$$\dot{Q} = \frac{\dot{q}\,0.415 \times 3 \times 10^{-2} \times 0.5}{6 \times 5 \times 10^{-4} \times 0.585}$$

$$\dot{Q} = 0.0036\,\text{m}^3/\text{s}.$$

The gas consumption is slightly more than half compared to using only one unit.

7.6 $$[\%\,\text{H}] = \left\{\left(\frac{D_{H_2O}}{D_{H_2}}\right)^{2/3} p_{H_2O}\right\}^{1/2} \Bigg/ K_H = \left\{\left(\frac{0.24}{0.63}\right)^{2/3} \times 10^{-2}\right\}^{1/2} \Bigg/ K_H$$

$$\ln K_H = \frac{5869}{973} + 3.282$$

$$K_H = 11.1 \times 10^3$$

$$[\%\,\text{H}] = 0.07 \times 10^{-4} = 0.07\,\text{p.p.m.}$$

ANSWERS FOR CHAPTER 9

9.1 $$k_t = 0.15\,D_{Fe}^{2/3}\left(\frac{g\,\Delta\rho}{\rho\,\nu}\right)^{1/3}$$

$$= 0.15 \times 5^{2/3} \times 10^{-6}\left(\frac{9.81 \times 353}{3200}\right)^{1/3}\frac{1}{10^{-2}}$$

$$= 4.5 \times 10^{-5}\,\text{m/s}.$$

9.2 A mass balance for Si gives

$$75 = 100x + 58\,(1 - x)$$

where x is the fraction that is pure Si. Then

$$x = \frac{17}{42} \qquad \text{and} \qquad 1 - x = \frac{25}{42}.$$

The relative amounts of Si and eutectic are $\frac{17}{25} = 0.68$.

9.3 Equations (9.20) and (9.21) still apply but eqn (9.24) is replaced by

$$d(\bar{c}_c f_c) + c_l \, df = 0. \tag{II}$$

These equations together give

$$c_l = \bar{c}_s / k$$

$$f_s = \left(\frac{1}{\beta} - 1\right) f' \qquad \text{and} \qquad f' = f_c \beta$$

Then from eqn(I) $\bar{c}_c f_c = \bar{c}_s \left\{ (1/\beta - 1) + \frac{1}{k} \right\} \beta f_c$

so that $\bar{c}_s / k = \bar{c}_c / b$ where

$$b = (1 - \beta) k + \beta.$$

Therefore eqn (II) becomes, on using $df + df_c = 0$:

$$d(\bar{c}_c f_c) = \bar{c}_c \, df_c / b$$

or

$$\frac{d\bar{c}_c}{\bar{c}_c} = \left(\frac{1}{b} - 1\right) \frac{df_c}{f_c}. \tag{III}$$

Initially $f_c = 1$. Then $\bar{c}_c = c_{in}$ where c_{in} is the initial concentration. Then integration of eqn (III) gives

$$\frac{\bar{c}_c}{c_{in}} = f_c^{1/b - 1}.$$

Thus $\bar{c}_c / c_{in} = 0.5^{1/0.19 - 1} = 0.052$. Wetting has drastically increased the refining ratio.

9.4 Equation (9.30) gives

$$\frac{bl}{0.9} = \exp(bl) - 1.$$

The solution is $bl = 0.207$. Then

$$A = \frac{bL}{\exp(bL) - 1} = \frac{20.7}{\exp(20.7) - 1} = 2.1 \times 10^{-8}$$

$$\frac{c_s}{c_{in}} = 2.1 \times 10^{-8} \exp(20.7 \, x/L)$$

$$\frac{c_s}{c_{in}} = 2.1 \times 10^{-8} \exp(20.7 \times 0.8) = 0.33.$$

REFERENCES

1. Nilmani, M., Engh, T.A., Gray, N.B., and Floyd, S.M. (1990). Plant and laboratory investigations on injection and related phenomena in some smelting and refining processes. *Scand. J. Metallurgy*, **19**, 278–87.
2. Shirota, Y., Marukawa, K., Hiroki, N., and Ikemiya, H. (1987). Clogging-free nozzle conditions and metallurgical effects of powder injection practice in hot metal treatment. *Trans. ISIJ*, **27**, 18–24.

INDEX

The manufacturer's authorised representative in the EU for product safety is Oxford University Press España S.A. of el Parque Empresarial San Fernando de Henares, Avenida de Castilla, 2 – 28830 Madrid (www.oup.es/en or product.safety@oup.com). OUP España S.A. also acts as importer into Spain of products made by the manufacturer.

www.ingramcontent.com/pod-product-compliance
Lightning Source LLC
LaVergne TN
LVHW011858200926
848484LV00002B/479
* 9 7 8 0 1 9 8 5 6 3 3 7 2 *